INTERMEDIATE ALGEBRA

Hayk Yegoryan

MIT & Kellogg School of Management
McKinsey Co.

Rubik Yegoryan

Elite Pre-University

Cover image © shutterstock.com

www.innovativeinkpublishing.com
Send all inquiries to:
4050 Westmark Drive
Dubuque, IA 52004-1840

ACKNOWLEDGMENTS

We would like to thank our colleagues and students from Woodbury University,
MIT, Kellogg School of Management, as well as from UCLA,
UC San Diego, USC, CSUN, Glendale and Valley Community Colleges
for discussions, comments, recommendations and suggestions.

Our deep thanks to Kristina Yegoryan, and Marina Hovhannisyan , who assisted in
the editing process.

Hayk Yegoryan

Rubik Yegoryan

In Memory of Dr. Rubik Yegoryan's parents,

Sh. Yegoryan *and* **H. Yegoryan,**

In Memory of Dr. Rubik Yegoryan's wife parents,

R. Hovhannisyan *and* **L. Hovhannisyan**

&

In Memory of Dr. Rubik Yegoryan's brother,

Dr. Eduard Yegoryan *- Author of the Declaration of Independence
and Father of the Constitution of the Republic of Armenia*

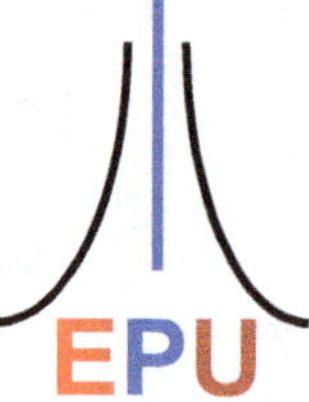

ELITE PRE - UNIVERSITY

BRIEF CONTENTS

PART 1 Numbers, Variables, and Expressions 1

PART 2 Equations, Inequalities, and Graphs 129

PART 3 Linear Systems 329

PART 4 Functions and Graphs 371

EXTRA PARTS

5 Bank of Tests 417
6 Bank of Extra Exercises 471
7 Bank of Nonstandard Problems 487

CONTENTS

PART 1 NUMBERS, VARIABLES, AND EXPRESSIONS 1

Chapter 1. Real Numbers and Variables 2

1.1.   Real Number Set and Number Line 3
1.2.   Operations and Properties 30
1.3.   The Order of Operations (PEMDAS + L→ R) 53

Chapter 1 Classwork 58
Chapter 1 Homework 59

Chapter 2. Powers, Polynomials, and Expressions 60

2.1.   Powers and Properties of Powers 61
2.2.   Polynomials and Operations 66
2.3.   Basic Algebraic Identities 82
2.4.   Factoring of Polynomials 85
2.5.   Fractional Expressions and Operations 97

Chapter 2 Classwork 108
Chapter 2 Homework 109

Chapter 3. Power Roots 110

3.1.   Power Roots and Their Properties 111
3.2.   Power Roots and Fractional Powers 114
3.3.   Operations with Power Roots 117
3.4.   Pythagorean Theorem and Distance Formula 120

Chapter 3 Classwork 124
Chapter 3 Homework 125

Part 1. Pre-Test 126

PART 2 EQUATIONS, INEQUALITIES, AND GRAPHS 129

Chapter 4. Equations and Inequalities 130

4.1. Linear Equations and Inequalities 131
4.2. Absolute Value Equations and Inequalities 145
4.3. Quadratic Equations and Inequalities 162
4.4. Polynomial Equations and Inequalities 228
4.5. Fractional equations and Inequalities 237
4.6. Radical Equations and Inequalities 258

Chapter 4 Classwork 272
Chapter 4 Homework 274

Chapter 5. Graphs of Lines and Inequalities 276

5.1. System of Coordinates 277
5.2. Equation of a Line 281
5.3. Graphs of Lines 293
5.4. Parallel, Coincident, Perpendicular, and Intersecting Lines 308
5.5. Linear Inequalities on a Coordinate Plane 312

Chapter 5 Classwork 317
Chapter 5 Homework 318

Part 2. Pre-Test 325

PART 3 LINEAR SYSTEMS 329

Chapter 6. Systems of Equations and Inequalities 330

6.1. Systems of Linear Equations 331
6.2. Cramer's Rules 351
6.3. Systems of Linear Inequalities 360

Chapter 6 Classwork 368
Chapter 6 Homework 369

Part 3. Pre-Test 370

PART 4 FUNCTIONS AND GRAPHS 371

Chapter 7. Algebraic Functions 372

7.1. Functions 373
7.2. Transformations of Functions 395
7.3. Polynomial and Fractional Functions 399

Chapter 7 Classwork 413
Chapter 7 Homework 414

Part 4. Pre-Test 415

EXTRA PARTS

5 Tests 417
6 Bank of Extra Exercises 471
7 Bank of Nonstandard Problems 487

PART 1

NUMBERS, VARIABLES, AND EXPRESSIONS

Chapter 1. Real Numbers and Variables

1.1. Real Number Set and Number Line
1.2. Operations and Properties
1.3. The Order of Operations (PEMDAS + L→ R)

Chapter 1   Classwork
Chapter 1   Homework

Chapter 2. Powers, Polynomials, and Expressions

2.1. Powers and Properties of Powers
2.2. Polynomials and Operations
2.3. Basic Algebraic Identities
2.4. Factoring of Polynomials
2.5. Fractional Expressions and Operations

Chapter 2   Classwork
Chapter 2   Homework

Chapter 3. Power Roots

3.1. Power Roots and Their Properties
3.2. Power Roots and Fractional Powers
3.3. Operations with Power Roots
3.4. Pythagorean Theorem and Distance Formula

Chapter 3   Classwork
Chapter 3   Homework

Part 1. Pre-Test

Chapter 1. Real Numbers and Variables

1.1. **Real Number Set and Number Line**
1.2. **Operations and Properties**
1.3. **The Order of Operations (PEMDAS + L$\rightarrow$ R)**

Chapter 1   Classwork
Chapter 1   Homework

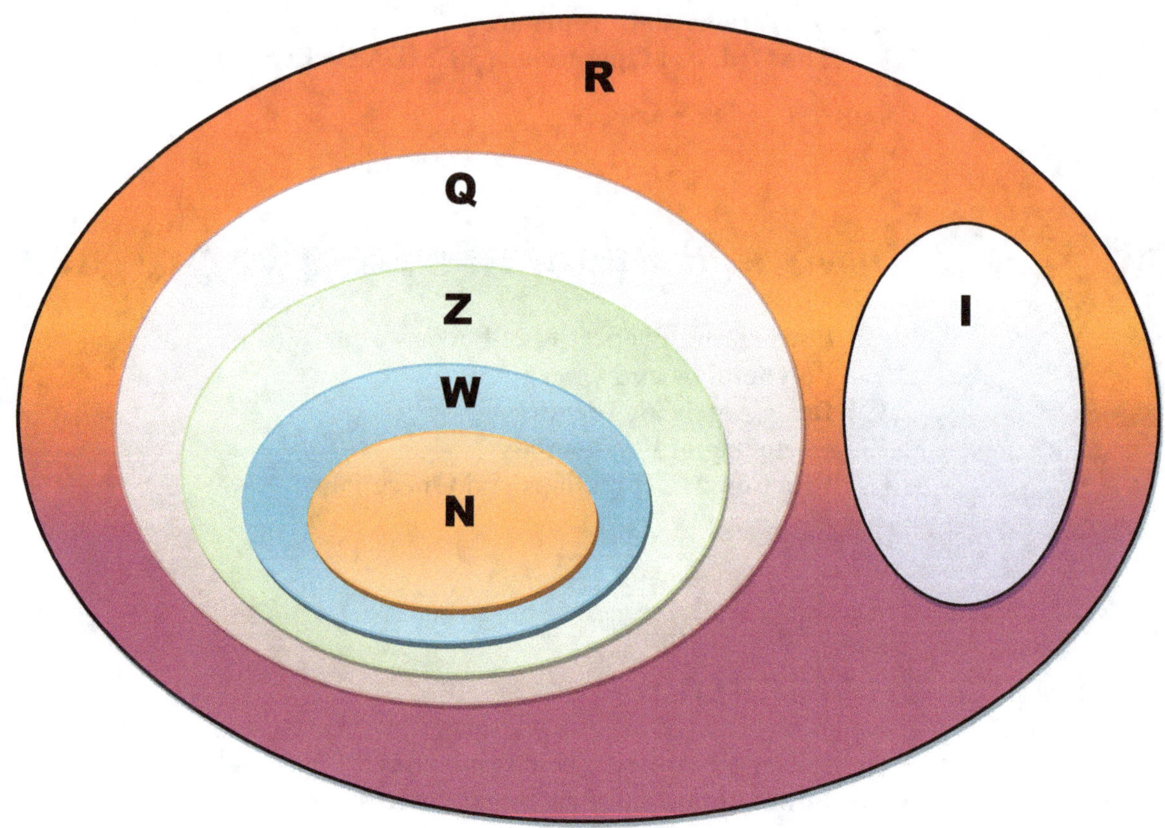

$$2,\ 0,\ -4,\ 0.5,\ \frac{1}{2},\ -\frac{3}{4},\ \sqrt{2},\ -\sqrt[3]{6},\ x, y, -z, t, \dots$$

$$N \subset W \subset Z \subset Q \subset R \ \& \ I \subset R$$

1.1. Real Number Set and Number Line

- ■ **THE REAL NUMBER SET AND ITS SUBSETS**
- ■ **THE NUMBER LINE**
- ■ **FACTORS, PRIME FACTORS, AND PRIME FACTORIZATION**
- ■ **GCF/GCD AND LCM/LCD**
- ■ **SETS AND VENN DIAGRAMS**

■ THE REAL NUMBER SET AND ITS SUBSETS

In this book we will study numbers, number sets, variables (letters which represent one or more numbers), algebraic expressions (combinations of numbers and variables), operations with numbers and expressions, as well as different types of equations and inequalities.

Below we study the Real Number Set **R** and its subsets: the Natural Number Set **N**, the Whole Number Set **W**, the Set of Integers **Z**, the Rational Number Set **Q**, and the Irrational Number Set **I**.

IMPORTANT DEFINITIONS

▶ The **set of natural numbers** is the infinite set $\mathbf{N} = \{1, 2, 3, 4, 5, ...\}$. *

▶ The **set of whole numbers** is the infinite set $\mathbf{W} = \{0, 1, 2, 3, 4, 5, ...\}$.

▶ The **set of integers** is the infinite set $\mathbf{Z} = \{..., -3, -2, -1, 0, 1, 2, 3, ...\}$.

▶ The **set of rational numbers** is the infinite set $\mathbf{Q} = \left\{ \frac{m}{n} \right\}$, where $\mathbf{n} \neq 0$ and $\mathbf{m, n} \in \mathbf{Z}$.

    Fractions, Mixed Numbers, and **Integers** are rational numbers (any integer **n** we can write as $\frac{n}{1}$).

    Terminating decimals and **repeating decimals** are also rational numbers ($0.25 = \frac{25}{100}$; $0.\overline{12} = \frac{12}{99}$).

▶ The **set of rational numbers** is the infinite set $\mathbf{Q} = \{$**all terminating** and **all repeating decimals**$\}$.

    Any terminating or any repeating decimal can be represented in fractional form and any fraction can be converted to a terminating or to a repeating decimal.
    There is a one-to-one relation between fractions and terminating and repeating decimals.

▶ The **set of irrational numbers** is the set **I** = $\{$all **non-terminating** and all **non-repeating** decimals$\}$.

    Decimals like **1.0200200020000200002...; 0.51511511151111511111...**; and so on are non-terminating and non-repeating decimals. They are irrational numbers.
    Any number such as $\sqrt{2}$; $\sqrt{3}$; $\sqrt{5}$; $\sqrt{7}$; $\sqrt{11}$; $\sqrt{13}$; and so on cannot be represented as a fraction, so all these numbers are irrational numbers. The set of irrational numbers is an infinite set.

▶ The **set of real numbers** is the infinite set $\mathbf{R} = \{$all **rational** and all **irrational** numbers$\} = \mathbf{Q} \cup \mathbf{I}$

▶ The **set of real numbers** is the infinite set $\mathbf{R} = \{$**all decimals**$\}$.

* Three dots mean that the numbers are continued infinitely in the indicated direction.

From the definitions of sets **N**, **W**, **Z**, **Q**, **I**, and **R** we can conclude the following:

❶ The set of Natural Numbers **N** is a subset of the Whole Number set **W** (**N** $\subset$ **W**): any element of **N** is also an element of **W**.

❷ The Whole Number Set **W** is a subset of the Integer Set **Z** (**W** $\subset$ **Z**).

❸ The set of Integers **Z** is a subset of the Rational Number Set **Q** (**Z** $\subset$ **Q**).

❹ The set of Rational Numbers **Q** is a subset of the Real Number Set **R** (**Q** $\subset$ **R**).

❺ And the set of Irrational Numbers **I** is also a subset of the Real Number Set **R** (**I** $\subset$ **R**).

The figure below shows the general relationship between the Natural Number Set (**N**), Whole Number Set (**W**), Integer Set (**Z**), Rational Number Set (**Q**), Irrational Number Set (**I**), and Real Number Set (**R**): **N** $\subset$ **W** $\subset$ **Z** $\subset$ **Q** $\subset$ **R** and **I** $\subset$ **R**.

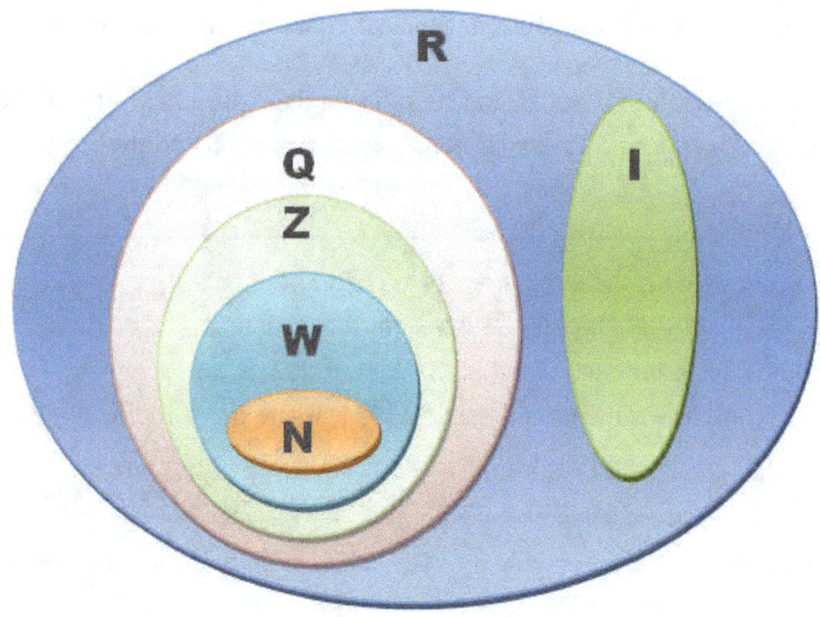

Figure 1.1.1 The graph of relations N $\subset$ W $\subset$ Z $\subset$ Q $\subset$ R and I $\subset$ R

We will also use some of definitions for the **special number subsets** of the set **R**.

Special subsets of the **Real Number Set R** are sets of **Positive Integers**, **Negative Integers**, **EVEN Numbers**, **Odd Numbers**, **Prime Numbers** and **Composite Numbers**.

The sets of **the positive integers** and **the negative integers**, definitions for which are below, are the first two groups of **special numbers**.

The **set of** *positive integers* is **Z⁺** = {1, 2, 3, 4, 5, 6, 7, 8, 9, 10, 11, 12,...}.
The **set of** *negative integers* is **Z⁻** = {-1, -2, -3, -4, -5, -6, -7, -8, -9, -10, -11, -12,...}.

The number **0** is a *neutral number* (**neither positive nor negative**).
Positive and negative numbers are *opposites* in relation to **0**.

Some Examples:

1. The numbers **2, 3, 8, 15, 21, 111,** and **999** are positive integers.
2. The numbers **–2, –1, –11, –111,** and **888** are negative integers.
3. The number **–7** is opposite to **7**, and **7** is opposite to **–7**.
   The number **– (– 11)** is opposite to **– 11**, so is equal to **11**.

Other types of **special numbers** are **even** and **odd numbers**.

The **set of** *even numbers* is **Z²ᵏ** = {...,-12,-10,-8,-6,-4,-2, 0, 2, 4, 6, 8, 10, 12 ...}.
An *even number* is an integer which is divisible by two and can be represented in the form **2k.**

The **set of** *odd numbers* is **Z²ᵏ⁺¹** = {...,-13,-11,-9,-7,-5,-3,-1, 1, 3, 5, 7, 9, 11, 13 ...}.
An *odd number* is an integer which is not an even number, so it is not divisible by two and can be represented in the form **2k + 1** (**k** is any integer).

Some Examples:

1. Positive or negative numbers whose last digits are **0, 2, 4, 6,** or **8** are even numbers.
   For example the integers **0, 2, 4, 6, 8** or the numbers **–14, –28, –116, 130,** and **1352** are even numbers.
2. Positive or negative integers whose last digits are **1, 3, 5, 7,** or **9** are odd numbers:
   For example the integers **1, 3, 5, 7, 9** or the numbers **–21, –33, –645, 107,** and **4829** are odd numbers.
3. The sum/difference of two even numbers is an even number (**2 + 4 = 6, 8 – 4 = 4**).
   The sum/difference of two odd numbers is an even number (**3 + 5 = 8, 9 – 3 = 6**).
   The sum/difference of one even and one odd number is an odd number (**2 + 3 = 5, 8 – 3 = 5**).
   The product of two even numbers is an even number (**4 · 6 = 24**).
   The product of two odd numbers is an odd number (**–3 · 9 = –27**).
   The product of one even and one odd number is an even number (**8 · 5 = 40**).

The next sets of **special numbers** are sets of prime and composite numbers.

■ A *prime number* is a **positive integer** that has **exactly two different factors**: **1** and **itself**.

The number **1** is not a prime number because it does not have two **different factors** (1 = 1·1).
The number **2** is the smallest prime number. It has exactly two different factors **1** and **2** (2 = 1·2).
The number **2** is the only even prime number.

The **set of** *prime numbers* is **P** = {2, 3, 5, 7, 11, 13, 17, 19, 23, 29, 31, 37, 41, 43, 47, 53,...}.

■ A *composite number* (non-prime number) is a **positive** integer that has **more than two
different** <u>factors</u>.

Any number that is not a prime number is a composite number, with the exception of **1**.
The number **1** is not a composite number because its factors are the same (1 = 1·1).

The **set of** *composite numbers* is **C** = {4, 6, 8, 9, 10, 12, 14, 15, 16, 18, 20, 21, 22, 24, 25,...}.

■ The *number 1* is neither a prime nor a composite number: it has only itself as a factor.

■ Two or more numbers are *relatively prime numbers* if the only common factor is **1**.

Some Examples:

1. The numbers **3, 5, 7,** and **11** are prime numbers because $3 = 1·3$; $5 = 1·5$; $7 = 1·7$; $11 = 1·11$.
2. The number **6** is a composite number because it has four factors (**1, 2, 3,** and **6**).
    The numbers **8, 10, 12** are also composite numbers because they have more than two
    different factors : $F(8) = 1,2,4,8$; $F(10) = 1,2,5,10$ and $F(12) = 1,2,3,4,6,12$.
3. The numbers **4** and **21** are relatively prime numbers because the **factors** of 4 are **1,2,4**
    ($F(4) = 1,2,4$), and the **factors** of 21 are **1,3,7,21** ($F(21) = 1,3,7,21$), and there are no
    common factors except **1.**
    The set {4,9,25,49} is a set of relatively prime numbers, because there are no common
    factors except 1 ($F(4) = 1,2,4$; $F(9) = 1,3,9$; $F(25) = 1,5,25$; $F(49) = 1,7,49$).
4. Any set of numbers with two or more even numbers cannot be a set of relative prime
    numbers because even numbers have **2** as a common factor other than one.

There are some relations between these sets.
For example, $\mathbf{Z}^+ \subset \mathbf{Z}$, $\mathbf{Z}^- \subset \mathbf{Z}$, $\mathbf{Z}^{2k} \subset \mathbf{Z}$, $\mathbf{Z}^{2k+1} \subset \mathbf{Z}$, $\mathbf{P} \subset \mathbf{Z}$, $\mathbf{C} \subset \mathbf{Z}$, $\mathbf{P} \subset \mathbf{N}$, and $\mathbf{C} \subset \mathbf{N}$.

List all natural numbers, whole numbers, and integers between the given pairs of numbers.

-7 and 10	0 and 11	10 and 25	-7 and 0	-10 and -3

Write the opposite of each number.

-8;   9;   0;   11;   -25;   100;   -28;   17;   20;   99;   -125;   1

List all even, odd, prime, and composite numbers.

-17;   7;   0;   19;   -25;   11;   -99;   2;   1;   29;   -18;   39;   51

Find the sum of all the even integers between the given numbers.

-8 and 7	0 and 11	11 and 20	-9 and 0	-8 and -1

Find the sum of all the odd integers between numbers.

-9 and 6	0 and 7	8 and 20	-6 and 0	-5 and -4

Find the product of all the prime numbers between the given numbers.

-90 and 6	0 and 12	11 and 18	-16 and 3	-5 and -4

Find two prime numbers whose sum is a prime number.

How many prime-even numbers are there?

CHALLENGE

Write the missing positive integers inside the squares.

□7 + □ = □□1	□□ + □□ = □97	□□ ÷ 9 = □0
□8 + □ = □□5	□2 · □ = 8 + □	□□6 − □ = □8
□□ · □ + □ = 11	□□ ÷ □ − □ = 98	□□ · □ − □ = 1

Put either a + or − between the numbers **0, 1, 2, 3, 4, 5, 6, 7, 8, 9** to get **1** as the result.

0   1   2   3   4   5   6   7   8   9 = 1

Put the integers **1, 2, 3, 4, 5, 6, 7, 8, 9** (use each only once) in the squares to get a true statement.

□ + □ = □ − □ = □ · □ = □□ ÷ □

□ + □ = □ + □ = □ + □ = □ + □ = 2 · □

□ ÷ □ + □ = □ · □ = □□ = □□ = □ + □ − □

$$
\begin{array}{c}
□ - □ \\
\| \\
□ = □ + □ = □ + □ \\
\| \\
□ - □
\end{array}
$$

Put the integers **1, 2, 3, 4, 5, 6, 7, 8, 9** (use each only once) in the squares to get a **magic square**. In the **magic square** each row, column, and diagonal has the same sum, called the **magic sum**.

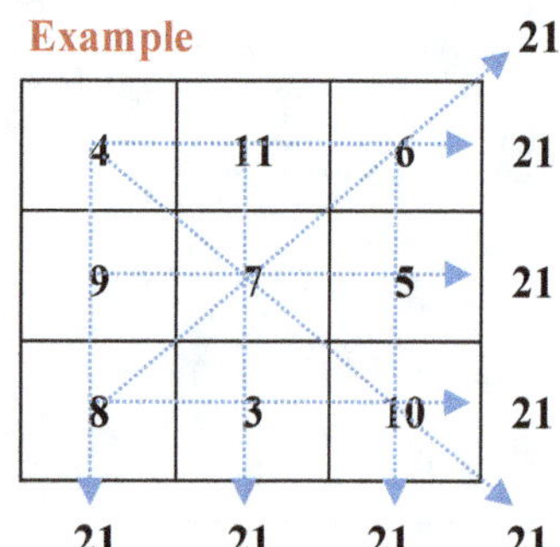

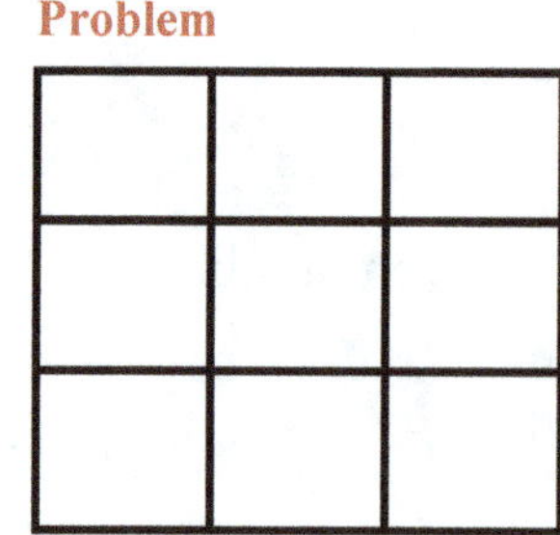

CHALLENGE

Complete each **magic square**.

4	11	
	7	5
		10

Sum = ☐

	7	6
9	5	
	3	

Sum = ☐

6		8
	5	3
		4

Sum = ☐

	13	
11	9	7
		12

Sum = ☐

4		8
	7	
6	5	

Sum = ☐

8	6	
3	10	5

Sum = ☐

2.1	9.1	4.1
7.1		
	1.1	

Sum = ☐

15		17
20		
19		21

Sum = ☐

	17	
15	13	11
14		

Sum = ☐

■ THE NUMBER LINE

With the **Number Line** we can show the places of **Real Numbers** on a line, find the opposite numbers of the given numbers, and compare and order the **Real Numbers**. On the Number Line we can also show the intervals of solutions for inequalities. The Number line is a very important instrument for understanding many ideas, as well as for solving many problems in Mathematics, Physics, other Sciences, and also in life.

We will use the following general definition of the **Number Line**.

A **Number Line** is a horizontal **line** on which **real numbers** are placed on the **line** according to their values.

There are **one-to-one** relations between the Real Number Set and the Number Line.

Each **real number** has one corresponding point on the **number line**, and each point on the number line has one corresponding number in the **Real Number Set R**.

The number corresponding to the point on the number line is called the **coordinate**. The point corresponding to the number **0 (zero)** is called the **origin** of the number line. A number line is divided into two symmetric halves at the **origin**. All **positive numbers** lie on the right side of **0**, and all **negative numbers** lie on the left side of **0**. The Number Line continues infinitely in the **positive** and **negative directions**. For any two points on the number line, the number corresponding to the rightmost point is always greater than the number corresponding to the leftmost point.

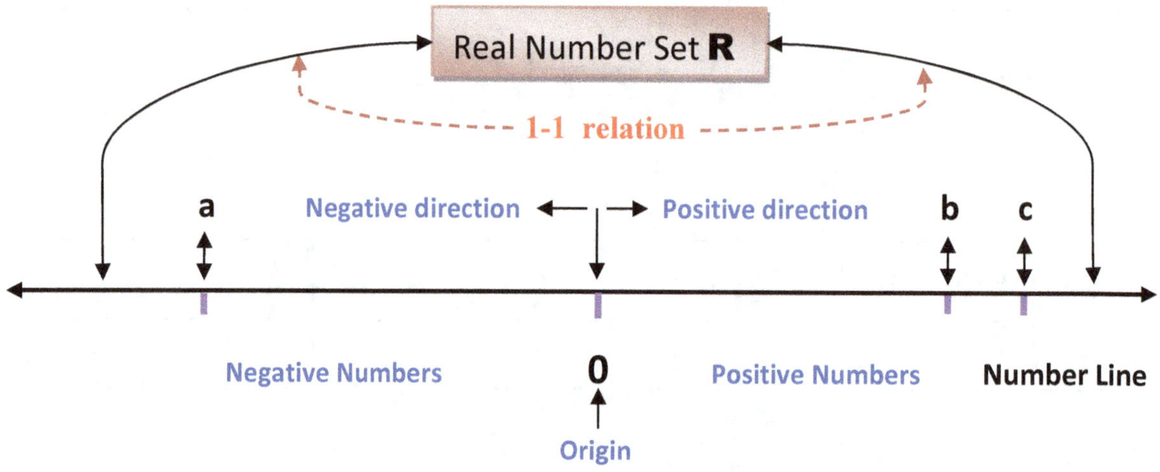

The number **c** is greater than **b** and **b** is greater than **a (c > b > a)**, because the corresponding point of **c** lies further right on the line than the corresponding point of **b** and the corresponding point of **b** lies further right on the line than the corresponding point of **a).**

INTEGERS ON THE NUMBER LINE

To draw a **Number Line** we draw a horizontal line with several dashes on it and order the integers below the line: negative integers, **0**, and positive integers (integer coordinates).

On the Number Line the numbers increase in value going from left to right.
Beginning with **0**, numbers increase in value to the right **(0,1,2, 3,...)** and decrease in value to the left **(...–3, –2, –1,0)**. In the picture below the Number Line labeled with integers **–9, –8, –7, –6, –5, –4, –3, –2, –1, 0, 1, 2, 3, 4, 5, 6, 7, 8, 9** is increasing in order from left to right.

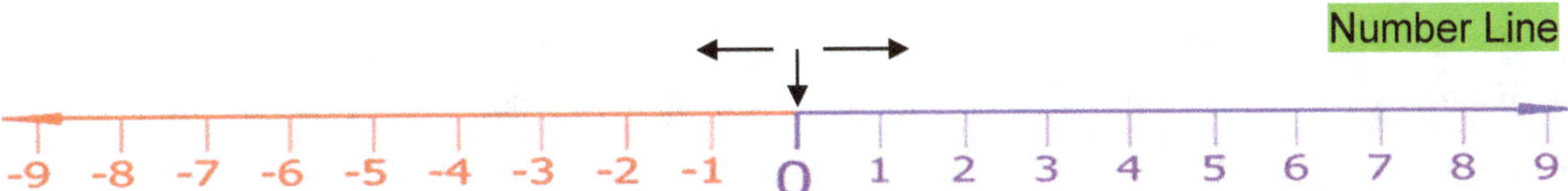

As you move from **left to right** on a number line, the numbers get larger: $-9 < -7 < 0 < 7 < 9$.

As you move from **right to left** on the number line, the numbers get smaller: $5 > 0 > -1 > -9$.

For any two points on the number line, the integer corresponding to right-place point is greater than the integer corresponding to left-place point: $-2 > -3$, $1 > -1$, $7 > 3$, and so on.

RATIONAL NUMBERS ON THE NUMBER LINE

Sometimes we also locate and label the points on the number line associated with **Rational Numbers** (fractions, terminating and repeating decimals).
The **Rational Number Set** includes all **Proper Fractions** ($\frac{m}{n}$, where **m < n**, and **n ≠ 0**; for example $\frac{2}{5}$), all **Improper Fractions** ($\frac{m}{n}$, where **m > n** and **n ≠ 0**; for example $\frac{7}{5}$), all **Integers**, all **Mixed Numbers** (**H**$\frac{m}{n}$, where **H** is the integer and is called the whole part of the mixed number, $\frac{m}{n}$ is called the fractional part of the mixed number, **m < n**, and **n ≠ 0**; for example, $9\frac{2}{5}$), all **Terminating Decimals** (numbers with finite digits after the decimal point, for example **8.456**), and all **Repeating Decimals** (numbers with infinite, but with any repeating groups of digits after the decimal point, for example **12.04123123123...**).

 **Proper Fractions on the Number Line**

Any proper fraction $\dfrac{m}{n}$ ($m < n$, $n \neq 0$, $n > 0$), is always greater than –1 and less than **1**. So, all points corresponding to **proper fractions** are located between points corresponding to –1 and **1**.

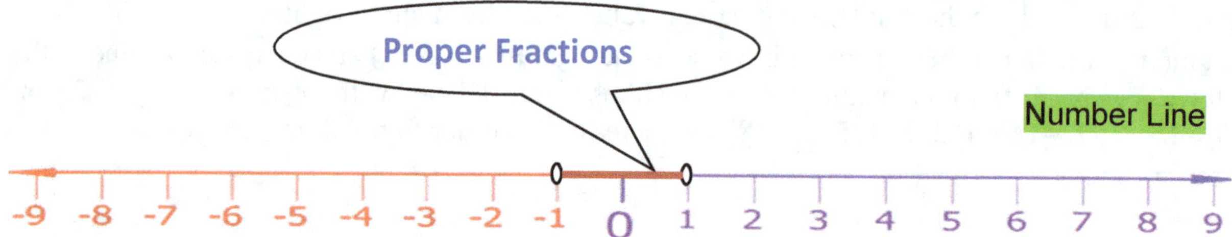

For example, the figure below shows locations of corresponding points of proper fractions
$-\dfrac{1}{4}; -\dfrac{1}{2}; -\dfrac{3}{4}; \dfrac{1}{3}; \dfrac{2}{3}; \dfrac{1}{6}; \dfrac{2}{6}; \dfrac{3}{6}; \dfrac{4}{6}; \dfrac{5}{6};$ on the number line.

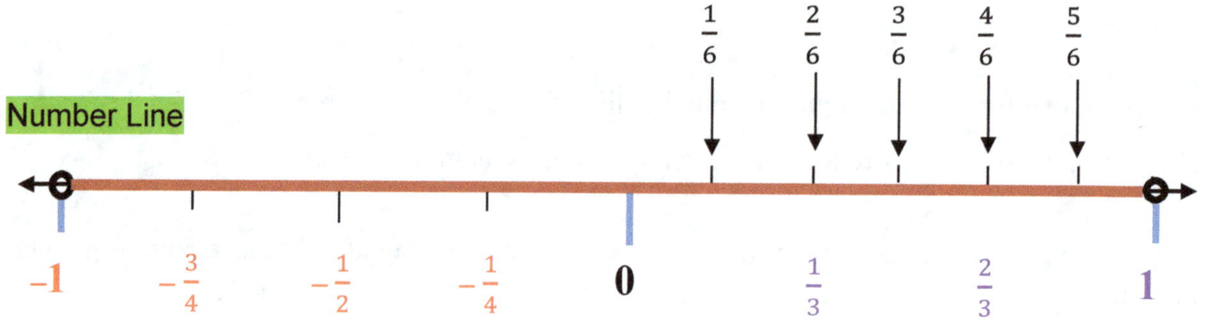

For finding the location of the point corresponding to the proper positive fraction $\dfrac{m}{n}$, on the number line, we will divide the segment between **0** and **1** by **n** parts and take **m** parts from **0**.

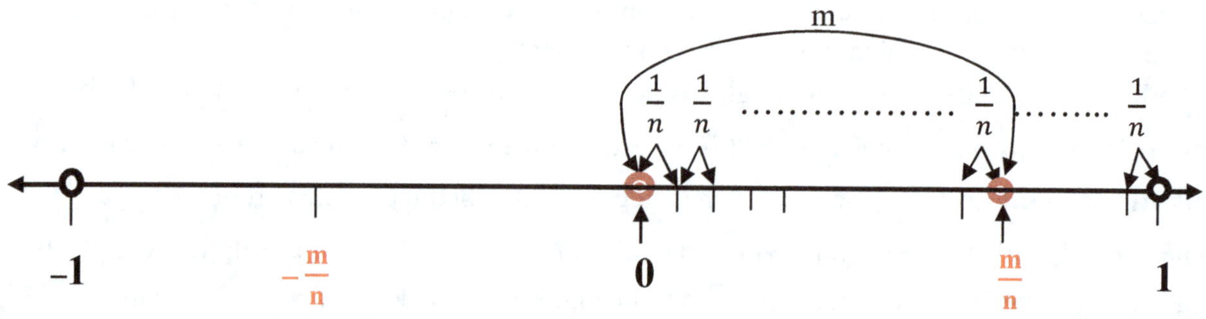

The point associated with $-\dfrac{m}{n}$ is the point symmetric to the point associated with $\dfrac{m}{n}$ in relation to the **origin** (**"0 point"** of the number line).

 **Improper Fractions and Mixed Numbers on the Number Line**

Any **Improper Fraction** can be converted to a mixed number (for example, $\frac{3}{2} = 1\frac{1}{2}$; $\frac{8}{3} = 2\frac{2}{3}$).

Any **Mixed Number** can be represented as a sum of an integer and proper fraction with the same sign (for example, $-1\frac{1}{2} = -1 - \frac{1}{2}$; $2\frac{2}{3} = 2 + \frac{2}{3}$).

To locate and label the points on the number line associated with a **positive mixed number** $k\frac{m}{n}$ (**k, m, n** are positive integers, and **m < n**, and **n ≠ 0**) we find the location of the point associated with positive integer **k**, and then find the location of the point associated with the positive proper fraction $\frac{m}{n}$, between the points associated with integers **k** and **k + 1**.

To locate and label the point on the real number line associated with a **negative mixed number** $-k\frac{m}{n}$ (**k, m, n** are positive integers, and **m < n**, and **n ≠ 0**) we mark the point on the number line which is the symmetric point to the point associated with a positive mixed number $k\frac{m}{n}$ (**k, m, n** are integers, and **m < n**, and **n ≠ 0**) in relation to the **origin** of the number line.

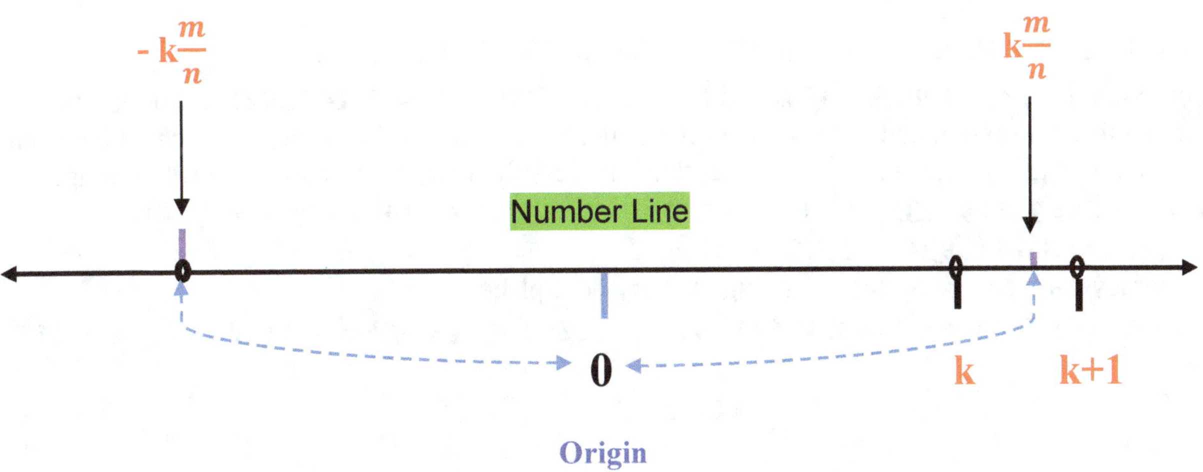

 **Terminating and Repeating Decimals on the Number Line**

Any **Terminating Decimal** can be converted into a **Proper Fraction** or into a **Mixed Number** (for example, $0.25 = \frac{1}{4}$; $-2.5 = -2\frac{1}{2}$). So, to locate and label the points on the number line associated with positive or negative **terminating decimals** we use the method for locating and labeling the points on the number line associated with corresponding positive or negative **proper fractions** or **mixed numbers**.

$\mathbf{A}$ny **Repeating Decimal** can be converted into a **Proper Fraction** or into a **Mixed Number** (for example, $1.3333\ldots = 1.\overline{3} = 1\frac{3}{9} = 1\frac{1}{3}$; $2.6666\ldots = 2.\overline{6} = 2\frac{6}{9} = 2\frac{2}{3}$). So, to locate and label the points on the real number line associated with positive or negative **repeating decimals** we use the same method for locating and labeling the points on the real number line associated with corresponding positive or negative **proper fractions** or **mixed numbers**.

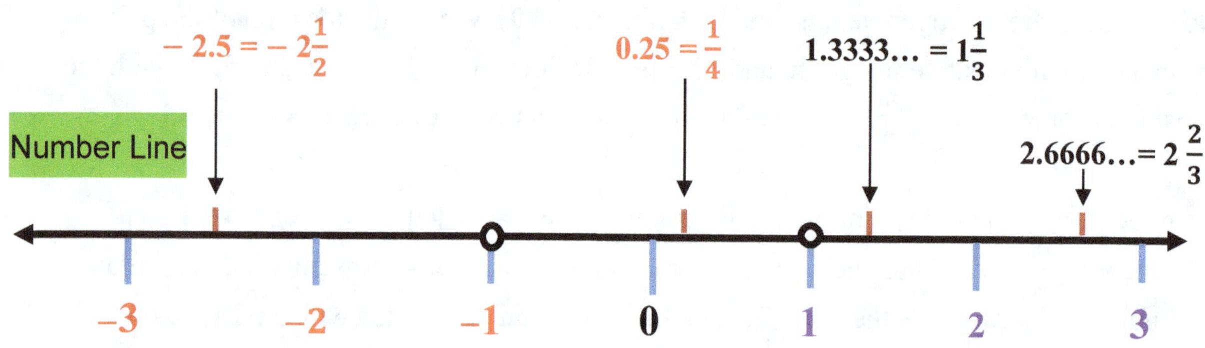

IRRATIONAL NUMBERS ON THE NUMBER LINE

$\mathbf{A}$ny **Irrational Number** is a **Non-Terminating** or **Non-Repeating Decimal.**
For example, $\sqrt{2} = 1.4142135\ldots$; $\sqrt{3} = 1.7320508\ldots$; $\pi = 3.1415926\ldots$ So, points on the number line associated with any positive or negative **irrational number** are located between two points with rational numbers, greater than and less than that irrational number. For example, $1.4 < \sqrt{2} = 1.4142135\ldots < 1.5$; or $1.7 < \sqrt{3} = 1.7320508\ldots < 1.8$; or $3.1 < \pi = 3.1415926\ldots < 3.2$.
Accuracy can be increased by rounding at every next place.
For example, $1.41 < \sqrt{2} = 1.4142135\ldots < 1.42$; $1.414 < \sqrt{2} = 1.4142135\ldots < 1.415$.

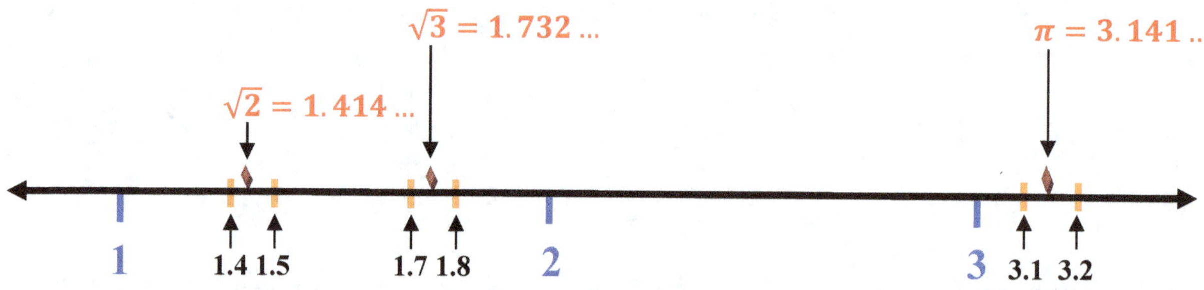

Plot **-7, -2.5, -1**$\frac{1}{2}$**, 0,** $\frac{1}{5}$**,** $\frac{11}{10}$**,** $\sqrt{3}$**, 3, 4**$\frac{1}{5}$**,** $\sqrt{24}$**,** and **6.1** real numbers on a number line.

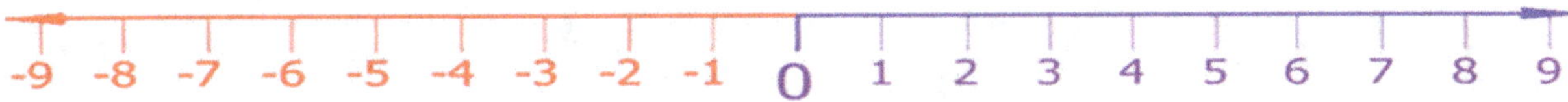

Plot **3**$\frac{3}{10}$**, -5, 3.7,** and $\frac{7}{2}$ numbers on a number line and write in order from list to greatest.

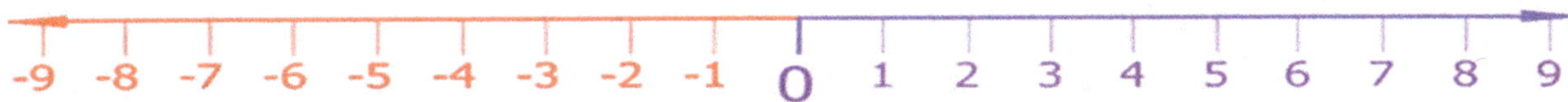

Match the numbers **6.7, -4**$\frac{1}{10}$**, 2.5, -1,** and **-7**$\frac{8}{9}$ with their positions on the number line.

Graph the numbers **7, -2**$\frac{1}{2}$**, 2, 5, -1,** and **-3** on a number line. Then write 3 inequalities that compare any two numbers from this number set.

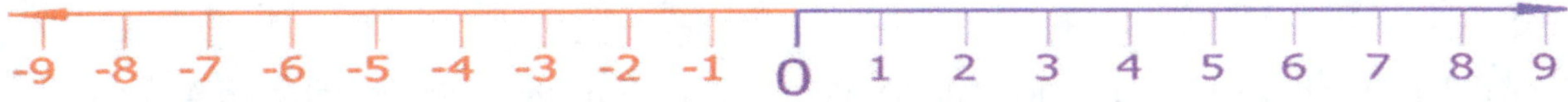

Graph the numbers **- 1**$\frac{1}{2}$**, -**$\frac{3}{2}$**, -**$\frac{9}{6}$**,** $\frac{1}{4}$**,** $\frac{2}{8}$**, 3,** $\frac{6}{2}$**,** on a number line.

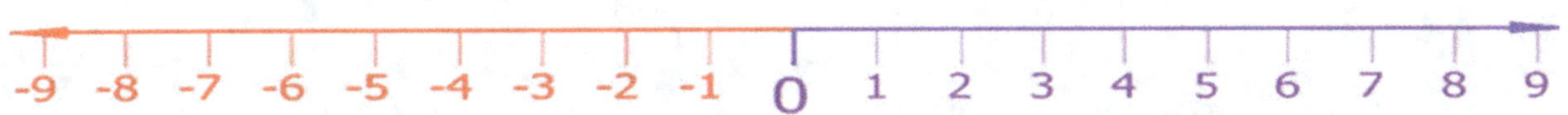

CHALLENGE

Which point has a coordinate $\left(-1\frac{2}{3}\right)^2$?

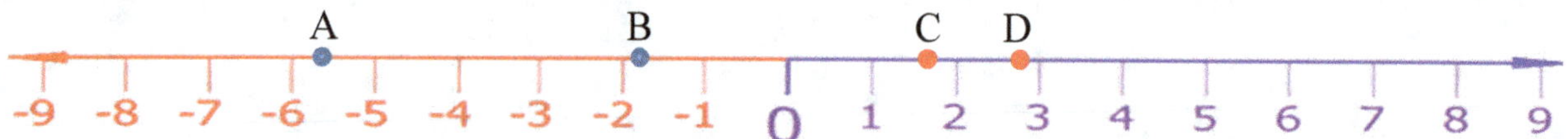

What is the value of $\frac{AC}{BD}$?

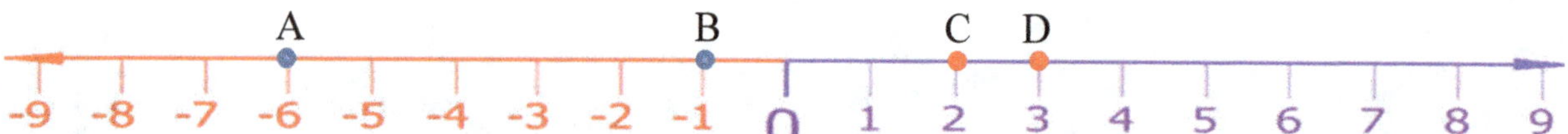

The tick marks have equal distances on the number line. Find the value of **x**.

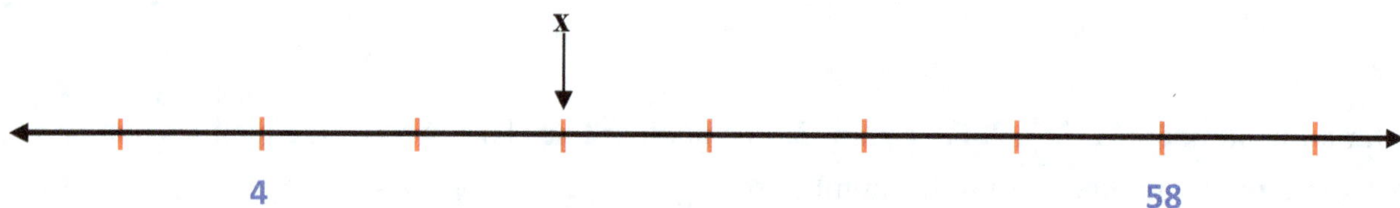

The tick marks have equal distances on the number line. Find the distance between **A** and **B**.

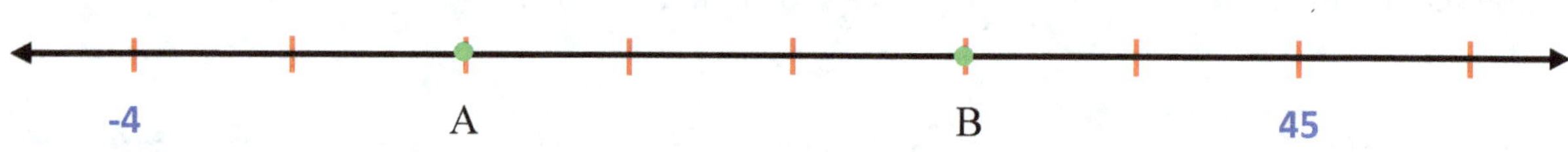

FACTORS, PRIME FACTORS, AND PRIME FACTORIZATION

Knowledge of the definitions of a factor, a prime factor, and the factors of the number is important throughout Algebra. There are also two important operations: factoring of a number and prime factorization of a number. We will use the following definitions.

DEFINITIONS

▶ A *factor* of a given number is a number that can divide the given number evenly without a remainder, or **a** is a factor for **b** if there exists a nonzero **c**, such that **b = ac**.

For example, **3** is a factor of **6** because **6÷3 = 2** or because **6 = 3·2**. By the same reason **2** is also a factor for **6**. Two other factors of **6** are **1** and **6**.

▶ The set of *all factors* of a given number is the set of numbers that each can divide the given number evenly.

For example, all factors of **35** are **1, 5, 7,** and **35**. All factors of **24** are **1, 2, 3, 4, 6, 8, 12,** and **24**.

▶ A *prime factor* is a factor which is also a prime number.

For example, **5** and **7** are prime factors of **35**. Numbers **2** and **3** are prime factors of **48**.

▶ The *factoring* of a given number is a representation of the number as a product of its any two or more factors.

For example, **15 = 5·3 = 15·1 = 5· 3 ·1**, or **75 = 5·5·3 = 25·3 = 5·15 = 75·1**.

▶ The *prime factorization* of a given number is a representation of the number as a product in only prime factors.

For example, prime factorization of **35** is **35 = 5·7** and prime factorization of **48** is **48=2·2·2·2·3**.

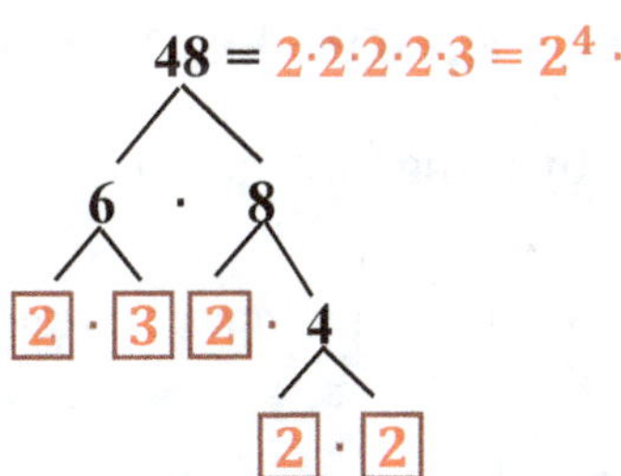

▶ Two numbers are *relatively prime* if those numbers have no common factors except for **1**.

For example, numbers **35** (**35 = 5·7**) and **48** (**48 = 2·2·2·2·3**) are relatively prime numbers.

Examples Find all factors and prime factors of numbers **4, 11,** and **12**.

Solution We will use definitions

■ $F(4)\ = 1, 2, 4$ ■ $PF(4)\ = 2$
■ $F(11) = 1, 11$ ■ $PF(11) = 11$
■ $F(12) = 1, 2, 3, 4, 6, 12$ ■ $PF(12) = 2, 3$

GCF/GCD AND LCM/LCD

There are different methods to find **GCF/GCD** and **LCM/LCD** of numbers.

Prime factorization method

- $12 = 2 \cdot 2 \cdot 3$ & $18 = 2 \cdot 3 \cdot 3$

  GCF/GCD$(12,18) = 2 \cdot 3 = 6$

  LCM$(12,18) = \underbrace{2 \cdot 3 \cdot 3}_{\substack{\text{geatest} \\ \text{number} \\ 18}} \underbrace{\cdot 2}_{\substack{\text{factor} \\ \text{from} \\ \text{other} \\ \text{number} \\ 12}} = 18 \cdot 2 = 36$

Factor method

- $F(12) = 1, 2, 3, 4, 6, 12$

  $F(18) = 1, 2, 3, 6, 9, 18$

  GCF/GCD$(12,18) = 6$

Prime factorization method

- $\dfrac{1}{12}$ & $\dfrac{1}{18}$

  LCD $\left(\dfrac{1}{12}, \dfrac{1}{18} \right) = $ LCM$(12,18) = 36$

Multiples method

- $M(12) = 12, 24, 36, 48, 60, 72, 84, 96, 108, ..$

  $M(18) = 18, 36, 54, 72, 90, 108$

  LCM$(12,18) = 36$

Examples Find the **GCF** and the **LCM** for the given numbers: **6 & 8**; **10 & 15**; and **12 & 18** by using any method.

Solution We will use the prime factorization method.

- $6 = 2 \cdot 3$ & $8 = 2 \cdot 2 \cdot 2$

  GCF$(6,8) = 2$

  LCM$(6,8) = \underbrace{2 \cdot 2 \cdot 2}_{\substack{\text{geatest} \\ \text{number} \\ 8}} \underbrace{\cdot 3}_{\substack{\text{factor} \\ \text{from} \\ \text{other} \\ \text{number} \\ 6}} = 24$

- $10 = 5 \cdot 2$ & $15 = 5 \cdot 3$

  GCF$(10,15) = 5$

  LCM$(10,15) = 5 \cdot 3 \cdot 2 = 30$

- $12 = 3 \cdot 2 \cdot 2$ & $18 = 3 \cdot 2 \cdot 3$

  GCF$(12,18) = 3 \cdot 2 = 6$

  LCM$(12,18) = 3 \cdot 2 \cdot 3 \cdot 2 = 36$

Write **all factors** and all **prime factors** of the numbers.

F(8) = PF(8) =	F(10) = PF(10) =	F(13) = PF(13) =	F(15) = PF(15) =
F(21) = PF(21) =	F(23) = PF(23) =	F(25) = PF(25) =	F(27) = PF(27) =
F(32) = PF(32) =	F(36) = PF(36) =	F(45) = PF(45) =	F(48) = PF(48) =

Write the missing numbers in the **prime factorization** of each number.

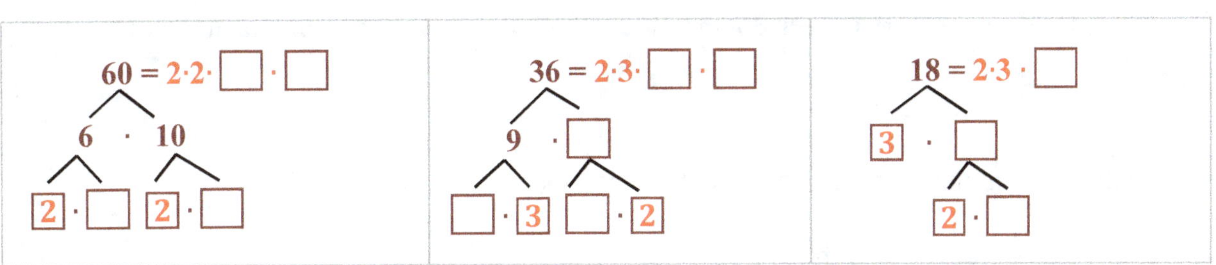

Which sets are sets of **relatively prime** numbers?

{4, 9}	{2, 6}	{4, 5, 6}	{7, 28, 35}	{9, 11, 25, 32}

Find the **GCF/GCD** or **LCM/LCD** by using the prime factorization and multiple methods.

2 & 3	15 & 25	$\frac{1}{8}$ & $\frac{1}{16}$
GCF/GCD(2,3) = LCM(2,3) =	GCF/GCD(15,25) = LCM(15,25) =	LCD $\left(\frac{1}{8}, \frac{1}{16}\right)$ =
2 & 4 & 6	**3 & 6 & 12**	**4 & 5 & 8**
GCF/GCD(2,4,6) = LCM(2,4,6) =	GCF/GCD(3,6,12) = LCM(3,6,12) =	GCF/GCD(4,5,8) = LCM(4,5,8) =

CHALLENGE

What are two factors of **24**, the product of which is **36**?

Is the perimeter of a triangle a prime number if the measures of the sides of that triangle are three consecutive numbers?

Find a composite number between **30** and **40** of which the sum of its prime factors is **12**.

Write the missing numbers for the prime factorization of each number.

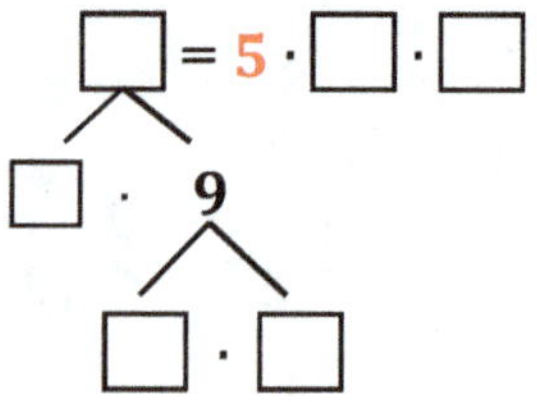

$$\square = 5 \cdot \square \cdot \square$$
$$\square \cdot 9$$
$$\square \cdot \square$$

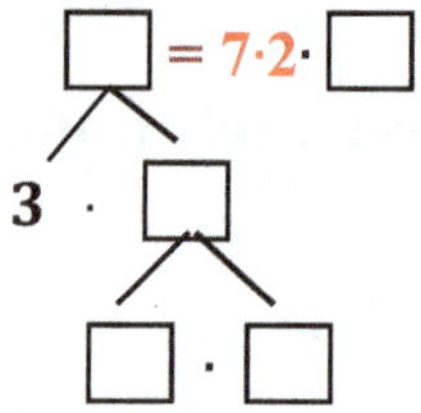

$$\square = 7 \cdot 2 \cdot \square$$
$$3 \cdot \square$$
$$\square \cdot \square$$

The **GCF** of **80** and a number **x** is **16**. The sum of digits of the number **x** is **13**. Find the minimum possible value of **x**.

The **GCF(x, y) = 1**. What is the **LCM(x, y)** of the same two numbers **x** and **y**?

The **GCF(x, y) = 5**, the **LCM(x, y) = 30**, and **x < 20, y < 20**. Find two numbers **x** and **y**.

■ SETS AND VENN DIAGRAMS

DEFINITION

In algebra there are many number sets. So it is useful for one to get acquainted with what is known as the **General Definition of Sets**.
We will use the following **general definitions of a set**.

> ▸ A *set* is a well-defined collection of objects called members or elements.
>
> ▸ A *set* is a group of elements.
>
> There are sets, which have a *finite number of elements*.
> There are also sets which have an *infinite number of elements*.
> There is also a type of sets, called *empty sets*, which have no elements. The symbol for empty set is $\emptyset$.
>
> Usually, a set is indicated by braces { }. For example, the **set of natural numbers** is the infinite set $N = \{1, 2, 3, 4, 5,...\}$.

Some Examples:

1. Books on the table, pencils in the pencil-box, passengers in an airplane, and face-colored cubes in the Rubik's cube are all examples of sets with a **finite number of elements**.
2. The sets of natural numbers, whole numbers, integers, rational numbers, irrational and real numbers are all examples of sets with an **infinite number of elements**.
3. The set of molecules in a perfect vacuum is an example of an **empty set**.

Sets are usually denoted by capital letters. The elements are not always indicated.
For example, the set **A**, the set **B**, the set $C = \{2,4,6\}$, or the set of natural numbers
$N = \{1, 2, 3, 4, 5,...\}$, all represent different sets. The elements of **A** and **B** are not indicated.
The notation $C = \{2, 4, 6\}$ means that the set **C** has three elements: **2, 4,** and **6**. The set
$N = \{1,2,3,4,...\}$ has an infinite number of elements (indicated by the **...** symbol).

We will use two different **notations**: *roster notation* (if we have listed elements inside braces) and *set-builder notation* (if we have ruled elements inside braces of the given set).
Roster notation for the set of natural numbers is $N = \{1,2,3,4,...\}$.
Set-builder notation for natural numbers greater than **100** is $\{n/n>100 \text{ and } n\in N\}$.
Set-builder notation for the inequality $2 < x \leq 5$ is $\{x|2 < x \leq 5\}$.

SYMBOLS

Some of the symbols used in set theory are presented below with their respective meanings and examples. When we discuss the number sets or sets of solutions of inequalities, we use the following symbolism.

Symbolism	How to Read	Meaning	Example
$A = \{a,b,c,d,e\}$	A is the set with a finite number of elements a,b,c,d,e	a set of elements, or a group of elements, or a collection of elements	$A = \{-1,0,1,2,3\}$
$B = \{b_1,b_2,b_3,\ldots\}$	B is a set with an infinite number of elements b_1,b_2,b_3, and so on	a set of an infinite number of elements	$B = \{5,10,15,\ldots\}$
$C = \{x\mid a<x\le b\}$	C is a set of numbers x which are greater than a and less than or equal to b	a set of elements as an interval on the number line	$C = \{x\mid 2< x\le 9\}$
$D = \varnothing$	D is a null set, or an empty set	a set that does not contain any elements	$D = \{\,\}=\varnothing$
$a \in A$ $b_1 \in B$ $x \in C$	a is an element of set A b_1 is an element of set B x is an element of set C	a is in set A b_1 is in set B x is in set C	$-1\in Z,\ 0\in W$ $5\in N,\ 10\in N$ $x\in R$
/	is not	$\notin$ is not an element of $\neq$ is not equal to $\not\subset$ is not a subset of	$-4\notin N$ $-1\neq 0$ $R \not\subset N$

Examples

EXAMPLE 1: Write the set of all odd integers in
  a). **roster notation** and in
  b). **set-builder notation**

Solution

a). **Roster notation**

$$Z^{2k+1} = \{\ldots,- 5, - 3, -1, 1, 3, 5,\ldots\} \qquad \text{Listed elements}$$

b). Set-builder notation

$$Z^{2k+1} = \{n | n = 2k+1, \quad k = 0, \pm 1, \pm 2, \pm 3, \dots \}$$ Ruled elements

Three dots indicate that the numbers continue.

EXAMPLE 2: Write the set of all even integers in
a). **roster notation** and in
b). **set-builder notation**

Solution

a). **Roster notation**

$$Z^{2k} = \{\dots, -6, -4, -2, 0, 2, 4, 6, \dots\}$$ Listed elements

b). **Set-builder notation**

$$Z^{2k} = \{n | n = 2k, \quad k = 0, \pm 1, \pm 2, \pm 3, \dots \}$$ Ruled elements

VENN DIAGRAMS AND OPERATIONS WITH SETS

We use graphical diagrams - Venn Diagrams - to show **sets**, **set relationships** and **set operations**. For example, the **Venn Diagram** of a set of digits is shown below.

$$
\begin{array}{ccccc}
0 & 2 & 4 & 6 & \\
1 & 3 & 5 & 7 & 9
\end{array}
$$

In Algebra, when you want to solve inequalities or when you need to find the domains of functions, you have to use two operations between sets: *union* of sets and *intersection* of sets.

The Venn Diagrams of these operations, corresponding symbols, and meanings are shown below.

Operation	Symbol	Venn Diagrams	Meaning / Example
A union B $x \in \mathbf{A} \cup \mathbf{B}$ if and only if $x \in \mathbf{A}$ or $x \in \mathbf{B}$	$\mathbf{A} \cup \mathbf{B}$		The union of two sets **A** and **B**, denoted as **A ∪ B**, is a set of elements of which are in **A or** in **B**. **Example:** A = {-3, -2, -1, 0, 1} B = {-1, 0, 1, 2, 3} A ∪ B = {-3, -2, -1, 0, 1, 2, 3}
A intersection B $x \in \mathbf{A} \cap \mathbf{B}$ if and only if $x \in \mathbf{A}$ and $x \in \mathbf{B}$	$\mathbf{A} \cap \mathbf{B}$		The intersection of two sets **A** and **B**, denoted as **A ∩ B**, is a set of elements of which are in both sets **A and B** at the same time. **Example:** A = {-3, -2, -1, 0, 1} B = {-1, 0, 1, 2, 3} A ∩ B = { -1, 0, 1}

Examples

EXAMPLE 1: Find **A ∪ B** and **A ∩ B** if A = {-5, -3, -1, 0, 1, 3, 5} and B = {- 4, -2, 0, 2, 4}

Solution

$$A \cup B = \{-5, -4, -3, -2, -1, 0, 1, 2, 3, 4, 5\}; \quad A \cap B = \{0\}$$

EXAMPLE 2: Find **A ∪ (B ∩ C)** if A = {-3, -1, 0, 1, 3 } , B = {-2, 0, 2}, and C = {- 4, 0, 2, 5}

Solution

$$A \cup (B \cap C) = \{-3, -1, 0, 1, 3\} \cup \{0, 2\} = \{-3, -1, 0, 1, 2, 3\}$$

Solution

A ∪ B = 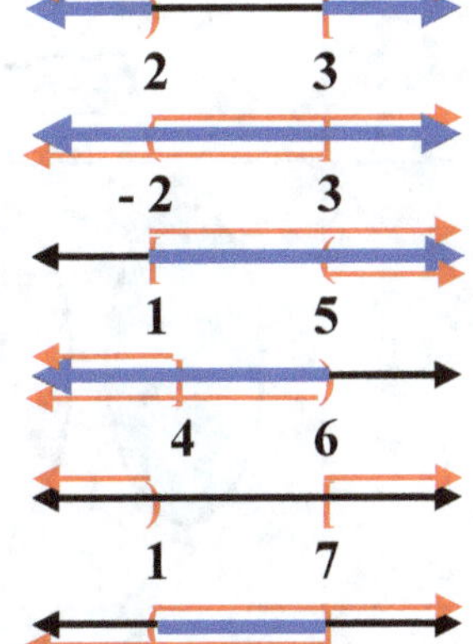

A ∩ B = ∅

EXAMPLE 4: Find

a) $x < 2 \;\cup\; x \geq 3$
b) $x > -2 \;\cup\; x < 3$
c) $x \geq 1 \;\cup\; x > 5$
d) $x \leq 4 \;\cup\; x < 6$
e) $x < 1 \;\cap\; x \geq 7$
f) $x > 2 \;\cap\; x < 4$
g) $x \geq 2 \;\cap\; x > 9$
h) $x \leq 3 \;\cap\; x < 8$

Solution

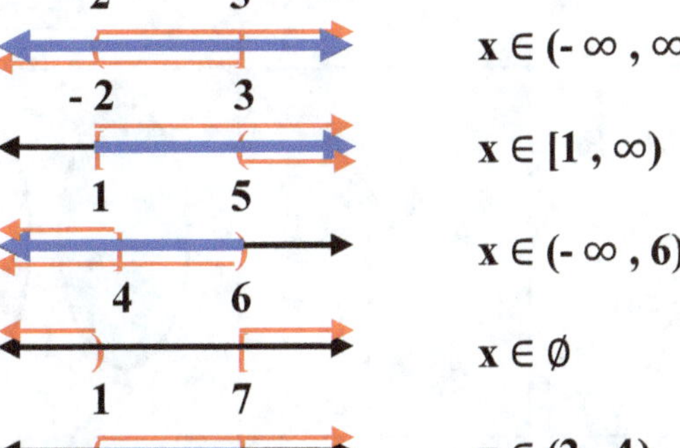

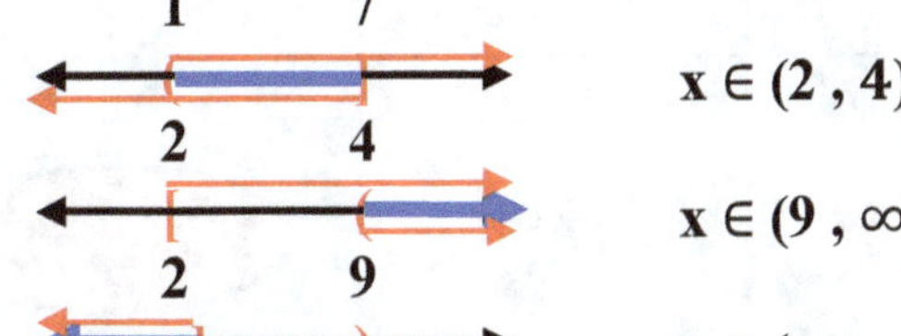

a) $x < 2 \;\cup\; x \geq 3$ $x \in (-\infty, 2) \cup [3, \infty)$

b) $x > -2 \;\cup\; x < 3$ $x \in (-\infty, \infty)$

c) $x \geq 1 \;\cup\; x > 5$ $x \in [1, \infty)$

d) $x \leq 4 \;\cup\; x < 6$ $x \in (-\infty, 6)$

e) $x < 1 \;\cap\; x \geq 7$ $x \in \emptyset$

f) $x > 2 \;\cap\; x < 4$ $x \in (2, 4)$

g) $x \geq 2 \;\cap\; x > 9$ $x \in (9, \infty)$

h) $x \leq 3 \;\cap\; x < 8$ $x \in (-\infty, 3]$

RELATIONS OF SETS

The **Venn Diagrams** of relations between sets, corresponding symbols, and meanings are shown below.

Relation of Sets	Symbolism	Venn Diagram	Meaning / Example
A is equal to B	**A = B**		**A = B** if both sets **A** and **B** contain exactly the same elements. Example: A = {-1,0,1,2}; B = {-1,0,1,2}
A is subset of B or **B is superset of A** A superset is the exact opposite of a subset: if set **A** is a subset of set **B**, then **B** is a superset of **A**.	$A \subset B$ or $B \supset A$ We also use the other subset/superset symbols $\subseteq$ and $\supseteq$.	$A \subset B$ or $B \supset A$ $A \not\subset B$	The set **A** is a subset of set **B** if every element of set **A** is also the element of the set **B**. or The set **B** is a superset for set **A** if the set **A** is a subset of the set **B**. Example: A = {-1,0,1,2}; B = {-2,-1,0,1,2,4} $A \subset B$ or $B \supset A$ Note: Set **A** is **not** a subset of **B** ($A \not\subset B$) or set **B** is **not** a superset of **A** if an element of **A** that is not element of **B** exists.
A separation B or **A disjoint B**	$A \cap B = \varnothing$		The sets **A** and **B** are separated or disjointed sets if they have no common elements. Example: A = {-3,-2,5,6,7}; B = {-1,0,1,2, 3,4} $A \cap B = \varnothing$
A^c complements A	A^c		A^c is a complement set of set **A** if the elements of A^c are all elements which are not in the set **A**. Example: A = {1,2,3}, A^c = {4,5,6,...}

EXAMPLE 1: Find some relations between $A = \{-5, -3, -1, 0, 1, 3, 5\}$; $B = \{0, 1, 3, 5\}$;
and $C = \{-5, -4, -3, -2, -1, 0, 1, 2, 3, 4, 5\}$.

Solution

$$B \subset A \quad (\text{or } A \supset B)$$
$$B \subset C \quad (\text{or } C \supset B)$$
$$A \subset C \quad (\text{or } C \supset A)$$
$$A \cup B = A \subset C \quad (\text{or } C \supset A \cup B)$$
$$A \cap B = B \subset C \quad (\text{or } C \supset A \cap B)$$
$$A \cap B \subset A \cup B$$

EXAMPLE 2: Find the relations between Rational Number Set **Q,** Irrational Number Set **I,** and
Real Number Set **R.**

Solution

$$Q \subset R; \quad I \subset R; \quad I \not\subset Q; \quad Q \not\subset I$$

EXAMPLE 3: Find the relations between EVEN Number Set $\mathbf{Z^{2k}}$, Odd Number Set $\mathbf{Z^{2k+1}}$, Prime
Number Set **P,** Composite Number Set **C,** and Set of Integers **Z.**

Solution

$$Z^{2k} \subset Z; \quad Z^{2k+1} \subset Z; \quad Z^{2k} \not\subset Z^{2k+1}; \quad P \subset Z; \quad P \not\subset C$$

EXAMPLE 4: Find results of operations between sets $\mathbf{N, W, Z, Q, I, Z^{2k}, Z^{2k+1}, P, C}$ and **R.**

Solution

There are results of some operations. For example,

$$N \cup \{0\} = W, \; N \cup W = W, \; N \cap W = N;$$
$$W \cup Z = Z, \; W \cap Z = W, \; Z \cup Z = Z, \; Z \cap Z = Z;$$
$$W \cup Q = Q, \; W \cap Q = W, \; Z \cup Q = Q, \; Z \cap Q = Z;$$
$$Q \cup Q = Q, \; Q \cap Q = Q, \; Q \cup I = R, \; Q \cap I = \varnothing;$$
$$R \cup I = R, \; I \cap R = I, \; R \cup Q = R, \; Q \cap R = Q, \; Z^{2k} \cup Z^{2k+1} = Z, \; Z^{2k} \cap Z^{2k+1} = \varnothing;$$
$$P \cap C = \varnothing, \; P \cup C = \{2, 3, 4, 5, 6, 7, ...\}, \; P \cap Z^{2k} = \{2\}, \; P \cup Z^{2k+1} = Z^{2k+1}$$

Write a set of all whole numbers greater than **71** and less than **90** in

a) roster notation

b) set-builder notation

Let set $A = \{-4, -3, -2, -1,\ 0,\ 1,\ 2,\ 3,\ 5\}$ and set $B = \{-3, -1,\ 1,\ 3,\ 5\}$.

Find

a) $A \cup B$

b) $A \cap B$

c) $A - B$

d) $B - A$

e) Is $A \subset B$?

f) Is $B \subset A$?

g) Are sets A and B disjoint/separate sets?

Let set $A = \{1,\ 2,\ 3,\ 4,\ 5\}$ and set $B = \{1,\ 2,\ 3,\ 5,\ 6,\ 7\}$.

Find

a) $(A \cup B) \cup B$

b) $(A \cup B) \cap B$

c) $(A \cap B) \cup B$

d) $(A \cap B) \cap B$

e) $(A \cup B) \cup \emptyset$

f) $(A \cup B) \cap \emptyset$

CHALLENGE

Write set **A** using **roster notation** and **set-builder notation** if:

a) **A** is set of whole numbers less than or equal to **11**.
b) **A** is set of even numbers between **0** and **31**.
c) **A** is set of prime numbers between **1** and **50**.
d) **A** is set of prime numbers sum of which is **10**.

Let set $A = \{-4, -3, -2, -1,\ 0,\ 1,\ 2,\ 3,\ 5\}$ and set $B = \{-3, -1,\ 0,\ 1,\ 3\}$.

Find

a) $(A - B) \cup A$

b) $(A - B) \cap A$

c) $(A - B) \cup B$

d) $(A - B) \cap B$

e) $(A - B) \cup \varnothing$

f) $(A - B) \cap \varnothing$

g) $(A \cup B) \cup (A - B)$

h) $(A \cup B) \cap (A - B)$

i) $(A - B) \cup (A \cup B)$

j) $(A - B) \cap (A \cup B)$

k) $(A - B) \cup (A \cap B)$

l) $(A - B) \cap (A \cap B)$

m) $(A \cup B) \cup (A - B)$

n) $(A \cup B) \cap (A - B)$

1.2. Operations and Properties

- **GENERAL RULES OF BASIC OPERATIONS**
- **OPERATIONS WITH INTEGERS**
- **OPERATIONS WITH FRACTIONS AND MIX NUMBERS**
- **OPERATIONS WITH DECIMALS**
- **FRACTIONS, DECIMALS, AND PERCENTS**
- **USING VARIABLES**
- **PROPERTIES OF BASIC OPERATIONS**

■ GENERAL RULES OF BASIC OPERATIONS

In Algebra there are **4** basic operations between real numbers:
Addition (as **a + b**).
Subtraction (as **a - b**).
Multiplication (as **a x b, a · b,** or **ab**).
Division (as **a÷b, b)$\overline{\text{a}}$** , $\frac{a}{b}$ or **a/b**).

There are also **3** additional operations with real numbers: **Absolute Value** (as **|a|**),
Power (as **(a)^n**), and **Power Root** (as $\sqrt[n]{\text{a}}$).

There are the following general rules of basic operations with real numbers.

GENERAL RULES OF BASIC OPERATIONS

If we have addition (or subtraction) with two real numbers written with the same signs (both positive, or both negative), then we will add these two numbers and put the common sign in front of the result. **Examples: 2 + 3 = 5; -2 - 3 = -5**	If we have multiplication between two real numbers written with the same signs (both positive, or both negative), then we will multiply these two numbers and put a positive sign (or nothing) in front of the result. **Examples: 2 · 3 = 6; (-2) · (-3) = 6**	If we have division between two real numbers written with the same signs (both positive, or both negative), then we will divide these two numbers and put a positive sign (or nothing) in front of the result. **Examples: 6÷3= 2; (-6)÷(-3) = 2**
If we have addition (or subtraction) with two real numbers written with different signs (one positive, and one negative), then we will subtract the smaller number from the larger number and put the sign of the larger number in front of the result. **Examples: -2 + 3 = 1; 2 - 3 = -1**	If we have multiplication between two real numbers written with different signs (one positive, and the other negative), then we will multiply these two numbers and put a negative sign in front of the result. **Examples: 2·(-3) = - 6; (-2) ·3= - 6**	If we have division between two real numbers written with different signs (one positive, and the other negative), then we will divide these two numbers and put a negative sign in front of the result. **Examples: 6÷(-3) =-2; (-6)÷3=-2**

■ OPERATIONS WITH INTEGERS

The set of integers is an infinite set of numbers $\mathbf{Z} = \{\ldots -3, -2, -1, 0, 1, 2, 3 \ldots\}$.

We will use the following important rules for basic operations of integers which come from the general rules of basic operations with real numbers.

Addition and Subtraction of Integers	Multiplication and Division of Integers
■ $4 + 2 = 6$	■ $4 \cdot 2 = 8$
■ $-4 - 2 = -6$	■ $(-4) \cdot (-2) = 8$
■ $-4 + 2 = -2$	■ $(-4) \cdot (2) = -8$
■ $4 - 2 = 2$	■ $(4) \cdot (-2) = -8$
■ $4 + (+2) = 4 + 2 = 6 \quad [\ +(+2) = +2\ ]$	■ $4 \div 2 = 2$
■ $4 - (-2) = 4 + 2 = 6 \quad [\ -(-2) = +2\ ]$	■ $(-4) \div (-2) = 2$
■ $4 + (-2) = 4 - 2 = 2 \quad [\ +(-2) = -2\ ]$	■ $(-4) \div (2) = -2$
■ $4 - (+2) = 4 - 2 = 2 \quad [\ -(+2) = -2\ ]$	■ $(4) \div (-2) = -2$

Example 1 Evaluate the expression $(-1) \cdot 2 - (-(-3))$

Solution

$$(-1) \cdot 2 - (+3) = -2 - 3 = -5$$

Example 2 Evaluate the expression $(-5) \cdot (-1) - (-6) + (-3) \cdot (4)$

Solution

$$(-5) \cdot (-1) - (-6) + (-3) \cdot (4) = 5 + 6 - 12 = 11 - 12 = -1$$

Example 3 Evaluate the expression $(-4) \cdot 5 - 12 \div (-3) + (-2)$

Solution

$$(-4) \cdot 5 - 12 \div (-3) + (-2) = -20 + 4 - 2 = -16 - 2 = -18$$

Example 4 Evaluate the expression $(-2) \cdot (-3) - (-6) \div (-3) - (+5)$

Solution

$$(-2) \cdot (-3) - (-6) \div (-3) - (+5) = 6 - 2 - 5 = 4 - 5 = -1$$

ADDITION AND SUBTRACTION

Find the results of each operation.

$-8 + 4$	$-7 - 14$	$6 - 15$	$7 + 43$	$-10 - 3$
$-2 + 5$	$-3 - 25$	$2 - 18$	$-21 + 7$	$-65 + 15$
$-7 + (-2)$	$-6 - (-41)$	$77 - (+87)$	$12 + (-4)$	$-11 + (+3)$
$-17 - (+15)$	$-13 - (-16)$	$1 - (-18)$	$-24 + (-17)$	$-14 - (-12)$
$-18 + (+14)$	$-22 - (-13)$	$34 - (-12)$	$11 + (-15)$	$-17 - (+25)$
$-24 + (-21)$	$-23 - (-35)$	$1 - (-14) - 15$	$7 - (-24) - (+7)$	$-13 - (+7) - 3$
$-7 + 24 - (-62)$	$-6 - (-4) + (-7)$	$17 - 8 - (-9)$	$12 + 24 + (-10)$	$-11 + 13 - 21$
$-32 + 19 - 18$	$5 - (-17) - (+16)$	$31 + 5 - (+7)$	$16 - 6 - (-9)$	$-4 + 14 + (-14)$

MULTIPLICATION AND DIVISION

Find the results of each operation.

$(-2) \cdot (-9)$	$(-4) \cdot (8)$	$(2) \cdot (-5)$	$(7) \cdot (0)$	$7 \cdot 8$
$(-12) \div (-3)$	$(-54) \div (6)$	$(25) \div (-5)$	$(72) \div (8)$	$(-27) \div (-9)$
$(-8) \cdot (-2) \cdot (-5)$	$(-10) \cdot (7) \cdot (5)$	$(11) \cdot (-6) \cdot (9)$	$(-9) \cdot (-3) \cdot (10)$	$(-10) \cdot (-8) \cdot (-4)$
$(-8) \div (2) \div (-4)$	$(-48) \div (6) \cdot 2$	$(55) \div (5) \div (-11)$	$(63) \div (9) \div (7)$	$(-45) \div (-9) \cdot 5$
$(-20) \cdot (-3) \div 6$	$(-7) \cdot (5) \div (-5)$	$(15) \cdot (-3) \div (-9)$	$(55) \div (-5) \cdot (2)$	$(-8) \cdot (-2) \div (-4)$
$(-36) \div (-9) \cdot 7$	$(-36) \div (4) \cdot 5$	$(25) \div (-5) \cdot (-2)$	$(-2) \cdot (27) \div (9)$	$(-24) \div (-6) \cdot 2$

CHALLENGE

Complete each number square so the rows, columns, and diagonals have the same sum.

		5
	−1	
−7		−3

Sum =

	−5	
−2	−9	−4

Sum =

	−9	1
	−1	
−3		

Sum =

−8		1	7
6	0		−7
	2		
	3		4

Sum =

			6
5		−2	
		1	−7
−6	2	−5	3

Sum =

−7			8
7		0	
6		3	
−4		−3	5

Sum =

Complete each number triangle if sides have the same sum.

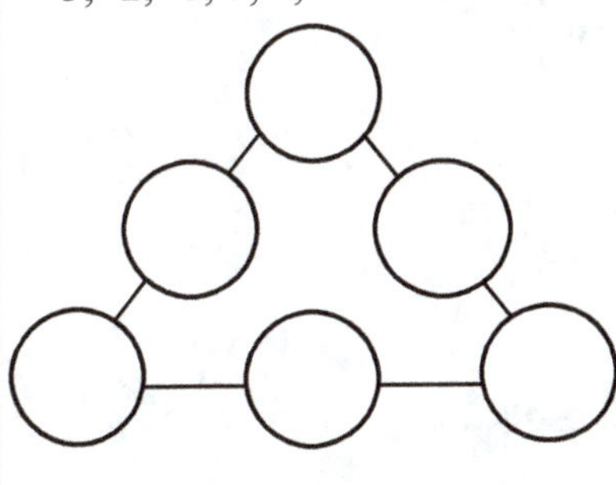

Arrange the integers
-3, -2, -1, 1, 2, 3 in the circles

Sum =

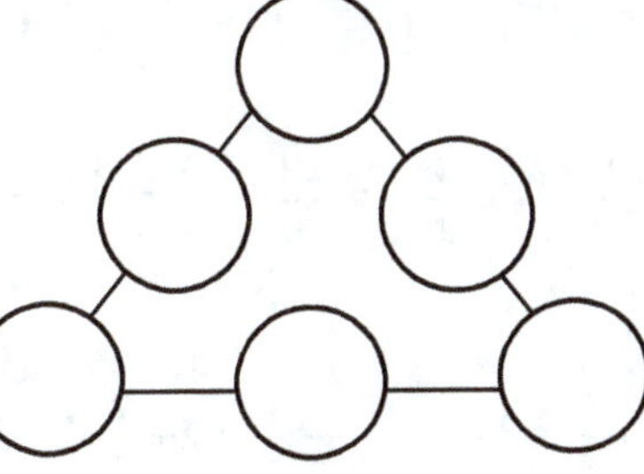

Arrange the integers
-1, 0, 1, 2, 3, 4 in the circles

Sum =

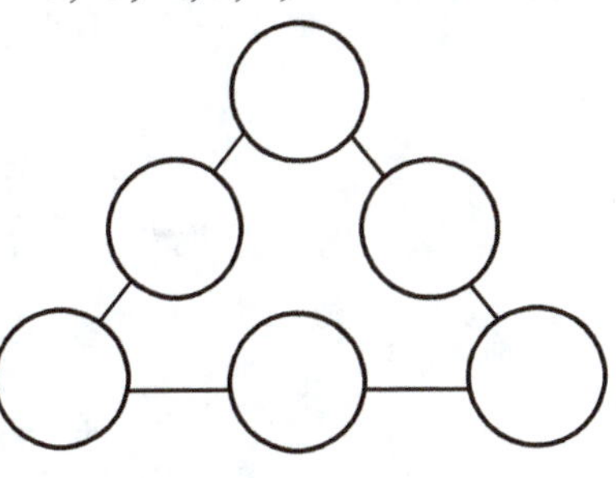

Arrange the integers
-5, -3, -2, 0, 1, 3 in the circles

Sum =

OPERATIONS WITH FRACTIONS AND MIX NUMBERS

The set of Fractions is the set of numbers $\mathbf{Q} = \left\{\frac{n}{m}\right\}$, where $m \neq 0$ and $n, m \in \mathbf{Z}$.

The Integer set is a subset for the Fraction set: $\mathbf{Z} \subset \mathbf{Q} \left(4 = \frac{4}{1}, \ -3 = \frac{-3}{1}, \ 0 = \frac{0}{1}\right)$.

The set of Mixed Numbers $\mathbf{M} = \left\{p\frac{n}{m}, \ m \neq 0 \text{ and } p, n, m \in \mathbf{Z}\right\}$ is also a subset of $\mathbf{Q}$: $\mathbf{M} \subset \mathbf{Q}$.

We will use the following important rules for basic operations of fractions and mixed numbers.

Adding and subtracting with like denominators

- $\frac{1}{3} + \frac{2}{3} = \frac{3}{3} = 1$ **(LCD = 3)** ;

- $\frac{1}{4} - \frac{3}{4} = \frac{-2}{4} = -\frac{1}{2}$ **(LCD = 4)**

- $\frac{1}{5} + 2\frac{3}{5} = 2\frac{4}{5}$ **(LCD = 5)** ;

- $3\frac{1}{5} - 1\frac{4}{5} = \frac{16}{5} - \frac{9}{5} = \frac{7}{5} = 1\frac{2}{5}$ **(LCD = 5)**

- $1 + \frac{2}{3} = 1\frac{2}{3}$

- $\frac{1}{8} - 1 = \frac{1}{8} - \frac{8}{8} = \frac{1-8}{8} = \frac{-7}{8}$ **(LCD = 8)**

Adding and subtracting with unlike denominators

- $\frac{1}{8} + 2\frac{3}{4} = \frac{1}{8} + 2\frac{3 \cdot 2}{4 \cdot 2} = 2\frac{7}{8}$ **(LCD = 8)**

- $3\frac{5}{8} - 1\frac{1}{3} = 3\frac{5 \cdot 3}{8 \cdot 3} - 1\frac{1 \cdot 8}{3 \cdot 8} = 2\frac{7}{24}$ **(LCD = 24)**

Two ways to add and subtract mixed numbers

- $8\frac{1}{3} + 2\frac{3}{5} =$

  $\frac{25}{3} + \frac{13}{5} = \frac{25 \cdot 5}{3 \cdot 5} + \frac{13 \cdot 3}{5 \cdot 3} = \frac{125}{15} + \frac{39}{15} = \frac{164}{15} = 10\frac{14}{15}$

  $10 + \frac{1}{3} + \frac{3}{5} = 10 + \frac{1 \cdot 5}{3 \cdot 5} + \frac{3 \cdot 3}{5 \cdot 3} = 10 + \frac{14}{15} = 10\frac{14}{15}$

  (LCD = 15)

- $3\frac{2}{3} - 1\frac{1}{2} =$

  $\frac{11}{3} - \frac{3}{2} = \frac{11 \cdot 2}{3 \cdot 2} - \frac{3 \cdot 3}{2 \cdot 3} = \frac{22}{6} - \frac{9}{6} = \frac{13}{6} = 2\frac{1}{6}$

  $3 + \frac{2}{3} - 1 - \frac{1}{2} = 3 - 1 + \frac{2}{3} - \frac{1}{2} = 2 + \frac{2 \cdot 2}{3 \cdot 2} - \frac{1 \cdot 3}{2 \cdot 3} = 2 + \frac{1}{6} = 2\frac{1}{6}$

  (LCD = 6)

Multiplication

$$\blacksquare\ \frac{1}{2}\cdot\frac{3}{5}=\frac{1\cdot3}{2\cdot5}=\frac{3}{10}\qquad\blacksquare\ \frac{\overset{1}{\cancel{2}}}{\underset{1}{\cancel{3}}}\cdot\frac{\overset{1}{\cancel{3}}}{\underset{2}{\cancel{4}}}=\frac{1}{2}\qquad\blacksquare\ 2\cdot\frac{1}{4}=\frac{\overset{1}{\cancel{2}}}{1}\cdot\frac{1}{\underset{2}{\cancel{4}}}=\frac{1}{2}\qquad\blacksquare\ 2\frac{1}{4}\cdot1\frac{1}{3}=\frac{\overset{3}{\cancel{9}}}{\underset{1}{\cancel{4}}}\cdot\frac{\overset{1}{\cancel{4}}}{\underset{1}{\cancel{3}}}=3\qquad\blacksquare\ 4\cdot3\frac{1}{4}=\frac{4}{1}\cdot\frac{13}{\cancel{4}}=13$$

Division

$$\blacksquare\ \frac{1}{2}\div\frac{3}{5}=\frac{1\cdot5}{2\cdot3}=\frac{5}{6}\qquad\blacksquare\ 2\div\frac{2}{3}=\frac{\overset{1}{\cancel{2}}}{1}\cdot\frac{3}{\underset{1}{\cancel{2}}}=3\qquad\blacksquare\ 2\frac{1}{4}\div4\frac{1}{2}=\frac{9}{4}\div\frac{9}{2}=\frac{\overset{1}{\cancel{9}}}{\underset{2}{\cancel{4}}}\cdot\frac{\overset{1}{\cancel{2}}}{\underset{1}{\cancel{9}}}=\frac{1}{2}\qquad\blacksquare\ 5\div1\frac{1}{4}=\frac{5}{1}\cdot\frac{4}{\underset{1}{\cancel{5}}}=4$$

Example 1 Evaluate the expression $\dfrac{1}{7}+\dfrac{5}{7}-\dfrac{4}{7}+\dfrac{1}{5}$

Solution

$$\frac{1}{7}+\frac{5}{7}-\frac{4}{7}=\frac{1+5-4}{7}+\frac{1}{5}=\frac{2}{7}+\frac{1}{5}=\frac{10+7}{35}=\frac{17}{35}$$

Example 2 Evaluate the expression $-\dfrac{1}{7}+3\dfrac{2}{5}-2$

Solution

$$-\frac{1}{7}+3\frac{2}{5}-2=-\frac{1}{7}+\frac{17}{5}-\frac{2}{1}=\frac{-5+114-70}{35}=\frac{39}{35}=1\frac{4}{35}$$

Example 3 Evaluate the expression $2\dfrac{1}{2}\cdot1\dfrac{1}{5}\cdot\left(-\dfrac{1}{3}\right)\cdot5$

Solution

$$2\frac{1}{2}\cdot1\frac{1}{5}\cdot\left(-\frac{1}{3}\right)\cdot5=\frac{5}{2}\cdot\frac{6}{5}\cdot\left(-\frac{1}{3}\right)\cdot\frac{5}{1}=-\frac{5}{2}\cdot\frac{6}{5}\cdot\frac{1}{3}\cdot\frac{5}{1}=-\frac{150}{30}=-5$$

Example 4 Evaluate the expression $3\dfrac{1}{2}\div1\dfrac{1}{2}\cdot\dfrac{3}{5}\div2$

Solution

$$3\frac{1}{2}\div1\frac{1}{2}\cdot\frac{3}{5}\div2=\frac{7}{2}\div\frac{3}{2}\cdot\frac{3}{5}\div\frac{2}{1}=\frac{7}{2}\cdot\frac{2}{3}\cdot\frac{3}{5}\cdot\frac{1}{2}=\frac{7}{10}$$

ADDITION AND SUBTRACTION

Find the result of each operation.

$\frac{1}{2}+\frac{1}{2}$	$\frac{3}{5}-\frac{1}{5}$	$\frac{1}{6}+\frac{5}{6}$	$\frac{3}{5}-\frac{4}{5}$	$\frac{1}{7}+\frac{5}{7}-\frac{4}{7}$
$\frac{1}{2}+\frac{1}{3}$	$\frac{3}{5}-\frac{1}{2}$	$\frac{1}{8}+\frac{2}{3}$	$\frac{2}{7}-\frac{1}{2}$	$\frac{1}{2}+\frac{3}{4}-\frac{2}{3}$
$1+\frac{3}{5}$	$\frac{3}{4}+2$	$3-\frac{5}{6}$	$\frac{1}{5}-4-\frac{1}{2}$	$\frac{1}{3}+\frac{1}{4}-5$
$1\frac{1}{6}+\frac{1}{6}$	$\frac{3}{8}+3\frac{1}{4}$	$2\frac{5}{12}-\frac{2}{4}+3$	$2\frac{1}{2}-\frac{5}{6}-\frac{1}{3}$	$4\frac{1}{3}-\frac{1}{9}+1$
$2\frac{4}{5}+\frac{3}{5}-\frac{1}{4}$	$3\frac{7}{10}-2\frac{1}{5}-\frac{3}{5}$	$4\frac{1}{5}-\frac{5}{6}+1$	$-2\frac{1}{4}+1\frac{2}{5}-1\frac{1}{3}$	$-2\frac{1}{7}-3\frac{2}{5}-2$
$2\frac{4}{5}+4\frac{1}{2}-1\frac{1}{3}$	$5\frac{1}{4}-3\frac{1}{7}-2\frac{2}{3}$	$3\frac{1}{3}-1\frac{5}{6}+5$	$3\frac{1}{2}+2\frac{1}{4}-4\frac{1}{5}$	$-4\frac{1}{6}-1\frac{2}{3}-5$

MULTIPLICATION AND DIVISION

Find the result of each operation.

$\left(-\frac{2}{3}\right)\cdot\left(-\frac{9}{14}\right)$	$\frac{3}{4}\cdot\left(-\frac{8}{9}\right)$	$(-3)\cdot\left(-\frac{5}{9}\right)\cdot3$	$(-4)\cdot3\frac{1}{2}$	$1\frac{1}{2}\cdot2\frac{1}{4}\cdot(-8)$
$(-8)\div\frac{4}{7}$	$\frac{3}{25}\div\frac{9}{75}$	$\left(-\frac{1}{3}\right)\div(-2)$	$\frac{3}{4}\div2\frac{1}{4}$	$\left(-\frac{1}{2}\div2\frac{1}{4}\right)\div\frac{2}{9}$
$\left(-\frac{2}{3}\right)\div\left(-\frac{1}{6}\right)$	$\frac{2}{5}\cdot\left(-\frac{3}{4}\right)\div\frac{3}{10}$	$(-2)\div\left(-\frac{2}{3}\right)\cdot\frac{1}{3}$	$(-4)\cdot3\frac{1}{2}\div7$	$2\frac{1}{2}\div1\frac{1}{5}\cdot(-3)$
$\frac{1}{7}\div\left(-\frac{3}{14}\right)$	$\frac{2}{5}\div\left(-\frac{1}{5}\right)\div\frac{1}{5}$	$(-2)\div\left(-\frac{1}{3}\right)\cdot\frac{1}{6}$	$(-2)\cdot3\frac{1}{2}\div7$	$3\frac{1}{2}\div1\frac{1}{2}\cdot\frac{3}{5}$

CHALLENGE

Find the missing integers.

$\dfrac{1}{\square} - \dfrac{2}{7} = \dfrac{\square}{14}$	$\dfrac{\square}{7} - \dfrac{1}{\square} = \dfrac{5}{14}$	$\dfrac{1}{3} - \dfrac{1}{\square} = \dfrac{1}{\square}$

Find the missing integers if all of the fractions are in proper and simplest form.

$70 \cdot \left(\dfrac{\square}{5}\right) \cdot \dfrac{2}{7} = 12$	$\dfrac{\square}{7} \div \dfrac{2}{\square} = \dfrac{3}{7}$	$\left(\dfrac{\square}{6}\right) \cdot \dfrac{2}{5} \div \left(\dfrac{\square}{5}\right) = \dfrac{1}{9}$

Put the digits 0, 1, 2, 3, 4, 5, 6, 7, 8, 9 (use each only once) in the boxes to get a true statement.
Find all the possibilities.

$$\square\,\square\,\dfrac{\square}{\square} + \square\,\square\,\dfrac{\square\,\square}{\square\,\square} = 100$$

Put the digits 0, 1, 2, 3, 4, 5, 6, 7, 8, 9 (use each only once) in the boxes to get a true statement.
Find all the possibilities.

$$\dfrac{\square\,\square}{\square\,\square} + \dfrac{\square\,\square\,\square}{\square\,\square\,\square} = 1$$

$$\dfrac{1}{2} + \dfrac{1}{2} = 1$$

■ OPERATIONS WITH DECIMALS

Whole Part   Decimal Part

A decimal number is a number that contains a decimal point: $\underbrace{2345675}\,.\,\underbrace{12587695}$.

Decimal Number

The number before the decimal point is the whole part of the decimal number.

The number after the decimal point is the decimal part of the decimal number.

We will use the following important rules for basic operations between decimals.

IMPORTANT RULES

Addition and Subtraction of Decimals	Multiplication and Division of Decimals

Addition and Subtraction of Decimals

■ $10.1 + 0.5$ ■ $15.2 + 1.576$ ■ $89 + 5.25$

$$
\begin{array}{r} 10.1 \\ +\ 0.5 \\ \hline 10.6 \end{array}
\qquad
\begin{array}{r} 15.200 \\ +\ 1.576 \\ \hline 16.776 \end{array}
\qquad
\begin{array}{r} ^{1} \\ 89.00 \\ +\ 5.25 \\ \hline 94.25 \end{array}
$$

■ $16.7 - 3.9$ ■ $19.6 - 3.258$ ■ $23 - 6.97$ ■ $11.3 - 7$

$$
\begin{array}{r} 16.7 \\ -\ 3.9 \\ \hline 12.8 \end{array}
\qquad
\begin{array}{r} 19.600 \\ -\ 3.258 \\ \hline 16.342 \end{array}
\qquad
\begin{array}{r} 23.00 \\ -\ 6.97 \\ \hline 16.03 \end{array}
\qquad
\begin{array}{r} 11.3 \\ -\ 7.0 \\ \hline 4.3 \end{array}
$$

Multiplication and Division of Decimals

■ $10.1 \cdot 0.5$ ■ $5.2 \cdot 1.57$ ■ $82 \cdot 5.25$

$$
\begin{array}{r} 10.1 \\ \times\ 0.5 \\ \hline 5.05 \end{array}
\qquad
\begin{array}{r} 1.57 \\ \times\ 5.2 \\ \hline 314 \\ +785 \\ \hline 8.164 \end{array}
\qquad
\begin{array}{r} 5.25 \\ \times\ 82 \\ \hline 1050 \\ +4200 \\ \hline 430.50 \end{array}
$$

■ $16.2 \div 0.3$ ■ $19.8 \div 9$

$$
\begin{array}{r} 54 \\ 0.3\,)\overline{16.2} \\ -15 \\ \hline 12 \\ -12 \\ \hline 0 \end{array}
\qquad
\begin{array}{r} 2.2 \\ 9\,)\overline{19.8} \\ -18 \\ \hline 18 \\ -18 \\ \hline 0 \end{array}
$$

Example 1   Evaluate the expression $7.9 + 2 - 3.2$

Solution

$$7.9 + 2 - 3.2 = 9.9 - 3.2 = 6.7$$

Example 2   Evaluate the expression $10.3 \cdot 0.5 \div 5$

Solution

$$
10.3 \cdot 0.5 = \begin{array}{r} 10.3 \\ \times\ 0.5 \\ \hline 5.15 \end{array}
\qquad
\begin{array}{r} 1.03 \\ 5\,)\overline{5.15} \\ -5 \\ \hline 15 \\ -15 \\ \hline 0 \end{array}
$$

Find the result of each operation.

$0.6 + 0.3$	$1.24 + 5.15$	$12.7 + 1.345$	$9.99 + 0.117$	$10.25 + 8.75$
15.8 $+ 9.56$	1.14 $+ 6.09$	14.99 $+ 5.111$	4.57 $+ 5.432$	2.672 4.5 $+ 3.321$
$5.28 - 1.26$	$7.4 - 3.1$	$49.3 - 11.5$	$7.7 - 5.8$	$2.175 - 0.582$
10.87 $- 9.56$	9.14 $- 3.09$	52.002 $- 5.789$	8.4 $- 3.435$	22.67 $- 3.999$
$6.3 \cdot 2.4$	$3.3 \cdot 6.25$	$7.3 \cdot 23$	$10.3 \cdot 0.5$	$11.5 \cdot 4.2$
$5 \cdot 14.2$	$5.11 \cdot 3$	$7.34 \cdot 5.5$	$26.34 \cdot 31$	$6.48 \cdot 2.3$
$7.65 \div 5$	$15 \div 0.05$	$6.955 \div 1.3$	$64.45 \div 12.89$	$7.696 \div 1.6$
$4.75 \div 2.5$	$9.18 \div 3.6$	$72.3 \div 6$	$8.1 \div 0.9$	$17.25 \div 3.45$
$(5.9 - 1.6) \cdot 5$	$3.5 + 3.24 \div 9$	$(2.6 + 3.9) \cdot 2$	$3.5 \div (2.6 + 2.4)$	$8.1 \div 0.9 \cdot 1.6$

CHALLENGE

Find the missing integers.

$\square.6 + 3.\square = 5$	$\square.\square7 + 5.4\square = 9$	$\square.39 + 8.\square = 11.09$
$\square.5 - 2.\square = 4$	$\square.\square6 - 2.8\square = 3$	$\square.57 - 4.\square\square = 1.6$
$\square.6 \times 0.\square = 0.8$	$\square.\square8 \times 5 = 5.9$	$\square.39 + 8.\square = 11.09$
$\square.5 - 2.\square = 4$	$\square.\square6 - 2.8\square = 3$	$\square.57 - 4.\square\square = 1.6$

Find the missing integers if all fractions are proper fractions in simplest form.

$\left(\dfrac{\square}{6}\right) \cdot \square.6 = 0.1$	$4.\square \cdot 3 = 1\square.5$	$70 \cdot (\square.5) \cdot \dfrac{2}{7} = 10$

Find the missing integers if all fractions are proper fractions in simplest form.

$8.\square \div 0.\square = 9$	$11.9 \div \square = \square.7$	$\left(\dfrac{\square}{6}\right) \cdot \square.6 = 0.1$

■ FRACTIONS, DECIMALS, AND PERCENTS

CONVERTING RULES

The word "percent" come from the Latin *per centum* ("thoroughly hundred") and means "out of a hundred".

We will use the following definition of a percent.

DEFINITIONS

The **x percent** (**x%**) is a ratio that compares the number **x** to **100**: $x\% = \dfrac{x}{100}$.

Any percent number with symbol **%** we can also write as a fraction or as a decimal.

Percent to fraction conversion comes from the definition $x\% = \dfrac{x}{100}$.

For example, $25\% = \dfrac{25}{100} = \dfrac{1}{4}$; $235\% = \dfrac{235}{100} = \dfrac{47}{20}$; $8\% = \dfrac{8}{100} = \dfrac{2}{25}$; $0.5\% = \dfrac{0.5}{100} = \dfrac{5}{1000} = \dfrac{1}{200}$.

Percent to decimal conversion is very easy; we have to move the decimal point two places to the left. For example, $25\% = 0.25$; $235\% = 2.35$; $8\% = 0.08$; $0.5\% = 0.005$.

We will use the following converting rules for decimals, fractions, and percents.

EXAMPLES FOR CONVERTING RULES

■ $20\% = \dfrac{20}{100} = 0.20$	■ $\dfrac{1}{4} = \dfrac{1 \cdot 25}{4 \cdot 25} = \dfrac{25}{100} = 0.25 = 25\%$	■ $0.18 = 18\% = \dfrac{18}{100}$
■ $2\% = \dfrac{2}{100} = 0.02$	■ $\dfrac{3}{10} = \dfrac{3 \cdot 10}{10 \cdot 10} = \dfrac{30}{100} = 0.3 = 30\%$	■ $0.04 = 4\% = \dfrac{4}{100}$
■ $152\% = \dfrac{152}{100} = 1.52$	■ $\dfrac{1}{8} = 0.125 = 12.5\%$	■ $1.7 = 170\% = \dfrac{170}{100}$
■ $0.8\% = \dfrac{0.8}{100} = 0.008$	■ $\dfrac{1}{3} \approx 0.333 = 33.3\%$	■ $0.007 = 0.7\% = \dfrac{0.7}{100}$

Example 1 Find the corresponding fraction and decimal for **75%**

Solution

$$75\% = \dfrac{75}{100} = \dfrac{3}{4} = 0.75$$

Example 2 Find the corresponding percent and decimal for $\dfrac{3}{10}$

Solution

$$\dfrac{3}{10} = 10\overline{)3} = 10\overline{)3.0} = 30\,\%$$

with the long division showing 0.3, -30, remainder 0.

Example 3 Find the corresponding percent and fraction for **0.08**

Solution

$$0.08 = 8\% = \dfrac{8}{100} = \dfrac{2}{25}$$

Example 4 Find the two other relations for the given number.

25%		
	$\dfrac{1}{5}$	
		0.5

Solution

25%	$\dfrac{25}{100}$	0.25
20%	$\dfrac{1}{5}$	0.2
50%	$\dfrac{50}{100}$	0.5

OPI FORMULA

We will use the **OPI formula** (created by Hayk Yegoryan in 2005) to solve many math and life problems related with percents.

OPI FORMULA

$$\boxed{\textbf{O}\text{f number}} \cdot \boxed{\text{Decimal equivalent of } \textbf{P}\text{ercent number}} = \boxed{\textbf{I}\text{s number}}$$

Examples

- What is **20%** of **160**?

  $20\% \rightarrow 0.20$

  $$\boxed{\begin{matrix}O\\160\end{matrix}} \cdot \boxed{\begin{matrix}P\\0.20\end{matrix}} = \boxed{\begin{matrix}I\\x\end{matrix}}$$

  $x = 160 \cdot 0.20 = \mathbf{32}$

- **30%** of what number is **15**?

  $30\% \rightarrow 0.30$

  $$\boxed{\begin{matrix}O\\x\end{matrix}} \cdot \boxed{\begin{matrix}P\\0.30\end{matrix}} = \boxed{\begin{matrix}I\\15\end{matrix}}$$

  $x = 15 \div 0.30 = \mathbf{50}$

- **20** is what percent of **80**?

  $0.25 \rightarrow 25\%$

  $$\boxed{\begin{matrix}O\\80\end{matrix}} \cdot \boxed{\begin{matrix}P\\x\end{matrix}} = \boxed{\begin{matrix}I\\20\end{matrix}}$$

  $x = 20 \div 80 = 0.25 = \mathbf{25\%}$

- What is **5%** of **200**?

  $5\% \rightarrow 0.05$

  $$\boxed{\begin{matrix}O\\200\end{matrix}} \cdot \boxed{\begin{matrix}P\\0.05\end{matrix}} = \boxed{\begin{matrix}I\\x\end{matrix}}$$

  $x = 200 \cdot 0.05 = \mathbf{10}$

- **0.5%** of what number is **10**?

  $0.5\% \rightarrow 0.005$

  $$\boxed{\begin{matrix}O\\x\end{matrix}} \cdot \boxed{\begin{matrix}P\\0.005\end{matrix}} = \boxed{\begin{matrix}I\\10\end{matrix}}$$

  $x = 10 \div 0.005 = \mathbf{2,000}$

- **100** is what percent of **20**?

  $5 \rightarrow 500\%$

  $$\boxed{\begin{matrix}O\\20\end{matrix}} \cdot \boxed{\begin{matrix}P\\x\end{matrix}} = \boxed{\begin{matrix}I\\100\end{matrix}}$$

  $x = 100 \div 20 = 5 = \mathbf{500\%}$

- What is **120%** of **60**?

  $120\% \rightarrow 1.20$

  $$\boxed{\begin{matrix}O\\60\end{matrix}} \cdot \boxed{\begin{matrix}P\\1.20\end{matrix}} = \boxed{\begin{matrix}I\\x\end{matrix}}$$

  $x = 60 \cdot 1.20 = \mathbf{72}$

- **2%** of what number is **50**?

  $2\% \rightarrow 0.02$

  $$\boxed{\begin{matrix}O\\x\end{matrix}} \cdot \boxed{\begin{matrix}P\\0.02\end{matrix}} = \boxed{\begin{matrix}I\\50\end{matrix}}$$

  $x = 50 \div 0.02 = \mathbf{2500}$

- **2** is what percent of **200**?

  $0.01 \rightarrow 1\%$

  $$\boxed{\begin{matrix}O\\200\end{matrix}} \cdot \boxed{\begin{matrix}P\\x\end{matrix}} = \boxed{\begin{matrix}I\\2\end{matrix}}$$

  $x = 2 \div 200 = 0.01 = \mathbf{1\%}$

- What is **25%** of **400**?

  $25\% \rightarrow 0.25$

  $$\boxed{\begin{matrix}O\\400\end{matrix}} \cdot \boxed{\begin{matrix}P\\0.25\end{matrix}} = \boxed{\begin{matrix}I\\x\end{matrix}}$$

  $x = 400 \cdot 0.25 = \mathbf{100}$

- **0.01%** of what number is **2**?

  $0.01\% \rightarrow 0.0001$

  $$\boxed{\begin{matrix}O\\x\end{matrix}} \cdot \boxed{\begin{matrix}P\\0.0001\end{matrix}} = \boxed{\begin{matrix}I\\2\end{matrix}}$$

  $x = 2 \div 0.0001 = \mathbf{20,000}$

- **80** is what percent of **4**?

  $20 \rightarrow 2,000\%$

  $$\boxed{\begin{matrix}O\\4\end{matrix}} \cdot \boxed{\begin{matrix}P\\x\end{matrix}} = \boxed{\begin{matrix}I\\80\end{matrix}}$$

  $x = 80 \div 4 = 20 = \mathbf{2,000\%}$

Find the two other relations for the given number.

50%	78%	9.7%	0.08%	125%
35%	6.3%	0.33%	260%	100%
$\frac{1}{2}$	$\frac{1}{4}$	$\frac{3}{8}$	$\frac{1}{10}$	$\frac{3}{25}$
$\frac{2}{3}$	$\frac{6}{8}$	$\frac{6}{4}$	$\frac{7}{5}$	$\frac{9}{4}$
0.2	0.89	0.04	0.005	9.5

Find the two other relations for the given number.

0.5%		
	$\frac{1}{100}$	
		10.5

0.05%		
	$\frac{1}{3}$	
		1.0

20.5%		
	$\frac{1}{10}$	
		0.52

45%		
	$\frac{1}{6}$	
		2.85

Find the unknown number or percent using the **OPI** formula.

What is 40% of 120?	What is 7% of 90?	What is 70% of 120?
20% of what number is 30?	5% of what number is 40?	0.4% of what number is 50?
60 is what percent of 200?	5 is what percent of 50?	150 is what percent of 300?
What is 15% of 80?	50 is what percent of 600?	15 is what percent of 60?

CHALLENGE

Is **20%** of a number, which is **60%** of **350**, the same as **12%** of **350**?

David solves **30** problems out of **50**. What percent of **50** problems does he solve? How many more problems does he have to solve to receive **80%**?

8% of a number is **x**, **50%** of the same number is **x + 21**, and that number is a two digit number divisible by **5**. What is the value of **x**? What is the number?

All missing numbers are positive whole numbers greater than **1**. Write the missing numbers. Write all the possible solutions.

$$(\Box \% \text{ of } 100) \cdot (\Box \% \text{ of } 100) = 35? \qquad (\Box \% \text{ of } 25) \cdot (\Box \% \text{ of } 70) = 35?$$

All missing numbers are positive whole numbers greater than **1**. Write the missing numbers. Write all the possible solutions.

$$(50 \% \text{ of } \Box) \cdot (50 \% \text{ of } \Box) = 21? \qquad (25 \% \text{ of } \Box) \cdot (25 \% \text{ of } \Box) = 15?$$

All missing numbers are positive whole numbers greater than **1**. Write the missing numbers. Write all the possible solutions.

$$(\Box \% \text{ of } 100) \cdot (50 \% \text{ of } \Box) = 6? \qquad (\Box \% \text{ of } 100) \cdot (25 \% \text{ of } \Box) = 10?$$

USING VARIABLES

In algebra we use numbers, number expressions (or numerical expressions), variables (letters), variable expressions, algebraic expressions, equations, inequalities, and functions.

A **number** is any real or complex number like $2, \frac{1}{4}, 3.7, \sqrt{2}, 4 - 2i$, and so on.

A **number (numerical) expression** consists of real or complex numbers and operations (addition, subtraction, multiplication, division, absolute value, power, and power root).
For example, $5 + 2, 3 - 20 \div 4, |2^2 - 10|, \sqrt{44} + 2, 7 + 6 \cdot (3 - 20 \div 5)$ are numerical expressions.

A **variable** is a letter from any alphabet, like $x, y, z, a, b, c, d, \alpha, \beta, \delta, \varepsilon$, etc. and is a symbol for unknown numbers.

When a letter is used as a variable to represent a set (or range) of numbers, the numbers of that set are called **values** of the variable.

A **variable expression** consists of variables, and combinations (operations) of variables (addition, subtraction, multiplication, division, absolute value, power, and power root).
For example, $x + y, |xy - z|, x^2 - x, \frac{2x + y}{x + y}, \sqrt{4x^2 yz^8}$, and $\frac{2x + y}{x + y} \cdot \frac{6x + 6y}{4x^2 - y^2}$, are variable expressions.

An **algebraic expression** consists of a number or numbers, a variable or variables, and combinations (addition, subtraction, multiplication, division, absolute value, power, and power root) of numbers and variables.
In another words, an **algebraic expression** consists of one or more numbers and variables with or without one or more algebraic operations ($+, -, \times, \div, |a|, a^n, \sqrt[n]{a}$).
For example, $4, 3 + 6, y, z, x + y, 2|xy - z|, x^2 - x + 1, \frac{2 + y}{x + y}, \sqrt{4x^2 yz^8}$, are algebraic expressions.

In algebraic expressions there are **variables**, **coefficients**, and **constants**.
For example, in expression $4x + 2$, **x** is the **variable**, 4 is the **coefficient** of the variable **x**, and 2 is the **constant**.

An **equation** is a relation between any two algebraic expressions with an equal sign: $E_1 = E_2$.
For example, $x + 1 = 2, |xy - z| = 5x + 1, x^2 - x - 2 = 0$ are equations.

An **inequality** is a relation between any two algebraic expressions with any inequality sign: $E_1 < E_2; E_1 > E_2; E_1 \leq E_2$ and $E_1 \geq E_2$.
For example, $2x + 1 < 3, |x - y| > 2, x^2 - x - 2 \leq 0, 2x^2 - 3x - 5 \geq 0$ are inequalities.

A **function** is a relation between two or more variables with an equal sign.
For example, $y = 2x + 1, y = |x - 3|, y = x^2 - x - 2 = 0$, and $x^2 + y^2 = 9$ are functions.

In algebra there are many questions related with variables and expressions.
For example, evaluate the expression $4xy + 2$ if $x = 1, y = 3$; simplify the expression $\frac{2xy + y}{3y}$; solve the equation $x - 2 = 4$; solve the inequality $3x + 4 < 0$; graph the given one-variable function $f(x) = 2x + 1$; etc.

 Identify the variables, coefficients and constants in each expression.

1. $2x + 3$ 2. $1 - 2x + 3y^2$ 3. $x + 3xyz + 5y^2$

Solution

1. in $2x + 3$
   x is the variable
   2 is the coefficient of x
   3 is the constant

2. in $1 - 2x - 3y^2$
   x and y are variables
   -2 is the coefficient of x
   3 is the coefficient of y^2
   1 is the constant

3. $x + 3xyz + 5y^2$
   x, y and z are variables
   1 is the coefficient of x
   3 is the coefficient of xyz
   5 is the coefficient of y^2
   there is no constant

 Evaluate the expression $7 + 2x - 12$ for $x = 3$.

Solution

$$7 + 2x - 12 = 7 + 2 \cdot 3 - 12 = 7 + 6 - 12 = 13 - 12 = 1$$

 A plane travels at a speed **500** miles per hour for **4** hours. Find the traveled distance using distance formula $d = vt$, where v is the speed and t is the time.

Solution

$$d = vt = 500 \cdot 4 = 2000 \text{ miles}$$

 Find the perimeter of the triangle using perimeter formula $P = a + b + c$, where $a = 3$in, $b = 4$in, and $c = 5$in are lengths of sides of the triangle.

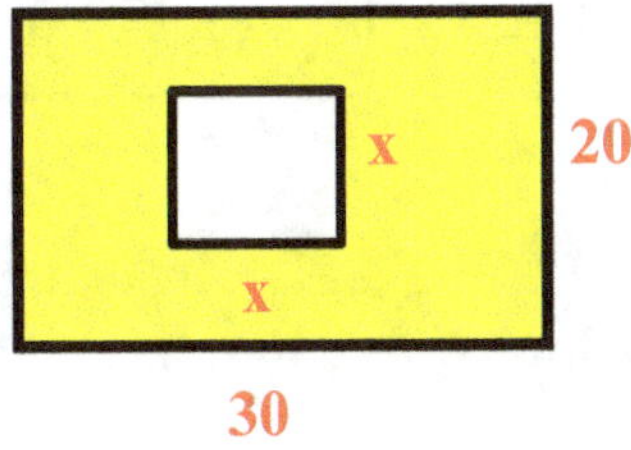

Solution

$$P = a + b + c = 3 \text{ in} + 4 \text{ in} + 5 \text{ in} = 12\text{in}$$

 Write the expression that represents the area of the shaded region.

Solution

$$A = 30 \cdot 20 - x \cdot x = 600 - x^2$$

■ PROPERTIES OF BASIC OPERATIONS

Addition and multiplication operations have the following properties.

Property of Addition	Name of Property	Examples
Changing the order of addends does not change the sum.	Commutative property of Addition $a + b = b + a$	$2 + 3 = 3 + 2 = 5$ $7 + (-3) = (-3) + 7 = 4$ $(-4) + (-3) = (-3) + (-4) = -7$
Changing the grouping of the addends does not change the sum.	Associative property of Addition $(a + b) + c = a + (b + c)$	$(2 + 3) + 4 = 2 + (3 + 4) = 9$ ♦ $[7+(-3)]+1 = 7+[(-3)+1] = 5$
The sum of a number and zero is that number.	Identity property of Addition $a + 0 = a$	$8 + 0 = 0 + 8 = 8 \; ; \; 0 + 0 = 0$ $(-3) + 0 = 0 + (-3) = -3$
The sum of a number and its opposite number is zero.	Inverse property of Addition $a + (-a) = 0$	$9 + (-9) = (-9) + 9 = 0$ $(-3) + 3 = 3 + (-3) = 0$

♦ In the future we will use **3** types of **grouping symbols: ( ) - parentheses; [ ] - brackets; { } - braces.**

Property of Multiplication	Name of Property	Examples
Changing the order of factors does not change the product.	Commutative property of Multiplication $a \cdot b = b \cdot a$	$2 \cdot 3 = 3 \cdot 2 = 6$ $7 \cdot (-3) = (-3) \cdot 7 = -21$ $(-4) \cdot (-3) = (-3) \cdot (-4) = 12$
Changing the grouping of the factors does not change the product.	Associative property of Multiplication $(a \cdot b) \cdot c = a \cdot (b \cdot c)$	$(2 \cdot 3) \cdot 4 = 2 \cdot (3 \cdot 4) = 24$ $[7 \cdot (-3)] \cdot 1 = 7 \cdot [(-3) \cdot 1] = -21$
The product of a number and one is that number.	Identity property of Multiplication $a \cdot 1 = 1 \cdot a = a$	$8 \cdot 1 = 1 \cdot 8 = 8 \; ; \; 1 \cdot 1 = 1 \cdot 1 = 1$ $(-3) \cdot 1 = 1 \cdot (-3) = -3$
The product of a number and its multiplicative inverse is one.	Inverse property of Multiplication $a \cdot (1/a) = (1/a) \cdot a = 1$	$9 \cdot (1/9) = (1/9) \cdot 9 = 1$ $(-1/3) \cdot (-3) = (-3) \cdot (-1/3) = 1$
The product of a number and 0 is zero.	"Zero" property of Multiplication $a \cdot 0 = 0 \cdot a = 0$	$9 \cdot 0 = 0 \cdot 9 = 0$ $(-3) \cdot 0 = 0 \cdot (-3) = 0$

The product of a number and sum of numbers has the following important property.

Property	Name of Property	Examples
We have to multiply a number with each addend inside the parentheses.	Distributive property $a \cdot (b + c) = a \cdot b + a \cdot c$	$2 \cdot (3 + 5x) = 2 \cdot 3 + 2 \cdot 5x = 6 + 10x$ $(x - 4) \cdot 3 = 3x + 3 \cdot (-4) = 3x - 12$

Examples

EXAMPLE 1: Which property do you see below?

$$y + 4 = 4 + y$$

Solution

$$y + 4 = 4 + y \quad \text{commutative property of addition}$$

EXAMPLE 2: Which properties do you see below?

$$x(y + z) = zx + yx$$

Solution

$$x(y + z) = xy + xz \qquad \text{distributive property}$$
$$= xz + xy \qquad \text{commutative property of addition}$$
$$= zx + yx \qquad \text{commutative property of multiplication}$$

EXAMPLE 3: Which properties can be used to simplify the expression $5(3 - xy) + 5yx$?

Solution

$$5(3 - xy) + 5yx = 15 - 5xy + 5yx \quad \text{distributive property}$$
$$= 15 - 5yx + 5yx \quad \text{commutative property of multiplication}$$
$$= 15 + 0 \qquad \text{inverse property of addition}$$
$$= 15 \qquad \text{identity property of addition}$$

EXAMPLE 4: Which properties do you see below?

$$(a + b)c - bc = ca$$

Solution

$$(a + b)c - bc = ac + bc - bc \qquad \text{distributive property}$$
$$= ac + 0 \qquad \text{inverse property of addition}$$
$$= ac \qquad \text{identity property of addition}$$
$$= ca \qquad \text{commutative property of multiplication}$$

EXAMPLE 5: Which properties can be used to simplify $7[4(x + 9) - 2(2x + 18)]$?

Solution

$$7[4(x + 9) - 2(2x + 18)] = \qquad \text{distributive property}$$
$$7[4x + 36 - 4x - 36] = \qquad \text{commutative property of addition}$$
$$7[4x - 4x + 36 - 36] = \qquad \text{inverse property of addition}$$
$$7[0 + 0] = \qquad \text{identity property of addition}$$
$$7[0] = \qquad \text{identity property of addition}$$
$$0 \qquad \text{zero property of multiplication}$$

EXAMPLE 6: Simplify the expression $(-2 - 5x)6 - 10(5 - 3x) + 3 \cdot 1$.

Solution

$$(-2 - 5x)6 - 10(5 - 3x) + 3 \cdot 1 = -12 - 30x - 50 + 30x + 3$$
$$= -12 - 30x + 30x - 50 + 3 = -12 - 50 + 3 = -59$$

Evaluate each expression for **x = 3, y = − 2, z = 5**.

$x + y$	$x - y - z$	$x - (y - z)$
$xy - z$	$xy - 2x - z$	$5 - 2(x - y)$
$1 + |x - z|$	$|y + z - x| - 2$	$x + |y - z|$
$\dfrac{y + |y - z|)}{x}$	$1 - x^2$	$\dfrac{x}{z - y}$

Write the phrase as a variable expression or relation.

8 times a number x	y is at list 20	quotient of 2 and a number
sum of x and the square of y	9 minus z is 5	x is less than or equal 8

Write the expression that represents the area of the shaded region.

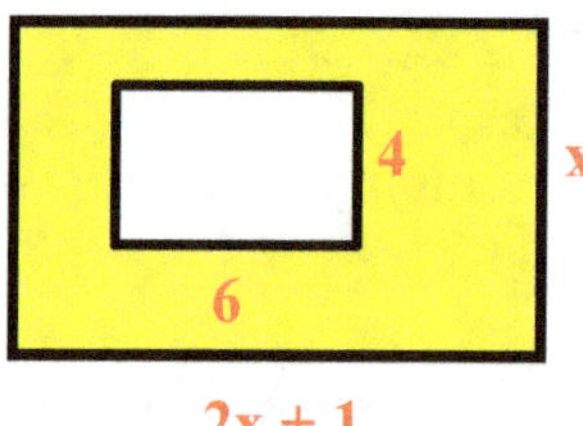

Which properties do you see below?

$5 + 0 = 5$	$-3 + 7 = 7 - 3$	$xy + 1 = 1 + yx$
$7(x + y) = 7x + 7y$	$5(x + y) + xz = zx + 5x + 5y$	$ab \cdot \dfrac{1}{ba} = 1$
$-3 + 0 = -3$	$2(x + 3y) - 2x = 6xy$	$x(y + 1) = x + yx$

CHALLENGE

Evaluate each expression for **x = 1, y = − 2, z = 2**.

$2x^2yz - yz$	$2(x-3)y(2-z)$	$3x\,(y-z)$
$3\lvert y - \lvert x - z \rvert\rvert$	$\lvert y\lvert yz - x\rvert\rvert$	$\lvert(\lvert x - \lvert y - z\rvert\rvert) - 3\rvert$
$1 + \dfrac{x\lvert y - z\rvert}{\lvert z - y\rvert}$	$\dfrac{3xzy^2}{z - y}$	$\dfrac{xyz}{z-y}\,\lvert y + z + 1\rvert$

Write the expression for the area of each shaded region.
The side of each big square is **2x** and the radius of each big circle is **v**.

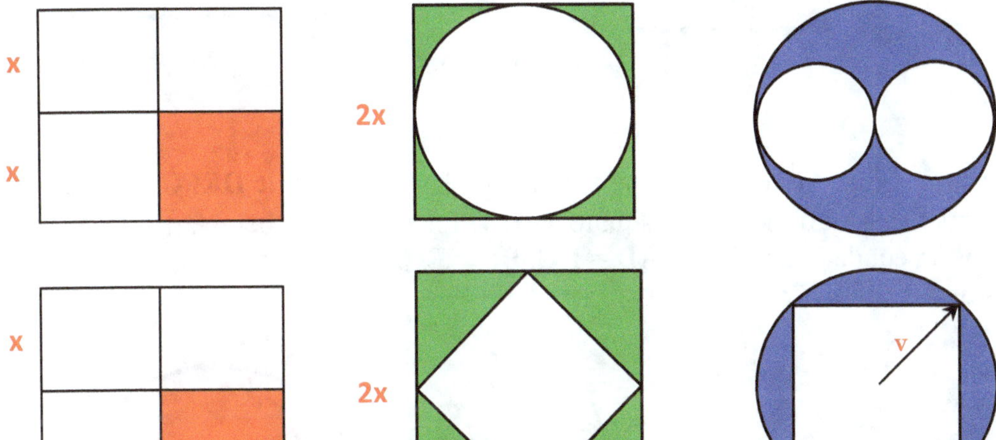

Write the expression that represents the area of the shaded region.

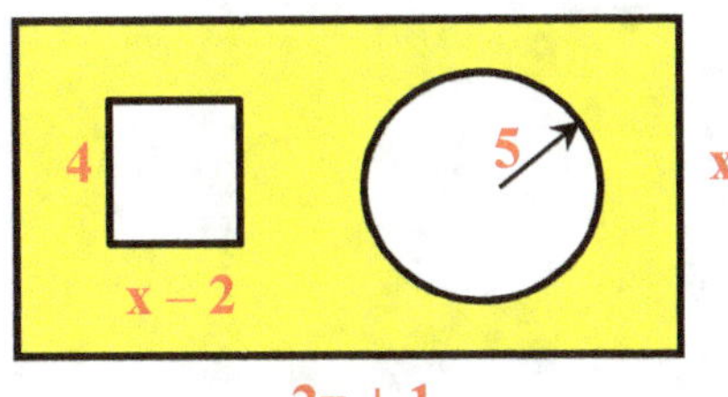

CHALLENGE

Write the expression for the sum of the perimeters of the two squares and the expression for the sum of the two equilateral triangles if **DE = x** and **MN = y**.

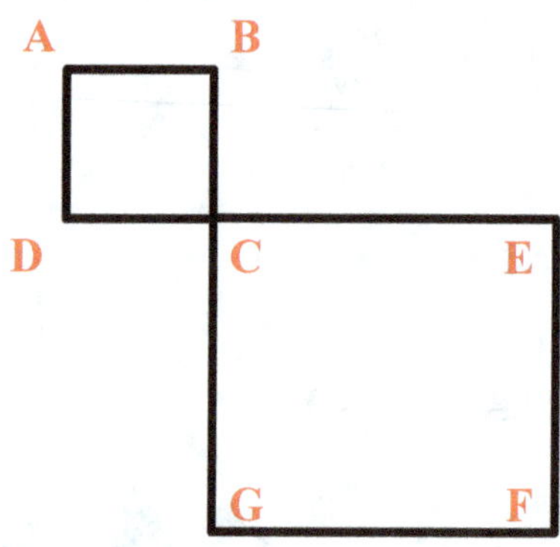
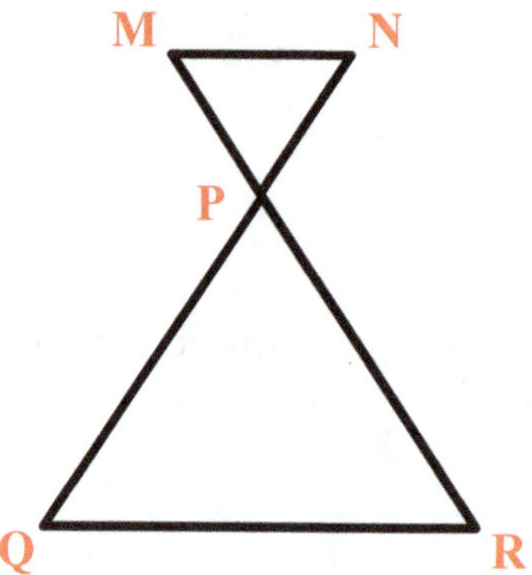

Write the expression that represents the area of the shaded rectangle **DEFG** and the expression that represents the area of the shaded square **MNPQ** for the figures below.
The △**ABC** is an equilateral triangle, **AD = DC**, **BE = EC**, and **AC = x**.
The radius of the circle is **r**.

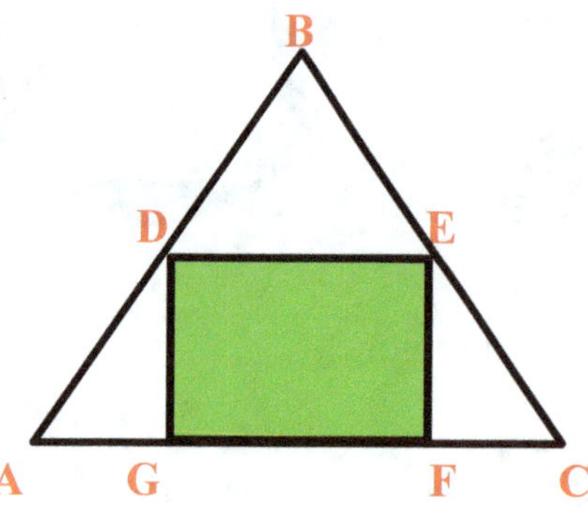
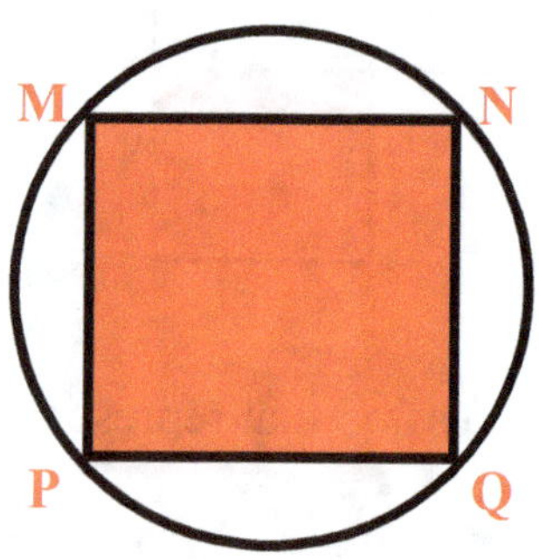

Which properties do you see below?

$x(yz + y) + (4 - y)x = 4x + yxz$	$6(x + \frac{1}{6}y) = y + 6x$	$x + z(y + 1) = 1 + yz$

1.3. The Order of Operations (PEMDAS + L→R)

- PEMDAS + L→R
- USING PEMDAS + L→R

■ PEMDAS + L→R

If we have an **algebraic expression** with real numbers, with more than two operations, and with **grouping symbols: ( )-parentheses, [ ]-brackets and { }-braces**, then we use the PEMDAS + L → R (PEMDAS plus "from Left to Right") rule for ordering operations, and simplifying expressions. **PEMDAS** stands for Parentheses, Exponents, Multiplication, Division, Addition, and Subtraction.

PEMDAS + (L → R)

In expressions inside the grouping symbols and in expressions without grouping symbols, multiplication and division have first priority. Addition and subtraction have second priority.

If an expression includes only **multiplication** and **division** operations, like **12 ÷ 4 x 8**, then we use the **"from left to right (L → R)"** rule: **12 ÷ 4 x 8 = 3 x 8 = 24**.
Multiplication signs (x) **and division signs** (÷) **have the same priority.** The fact that the "M" in **PEMDAS** comes before the "D" does not mean that multiplication is of higher priority.
If we have a division operation before a multiplication operation, then we have to divide first and then multiply.

If an expression includes only **addition** or **subtraction** operations, like **4 + 5 − 8**, then we use the **"from left to right (L → R)"** rule: **4 + 5 − 8 = 9 − 8 = 1**.
The signs + and − have the same priority. The fact that the "A" in **PEMDAS** comes before the "S" does not mean that addition is of higher priority. So, for example, to evaluate the expression **5 − 8 + 2** we use the rule **from left to right (L → R): 5 − 8 + 2 = − 3 + 2 = − 1**.

If we have an expression with **multiplication, division, addition,** and **subtraction**, then we start from multiplication or division using the rule **from left to right (L → R)** and afterwards, use addition and subtraction for the remaining and resultant terms from **left to right (L → R)** again:
4 + 6 ÷ 3 − 7 + 3 x 2 = 4 + 2 − 7 + 3 x 2 = 4 + 2 − 7 + 6 = 6 − 7 + 6 = − 1 + 6 = 5.

If we have any expression with **many operations** and **many grouping symbols: (), [ ] and { },** then we first clear the parentheses (), **working** from **left to right**; then [ ], then { }. **Inside or outside of grouping symbols** we take care of the exponents (powers). Then we perform all multiplication and division from **left to right**. And then we perform all addition and subtraction from **left to right**.

■ USING PEMDAS + L→R

Example 1 Evaluate the expression: $2 + 40 \div 8$

Solution

$$2 + \underset{5}{\underline{40 \div 8}} = 2 + 5 = 7$$

Example 2 Evaluate the expression: $7 + 6 \cdot 3 - 20 \div 5 - 2$

Solution

$$7 + 6 \cdot 3 - 20 \div 5 - 2 = 7 + \underset{18}{\underline{6 \cdot 3}} - 20 \div 5 - 2 = 7 + 18 - \underset{4}{\underline{20 \div 5}} - 2 =$$

$$7 + 18 - 4 - 2 = \underset{25}{\underline{7 + 18}} - 4 - 2 = 25 - 4 - 2 = \underset{21}{\underline{25 - 4}} - 2 = 21 - 2 =$$

$$\underset{19}{\underline{21 - 2}} = 19$$

Example 3 Evaluate the expression: $30 \div 6 \cdot [\, 28 - 2 \,(7 + 2 \cdot 3)\,]$

Solution

$$30 \div 6 \cdot [\, 28 - 2 \,(7 + 2 \cdot 3)\,] = 30 \div 6 \cdot [\, 28 - 2 \,(7 + \underset{6}{\underline{2 \cdot 3}})\,] =$$

$$30 \div 6 \cdot [\, 28 - 2 \,(7 + 6)\,] = 30 \div 6 \cdot [\, 28 - 2 \,(\underset{13}{\underline{7 + 6}})] = 30 \div 6 \cdot [\, 28 - 2 \cdot 13] =$$

$$30 \div 6 \cdot [\, 28 - \underset{26}{\underline{2 \cdot 13}} \,] = 30 \div 6 \cdot [\, 28 - 26] = 30 \div 6 \cdot [\, \underset{2}{\underline{28 - 26}} \,] =$$

$$30 \div 6 \cdot 2 = \underset{5}{\underline{30 \div 6}} \cdot 2 = 5 \cdot 2 = \underset{10}{\underline{5 \cdot 2}} = 10$$

Example 4 Evaluate the expression: $4\{2+3^2 - 3(5 - 3) - 4[4+2(3+2) - 8 \div 2] - 5\} + 2(3+1)$

Solution

$$4\{2 + 3^2 - 3(5 - 3) - 4[4 + 2(3 + 2) - 8 \div 2] - 5\} + 2(3 + 1) =$$
$$4\{2 + 3^2 - 3(\underset{2}{\underline{5 - 3}}) - 4[4 + 2(\underset{5}{\underline{3 + 2}}) - 8 \div 2] - 5\} + 2(\underset{4}{\underline{3 + 1}}) =$$
$$4\{2 + 3^2 - 3 \cdot 2 - 4[4 + 2 \cdot 5 - 8 \div 2] - 5\} + 2 \cdot 4 =$$
$$4\{2 + 9 - 3 \cdot 2 - 4[4 + 2 \cdot 5 - 8 \div 2] - 5\} + 2 \cdot 4 =$$
$$4\{2 + 9 - 3 \cdot 2 - 4[4 + \underset{10}{\underline{2 \cdot 5}}) - \underset{4}{\underline{8 \div 2}} \,] - 5\} + 2 \cdot 4 =$$
$$4\{2 + 9 - 3 \cdot 2 - 4[\underset{10}{\underline{4 + 10 - 4}} \,] - 5\} + 2 \cdot 4 =$$
$$4\{2 + 9 - \underset{6}{\underline{3 \cdot 2}} - \underset{40}{\underline{4 \cdot 10}} - 5\} + 2 \cdot 4 = 4\{\underset{-40}{\underline{2 + 9 - 6 - 40 - 5}}\} + 2 \cdot 4 =$$
$$4\{-40)\} + 2 \cdot 4 = \underset{-160}{\underline{4 \cdot (-40)}} + 2 \cdot 4 = -160 + \underset{8}{\underline{2 \cdot 4}} = -160 + 8 = -152$$

Use the **PEMDAS + L→R** rule and evaluate each expression.

$1 + 2 - 3$	$1 + 2 - 3 + 2^2$	$1 + 2 - 3 + 4 - 5$
$4 + 5 - (3 + 4)$	$1 - (5 + 6 - 3^2)$	$2 - [3 + 4 - (5 + 6)] - 7$
$1 + 2 - 3 + 4 - 5 + 6 - 7$	$1 + 2 - (3 + 4) - 5 + 6 - 7$	$1 + 2 - 3 + 4 - (5 + 6) - 7$
$1 + 2 - (3 + 4) - (5 + 6) - 3^2$	$1 + 2 - [3 + 4 - (5 + 6 - 7)]$	$1 + \{2 - [3 + 4 - (5 + 6)]\} - 7$
$32 \div 8 \cdot 3$	$2 \cdot 4 \cdot 6 \div 8$	$6 \div 3 \cdot 4 \div 2$
$9 \div 3 \cdot 3 \div 3^2$	$2^5 \div 8 \cdot 4 \div 2 \div 2$	$2 \cdot 4 \cdot 8 \div 16 \cdot 2 \div 8$
$36 \div (2 \cdot 9) \div 2$	$2 \cdot 4 \cdot (16 \div 2^3 \cdot 2)$	$2^6 \div 8 \cdot 4 \div (4 \cdot 2) \cdot 2$
$64 \div (8 \cdot 4 \div 4) \cdot (4 \div 2)$	$64 \div [8 \cdot (4 \div 4 \cdot 2)] \cdot 5$	$64 \div \{[8 \cdot (4 \div 4 \cdot 2)] \cdot 4\}$
$7 + 4 - (2 + 5)$	$3 + 24 \div 8$	$2 \cdot 4 - 16 \div 2$
$1 + 6 \cdot 0 \div 2 + 5$	$(4 - 7) \div 3 + 1$	$4 - (9 \div 3 + 1) + 8$
$1 - 9 \div 3 + 5 \cdot 2$	$4 + 3^2 - (7 + 3) + 1$	$8 \cdot 6 - (8 + 1) \div 3 + 2^4 \div 4$
$4 - 3^2 + 2(1 + 5) \div 4$	$0.2 + 5(1 \div 10) - 0.4 \div 2$	$2 \cdot (1/2 + 3/4) - 2.5 + 4 \div 2$
$10 - 27 \div 3 \div 9 - 8$	$2 + 3(16 \div 2^2 \cdot 1) - 4 \div 2$	$2 \cdot \{[4(2 + 16 \div 8) - 6] - 2\}$

Evaluate each expression.

$4 + 2 \cdot 5 - 7$	$6 + 8 \div 2$	$4(6+1) \div 7 \cdot 5 - 20$	$9 + [2(2^2 + 1) \div 5 - 1]$
$6 + 0.3 \cdot 2$	$1.2 \cdot 5 + 5.5 \div 5$	$(1.7 + 1.3) \cdot 5 - 7$	$4.5 - 1.1 \cdot (5 - 3)$
$\frac{1}{3} - \frac{1}{2} + \frac{1}{4}$	$\frac{1}{2} - (\frac{1}{2} - \frac{1}{4}) \cdot 2$	$\frac{1}{2} \div \frac{1}{4} \cdot \frac{1}{3}$	$1.5 - \frac{1}{3} \div (\frac{1}{2} - \frac{1}{3}) + 2$
$3^2 + 2(5-2) \div 3 - 1$	$9 - [1 + \frac{5 \cdot (6-3)}{15}] + 1$	$-\frac{3 \cdot (6-2)}{4} + 8 \div 4 + 1$	$-[3 + \frac{2 \cdot (7-2)}{5}] + 8$

Compare. Use >, <, or = to complete each statement.

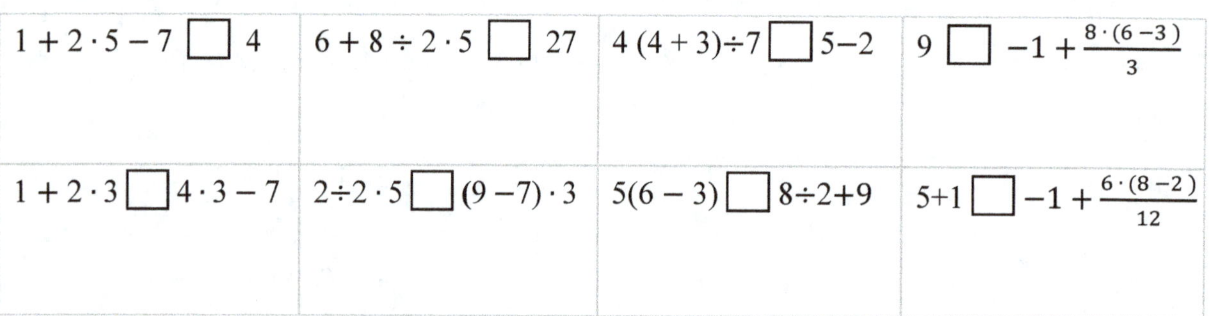

$1 + 2 \cdot 5 - 7 \ \square\ 4$	$6 + 8 \div 2 \cdot 5 \ \square\ 27$	$4(4+3) \div 7 \ \square\ 5-2$	$9 \ \square\ -1 + \frac{8 \cdot (6-3)}{3}$
$1 + 2 \cdot 3 \ \square\ 4 \cdot 3 - 7$	$2 \div 2 \cdot 5 \ \square\ (9-7) \cdot 3$	$5(6-3) \ \square\ 8 \div 2 + 9$	$5 + 1 \ \square\ -1 + \frac{6 \cdot (8-2)}{12}$

Add parentheses to make a true statement.

$6 + 2 \cdot 5 - 7 = 2$	$6 + 8 \div 2 \cdot 3 = 30$	$3 \cdot 4 + 6 \div 3 - 1 = 15$	$2 + 8 \div 6 - 1 = 2$
$3 \cdot 2^3 - 4 - 3 = 1$	$8 + 2^3 \div 4 - 3 = 1$	$2 \cdot 4^2 - 3^2 + 2^3 = 30$	$4 \cdot 3^2 - 5^2 + 2^2 = 7$

CHALLENGE

Write the missing integers.

$$[(\,\square\, + 2\,) \div 6 \cdot 3 + 1] \div 4 \cdot 5 = 5$$

$$\square = 2 \cdot \{[(\,\square\, + 1) - 2\,] \div 3 + 1\}$$

$$2 \cdot \square = 3 \cdot \{[(\,\square\, + 2) - 1\,] \div 4 + 1\}$$

Put the signs +, −, · , and ÷ between the numbers, so that the following equations are true.

$$2\ \square\ 3\ \square\ 4\ \square\ 5\ \square\ 6 = 1$$

$$7\ \square\ 8\ \square\ 9\ \square\ 10\ \square\ 11 = 100$$

$$12\ \square\ 13\ \square\ 14\ \square\ 15\ \square\ 16 = -70$$

1. In the set $\{-20, -8.3, -3.77, -\frac{1}{5}, 0, \sqrt{5}, 3, \pi, 5\frac{1}{4}, 7, 9.\overline{3}, 11.5\}$ list all of the:

   - Natural Numbers
   - Whole Numbers
   - Integers
   - EVEN Numbers
   - Odd numbers
   - Prime Numbers
   - Rational Numbers
   - Irrational Numbers
   - Real Numbers

2. Evaluate the numerical expression

   $$-4 + 2 \cdot (-3) \cdot (-2)$$

3. Let $A = \{1, 3, 5, 7\}$, $B = \{-1, 0, 1, 2, 3, 4, 5, 6\}$. Find each of the following.

   a) $A \cup B$
   b) $A \cap B$
   c) $A \cap A$
   d) $(B \cup B) \cap A$
   e) $(A \cap B) \cup A$
   f) $\{x/ \; x \in A, \; x \notin B\}$
   g) $\{x/ \; x \in B, \; x \geq 3\}$

4. Evaluate the numerical expression.

   $$\frac{|3 - 7| + |-(3 - 5) - 8 \div 2|}{|8 - 6| + |-9 \div 3 + (7 - 8)|}$$

5. Evaluate the variable expression for $x = 1$, $y = 2$.

   $$\frac{2x + 4y}{xy} + 1$$

6. Use PEMDAS + L→R to evaluate the expression.

   $$(-7) \cdot \{[-(-1)^2 + 2] + 4\} - 15 \div (-3) + 7$$

1. In the set $\{-11, -5.2, -2, -\frac{1}{2}, 0, \sqrt{3}, 2\pi, 9\frac{1}{3}, 11, 15.\overline{6}, 19, 21, 22, 28, 29\}$ list all of the:

   - Natural Numbers
   - Whole Numbers
   - Integers
   - EVEN Numbers
   - Odd numbers
   - Prime Numbers
   - Rational Numbers
   - Irrational Numbers
   - Real Numbers

2. Evaluate the numerical expression.

   $$43 - 7 + 5 \cdot (-3) \cdot (-1) \cdot (-2)$$

3. Let $A = \{1, 2, 3, 7\}$, $B = \{-1, 0, 1, 2, 3, 4, 7\}$, and $C = \{0, 4, 7\}$. Find each of the following.

   a) $A \cup B$
   b) $A \cap C$
   c) $B \cap C$
   d) $(A \cup B) \cap C$
   e) $(A \cap B) \cup C$
   f) $\{x/\ x \in A,\ x \notin C\}$
   g) $\{x/\ x \in C,\ x \geq 3\}$

4. Evaluate the numerical expression.

   $$\frac{|1 - 3| + |(1 - 4) + 9 \div 3| + 2}{|5 - 3| + |-6 \div 3 + (8 - 4)|}$$

5. Evaluate the variable expression for $x = 2$, $y = 4$.

   $$\frac{4(x + y)}{xy} + x - 2(-y)^2$$

6. Use PEMDAS + L→R to evaluate the expression.

   $$(-4) \cdot \{[(-2)^3 + 1] - 3\} - 9 \div (-3) - 5$$

Chapter 2. Powers, Polynomials, and Expressions

2.1. Powers and Properties of Powers
2.2. Polynomials and Operations
2.3. Basic Algebraic Identities
2.4. Factoring of Polynomials
2.5. Fractional Expressions and Operations

Chapter 2   Classwork
Chapter 2   Homework

$$a^2 - b^2 = (a - b)(a + b)$$

$$(a + b)^2 = a^2 + 2ab + b^2$$
$$(a - b)^2 = a^2 - 2ab + b^2$$

$$a^3 - b^3 = (a - b)(a^2 + ab + b^2)$$
$$a^3 + b^3 = (a + b)(a^2 - ab + b^2)$$

$$(a - b)^3 = a^3 - 3a^2b + 3ab^2 - b^3$$
$$(a - b)^3 = a^3 - 3a^2b + 3ab^2 - b^3$$

2.1. Powers and Properties of Powers

- **DEFINITIONS**
- **PROPERTIES OF POWERS**
- **SCIENTIFIC NOTATION**

■ DEFINITIONS

A product where all factors are the same number is called a **power**.

We will use the following general definition of the nth **power**, and two special definitions for the **second** and **third** powers of a number.

IMPORTANT DEFINITIONS

General Definition ▶

A power with a natural exponent **n** (or n^{th} power) of a number a is:

$$\underbrace{a^n}_{\text{power}} = \underbrace{a \cdot a \cdot a \cdots a}_{n \text{ times}}$$

The number a is called the **base**; the number **n** is called the **exponent**; and a^n is called the "n^{th} power of a" (or "a to the nth power").
The exponent tells how many times the base a is repeated as a factor in multiplication.

Special Definitions ▶

A **square** of a number a is a^2 (**a to the second power**).
A **cube** of a number a is a^3 (**a to the third power**).

Examples Evaluate the expressions: $1. 2^3$; $2. 0^7$; $3. 0.5^2$; $4. -5^2$; $5. \left(-\frac{1}{2}\right)^3$; $6. (-5)^2$; $6. (3a)^4$.

Solution

1. $2^3 = 2 \cdot 2 \cdot 2 = 8$
2. $0^7 = 0 \cdot 0 \cdot 0 \cdot 0 \cdot 0 \cdot 0 \cdot 0$
3. $0.5^2 = 0.5 \cdot 0.5 = 0.25$
4. $-5^2 = -5 \cdot 5 = -25$

5. $\left(-\frac{1}{2}\right)^3 = \left(-\frac{1}{2}\right) \cdot \left(-\frac{1}{2}\right) \cdot \left(-\frac{1}{2}\right) = -\frac{1}{8}$
6. $(-5)^2 = (-5) \cdot (-5) = 25$
7. $(3a)^4 = (3a) \cdot (3a) \cdot (3a) \cdot (3a) = 3 \cdot 3 \cdot 3 \cdot 3 \cdot a \cdot a \cdot a \cdot a = 3^4 a^4$

■ PROPERTIES OF POWERS

We will use the following properties of integer powers (below $\mathbf{m, n \in Z}$; and $\mathbf{a, b \in R}$).

PROPERTY OF INTEGER POWERS	EXAMPLE
1. $a^m \cdot a^n = a^{m+n}$	1. $2^3 \cdot 2^4 = 2^7$
2. $a^m / a^n = a^{m-n} = 1/a^{n-m}$	2. $3^5 / 3^7 = 3^{-2} = 1/3^2$
3. $a^0 = 1$	3. $5^0 = 1, 7^0 = 1, (-8)^0 = 1$
4. $a^1 = a$	4. $7^1 = 7, x^1 = x, (-9)^1 = -9$
5. $(a^m)^n = a^{mn}$	5. $(2^3)^4 = 2^{12}$
6. $(ab)^n = a^n b^n$	6. $(2x)^5 = 2^5 x^5$
7. $(a/b)^n = a^n/b^n$	7. $(5/x)^2 = 5^2/x^2$
8. $a^{-n} = 1/a^n$	8. $10^{-2} = 1/10^2$
9. $1/a^{-n} = a^n$	9. $1/3^{-2} = 3^2$
10. $(a/b)^{-n} = (b/a)^n = b^n/a^n$	10. $(x/3)^{-2} = (3/x)^2 = 3^2/x^2$

Examples

EXAMPLE 1: Simplify the expression: $(x^2 x^{-4})^{-3}$

Solution

$$(x^2 x^{-4})^{-3} = (x^{-2})^{-3} = x^6$$

EXAMPLE 2: Simplify the expression: $4\left(-\dfrac{3x^{-4}y^{-3}}{2x^2 y}\right)^2 \cdot \left(-\dfrac{y^{-3}}{3^{-1}x^4}\right)^{-3}$.

Solution

$$4\left(-\frac{3x^{-4}y^{-3}}{2x^2 y}\right)^2 \cdot \left(-\frac{y^{-3}}{3^{-1}x^4}\right)^{-3} = 4\frac{9x^{-8}y^{-6}}{4x^4 y^2} \cdot \left(-\frac{y^9}{3^3 x^{-12}}\right) = \frac{9}{x^{12}y^8}\left(-\frac{y^9}{27x^{-12}}\right) = -\frac{y}{3}$$

Find the result of each operation.

$(-5)^3 \cdot (-5)^6$	$(3^2 x^3) \cdot (3^4 x^5)$	$\dfrac{8^5}{16^2}\left(\dfrac{4}{2}\right)^3$	$\dfrac{x^2 y^5}{x^2 y^6}$	$\dfrac{x^4 y^7}{x^3 y^9} \cdot \dfrac{x^{-3} y^{-2}}{x^{-5} y^{-6}}$
$y^3 y^0 \cdot 3^0$	$(x^2)^3$	xy^1	$5^1 \cdot 2^2$	$\dfrac{5^0 4^5}{2^3 6^{-4}}$
$(2^2)^3$	$(3x^2 y)^2$	$((5y^2 z)^3)^2$	$(2x^3 (3y^3 z)^2)^3$	$\left(\dfrac{2(3y^2 z)^2}{6}\right)^3$
$\left(\dfrac{3x}{2z}\right)^2$	$\left(\left(\dfrac{2x^2}{3y}\right)^2\right)^3$	$5x^{-2}$	$(5x)^{-2}$	$\left(\dfrac{1}{3} x^2 y^{-2}\right)^{-2}$
$\dfrac{5^2}{3x^{-2}}$	$\dfrac{14y^2}{2x^{-3}}$	$\dfrac{z^{-3}}{3x^{-2} y^{-5}}$	$\left(\left(\dfrac{x^{-2}}{2y}\right)^{-2}\right)^3$	$\left(\left(\dfrac{x^{-2}}{y}\right)^2\right)^3 \cdot \dfrac{y^6}{x^{12}}$
$\dfrac{(x^4 y^{-2})^3}{(xy^3)^2}$	$\dfrac{y^2 (2xy^{-3})^2}{4x^2 y^{-4} z^{-1}}$	$\left(\left(\dfrac{3x^6}{4y^3}\right)^3\right)^0$	$(2x^2 y^{-3})^2 \dfrac{y^6}{4x^4}$	$\left(\left(\left(\dfrac{2x^2}{y^{-2}}\right)^5\right)^3\right)^2$

■ SCIENTIFIC NOTATION

For writing big numbers in astronomy, writing small numbers in elementary particle physics, and in many other areas, scientists use the scientific notation of numbers.

DEFINITION

A number is in **scientific notation** when it is written in the form

$$A \cdot 10^n \quad \text{where} \quad A \in [1,10) \ \& \ n \in Z \ *)$$

*) **A** is a decimal number $1 \leq A < 10$ and **n** is an integer.

To rewrite a number from standard notation into scientific notation we have to move the decimal point to the position where we have only one nonzero digit before the decimal point. For example, $21{,}000 = 2\,1\,0\,0\,0.$

To rewrite a *large* number from standard notation into scientific notation we use positive powers of 10^n ($n \geq 1$). For example, $345{,}000 = 3\,4\,5\,0\,0\,0. = 3.45 \cdot 10^5$ (the exponent **5** is the number of places we moved the decimal point).

To rewrite a *small* number **a** ($0 < a < 1$) from standard notation into scientific notation we use negative powers of 10^n ($n \leq -1$). For example, $0.00045 = 0.0\,0\,0\,4\,5 = 4.5 \cdot 10^{-4}$.

The exponent **4** is the number of places we moved the decimal point.

Examples

EXAMPLE 1: Write the product in scientific notation: $8.5 \cdot 257 \cdot 10^5$

Solution

$$8.5 \cdot 257 \cdot 10^5 = 2184.5 \cdot 10^5 = 2.1845 \cdot 10^3 \cdot 10^5 = 2.1845 \cdot 10^8$$

EXAMPLE 2: Write the quotient in scientific notation: $\dfrac{2.5 \cdot 10^5}{625 \cdot 10^3}$

Solution

$$\frac{2.5 \cdot 10^5}{6250 \cdot 10^3} = 0.0004 \cdot 10^2 = 4 \cdot 10^{-4} \cdot 10^2 = 4 \cdot 10^{-2}$$

$\mathbf{W}$rite each number or expression in scientific notation.

450,000	65,000,000	124	9
0.08	4.02	0.00000432	0.0000000005
135.8	0.002	84.5	0.2
$456 \cdot 100$	$0.02 \cdot 0.002$	$8 \cdot 0.0004$	$125 \div 0.0005$
$0.08 \div 0.0002$	$8 \cdot 0.002$	$5 \cdot 75$	$0.0025 \div 125$
$(25 \cdot 10^7)(8 \cdot 10^{-3})$	$(1.5 \cdot 10^4)(5 \cdot 10^6)$	$(87 \cdot 10^{-2})(10^{-3})$	$(0.5 \cdot 10^{-5})(10^{11})$
$60(5 \cdot 10^5)(4 \cdot 10^0)$	$(8.5 \cdot 10^4)(5.4 \cdot 10^6)$	$(12 \cdot 10^5)(0.8 \cdot 10^3)$	$(4.5 \cdot 10^4)(5.4 \cdot 10^2)$
$\dfrac{(1.5 \cdot 10^{27})(6 \cdot 10^{-5})}{(2.5 \cdot 10^8)(0.4 \cdot 10^{-15})}$	$\dfrac{(1.8 \cdot 10^7)(0.5 \cdot 10^{-7})}{(2 \cdot 10^4)(0.5 \cdot 10^5)}$	$\dfrac{(1.8 \cdot 10^{-10})(0.7 \cdot 10^3)}{(12 \cdot 10^9)(6 \cdot 10^{-11})}$	$\dfrac{(200 \cdot 10^{-3})(14 \cdot 10^5)}{(0.05 \cdot 10^7)(80 \cdot 10^{-5})}$
$\dfrac{(1.3 \cdot 10^{-9})(2 \cdot 10^6)}{(0.4 \cdot 10^4)(6 \cdot 10^{-2})}$	$\dfrac{(9 \cdot 10^6)(0.05 \cdot 10^{-4})}{(5 \cdot 10^5)(4 \cdot 10^{-6})}$	$\dfrac{(17 \cdot 10^{-5})(2.5 \cdot 10^{-3})}{(23 \cdot 10^6)(5 \cdot 10^{-1})}$	$\dfrac{(15 \cdot 10^{-5})(20 \cdot 10^7)}{(0.5 \cdot 10^6)(400 \cdot 10^{-2})}$

$\mathbf{W}$rite each number or expression in standard (or in expanded) form.

$2.5 \cdot 10^7$	$1.4 \cdot 10^{-5}$	$1.23 \cdot 10^6$	$5.89 \cdot 10^{-2}$
$(1.5 \cdot 10^6)(10^{-3})$	$(1.8 \cdot 10^3)(5 \cdot 10^4)$	$(8.5 \cdot 10^2)(4 \cdot 10^5)$	$(0.5 \cdot 10^{-2})(10^2)$
$\dfrac{2.2 \cdot 10^5}{1.3 \cdot 10^7}$	$\dfrac{5 \cdot 10^8}{3 \cdot 10^5}$	$\dfrac{1.3 \cdot 10^{-5}}{(300)(4 \cdot 10^{-9})}$	$\dfrac{(2 \cdot 10^{-5})(1.4 \cdot 10^5)}{(5 \cdot 10^{-7})(8 \cdot 10^4)}$

2.2. Polynomials and Operations

- **DEFINITIONS AND CLASSIFYING OF POLYNOMIALS**
- **ADDITION AND SUBTRACTION OF POLYNOMIALS**
- **MULTIPLICATION OF POLYNOMIALS**
- **DIVISION OF POLYNOMIALS**

■ DEFINITIONS AND CLASSIFYING OF POLYNOMIALS

We will use the following important definitions.

IMPORTANT DEFINITIONS

▶ A **monomial** is a constant, or a variable, or the product of a constant and one or more variables with only **natural-number** exponents.

For example, 2, x, $3xy$, and $4xy^2$ are monomials, but $\frac{1}{x}$ and $\sqrt[3]{x+1}$ are not monomials.

▶ A **polynomial** is a monomial or the sum or difference of any finite number of monomials. Monomials are **terms of polynomials**. The numeric factor of a term is called the **coefficient**. The **separating signs** between terms of a polynomial are $+$ or $-$ signs.

For example, 3, y, $1+x$, $2x-3xy$, and $5y^2-3xy+1$ are polynomials. The numbers 5, 3, and 1 are coefficients of the polynomial $5y^2-3xy+1$.

▶ A **binomial** is a polynomial with **2** terms.
▶ A **trinomial** is a polynomial with **3** terms.

For example, the expressions $2x$ and xy are monomials, $1+x$ and $2x-3xy$ are binomials, and $5y^2-3xy+1$ is a trinomial.

▶ The **degree of a monomial** is the sum of exponents of all the variables of the monomial.
▶ The **degree of a polynomial** is the degree of the term with the highest power.

For example, the degree of monomial 2 is **0**; the degree of monomial x is **1**; the degree of binomial $2x-3xy$ is **2**; the degree of trinomial $5y^3-3xy+1$ is **3**.

▶ Two terms of a polynomial are **like terms** if those terms contain the same variables with the same exponents.

For example, 3 & 5, y & $6y$, $5y^2$ & $8y^2$ are like terms, but 2 & x, or $2x$ & $3y$, or $4x$ & $5x^2$ are not like terms.

▶ A polynomial with one variable is in **standard form** when all terms are in order from greatest to least by exponent of variable. The term with the highest degree is called the **leading term**. The coefficient of the leading term is called the **leading coefficient**.

For example, the polynomial $5x^3-3x^2+2x-4$ is in standard form. The leading term is $5x^3$ and the leading coefficient is **5**.

Examples Determine which expressions are polynomials.

3	$9x$	$x + y - z$	$1/x$	$\sqrt[3]{x+1} + x$
$5x^3 - y^7 + x^5y^8$	$4y^2 + 6y^4 - z^3$	$x + y + \dfrac{3}{x}$	$x^6z^7 + x^6y^9$	$x^5y^7 - x^6y^9 + \dfrac{2}{xy}$

Solution

3	$9x$	$x + y - z$	$1/x$	$\sqrt[3]{x+1} + x$
is a polynomial	is a polynomial	is a polynomial	is not a polynomial	is not a polynomial
$5x^3 - y^7 + x^5y^8$	$4y^2 + 6y^4 - z^3$	$x + y + \dfrac{3}{x}$	$x^6z^7 + x^6y^9$	$x^5y^7 - x^6y^9 + \dfrac{2}{xy}$
is a polynomial	is a polynomial	is not a polynomial	is a polynomial	is not a polynomial

Examples Classify each polynomial by number of terms and by degree.

4	$5x$	$3xy$	$6x^2y^3$	$9x^3y^5z$
$y^8z^3 + x^9y^{10}$	$9x^5 + 6y^9z^3$	$5 + 3x^5y^7$	$9x^8y^{12} + 4x^5y^3$	$23 + 5x^{12}y^{15}$
$x^7 + x^6y^5 + y^{10}$	$x^3 + y^7 + z^{13} + t^4$	$9 + 2x^3 + 5y^7 + y^3$	$3y^2 + x^6y^5 + y^4$	$xy + x^2y^4 + z^9$

Solution

4	$5x$	$3xy$	$6x^2y^3$	$9x^3y^5z$
monomial, $d = 0$	monomial, $d = 1$	monomial, $d = 2$	monomial, $d = 5$	monomial, $d = 9$
$y^8z^3 + x^9y^{10}$	$9x^5 + 6y^9z^3$	$5 + 3x^5y^7$	$9x^8y^{12} + 4x^5y^3$	$23 + 5x^{12}y^{15}$
binomial, $d = 19$	binomial, $d = 12$	binomial, $d = 12$	binomial, $d = 20$	binomial, $d = 27$
$x^7 + x^6y^5 + y^{10}$	$x^3 + y^7 + z^{13} + t^4$	$9 + 2x^3 + 5y^7 + y^3$	$3y^2 + x^6y^5 + y^4$	$xy + x^2y^4 + z^9$
trinomial, $d = 11$	four terms, $d = 13$	four terms, $d = 7$	trinomial, $d = 11$	trinomial, $d = 9$

Examples Write the polynomial in standard form.

$5 + 2x^2 - x$	$3x^2 - 6x + 3$	$4y^3 + 12y^5 + y^6$	$7y^3 + 3y^4 + 5$	$10y^5 + 4y^7 + 9z^6$
$1 - x^2$	$4x - 2x^2$	$5 - x^3 + x^4 + x^2$	$3x^5 + y^3 + x^7$	$y^2 - 8y + 3y^3$

Solution

$5 + 2x^2 - x$	$3x^2 - 6x + 3$	$4y^3 + 12y^5 + y^6$	$7y^3 + 3y^4 + 5$	$10y^5 + 4y^7 + 9z^6$
$2x^2 - x + 5$	$3x^2 - 6x + 3$	$y^6 + 12y^5 + 4y^3$	$3y^4 + 7y^3 + 5$	$4y^7 + 9z^6 + 10y^5$
$1 - x^2$	$4x - 2x^2$	$5 - x^3 + x^4 + x^2$	$3x^5 + y^3 + x^7$	$y^2 - 8y + 3y^3$
$-x^2 + 1$	$-2x^2 + 4x$	$x^4 - x^3 + x^2 + 5$	$x^7 + 3x^5 + y^3$	$3y^3 + y^2 - 8y$

67

We also will use the following group of definitions.

► A polynomial is written in **descending order** for the variable **x** if the term with the greatest exponent for **x** is the first term; the term with the next greatest exponent for **x** is the second term, and so on.

For example, the polynomial $6x^5 - 4x^3y + 2xy + 5$ is written in descending order.

► A polynomial is written in **ascending order** for the variable **x** if the term with the smallest exponent for **x** is the first term; the term with the next smallest exponent for **x** is the second term, and so on.

For example, the polynomial $3 + 4xy + 4x^3 - 2x^4y$ is written in ascending order.

► Replacing the variables in a polynomial with numbers and calculating the resulting number is called **evaluating the polynomial**.

For example, evaluation of the polynomial $3 + 4xy - x^2y$ for $x = 2$ and $y = 3$ gives the result **15** $(3 + 4xy - x^2y = 3 + 4 \cdot 2 \cdot 3 - 2^2 \cdot 3 = 3 + 24 - 12 = 15)$.

Examples Arrange the polynomials in descending order for **x**.

| $5x^3 - y^7 + x^5y$ | $x^2 + 6x^4 - z^3$ | $x + 2x^3y + 3x^2$ | $x^6z^7 + x^5y^9$ | $x^5y^7 - x^6 - x$ |

Solution

| $x^5y + 5x^3 - y^7$ | $6x^4 + x^2 - z^3$ | $2x^3y + 3x^2 + x$ | $x^6z^7 + x^5y^9$ | $-x^6 + x^5y^7 - x$ |

Examples Arrange the polynomials in ascending order for **x**.

| $5x^3 - y^7 + x^5y$ | $x^2 + 6x^4 - z^3$ | $x + 2x^3y + 3x^2$ | $x^6z^7 + x^5y^9$ | $x^5y^7 - x^6 - x$ |

Solution

| $-y^7 + 5x^3 + x^5y$ | $-z^3 + x^2 + 6x^4$ | $x + 3x^2 + 2x^3y$ | $x^5y^9 + x^6z^7$ | $-x + x^5y^7 - x^6$ |

Examples Evaluate each polynomial for **x = 1, y = 2, z = 3**.

| $x^3 - y^2 + x^2y$ | $x + 6x^2 - z^2$ | $x + zy + x^2$ | $x^2z^2 + xy^3$ | $xy^2 - x^2 - z$ |

Solution

| $1^3 - 2^2 + 1^2 \cdot 2$ | $1 + 6 \cdot 1^2 - 3^2$ | $1 + 3 \cdot 2 + 1^2$ | $1^2 \cdot 3^2 + 1 \cdot 2^3$ | $1 \cdot 2^2 - 1^2 - 3$ |
| $1 - 4 + 2 = -1$ | $1 + 6 - 9 = -2$ | $1 + 6 + 1 = 8$ | $9 + 8 = 17$ | $4 - 1 - 3 = 0$ |

Determine which expressions are polynomials.

8	$9 + x$	$3x + 4z - 5y$	$1 - \dfrac{1}{x}$	$\sqrt[3]{2 + x} + x$
$x^4 - 13y^3 + x^5z^2$	$\sqrt[4]{3y + 1} + x^2$	$4 - \dfrac{x}{x + 1}$	$x^5y^2 + x^2y^8$	$4x^5y^7 - x^6 + \dfrac{1}{xy}$

Identify the terms. Give the coefficient of each term. Rewrite the polynomial in standard form.

$2x^2 - x$	$4 + 2x^2 - 5x$	$2y^7 + y^5 + 4y^6$	$5y^2 + 3y^4 + 1$	$6y^5 + 4y^8 + 7z^2$
$1 - 2x^2 + x$	$x - 3x^2 - 5$	$6 - 2x^3 + x^4$	$x^5 + y^2 + x^8$	$y^6 - 2y + 3y^2$

Classify each polynomial by number of terms and by degree.

$5y$	$1 + x$	$5x + 2y - 1$	$2x^3 + y^3$	$5x^2y^3z^4$
$3x^5y^8 + 4y^7z^9$	$5x^3 + 3y^5 + 4z^2$	$x + 8x^5 + y^6$	$3x^8y^7 - 2x^4y^2$	$x + 4x^5 - y^{11}$

Classify each polynomial as a monomial, binomial or trinomial.

2	$1 + 2x$	$4 + x - 3y$	$x + 4x^2y^2$	$2x^5y^3z$
$5y^2 + z^2 + x^3y^2$	y^2z^3	$1 + x^2 - 5y^2$	$4x^3y^8 + 2x^2y^3$	$23 + 5x^{12}y^{15}$

Arrange the polynomials in descending order for **x**.

$x^7 + x^6y^5 + y^{10}$	$x^3 + y^7 + z^{13}$	$9 + 2x^3$	$2y^2 + x^6y^2 + y^3$	x^2y^3

Arrange the polynomials in ascending order for **x**.

$6x - x^2y$	$8x + x^3$	$5 - x^5 + x + x^6$	$4 + x^3 - x^4 + x^7$	$5x^2 - 5x + x^3$

Evaluate each polynomial for **x = 1, y = 2, z = −1**.

$x^3 - 2y^2 + x^2z$	$y + 6x^2 - z^2$	$x + zy + y^2$	$x^2 + xy^3z^2$	$2y^2 - x^2 - zx$

■ ADDITION AND SUBTRACTION OF POLYNOMIALS

We will use the **combining like terms method** to add or subtract two or more polynomials and represent the result in the simplest form if the given polynomials contain like terms.

But we cannot combine like terms if the polynomials do not contain any like terms.

Two terms are **like terms** if those terms contain the same variables with the same exponents. For example, **2 & 3**; **x & 2x**; or **$5y^2$ & $8y^2$** are like terms.

To combine two like terms we have to add or subtract coefficients of like terms: **2 + 3 = 5; 2 − 3 = −1; x + 2x = 3x; x − 2x = −x; $5y^2 + 8y^2 = 13y^2$; $5y^2 − 8y^2 = −3y^2$**.

Examples Combine like terms. Write the answer in simplest form.

$x + (4x − 5)$	$(7x + 4) + (3 − 6x)$	$(5xy + 8x) + (7x − 2xy + 3)$
$(4y^4 + 6z^2) + (z^2 − 4y^4) + 2$	$6y^3 + (x + y^3) + (4 + x − y^3)$	$(7y^2 + 6y) + (4y − 3) + 2$

Solution

$x + (4x − 5)$ $x + 4x − 5$ $5x − 5$	$(7x + 4) + (3 − 6x)$ $7x + 4 + 3 − 6x$ $x + 7$	$(5xy + 8x) + (7x − 2xy + 3)$ $5xy + 8x + 7x − 2xy + 3$ $3xy + 15x + 3$
$(4y^4 + 6z^2) + (z^2 − 4y^4) + 2$ $4y^4 + 6z^2 + z^2 − 4y^4 + 2$ $7z^2 + 2$	$6y^3 + (x + y^3) + (4 + x − y^3)$ $6y^3 + x + y^3 + 4 + x − y^3$ $6y^3 + 2x + 4$	$(7y^2 + 6y) + (4y − 3) + 2$ $7y^2 + 6y + 4y − 3 + 2$ $7y^2 + 10y − 1$

Examples Subtract polynomials by combining like terms.

$(y − 3) − (3x − 4)$	$(2x + 6) − (7 − x)$	$(5y + x) − (10x − y) − 7y$
$(y^2 + 5z^3) − (2z^3 − 4y^2) − 2$	$4y^2 − (2x + y^2) − (7 + x)$	$(y^2 + 6y − 4) − (9y − 2)$

Solution

$(y − 3) − (3x − 4)$ $y − 3 − 3x + 4$ $y − 3x + 1$	$(2x + 6) − (7 − x)$ $2x + 6 − 7 + x$ $3x − 1$	$(5y + x) − (10x − y) − 7y$ $5y + x − 10x + y − 7y$ $− y − 9x$
$(y^2 + 5z^3) − (2z^3 − 4y^2) − 2$ $y^2 + 5z^3 − 2z^3 + 4y^2 − 2$ $5y^2 + 3z^3 − 2$	$4y^2 − (2x + y^2) − (7 + x)$ $4y^2 − 2x − y^2 − 7 − x$ $3y^2 − 3x − 7$	$(y^2 + 6y − 4) − (9y − 2)$ $y^2 + 6y − 4 − 9y + 2$ $y^2 − 3y − 2$

$\mathbf{C}$ombine like terms in each expression and write the answer in standard form.

$8x + 4 - 3x - 4$	$x^4 - 7y^4 + 5x^4$	$x + 4z - 5x + z$	$1 - x^4 + 2x + 5$	$x - 3x^2 - 5x^2$

$\mathbf{A}$dd polynomials by combining like terms and write the answers in standard form.

$(x^2 - x) + (2x^2 - 5x)$	$(4 + 2x^2 - 5x) + (x - 3x^2)$	$(2y^5 + y^2) + (4y^2 - 3y^5) + y^5$
$(1 - 2x^2 + x) + (x - 3x^2 - 5)$	$(x^5 + y^2 + z) + (z - y^2 - 2x^5)$	$(6 - 2x^3 + x^4) + (2x^3 - x^4)$

$\mathbf{S}$ubtract polynomials by combining like terms and write the answers in standard form.

$(x^2 - x) - (2x^2 - 5x)$	$(4 + 2x^2 - 5x) - (x - 3x^2)$	$(2y^5 + y^2) - (4y^2 - 3y^5) + y^5$
$(1 - 2x^2 + x) - (x - 3x^2 - 5)$	$(x^5 + y^2 + z) - (z - y^2 + 2x^5)$	$(6 - 2x^3 + x^4) - (2x^3 - x^4)$

$\mathbf{F}$ind the perimeter of each figure and write the answer in standard form.

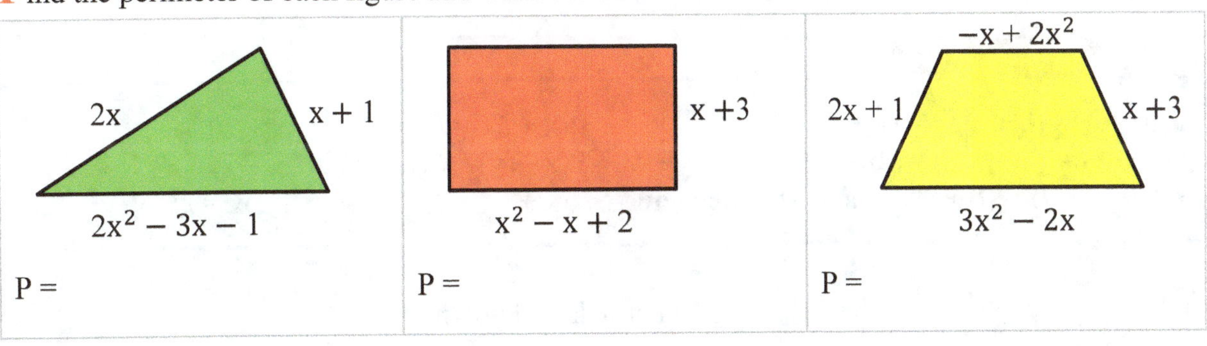

$\mathbf{F}$ind the area of each figure and write the answer in standard form.

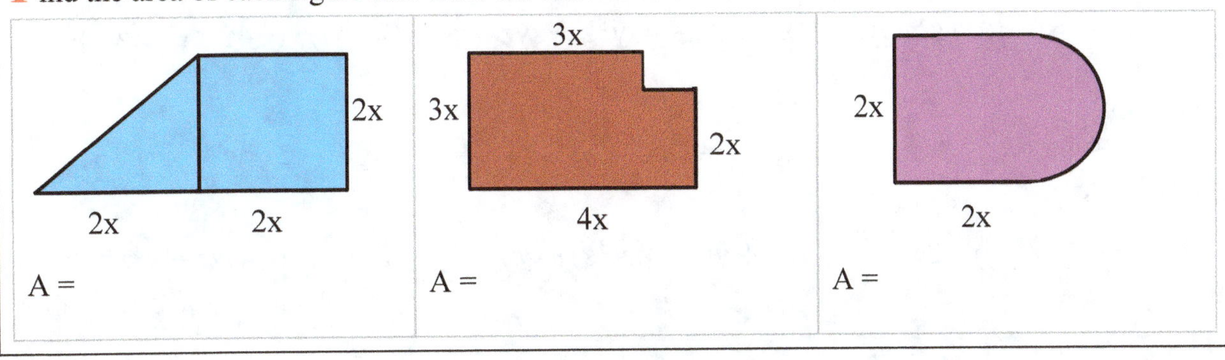

■ MULTIPLICATION OF POLYNOMIALS

MULTIPLYING MONOMIALS AND POLYNOMIALS

To multiply monomials we use the properties of multiplication and powers.

Examples Multiply the following monomials.

$x \cdot 4x$	$2x \cdot 3x^2$	$5xy \cdot (-4x^5y^7)$
$(4y^4 z^3) \cdot (2z^2 y^5)$	$6y^3 \cdot 2xy^3 \cdot (-4xy^3z) \cdot xyz$	$(5x^2y) \cdot (-4yzt) \cdot (-x^5zt^4)$

Solution

$x \cdot 4x$ $4x^2$	$2x \cdot 3x^2$ $6x^3$	$5xy \cdot (-4x^5y^7)$ $-20x^6y^8$
$(4y^4 z^3) \cdot (2z^2 y^5)$ $8y^9 z^5$	$6y^3 \cdot 2xy^3 \cdot (-4xy^3z) \cdot xyz$ $-48x^3 y^{10}z^2$	$(5x^2y) \cdot (-4yzt) \cdot (-x^5zt^4)$ $20x^7y^2z^2t^5$

To multiply monomials and polynomials with two or more terms we use the distributive property.

DISTRIBUTIVE PROPERTY

$a(b + c) = (b + c)a = ab + ac$ $a(b - c) = (b - c)a = ab - ac$ $a(b + c - d) = (b + c - d)a = ab + ac - ad$	$4(x + 5) = (x + 5)4 = 4x + 20$ $2(x - y) = (x - y)2 = 2x - 2y$ $x(x + y^2 - 2) = (x + y^2 - 2)x = x^2 + xy^2 - 2x$

Examples Multiply the following monomials and polynomials.

$x \cdot (x + 3)$	$2x \cdot (x^3 - 4x^2 + y)$	$5xy \cdot (-2x^6y^2 + x^2y^3)$
$(4y^4 z^3 - 3xz^3 y^2) \cdot (2z^2 y^5)$	$(x^2 + y^3 - 4xy^3z) \cdot xyz$	$(2x^2yt) \cdot (-4yzt - x^2zt^3)$

Solution

$x \cdot (x + 3)$ $x^2 + 3x$	$2x \cdot (x^3 - 4x^2 + y)$ $2x^4 - 8x^3 + 2xy$	$5xy \cdot (-2x^6y^2 + x^2y^3)$ $-10x^7y^3 + 5x^3y^4$
$(4y^4 z^3 - 3xz^3y^2) \cdot (2z^2 y^5)$ $8y^9 z^5 - 6xy^7z^5$	$(x^2 + y^3 - 4xy^3z) \cdot xyz$ $x^3yz + xy^4z - 4x^2y^4z^2$	$(2x^2yt) \cdot (-4yzt - x^2zt^3)$ $-8x^2y^2zt^2 - 2x^4yzt^4$

MULTIPLYING BINOMIALS AND POLYNOMIALS

To multiply two binomials we use the **FOIL** property.

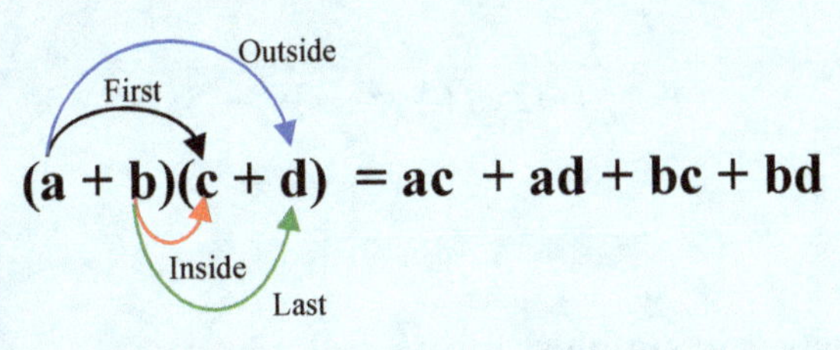

FOIL PROPERTY

$$(a + b)(c + d) = ac + ad + bc + bd$$

$(x + 2)(3x + 5) = 3x^2 + 5x + 6x + 10$

$(x - 2)(x - 2) = x^2 - 2x - 2x + 4$

$(x - y)(x + y) = x^2 + xy - yx - y^2$

Examples Multiply the following binomials. Then write the answer in simplest form.

| $(x + 3) \cdot (x + 3)$ | $(2x + 1) \cdot (x^3 - 4x^2)$ | $(x + y) \cdot (- x^2y^3 + x^3y^2)$ |

Solution

$(x + 3) \cdot (x + 3)$	$(2x + 1) \cdot (x^3 - 4x^2)$	$(x + y) \cdot (- x^2y^3 + x^3y^2)$
$x^2 + 3x + 3x + 9$	$2x^4 - 8x^3 + x^3 - 4x^2$	$- x^3y^3 + x^4y^2 - x^2y^4 + x^3y^3$
$x^2 + 6x + 9$	$2x^4 - 7x^3 - 4x^2$	$x^4y^2 - x^2y^4$

To multiply a binomial with a trinomial we use the following **FST** property.

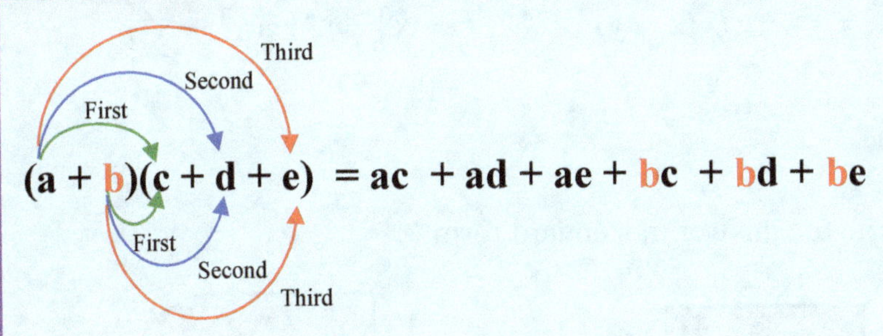

FST PROPERTY

$$(a + b)(c + d + e) = ac + ad + ae + bc + bd + be$$

$(x + 2)(x^2 + 2x + 1) =$
$x^3 + 2x^2 + x + 2x^2 + 4x + 2 =$
$x^3 + 4x^2 + 5x + 2$

$(x - y)(x + y - 2) =$
$x^2 + xy - 2x - yx - y^2 + 2y =$
$x^2 - 2x - y^2 + 2y$

Examples Multiply the following binomials and trinomials.

| $(x + y) \cdot (x - y + 2)$ | $(2x + 1) \cdot (x^3 - 4x^2 + x)$ | $(x + y) \cdot (x^2y^3 - x^3y^2 + xy^2)$ |

Solution

$(x + y) \cdot (x - y + 2)$	$(2x + 1) \cdot (x^3 - 4x^2 + x)$	$(x + y) \cdot (x^2y^3 - x^3y^2 + xy^2)$
$x^2 - xy + 2x + yx - y^2 + 2y$	$2x^4 - 8x^3 + 2x^2 + x^3 - 4x^2 + x$	$x^3y^3 - x^4y^2 + x^2y^2 + x^2y^4 - x^3y^3 + xy^3$
$x^2 + 2x - y^2 + 2y$	$2x^4 - 7x^3 - 2x^2 + x$	$- x^4y^2 + x^2y^2 + x^2y^4 + xy^3$

Multiply the following monomials and polynomials. Then write the answer in simplest form.

$2x(4x^2)$	$9(3 - 6x)$	$(4xy + 2x)(-3)$
$-4(2x - 5)$	$3x(3x + 4y^2 - 5)$	$(-2y)(4y^2 + 3y - 5x + 2)$

Multiply the following binomials. Then write the answer in simplest form.

$(y - 7)(y + 7)$	$(x + 1) \cdot (x^3 - x^2)$	$(x + y) \cdot (-x^2y^3 + x^3y^2)$
$(2y - 1)(2y + 1)$	$(x - y)(x^2 + y^2)$	$(x + 3z)(x + 3z)$

Multiply the following binomials and trinomials. Then write the answer in simplest form.

$(x - y) \cdot (x^2 + xy + y^2)$	$(x + y) \cdot (x^2 - xy + y^2)$	$(x + y) \cdot (x^3y^2 - x^2y^3 + xy)$
$(1 - 2x^2) \cdot (x - 3x^2 - 5)$	$(x^5 + y^2 + z) \cdot (z - y)$	$(1 - x^3 + x^4) \cdot (x^3 - x^4)$

Find the area of each figure and write the answer in standard form.

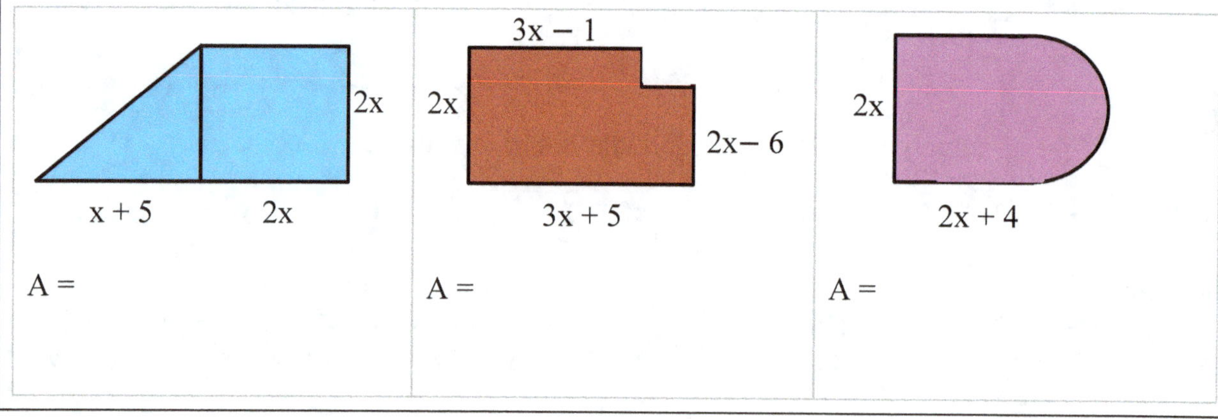

■ DIVISION OF POLYNOMIALS

DIVISION POLYNOMIALS BY MONOMIALS

To divide a polynomial by a monomial we use the **term by term division rule**.

$$\frac{a + b - c + d - e}{f} = \frac{a}{f} + \frac{b}{f} - \frac{c}{f} + \frac{d}{f} - \frac{e}{f} \qquad \frac{16x^4 + 8x^2 - 4x}{2x} = \frac{16x^4}{2x} + \frac{8x^2}{2x} - \frac{4x}{2x} = 8x^3 + 4x - 2$$

Examples Divide each polynomial by the given monomial.

$\dfrac{x + 3}{3x}$	$\dfrac{4x^2 - 2x}{x}$	$\dfrac{4x^4 - 8x^3 + 6x^2}{2x^2}$
$\dfrac{4y^4 z^3 - 16xz^3 y^2}{4z^2 y^2}$	$\dfrac{x^2y + y^3z - 4xy^3}{xyz}$	$\dfrac{-4xyzt - 8x^2yt^3}{2xyt}$
$\dfrac{6x^2 y^3 - 9xyz}{3xy}$	$\dfrac{xyz + x^2yz^2 - 3xy^3z}{xyz}$	$\dfrac{x^5 - 2x^4 + x^3 - 3x^2 + 5}{x}$

Solution

$\dfrac{x + 3}{3x}$	$\dfrac{4x^2 - 2x}{x}$	$\dfrac{4x^4 - 8x^3 + 6x^2}{2x^2}$
$\dfrac{x}{3x} + \dfrac{3}{3x}$	$\dfrac{4x^2}{x} - \dfrac{2x}{x}$	$\dfrac{4x^4}{2x^2} - \dfrac{8x^3}{2x^2} + \dfrac{6x^2}{2x^2}$
$\dfrac{1}{3} + \dfrac{1}{x}$	$4x - 2$	$2x^2 - 4x + 3$
$\dfrac{4y^4 z^3 - 16xz^3 y^2}{4z^2 y^2}$	$\dfrac{x^2y + y^3z - 4xy^3}{xyz}$	$\dfrac{-4xyzt - 8x^2yt^3}{2xyt}$
$\dfrac{4y^4 z^3}{4z^2 y^2} - \dfrac{16xz^3 y^2}{4z^2 y^2}$	$\dfrac{x^2y}{xyz} + \dfrac{y^3z}{xyz} - \dfrac{4xy^3}{xyz}$	$-\dfrac{4xyzt}{2xyt} - \dfrac{8x^2yt^3}{2xyt}$
$y^2 z - 4xz$	$\dfrac{x}{z} + \dfrac{y^2}{x} - \dfrac{4y^2}{z}$	$-2z - 4xt^2$
$\dfrac{6x^2 y^3 - 9xyz}{3xy}$	$\dfrac{xyz + x^2yz^2 - 3xy^3z}{xyz}$	$\dfrac{x^5 - 2x^4 + x^3 - 3x^2 + 5}{x}$
$\dfrac{6x^2 y^3}{3xy} - \dfrac{9xyz}{3xy}$	$\dfrac{xyz}{xyz} + \dfrac{x^2yz^2}{xyz} - \dfrac{3xy^3z}{xyz}$	$\dfrac{x^5}{x} - \dfrac{2x^4}{x} + \dfrac{x^3}{x} - \dfrac{3x^2}{x} + \dfrac{5}{x}$
$2xy^2 - 3z$	$1 + xz - 3y^2$	$x^4 - 2x^3 + x^2 - 3x + \dfrac{5}{x}$

LONG DIVISION

To divide a polynomial by a binomial we can use the **long division** method similar to long division of numbers.

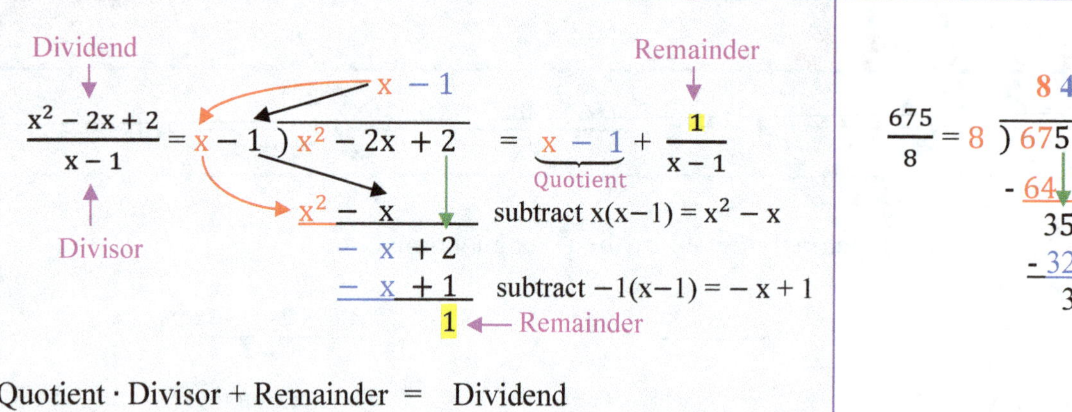

$$\text{Quotient} \cdot \text{Divisor} + \text{Remainder} = \text{Dividend}$$
$$(x - 1) \cdot (x - 1) + 1 = x^2 - 2x + 2$$

$$84 \cdot 8 + 3 = 675$$

Examples **D**ivide using long division.

$\dfrac{9x^2 - 7x + 5}{3x + 2}$	$\dfrac{x^3 - 1}{x - 1}$

Solution

$$\frac{9x^2 - 7x + 5}{3x + 2} = 3x + 2 \,\overline{)\,9x^2 - 7x + 5}$$

with quotient $3x - 4$:

$$9x^2 + 6x$$
$$-13x + 5$$
$$-12x - 8$$
$$- x + 13 \quad \text{Remainder}$$

$$\frac{9x^2 - 7x + 5}{3x + 2} = 3x - 4 + \frac{-x + 13}{3x + 2}$$

$$\text{Quotient} \cdot \text{Divisor} + \text{Remainder} = \text{Dividend}$$
$$(3x - 4) \cdot (3x + 2) + (-x + 13) = 9x^2 - 7x + 5$$

$$\frac{x^3 - 1}{x - 1} = x - 1 \,\overline{)\,x^3 + 0 \cdot x^2 + 0 \cdot x - 1}$$

with quotient $x^2 + x + 1$:

$$x^3 - x^2$$
$$x^2 + 0 \cdot x$$
$$x^2 - x$$
$$x - 1$$
$$x - 1$$
$$0$$

$$\frac{x^3 - 1}{x - 1} = x^2 + x + 1$$

$$\text{Quotient} \cdot \text{Divisor} + \text{Remainder} = \text{Dividend}$$
$$(x^2 + x + 1) \cdot (x - 1) = x^3 - 1$$

Synthetic Division

To divide a polynomial by a binomial we can also use the **synthetic division** method.
This method is shorter and easier for dividing a polynomial by a binomial in the form $x - k$.
Synthetic division method uses only the coefficients of terms with the variable and the constant term.

Synthetic division

For example, to divide $x^3 - 1$ by $x - 1$ we have to identify the value of **k** ($x - k = x - 1$, so $k = 1$), then the coefficients of the variable **x** with rule "power -1", and the constant term ($1x^3 + 0x^2 + 0x - 1$).

$$\frac{x^3 - 1}{x - 1} = \underline{1}\,|\,1 \quad 0 \quad 0 \quad -1$$
$$+\;0 \quad 1 \quad 1 \quad 1$$
$$1 \quad 1 \quad 1 \quad 0$$

Quotient $= 1x^2 + 1x + 1 = x^2 + x + 1$

$$\frac{x^3 - 1}{x - 1} = x^2 + x + 1$$

Long division

$$\frac{x^3 - 1}{x - 1} = x - 1 \,\overline{)\, x^3 + 0 \cdot x^2 + 0 \cdot x - 1}$$

$$\begin{array}{r} x^2 + x + 1 \\ \underline{x^3 - x^2} \\ x^2 + 0 \cdot x \\ \underline{x^2 - x} \\ x - 1 \\ \underline{x - 1} \\ 0 \end{array}$$

$$\frac{x^3 - 1}{x - 1} = x^2 + x + 1$$

Examples Divide using synthetic division.

$$\frac{2x^3 - 3x^2 - 8x + 4}{x + 2}$$

$$\frac{x^5 - 3x^4 + x^2 - 2x - 3}{x - 3}$$

Solution

$$\frac{2x^3 - 3x^2 - 8x + 4}{x + 2} = \underline{-2}\,|\,2 \quad -3 \quad -8 \quad 4$$
$$+\;0 \quad -4 \quad 14 \quad -12$$
$$2 \quad -7 \quad 6 \quad -8$$

$$\frac{x^5 - 3x^4 + x^2 - 2x - 3}{x - 3} = \underline{3}\,|\,1 \quad -3 \quad 0 \quad 1 \quad -2 \quad -3$$
$$+\;0 \quad 3 \quad 0 \quad 0 \quad 3 \quad 3$$
$$1 \quad 0 \quad 0 \quad 1 \quad 1 \quad 0$$

$$\frac{2x^3 - 3x^2 - 8x + 4}{x + 2} = 2x^2 - 7x + 6 + \frac{-8}{x + 2}$$

$$\frac{x^5 - 3x^4 + x^2 - 2x - 3}{x - 3} = 1x^4 + 1x + 1$$

$$\frac{2x^3 - 3x^2 - 8x + 4}{x + 2} = 2x^2 - 7x + 6 - \frac{8}{x + 2}$$

$$\frac{2x^3 - 3x^2 - 8x + 4}{x - 3} = x^4 + x + 1$$

$\mathbf{I}$f the divisor has a leading coefficient other than **1** ($ax - b$, $a \neq 1$) then we will factor out the leading coefficient of the divisor from the divisor and dividend to cancel and change the divisor into **x − k** form. After this procedure we will use synthetic division with the divisor in **x − k** form.

Examples $\mathbf{D}$ivide using synthetic division.

$\dfrac{4x^2 - 4x + 1}{2x - 1}$	$\dfrac{27x^3 + 8}{3x + 2}$
$\dfrac{3x^4 - 2x^3 + 5x^2 - 4x - 2}{3x + 1}$	$\dfrac{4x^4 - 5x^2 - 8x - 10}{2x - 3}$

Solution

$$\frac{4x^2 - 4x + 1}{2x - 1} = \frac{2\left(2x^2 - 2x + \frac{1}{2}\right)}{2\left(x - \frac{1}{2}\right)} =$$

$$= \frac{2x^2 - 2x + \frac{1}{2}}{x - \frac{1}{2}} = \quad \frac{1}{2}\bigg|\; 2 \quad -2 \quad \frac{1}{2}$$

$$+\; 0 \quad 1 \quad -\frac{1}{2}$$

$$\overline{\quad 2 \quad -1 \quad\;\; 0}$$

$$\frac{4x^2 - 4x + 1}{2x - 1} = 2x - 1$$

$$\frac{27x^3 + 8}{3x + 2} = \frac{3\left(9x^3 + 0x^2 + 0x + \frac{8}{3}\right)}{3\left(x + \frac{2}{3}\right)} =$$

$$= \frac{9x^3 + 0x^2 + 0x + \frac{8}{3}}{x + \frac{2}{3}} = \quad -\frac{2}{3}\bigg|\; 9 \quad 0 \quad 0 \quad \frac{8}{3}$$

$$+\; 0 \quad -6 \quad 4 \quad -\frac{8}{3}$$

$$\overline{\quad 9 \quad -6 \quad 4 \quad\;\; 0}$$

$$\frac{27x^3 + 8}{3x + 2} = 9x^2 - 6x + 4$$

$$\frac{3x^4 - 2x^3 + 5x^2 - 4x - 2}{3x + 1} =$$

$$\frac{3\left(x^4 - \frac{2}{3}x^3 + \frac{5}{3}x^2 - \frac{4}{3}x - \frac{2}{3}\right)}{3\left(x + \frac{1}{3}\right)} =$$

$$\frac{x^4 - \frac{2}{3}x^3 + \frac{5}{3}x^2 - \frac{4}{3}x - \frac{2}{3}}{x + \frac{1}{3}} =$$

$$-\frac{1}{3}\bigg|\; 1 \quad -\frac{2}{3} \quad \frac{5}{3} \quad -\frac{4}{3} \quad -\frac{2}{3}$$

$$+\; 0 \quad -\frac{1}{3} \quad \frac{1}{3} \quad -\frac{2}{3} \quad \frac{2}{3}$$

$$\overline{\quad 1 \quad -1 \quad 2 \quad -2 \quad\;\; 0}$$

$$\frac{3x^4 - 2x^3 + 5x^2 - 4x - 2}{3x + 1} = x^3 - x^2 + 2x - 2$$

$$\frac{4x^4 - 5x^2 - 8x - 10}{2x - 3} =$$

$$\frac{2\left(2x^4 + 0x^3 - \frac{5}{2}x^2 - 4x - 5\right)}{2\left(x - \frac{3}{2}\right)} =$$

$$\frac{2x^4 + 0x^3 - \frac{5}{2}x^2 - 4x - 5}{x - \frac{3}{2}} =$$

$$\frac{3}{2}\bigg|\; 2 \quad 0 \quad -\frac{5}{2} \quad -4 \quad -5$$

$$+\; 0 \quad 3 \quad \frac{9}{2} \quad 3 \quad -\frac{3}{2}$$

$$\overline{\quad 2 \quad 3 \quad 2 \quad -1 \quad -\frac{13}{2}}$$

$$\frac{4x^4 - 5x^2 - 8x - 10}{2x - 3} = 2x^3 + 3x^2 + 2x - 1 - \frac{13}{2x-3}$$

DIVISION BY FACTORING AND CANCELING

Sometimes to divide a polynomial by a polynomial we can use the **factoring and canceling method**. This method works if the dividend and the divisor are factorable polynomials with a minimum of one common factor other than **1** (or **–1**).

We will go back to the division of polynomials by factoring and canceling after studying factoring polynomials and the main identity formulas of algebra.

Now we will study some easy examples of division of polynomials by polynomials.

Examples Divide by using the factoring and canceling method.

$\dfrac{4x - 8}{x - 2}$	$\dfrac{x^2 - 3x}{x}$
$\dfrac{3x^2 - 9x}{6x^2 - 18x}$	$\dfrac{x^2 - 9}{x - 3}$
$\dfrac{x^2 - 9x}{2x^2 - 18x}$	$\dfrac{9x^2 - 4}{3x - 2}$
$\dfrac{2x^3 - 2x^2 + 4x}{x^2 - x + 2}$	$\dfrac{x^2 - 2x + x - 2}{x + 1}$

Solutions

$\dfrac{4x - 8}{x - 2} = \dfrac{4(x - 2)}{x - 2} = 4$	$\dfrac{x^2 - 3x}{x} = \dfrac{x(x - 3)}{x} = x - 3$
$\dfrac{3x^2 - 9x}{6x^2 - 18x} = \dfrac{3x(x - 3)}{6x(x - 3)} = \dfrac{1}{2}$	$\dfrac{x^2 - 9}{x - 3} = \dfrac{x^2 - 3^2}{x - 3} = \dfrac{(x - 3)(x + 3)}{x - 3} = x + 3$
$\dfrac{x^2 - 9x}{2x^2 - 18x} = \dfrac{x(x - 9)}{2x(x - 9)} = \dfrac{1}{2}$	$\dfrac{9x^2 - 4}{3x - 2} = \dfrac{(3x)^2 - (2)^2}{3x - 2} =$ $\dfrac{(3x - 2)(3x + 2)}{3x - 2} = 3x + 2$
$\dfrac{2x^3 - 2x^2 + 4x}{x^2 - x + 2} = \dfrac{2x(x^2 - x + 2)}{x^2 - x + 2} = 2x$	$\dfrac{x^2 - 2x + x - 2}{x + 1} = \dfrac{x(x - 2) + (x - 2)}{x + 1} =$ $\dfrac{(x - 2)(x + 1)}{x + 1} = x - 2$

Divide the following polynomials and monomials using **term by term division**.

$\dfrac{x^2 - 5x}{x}$	$\dfrac{4x^2 - 2x}{2x}$	$\dfrac{6x^2 - 3xy}{3x}$
$\dfrac{x^2y - 2xy^2}{xy}$	$\dfrac{x^3yz^3 - xy^2z - xy}{xyz}$	$\dfrac{4x^3y^2z^3 - 6xy^2z^3 + 4x^2y^2z}{2x^2y^2z^2}$

Divide the polynomials using **long division**.

$\dfrac{5x^2+7x - 6}{x + 2}$	$\dfrac{x^3 - x^2-10x +12}{x - 3}$	$\dfrac{x^4 - 16}{x + 2}$	$\dfrac{x^2 - 4x + 3}{x - 3}$

Divide using **synthetic division**.

$\dfrac{x^2 - 2x + 1}{x - 1}$	$\dfrac{8x^3 - 27}{2x - 3}$	$\dfrac{5x^2+7x - 6}{x + 2}$	$\dfrac{x^3 - 9x^2+10x +12}{x - 3}$

Divide using **factoring and canceling**.

$\dfrac{x^2 - 2x}{x - 2}$	$\dfrac{x^3 - 3x^2 - 4x +12}{x - 3}$	$\dfrac{3x^3y^2z^3 - 9xy^2z^3}{3xy}$

Write each polynomial in standard form. Name each by its degree and the number of its terms.

1. $2 + x$
2. $x^2 - x^3$
3. $-x + 4x^5$
4. $1 + 2x^2 - x^3$
5. $1 - x^3 + 8x$

6. $5 - 2x + x^2$
7. $3x^5 - 6x + x^2$
8. $x^4 - x + x^7$
9. $1 + x^4 - x + x^3 + 2x^5$
10. $x + x^2 - 3x + x^9 + 7x^6$

11. $y^2 + 5 - y$
12. $2z^2 + 5z^3 - z$
13. $x^7 + 2x^3 - x^5 - x + x^6$
14. $x^2 + x^3 - x^6 - 1 + x^5$
15. $x^4 + x^3 - x^8 + x^5 - x^2$

Find each sum or difference of polynomials. Write the answer in simplest and standard form.

1. $(x^2 - 2) + (x^2 + 2)$
2. $(2x^2 - x^3) - (3x^2 - 2x^3)$
3. $(x^2 - x + 4) - (4 + x^2)$
4. $(1 + x^2 - x^3) - (3 - x^3)$
5. $(4x^2 - x + 4) - 4(1 + x^2)$

6. $(x + 1) + (x^2 - x - 1)$
7. $(x^2 - x + 2) - (x^2 - x)$
8. $(2 + x - x^2) - (x - x^2)$
9. $(x^2 - x + 2) - (x^2 - x)$
10. $(y + xy - x^2) + (xy - y)$

11. $(2 + x + y) - (-x + y)$
12. $2(x + y) + (-2x + 3y)$
13. $3(x + xy) + (2x - 3xy)$
14. $5(xy + yz) + y(x + 3z)$
15. $(x + 4xy) + y(1 - 4x)$

Multiply or divide the following polynomials. Write the answer in simplest and standard form.

1. $(y - 2)(y + 2)$
2. $(3x - 2y)(2y + 3x)$
3. $\dfrac{6xy^2 + 3y^2}{3y}$
4. $\dfrac{xy^3 + y^3}{x + 1}$

5. $(x + 2) \cdot (x^2 - 2x + 4)$
6. $xy(x^2 - 2x + y^2 - y)$
7. $\dfrac{4x^3 + 8x}{2x}$
8. $\dfrac{x^2 - x}{x - 1}$

9. $(x + y) \cdot (x^2y^3 + x^3y^2)$
10. $(x - y) \cdot (x^2 + y^2)$
11. $\dfrac{x^2 - 2x + 4xy}{x}$
12. $\dfrac{x^2 - y^2}{x - y}$

Divide a polynomial by using **long division** and **synthetic division**.

1. $\dfrac{x^2 - 6x + 9}{x - 3}$
2. $\dfrac{2x^3 - 3x^2 + 4x + 6}{x - 2}$

3. $\dfrac{3x^2 - 4x + 4}{x - 1}$
4. $\dfrac{4x^2 - 16 + 16}{x - 2}$

5. $\dfrac{3y^2 - 12y + 12}{y - 2}$
6. $\dfrac{4y^2 - 4y + 2}{2y - 1}$

Simplify.

1. $5y - (10 + 4x) + (3y + x)$
2. $7x^2 - (4y + x)(4y - x)$
3. $\dfrac{4x^2 - 9y^2}{2x + 3y} - 2x + y$

4. $(5x + 11) - (12x + x^2) + x^2$
5. $(x + 11y - 3)x - 11xy + 3x$
6. $\dfrac{x^2 - 4y^2}{x + y}(x + y) - x^2 + 2y^2$

7. $x(xy^2 + x) + (-x^2y - y)y$
8. $(x + 1)(x^2 - x + 1) - x^3$
9. $\dfrac{x^2 - 3xy^2 + 2x}{x} - 2 + y^2$

2.3. Basic Algebraic Identities

- **BASIC IDENTITIES**
- **BINOMIAL FORMULA AND PASCAL'S TRIANGLE**

BASIC IDENTITIES

In the future we will use the following **Basic Algebraic Identities**.

BASIC ALGEBRAIC IDENTITIES	EXAMPLE
1. $a^2 - b^2 = (a - b)(a + b)$	1. $4x^2 - 9y^2 = (2x - 3y)(2x + 3y)$
2. $(a - b)^2 = a^2 - 2ab + b^2$	2. $(2x - 3)^2 = 4x^2 - 12x + 9$
3. $(a + b)^2 = a^2 + 2ab + b^2$	3. $(x + 2y)^2 = x^2 + 4xy + 4y^2$
4. $a^3 - b^3 = (a - b)(a^2 + ab + b^2)$	4. $8x^3 - 27y^3 = (2x - 3y)(4x^2 + 6xy + 9y^2)$
5. $a^3 + b^3 = (a + b)(a^2 - ab + b^2)$	5. $x^3 + (2y)^3 = (x + 2y)(x^2 - 2xy + 4y^2)$
6. $(a - b)^3 = a^3 - 3a^2b + 3ab^2 - b^3$	6. $(2x - y)^3 = 8x^3 - 12x^2y + 6xy^2 - y^3$
7. $(a + b)^3 = a^3 + 3a^2b + 3ab^2 + b^3$	7. $(z + 3)^3 = z^3 + 9z^2 + 27z + 27$

Additional Examples

EXAMPLE 1: Use the identity formula for the **difference of two squares** and factor: $81x^2 - 16$

Solution

$$81x^2 - 16 = (9x)^2 - 4^2 = (9x - 4)(9x + 4)$$

EXAMPLE 2: Use the identity formula for the **square of sum** and factor: $x^2 + 10x + 25$

Solution

$$x^2 + 10x + 25 = x^2 + 2 \cdot 5 \cdot x + 5^2 = (x + 5)^2 = (x + 5)(x + 5)$$

EXAMPLE 3: Use the identity formula for the **difference of two cubes** and factor: $27x^3 - 8y^3$

Solution

$$27x^3 - 8y^3 = (3x)^3 - (2y)^3 = (3x - 2y)(9x^2 + 6xy + 2y^2)$$

EXAMPLE 4: Use the identity formula for the **cube of difference** and factor: $8x^3 - 12x^2 + 6x - 1$

Solution

$$8x^3 - 12x^2 + 6x - 1 = (2x)^3 - 3 \cdot (2x)^2 \cdot 1 + 3 \cdot 2x \cdot 1^2 - 1^3 = (2x - 1)^3$$

■ BINOMIAL FORMULA AND PASCAL'S TRIANGLE

For a power of a binomial expression $(a + b)^n$, we will use the **Binomial Formula**.

$$(a + b)^n = C_0^n \cdot a^n b^0 + C_1^n \cdot a^{n-1} b^1 + \ldots + C_k^n \cdot a^{n-k} b^k + \ldots + C_{n-1}^n \cdot a^1 b^{n-1} + C_n^n \cdot a^0 b^n$$

The coefficients of this **Binomial Formula** are

$$C_0^n = \frac{n!}{0!(n-0)!} = \frac{n!}{n!} = 1, \qquad \text{where } n! = n(n-1)(n-2)(n-3)\ldots(2)(1) \text{ and } 1! = 1, \, 0! = 1$$

$$C_1^n = \frac{n!}{1!(n-1)!} = \frac{n!}{(n-1)!} = n$$

....

....

$$C_k^n = \frac{n!}{k!(n-k)!} = \frac{n(n-1)\ldots(n-k+1)(n-k)(n-k-1)\ldots(1)}{k(k-1)\ldots(1)\cdot(n-k)(n-k-1)\ldots(1)} = \frac{n(n-1)\ldots(n-k+1)}{k(k-1)\ldots(1)}$$

....

....

$$C_{n-1}^n = \frac{n!}{(n-1)![n-(n-1)]!} = \frac{n!}{(n-1)!1!} = n$$

$$C_n^n = \frac{n!}{n!(n-n)!} = \frac{n!}{n!0!} = 1$$

The coefficients $C_0^n, C_1^n, \ldots, C_k^n, \ldots, C_{n-1}^n, C_n^n$ of the **Binomial Formula** are the numbers from the nth row of **Pascal's triangle**.

Power	Expanding Binomial Expression	Pascal's Triangle
$(a + b)^0 =$	1	1
$(a + b)^1 =$	$a + b$	$1 \quad 1$
$(a + b)^2 =$	$a^2 + 2ab + a^2$	$1 \quad 2 \quad 1$
$(a + b)^3 =$	$a^3 + 3a^2b + 3a^2b + b^3$	$1 \quad 3 \quad 3 \quad 1$
$(a + b)^4 =$	$a^4 + 4a^3b + 6a^2b^2 + 4ab^3 + b^4$	$1 \quad 4 \quad 6 \quad 4 \quad 1$
$(a + b)^5 =$	$a^5 + 5a^4b + 10a^3b^2 + 10a^2b^3 + 5ab^4 + b^5$	$1 \quad 5 \quad 10 \quad 10 \quad 5 \quad 1$

Example   Use Pascal's triangle and expand the binomial expression $(x - 3)^4$.

Solution

$$(x - 3)^4 = 1x^4 + 4x^3(-3) + 6x^2(-3)^2 + 4x(-3)^3 + 1(-3)^4 = x^4 - 12x^3 + 54x^2 - 108x + 81$$

Use the corresponding identities.

$x^2 - 9$	$x^2 - 36y^2$	$81x^2 - 1$
$x^4 - 16$	$16x^4 - 81y^4$	$x^8 - 1$
$x^3 - 27y^3$	$x^3 + 64$	$27x^3 - 125$
$x^6 - 1$	$8x^3 - y^3$	$x^6 + 1$
$(x^2 - 1)^2$	$(2x - 3)^2$	$(3x + 2y)^2$
$(xy - 3)^2$	$(x^2 - 2y)^2$	$(x + 5x^2)^2$
$(x^2 - 2)^3$	$(x - 3y)^3$	$(3x + 4y)^3$
$(2xy - 1)^3$	$(x^2 + y)^3$	$(4x + 2x^2)^3$

Use Pascal's triangle and expand each binomial expression.

$(x - 3)^4$ 1 1 1 1 2 1 1 3 3 1 1 4 6 4 1	$(5 - 2y)^5$ 1 1 1 1 2 1 1 3 3 1 1 4 6 4 1 1 5 10 10 5 1	$(x + 1)^6$ 1 1 1 1 2 1 1 3 3 1 1 4 6 4 1 1 5 10 10 5 1 1 6 15 20 15 6 1
$(xy - 2z)^4$	$(x - 3y)^5$	$(3x + 4y)^4$

Find the following coefficients of the **Binomial Formula.**

C_7^9	C_8^8	C_1^4

2.4. Factoring of Polynomials

- **FACTORING MONOMIALS**
- **FACTORING OTHER POLYNOMIALS**

■ FACTORING MONOMIALS

To factor an expression means to rewrite the given expression in an equivalent form that is a product of two or more expressions. This is called **factorization** of the given expression. The factorization with only prime factors is called **prime factorization**.

To factor a monomial we should find two or more monomials whose product is the original monomial.

Examples

EXAMPLE 1: Factor the monomial $81x^3$.
Solution

$$81x^3 = (3)(27x^3)$$
$$81x^3 = (9)(9x^3)$$
$$81x^3 = (3x)(27x^2)$$
$$81x^3 = (9x)(9x^2)$$
$$81x^3 = (3x^2))(27x)$$
$$81x^3 = (9x^2)(9x)$$
$$...$$
$$81x^3 = (3)(3)(3)(3)(x)(x)(x) \quad \longleftarrow \text{ prime factorization of a monomial } 81x^3$$

EXAMPLE 2: Factor the monomial $2x$.
Solution

$$2x = (2x)(1) = (2)(x)$$

EXAMPLE 3: Find **2** different factorizations and the prime factorization for each monomial:
a) $12x^4$; b) $18x^2y^3$; c) $5xyz^2$
Solution

a) $12x^4 = (3x)(4x^3) = (2x)(4x^3) = (2)(2)(3)(x)(x)(x)(x)$
b) $18x^2y^3 = (3x^2)(6y^3) = (9x^2y^2)(2y) = (2)(3)(3)(x)(x)(y)(y)(y)$
c) $5xyz^2 = (5xyz)(z) = (z^2)(5xy) = (5)(x)(y)(z)(z)$

■ FACTORING OTHER POLYNOMIALS

For factoring binomials, trinomials, and other polynomials with more than three terms, we usually use **6** factoring methods: **1) factoring out** common factors; **2)** factoring by **grouping**; **3)** factoring by **"branch"** method; **4)** factoring by **"big X"** method; **5)** factoring by **identity formulas**; and **6) miscellaneous** factoring**.**

FACTORING BY " FACTOR OUT" METHOD

Factoring by factoring out common factors is the reverse of the distributive property.
To factor polynomials with two or more terms, we should write the factor common to each term with the greatest possible coefficient and with the variables to the greatest powers outside the parentheses, and inside the parentheses we should write the other non-common factors of each term with corresponding separating symbols +/− between original terms.

Examples

EXAMPLE 1: Factor the polynomial $3x^3 - 27x^2y$.
Solution

$$3x^3 - 27x^2y = 3 \cdot x^2 \cdot x - 3 \cdot 9 \cdot x^2 \cdot y = 3\,x^2(x - 9y)$$

EXAMPLE 2: Factor the polynomial: $5x^4 - 20x^3 - 5x^2$
Solution

$$5x^4 - 20x^3 - 5x^2 = 5 \cdot x^2 \cdot x^2 - 5 \cdot 4 \cdot x^2 \cdot x - 5 \cdot x^2 \cdot 1 = 5x^2(x^2 - 4x - 1)$$

FACTORING BY " GROUPING" METHOD

To factor some polynomials with four (or six, or eight, or other even number) terms, we can use grouping factoring method. Factoring by grouping requires using the factor out method two or more times for the given polynomial. Before factoring by grouping we should divide the given polynomial into proper groups.

Examples

EXAMPLE 1: Factor the polynomial: $2x^3 - 8x^2y + x - 4y$
Solution

$$2x^3 - 8x^2y + x - 4y = \underbrace{2x^3 - 8x^2y}_{\text{group 1}} + \underbrace{x - 4y}_{\text{group 2}} = \underbrace{2x^2(x - 4y)}_{\text{group 1}} + \underbrace{(x - 4y) \cdot 1}_{\text{group 2}}$$

$$= (x - 4y)(2x^2 + 1)$$

EXAMPLE 2: Factor the polynomial: $x^5 - 2x^4 + 2x^3 - 4x^2 - 3x + 6$

Solution

$$x^5 - 2x^4 + 2x^3 - 4x^2 - 3x + 6 = \underbrace{x^5 - 2x^4}_{\text{group 1}} + \underbrace{2x^3 - 4x^2}_{\text{group 2}} - \underbrace{3x + 6}_{\text{group 3}} =$$

$$\underbrace{x^4(x-2)}_{\text{group 1}} + \underbrace{2x^2(x-2)}_{\text{group 2}} - \underbrace{3(x-2)}_{\text{group 3}} = (x-2)(x^4 + 2x^2 - 3)$$

FACTORING BY " BRANCH" METHOD

We use the **"branch" factoring** method to factor the quadratic trinomials $x^2 + bx + c$ with $b, c \in \mathbf{R}$, if the quadratic expression $x^2 + bx + c$ is an **easily factorable expression**. For example, $x^2 + 3x + 2 = (x+2)(x+1)$; $x^2 - 3x - 4 = (x-4)(x+1)$; $x^2 - 4x + 3 = (x-3)(x-1)$ are **easily factorable expressions**. The expressions $x^2 + 3x - 2$; $x^2 - 3x + 4$; and $x^2 + 4x - 3$ are not easily factorable expressions. Each of these expressions is factorable like $(x - x_1)(x - x_2)$, where x_1 and x_2 are solutions of corresponding quadratic equations $(x^2 + 3x - 2 = 0$ or $x^2 - 3x + 4 = 0$ or $x^2 + 4x - 3 = 0)$, but it is not easy to find x_1 and x_2 mentally.

> **Theorem** ▶ If for the quadratic trinomial $x^2 + bx + c$ we can find two numbers c_1 and c_2 such that $c_1 \cdot c_2 = c$ and $c_1 + c_2 = b$, then we can present the quadratic trinomial $x^2 + bx + c$ in the factoring form $x^2 + bx + c = (x + c_1)(x + c_2)$.

Proof: In the trinomial $x^2 + bx + c$, let $c = c_1 \cdot c_2$ and $b = c_1 + c_2$. So we can write:

$$x^2 + bx + c =$$

$c_1 \cdot c_2 \quad \longleftarrow \quad c_1 \cdot c_2 = c , \ c_1 + c_2 = b$ (c_1 and c_2 are **branch factors**)

$$x^2 + (c_1 + c_2)x + c_1 \cdot c_2 =$$
$$x^2 + c_1 x + c_2 x + c_1 \cdot c_2 =$$
$$(x^2 + c_1 x) + (c_2 x + c_1 c_2) =$$
$$x(x + c_1) + c_2(x + c_1) =$$
$$(x + c_1)(x + c_2)$$

$$x^2 + bx + c = (x + c_1)(x + c_2) \quad \longleftarrow \quad \textbf{The quadratic trinomial is a factorable expression}$$

Examples

EXAMPLE 1: Factor the trinomial: $x^2 - 8x - 9$

Solution

$$x^2 - 8x - 9 = (x - 9)(x + 1)$$

$-9 \cdot 1 \quad \longleftarrow \quad -9 \cdot 1 = -9; \ -9 + 1 = -8$

EXAMPLE 2: Factor the trinomial: $x^2 - 5x + 6$

Solution

$$x^2 - 5x + 6 = (x - 3)(x - 2)$$

$$-3 \cdot (-2) \qquad \longleftarrow \qquad -3 \cdot (-2) = 6 \; ; \quad -3 + (-2) = -5$$

EXAMPLE 3: Factor the trinomial: $x^2 + 5x + 6$

Solution

$$x^2 + 5x + 6 = (x + 3)(x + 2)$$

$$3 \cdot 2 \qquad \longleftarrow \qquad 3 \cdot 2 = 6 \; ; \quad 3 + 2 = 5$$

FACTORING BY " BIG X" METHOD

We use the **"Big X" factoring** method to factor the standard quadratic trinomial with one variable $ax^2 + bx + c = 0$ $(a \neq 0, 1)$, if the trinomial $ax^2 + bx + c$ is a factorable expression.

Theorem ▶ If for quadratic trinomial $ax^2 + bx + c$ with $a \neq 0, 1$ we can find two numbers c_1 and c_2, such that $c_1 \cdot c_2 = ac$ and $c_1 + c_2 = b$,

$$
\begin{array}{c}
b \\
c_1 \quad \times \quad c_2 \qquad \longleftarrow \qquad c_1 \cdot c_2 = ac \, , \; c_1 + c_2 = b \\
ac
\end{array}
$$

then we can write that trinomial in the following factor form:

$$ax^2 + bx + c = \frac{1}{a} \cdot (ax + c_1) \cdot (ax + c_2)$$

Proof: For the trinomial $ax^2 + bx + c$, let there be c_1 and c_2 such that $c_1 \cdot c_2 = ac$ and $c_1 + c_2 = b$. So we can write

$$ax^2 + bx + c =$$
$$ax^2 + (c_1 + c_2)x + c_1 \cdot c_2 \, /a = \qquad \longleftarrow \qquad c = c_1 \cdot c_2 \, / \, a, \; b = c_1 + c_2$$
$$\frac{1}{a} [a^2 x^2 + ac_1 x + ac_2 x + c_1 c_2] =$$

$$\frac{1}{a} [(a^2 x^2 + ac_1 x) + (ac_2 x + c_1 c_2)] = \frac{1}{a} [ax(ax + c_1) + c_2(ax + c_1)] =$$

$$\frac{1}{a} \cdot (ax + c_1) \cdot (ax + c_2) \qquad \longleftarrow \qquad \text{The quadratic trinomial is a factorable expression}$$

EXAMPLE 1: Factor the trinomial: $6x^2 + 7x + 2$

Solution 1

$$6x^2 + 7x + 2$$

$$c_1 = 3 \quad \diagdown\!\!\!\diagup \quad 4 = c_2 \qquad \longleftarrow \qquad c_1 \cdot c_2 = 12 \,,\ c_1 + c_2 = 7 \ \Rightarrow\ c_1 = 3,\ c_2 = 4$$

with 7 above and 12 below the big X.

$$6x^2 + 7x + 2 = \frac{1}{6} \cdot (6x + 3) \cdot (6x + 4) = \frac{1}{6} \cdot 3(2x + 1) \cdot 2(3x + 2) = (2x + 1)(3x + 2)$$

$$6x^2 + 7x + 2 = (2x + 1)(3x + 2)$$

Check: $(2x + 1)(3x + 2) = 6x^2 + 4x + 3x + 2 = 6x^2 + 7x + 2$

We can also do the following to factor the same expression by using the big x method.

Solution 2

$$6 \times 2 = 12$$
$$6x^2 + 7x + 2$$

$$\boxed{\frac{1}{2x}} \leftarrow 3/6 \leftarrow 3 \quad \diagdown\!\!\!\diagup \quad 4 \rightarrow 4/6 \rightarrow \boxed{\frac{2}{3x}} \qquad \longleftarrow \quad 3 \cdot 4 = 12 \,,\ 3 + 4 = 7$$

with 7 above, 12 and 6×2 below the big X.

$$6x^2 + 7x + 2 = (2x + 1)\,(3x + 2)$$

Check: $(2x + 1)(3x + 2) = 6x^2 + 4x + 3x + 2 = 6x^2 + 7x + 2$

EXAMPLE 2: Factor the trinomial: $2x^2 - x - 1$

Solution

$$2x^2 - x - 1$$

$$c_1 = 1 \quad\diagdown\!\!\!\!\diagup\quad -2 = c_2 \quad\longleftarrow\quad c_1 \cdot c_2 = -2 \,, \ c_1 + c_2 = -1 \Rightarrow c_1 = 1, \ c_2 = -2$$

with -1 above and -2 below the cross.

$$2x^2 - x - 1 = \frac{1}{2} \cdot (2x - 2) \cdot (2x + 1) = \frac{1}{2} \cdot 2(x - 1) \cdot (2x + 1) = (x - 1)(2x + 1)$$

$$2x^2 - x - 1 = (2x + 1)(x - 1)$$

Check: $(2x + 1)(x - 1) = 2x^2 - 2x + x - 1 = 2x^2 - x - 1$

We can also do the following to factor the same expression by using the big x method

Solution 2

$$2x(-1) = -2$$
$$2x^2 - 1x - 1$$

$$\boxed{\frac{1}{2x}} \ \leftarrow\ 1/2 \leftarrow 1 \quad\diagdown\!\!\!\!\diagup\quad -2 \rightarrow -2/2 \rightarrow \boxed{\frac{-1}{1x}} \quad\longleftarrow\quad -2 \cdot 1 = -2 \,, \ -2 + 1 = 1$$

with -1 above and -2 below the cross, and $2x(-1)$ beneath.

$$2x^2 - x - 1 = (2x + 1)(1x - 1) = (2x + 1)(x - 1)$$

Check: $(2x + 1)(x - 1) = 2x^2 - 2x + x - 1 = 2x^2 - x - 1$

FACTORING BY " ALGEBRAIC IDENTITIES" FORMULAS

Sometimes we use the **"Algebraic Identities"** to factor expressions.
There are **7** important algebraic identities shown below.

1. $a^2 - b^2 = (a - b)(a + b)$
2. $(a + b)^2 = a^2 + 2ab + b^2$
3. $(a - b)^2 = a^2 - 2ab + b^2$
4. $a^3 + b^3 = (a + b)(a^2 - ab + b^2)$
5. $a^3 - b^3 = (a - b)(a^2 + ab + b^2)$
6. $(a + b)^3 = a^3 + 3a^2 b + 3a b^2 + b^3$
7. $(a - b)^3 = a^3 - 3a^2 b + 3a b^2 - b^3$

All these formulas work for any $a, b \in R$

Examples

EXAMPLE 1: Factor the polynomial: $x^2 - 4$
Solution

$$x^2 - 4 = x^2 - 2^2 = (x - 2)(x + 2)$$

EXAMPLE 2: Factor the polynomial: $25x^2 - 30x + 9$
Solution

$$25x^2 - 30x + 9 = (5x)^2 - 2 \cdot 3 \cdot 5x + 3^2 = (5x - 3)^2 = (5x - 3)(5x - 3)$$

EXAMPLE 3: Factor the polynomial: $x^2 + 4x + 4$
Solution

$$x^2 + 4x + 4 = x^2 + 2 \cdot 2 \cdot x + 2^2 = (x + 2)^2 = (x + 2)(x + 2)$$

EXAMPLE 4: Factor the polynomial: $8x^3 - 125y^3$
Solution

$$8x^3 - 125y^3 = (2x)^3 - (5y)^3 = (2x - 5y)(4x^2 + 10xy + 5y^2)$$

EXAMPLE 5: Use identity formula for **cube of difference** and factor: $27x^3 + 27x^2 + 9x + 1$
Solution

$$27x^3 + 27x^2 + 9x + 1 = (2x)^3 + 3 \cdot (3x)^2 \cdot 1 + 3 \cdot 3x \cdot 1^2 + 1^3 = (3x + 1)^3$$

MISCELLANEOUS FACTORING

Sometimes we use **combinations of different methods** to factor the expressions.

Examples

EXAMPLE 1: Factor $x^2 - 4x + 4 - y^2$

Solution

To factor this expression we use a combination of **2** identity formulas

$$x^2 - 4x + 4 - y^2 = x^2 - 4x + 4 - y^2 = (x - 2)^2 - y^2 = (x - 2 - y)(x - 2 + y)$$

EXAMPLE 2: Factor $x^4 - 13x^2 + 4$

Solution

We use a combination of addition properties and an identity formula

$$x^4 - 13x^2 + 4 = x^4 - 4x^2 - 9x^2 + 4 = x^4 - 4x^2 + 4 - 9x^2 =$$
$$= (x^2 - 2)^2 - (3x)^2 = (x^2 - 2 - 3x)(x^2 - 2 + 3x)$$

EXAMPLE 3: Factor $2x^5 + 6x^3 + 8x$

Solution

To factor this expression we use a combination of **2** factoring methods

$$2x^5 + 6x^3 + 8x = 2x(x^4 + 3x^2 + 4) = 2x(x^4 + 4x^2 - x^2 + 4) =$$
$$= 2x(x^4 + 4x^2 + 4 - x^2) = 2x[(x^2 + 2)^2 - x^2] =$$
$$= 2x(x^2 + 2 - x)(x^2 + 2 + x)$$

EXAMPLE 4: Factor $x^3 - 27y^3 - x^2 + 4xy - 3y^2$

Solution

To factor this expression we use a combination of **4** factoring methods: grouping factoring, identity formula, branch factoring, and factor out.

$$x^3 - 27y^3 - x^2 + 4xy - 3y^2 = (x^3 - 27y^3) - (x^2 - 4xy + 3y^2) =$$

$$-3 \cdot -1$$

$$= (x - 3y)(x^2 + 3xy + 9y^2) - (x - 3y)(x + 3y)$$
$$= (x - 3y)(x^2 + 3xy + 9y^2 - x - 3y)$$

Factor each monomial.

$81x^2yz^3 = 3{\cdot}3{\cdot}3{\cdot}3{\cdot}x{\cdot}x{\cdot}y{\cdot}z{\cdot}z{\cdot}z$	$8x$	$6x^2y$
$9xyz$	$18x^2yz^2$	$4x^2yz^3t$
$15ab^2c$	$12a^2bc^4$	$8x^3y^2a^4$

Factor each polynomial using the **factor out** method.

$3 - 12x$	$9x^2y + 6xy^2$	$x^2yz - 2xy^2 - 8xy$
$2x^2 + 4x$	$7x^3y^2 - 2xy$	$2x^2y^2z + 4xy^3z^2 - 12xyz$
$x^2y - 3xy$	$3x^4y + 6xy^5$	$-5a^4bc - 25a^2bc^3 + 10abc$

Factor each polynomial using **algebraic identities**.

$x^2 - 1$	$4x^2 - 9y^2$	$x^4 - 81$
$16 - y^2$	$x^3 - 8$	$x^6 - 1$
$x^2 - 4x + 4$	$4x^2 + 12xy + 9y^2$	$64 - 16y + y^2$
$y^3 + 27z^3$	$x^3 + 6x^2y + 12xy^2 + 8y^3$	$8x^3 - 12x^2y + 6xy^2 - y^3$

Factor each polynomial using the **grouping** method.

$2x^2 - 4x + x - 2$	$x^2 - 4xy + x - 4y$	$x^2yz - 2xy^2 - xz + 2y$
$3x^2 - 9xy - x + 3y$	$3x^2 - 27 - x + 3$	$x^3 - 8 + x - 2$
$4x^2 - 3x + 8x - 6$	$4x^2 - 4y^2 + 3x + 3y$	$5x^2 - 2xy + 10x - 4y$

Factor each polynomial using the **branch factoring** method.

$x^2 - 3x - 4 = (x - 4)(x + 1)$ $-4 \quad 1$	$x^2 + 6x - 7$
$x^2 + 7x + 10$	$x^2 - 10x - 11$
$x^2 + 3x - 10$	$x^2 + 6x - 40$
$x^2 - 8x + 15$	$x^2 + 2x - 15$
$x^2 + 9x + 20$	$x^2 + 8x + 7$
$x^2 - x - 20$	$x^2 + x - 12$
$x^2 + 3x - 4$	$x^2 - 6x - 7$
$x^2 - 7x - 18$	$x^2 + 12x + 11$
$x^2 + 5x - 14$	$x^2 + 6x - 27$
$x^2 + 8x + 16$	$x^2 + 2x - 24$
$x^2 - 10x + 25$	$x^2 + 4x - 45$

Factor each polynomial using the **big X factoring** method.

$2x^2 + 5x + 3 = (x + 1)(2x + 3)$ 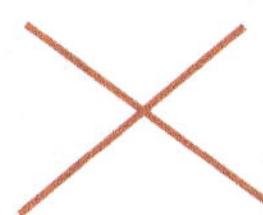 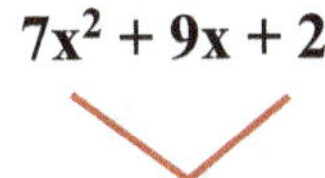 	$2x^2 + x - 3$
$3x^2 - x - 4$	$4x^2 - 2x - 6$ 
$7x^2 + 9x + 2$	$6x^2 + 8x + 2$
$5x^2 - 2x - 3$  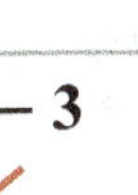	$3x^2 + 8x + 5$
$9x^2 - 6x - 3$ 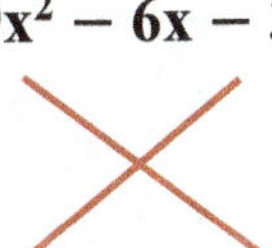	$5x^2 + 9x + 4$ 
$7x^2 + 5x - 2$ 	$6x^2 + 5x - 1$ 
$20x^2 + 22x + 2$ 	$4x^2 + 6x - 10$ 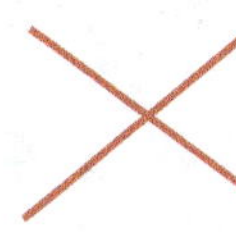
$12x^2 + 19x + 5$	$15x^2 - 37x - 8$ 

Factor each polynomial using **combinations of different methods**.

1. $4x^2 - 16 = 4(x-2)(x+2)$
2. $5x^4 - 5x$
3. $x^5 + x^8$
4. $x^5 - 8x^2$
5. $4x^8 - 9x^6$
6. $x^3y - xy^3$
7. $2x^3 - 8x^2 - 10x$
8. $x^5 - 10x^3 + 9x$
9. $x^4 - 5x^3 - x^2 + 5x$
10. $x^4 - 2x^3 - x^2 + 2x$
11. $x^4 - 27x$
12. $x^4 + 8x$
13. $3x^3 - 3$
14. $3x^3 - 2x^2 - 5x$
15. $2x^4 + 11x^3 + 5x^2$
16. $6yx^2 + 5xy - 6y$
17. $x^2 - 4x + 4 - y^2$
18. $x^3 - 2x^2y - x + 2y$
19. $x^4 - 8x^2 + 16 - 4y^2$
20. $x^5 - 3x^3 + 2x^2$
21. $x^5 - x^2 - x^3 + 1$

22. $2x^3 - 8x$
23. $x^4 - x$
24. $x^5 - x$
25. $x^7 - x$
26. $x^8 - 1$
27. $x^2y - 9y$
28. $x^3 - 2x^2 - 3x$
29. $x^5 - 3x^3 - 10x$
30. $x^4 - 5x^3 - x^2 + 5x$
31. $x^4 - 2x^3 + x^2$
32. $x^6 - 9x^4$
33. $x^4 - 5x^2 + 4$
34. $3x^3 - 2x^2 - 5x$
35. $2x^3 + 11x^2 + 5x$
36. $4x^4 + 5x^3 - 9x^2$
37. $2yx^2 + 5xy^2 + 2y^3$
38. $x^2 - 2x + 1 - y^2$
39. $x^2 - 2xy + y^2 - 1$
40. $x^4 - 6x^2 + 9 - 16y^2$
41. $x^4 - 3x^2 + 1$
42. $x^4 - 5x^2 + 4$

43. $x^3 - 6x^2 + 9x$
44. $x^3 - 6x^2 - 7x$
45. $x^3 - x^2 - x + 1$
46. $x^2 - 10x + 25 - x + 5$
47. $x^6 - 16x^2$
48. $x^4 - 5x^2 + 4$
49. $3x^5 + 9x^3 + 12x$
50. $x^9 + x^6 + x^3 + 1$
51. $x^{16} - 1$
52. $x^2y - 9y - x + 3$
53. $x^3 - 2x^2 - 4x + 8$
54. $x^5 - 3x^4 - x^3 + 3x^2$
55. $x^4 - 3x^3 + 3x^2 - x$
56. $2x^5 + 6x^4 + 6x^3 + 2x^2$
57. $x^3 - x^2 - 9x + 9$
58. $x^2y^2z + x^2y^2 - z - 1$
59. $x^2y - 9y + x + 3$
60. $x^2y - 2xy + y - 1$
61. $x^3y - xy^3 + x + y$
62. $x^4y - 8xy^4 + x^2 - 4y^2$
63. $x^3 - y^3 + x^2 + 2xy - y^2$

2.5. Fractional Expressions and Operations

- **DEFINITION AND EXCLUDED VALUES**
- **SIMPLIFYING FRACTIONAL EXPRESSIONS**
- **MULTIPLICATION OF FRACTIONAL EXPRESSIONS**
- **DIVISION OF FRACTIONAL EXPRESSIONS**
- **ADDITION AND SUBTRACTION OF FRACTIONAL EXPRESSIONS**

■ DEFINITION AND EXCLUDED VALUES

We will use the following definitions for **Fractional Expressions**.

DEFINITIONS

▶ A **fractional expression** is an expression $\dfrac{P}{Q}$ where P and Q are expressions and $Q \neq 0$.

For the value $Q = 0$ the expression $\dfrac{P}{Q}$ is undefined. The value $Q = 0$ is an **excluded value**.

For example, the expressions $\dfrac{1}{2}, \dfrac{3}{x}, \dfrac{x}{y}, \dfrac{x-3}{x+2}$, and $\dfrac{5y^2 - 3xy + 1}{2x - 3xy}$ are fractional expressions.

There is no excluded value for the expression $\dfrac{1}{2}$. The excluded value for the expression $\dfrac{3}{x}$ is $x = 0$.

The excluded value for the expression $\dfrac{x}{y}$ is $y = 0$. The excluded value for $\dfrac{x-3}{x+2}$ is $x = -2$ $(x + 2 = 0)$.

The excluded value for $\dfrac{5y^2 - 3xy + 1}{2x - 3xy}$ is $x = 0$ and $y = \dfrac{2}{3}$ $(2x - 3xy = 0)$.

▶ The expression $\dfrac{P(x)}{Q(x)}$ $(Q(x) \neq 0)$ is a **fractional expression with one variable x**, if the expressions $P(x)$ and $Q(x)$ contain only numbers and the variable **x**.

The value(s) of **x** for which $Q(x) = 0$ is (are) the **excluded value(s) of variable x**.

For example, the expressions $\dfrac{3}{x}, \dfrac{5}{x+2}, \dfrac{2x-4}{3x+5}$, and $\dfrac{2x^2 - 4x + 1}{2 - 4x}$ are fractional expressions with one variable **x**.

The excluded value for $\dfrac{3}{x}$ is $x = 0$. The excluded value for $\dfrac{5}{x+2}$ is $x = -2$ $(x + 2 = 0)$.

The excluded value for $\dfrac{2x-4}{3x+5}$ is $x = -\dfrac{5}{3}$ $(3x + 5 = 0)$.

The excluded value for $\dfrac{2x^2 - 4x + 1}{2 - 4x}$ is $x = \dfrac{1}{2}$ $(2 - 4x = 0)$.

 Classify each **fractional expression** as a **numerical, single-variable** or **multi-variable expression**.

$\dfrac{3}{2}$	$\dfrac{5}{3+x}$	$\dfrac{xy}{x+y}$	$\dfrac{2xyz}{xyz-4}$	$\dfrac{x-1}{x-3}$

Solution

$\dfrac{3}{2}$	$\dfrac{5}{3+x}$	$\dfrac{xy}{x+y}$	$\dfrac{2xyz}{xyz-4}$	$\dfrac{x-1}{x-3}$
numerical expression	single-variable expression	multi-variable expression	multi-variable expression	single-variable expression

Examples State the **excluded values** for each fractional expression.

$\dfrac{3+x}{2}$	$\dfrac{5}{3x}$	$\dfrac{x-5}{x+1}$	$\dfrac{2x}{x-2}$	$\dfrac{x-4}{6x-3}$
$\dfrac{4x-1}{x^2-1}$	$\dfrac{5+x}{5-x}$	$\dfrac{x-3}{4x^2-9}$	$\dfrac{3x+7}{x^2-2x-3}$	$\dfrac{5x}{2x^2-3x-5}$

Solution

$\dfrac{3+x}{2}$	$\dfrac{5}{3x}$	$\dfrac{x-5}{x+1}$	$\dfrac{2x}{x-2}$	$\dfrac{x-4}{6x-3}$
$2 \neq 0$	$3x=0$	$x+1=0$	$x-2=0$	$6x-3=0$
excluded value does not exist	excluded value $x=0$	excluded value $x=-1$	excluded value $x=2$	excluded value $x=\dfrac{1}{2}$
$\dfrac{4x-1}{x^2-1}$	$\dfrac{5+x}{5-x}$	$\dfrac{x-3}{4x^2-9}$	$\dfrac{3x+7}{x^2-2x-3}$	$\dfrac{5x}{2x^2-3x-5}$
$x^2-1=0$ $(x-1)(x+1)=0$	$5-x=0$ $x=5$	$4x^2-9=0$ $(2x-3)(2x+3)=0$	$x^2-2x-3=0$ $(x-3)(x+1)=0$	$2x^2-3x-5=0$ $(2x-5)(x+1)=0$
excluded values $x=-1,1$	excluded value $x=5$	excluded values $x=-\dfrac{3}{2},\dfrac{3}{2}$	excluded values $x=-1,3$	excluded values $x=-1,\dfrac{5}{2}$

Classify each **expression** as a **numerical**, **single-variable** or **multi-variable expression**.

$\dfrac{3 - 6}{2 \cdot 4 - 5}$	$\dfrac{5x}{y + x}$	$\dfrac{3x^4y + 6xy^5}{xy - 2}$	$\dfrac{1 - x^4}{x - 1}$	$\dfrac{x - y}{xy - 3}$

State the **excluded values** for each fractional expression.

$\dfrac{1 + x}{5}$	$\dfrac{2}{3x(x + 1)}$	$\dfrac{x - 1}{(x + 1)(x - 1)}$	$\dfrac{2x - 3}{4x - 2}$	$\dfrac{x + 5}{9x + 3}$
$\dfrac{x - 2}{x^2 - 4}$	$\dfrac{4 + 2x}{3x - x^2}$	$\dfrac{x - 2}{x^3 - 9x^2 - 10x}$	$\dfrac{x + 4}{x^4 - 1}$	$\dfrac{x + 2}{3x^2 - 4x - 7}$
$\dfrac{6 + 2x}{5x + 10}$	$\dfrac{7}{8x^2(3x + 6)}$	$\dfrac{x^2 - 1}{(x^2 - 25)(x - 3)}$	$\dfrac{6x + 2}{5x^2 - 45}$	$\dfrac{x + 6}{x(x + 3)(x - 2)}$
$\dfrac{x - 5}{(2x + 1)(x - 4)}$	$\dfrac{6 - x}{2x - 6x^2}$	$\dfrac{x + 3}{x^4 - 4x^3 - 12x^2}$	$\dfrac{x^3 + 9}{8x - 1}$	$\dfrac{x - 19}{5x^2 - 3x - 2}$
$\dfrac{3 - 9x^2}{x^2 - 16}$	$\dfrac{4x}{x^2 + 6x + 9}$	$\dfrac{3x - 2}{x^2 - 2x + 1}$	$\dfrac{4x + 8}{x^2 - 2x - 8}$	$\dfrac{x + 2}{x^3 - x^2 - 4x + 4}$

■ SIMPLIFYING FRACTIONAL EXPRESSIONS

Some fractional expressions have simplest form, but others you have to simplify and write in the simplest form. For example, each of the expressions $\frac{1}{2}$, $\frac{2}{3x}$, $\frac{x-2}{3+x}$, and $\frac{x^2-1}{4x}$ is in its simplest form and you can't simplify more, but the expressions $\frac{2}{4}$, $\frac{2x^2}{3x}$, $\frac{x-1}{x^2-1}$, and $\frac{x^2-1}{4x+4}$ can be simplified and rewritten in simplest forms: $\frac{2}{4}=\frac{1}{2}$, $\frac{2x^2}{3x}=\frac{2x}{3}$, $\frac{x-1}{x^2-1}=\frac{1}{x+1}$, and $\frac{x^2-1}{4x+4}=\frac{x-1}{4}$.

To simplify any non-simplest fractional expression and rewrite in the simplest form we have to factor the numerator and the denominator of that expression and then cancel all common factors.

Examples Simplify each **fractional expression** completely and write in simplest form.

$\dfrac{7}{2}$	$\dfrac{8}{12}$	$\dfrac{xy^3}{x+y}$	$\dfrac{2x^2y}{4xy^3}$
$\dfrac{x+y}{x+y}$	$\dfrac{5x+10y}{x^2+2xy}$	$\dfrac{x^3-8}{x^2-4}$	$\dfrac{x^2-1}{4x+4}$

Solution

$\dfrac{7}{2}=\dfrac{7}{2}$ **Is in simplest form**	$\dfrac{8}{12}=\dfrac{4\cdot2}{4\cdot3}=\dfrac{2}{3}$	$\dfrac{xy}{x+y}$ **Is in simplest form**	$\dfrac{2x^2y}{4xy^3}=\dfrac{2xxy}{2\,2xyyy}=\dfrac{x}{2y^2}$
$\dfrac{x+y}{x+y}=1$	$\dfrac{5x+10y}{x^2+2xy}=\dfrac{5(x+2y)}{x(x+2y)}$ $\dfrac{5(x+2y)}{x(x+2y)}=\dfrac{5}{x}$	$\dfrac{x^3-8}{x^2-4}=\dfrac{x^3-2^3}{x^2-2^2}$ $\dfrac{(x-2)(x^2+2x+4)}{(x-2)(x+2)}$ $\dfrac{x^2+2x+4}{x+2}$	$\dfrac{x^2-1}{4x+4}$ $\dfrac{(x-1)(x+1)}{4(x+1)}$ $\dfrac{x-1}{4}$

Simplify each **fractional expression** completely and write in simplest form.

$\dfrac{3}{27}$	$\dfrac{4x}{xy + x}$	$\dfrac{2x + 4y}{6xy(x + 2y)}$	$\dfrac{x^2 - 1}{x - 1}$	$\dfrac{x - 3y}{2xy - 6y^2}$
$\dfrac{1 + x}{5x + 5}$	$\dfrac{2 + 2x}{3x(x + 1)}$	$\dfrac{2x + 4}{x^2 - 4}$	$\dfrac{2x - 6}{4x - 12}$	$\dfrac{x^4y + xy^4}{xy(x + y)}$
$\dfrac{x - 2}{x^2 - 4}$	$\dfrac{x - 3}{3x - x^2}$	$\dfrac{1 - x^4}{x - 1}$	$\dfrac{x^2 + x}{x^3 - 9x^2 - 10x}$	$\dfrac{x^3 + 1}{3x^2 - 4x - 7}$
$\dfrac{z - 1}{z^2 - 11z + 10}$	$\dfrac{x^2 - 4}{x^2 + 4x - 12}$	$\dfrac{y^2 - 2y - 3}{y^2 + 3y + 2}$	$\dfrac{m^2 + 7m - 8}{m^2 + 16m + 64}$	$\dfrac{a^2 - 2a}{a^3 + a^2 - 6a}$
$\dfrac{x^3 - 1}{x^2 + x + 1}$	$\dfrac{x^2 - 3x + 2}{x^2 + 3x - 4}$	$\dfrac{x^2 - 2x - 8}{2x^2 + 5x + 2}$	$\dfrac{v^2 - 25}{2v^2 + 11v + 5}$	$\dfrac{3u^2 - 8u + 4}{3u^3 + 4u^2 - 4u}$
$\dfrac{3x^2 - 11x - 4}{2x^2 - 7x - 4}$	$\dfrac{54 - 2x^3}{18 + 2x^2 + 6x}$	$\dfrac{3x^3 - 11x^2 - 4x}{(2x + 1)(x - 4)}$	$\dfrac{x^4 + 27x}{x^3 + 3x^2 + 9x}$	$\dfrac{10x^2 + 4x}{5x^2 - 3x - 2}$
$\dfrac{x^2 - 1}{x^2 - 2x + 1}$	$\dfrac{4x^2 + 12x}{x^2 + 6x + 9}$	$\dfrac{x^3 + 2x^2 - 4x - 8}{x^2 + 4x + 4}$	$\dfrac{2x^2 + 3x - 2}{x^2 - 2x - 8}$	$\dfrac{x^2 - 4}{x^3 - x^2 - 4x + 4}$

■ MULTIPLICATION OF FRACTIONAL EXPRESSIONS

To multiply fractional expressions we have to multiply numerators and write the product as the numerator and multiply the denominators and write the result as the denominator of the resulting product of fractional expressions (for example, $\dfrac{2}{3x} \cdot \dfrac{x-2}{3+x} = \dfrac{2(x-2)}{3x(3+x)}$ or $\dfrac{2}{3x} \cdot \dfrac{x(x-2)}{4(3+x)} = \dfrac{2x(x-2)}{12x(3+x)}$).

Sometimes the product is a fraction in the simplest form (for example, the product $\dfrac{2(x-2)}{3x(3+x)}$ is in the simplest form). But sometimes the product is not in the simplest form and we can simplify and then write the product in its simplest form (for example, the product $\dfrac{2x(x-2)}{12x(3+x)}$ is not in the simplest form, and after simplifying we will get the following simplest form of the product $\dfrac{x-2}{6x(3+x)}$). We know that to simplify a non-simplest fractional expression and to rewrite it in the simplest form we have to factor the numerator and the denominator of that expression and then cancel all common factors of numerator and denominator.

Example 1 Multiply and write the product $\dfrac{3}{x} \cdot \dfrac{x+1}{x+2}$ in the simplest form.

Solution

$$\frac{3}{x} \cdot \frac{x+1}{x+2} = \frac{3(x+1)}{x(x+2)} = \frac{3(x+1)}{x(x+2)}$$

Example 2 Multiply and write the product $\dfrac{x^2-4}{x^2-x-6} \cdot \dfrac{x^2-9}{x-2}$ in the simplest form.

Solution

$$\frac{x^2-4}{x^2-x-6} \cdot \frac{x^2-9}{x-2} = \frac{(x^2-4)(x^2-9)}{(x^2-x-6)(x-2)} = \frac{(x-2)(x+2)(x-3)(x+3)}{(x-3)(x+2)(x-2)} = x+3$$

Example 3 Multiply and write the product $\dfrac{x^2-2x-3}{2x^2-5x-3} \cdot \dfrac{2x^2-x-1}{x^2-1}$ in the simplest form.

Solution

$$\frac{x^2-2x-3}{2x^2-5x-3} \cdot \frac{2x^2-x-1}{x^2-1} = \frac{(x-3)(x+1)(2x+1)(x-1)}{(x-3)(2x+1)(x+1)(x-1)} = 1$$

Simplify each **fractional expression** completely and write in simplest form.

$\dfrac{3}{2y} \cdot \dfrac{8xy}{9}$	$\dfrac{3x}{4y} \cdot \dfrac{8y}{6x}$	$\dfrac{x}{8y} \cdot \dfrac{6y^2}{3xz^2} \cdot \dfrac{8z^2}{2xy}$	$\dfrac{x^2}{y} \cdot \dfrac{3y^5}{x^4z^2} \cdot \dfrac{x^6z^4}{3xy}$
$\dfrac{4}{3y} \cdot \dfrac{9x^2y^2}{12x}$	$\dfrac{x^2-1}{x+1} \cdot \dfrac{4}{2x-2}$	$\dfrac{x^2-y^2}{x+y} \cdot \dfrac{4}{2x-2y}$	$\dfrac{x^3}{x^2yz} \cdot \dfrac{y^2z+yz}{y+1}$
$\dfrac{3x}{2} \cdot \dfrac{2x+4}{9x+18}$	$\dfrac{x+3}{2-x} \cdot \dfrac{3x-6}{4x+12}$	$\dfrac{x^2-1}{x+3} \cdot \dfrac{2x+6}{x+1}$	$\dfrac{x^2-4}{x^3-1} \cdot \dfrac{x^2+x+1}{x+2}$
$\dfrac{x^2-1}{x^2-4} \cdot \dfrac{12x+24}{6x+6}$	$\dfrac{x^2-9}{x^2-16} \cdot \dfrac{x^2+5x+4}{x^2-2x-3}$	$\dfrac{x^4-1}{x^2-5x-6} \cdot \dfrac{x^2-7x+6}{2x^2+2}$	$\dfrac{x^2+x}{x^2-3x-4} \cdot \dfrac{x^2-16}{2x^2+8x}$
$\dfrac{4x-4}{x^2-3x} \cdot \dfrac{x^3-2x^2-3x}{x^2-1}$	$\dfrac{2x+1}{x^2-x} \cdot \dfrac{3x^2-x-2}{6x^2+7x+2}$	$\dfrac{2y^2+5y+2}{2y^2-5y-3} \cdot \dfrac{y^2-2y-3}{y^2+3y+2}$	$\dfrac{z^2+7z-8}{z^2+16z+64} \cdot \dfrac{3z+24}{z^2-1}$
$\dfrac{x^3-1}{x^2+3x-4} \cdot \dfrac{2x+8}{x^2+x+1}$	$\dfrac{x^2-3x+2}{x^2+7x-8} \cdot \dfrac{x+8}{3x^2-4x-4}$	$\dfrac{x^2-2x-8}{2x^2+x} \cdot \dfrac{4x^2+4x+1}{2x^2+5x+2}$	$\dfrac{v^2-25}{2v^2+11v+5} \cdot \dfrac{2v^2-v-1}{v-5}$
$\dfrac{3x}{2} \cdot \dfrac{x^3+2x^2-4x-8}{x^3+4x^2+4x}$	$\dfrac{2x+6}{x+1} \cdot \dfrac{x^4-x^2}{2x^2+4x-6}$	$\dfrac{1}{x-2} \cdot \dfrac{x^3+2x^2-4x-8}{x^2+4x+4}$	$\dfrac{x-4}{x+2} \cdot \dfrac{2x^3+3x^2-2x}{2x^2-9x+4} \cdot \dfrac{2}{x}$

■ DIVISION OF FRACTIONAL EXPRESSIONS

Division of any two fractional expressions is the same as multiplication of the first fractional expression with the reciprocal of the second fractional expression.

For example, $\dfrac{2}{3x} \div \dfrac{x-2}{3+x} = \dfrac{2}{3x} \cdot \dfrac{3+x}{x-2} = \dfrac{2(3+x)}{3x(x-2)}$ or $\dfrac{x+2}{x-1} \div \dfrac{x+2}{x+1} = \dfrac{x+2}{x-1} \cdot \dfrac{x+1}{x+2} = \dfrac{(x+2)(x+1)}{(x-1)(x+2)} = \dfrac{x+1}{x-1}$.

Sometimes the quotient is a fraction in the simplest form (for example, the quotient $\dfrac{2(3+x)}{3x(x-2)}$ is in the simplest form). But sometimes the quotient is not in the simplest form and we can simplify and then rewrite it in the simplest form (for example, the quotient $\dfrac{(x+2)(x+1)}{(x-1)(x+2)}$ is not in the simplest form, and after simplifying we will get the following simplest form $\dfrac{x+1}{x-1}$).

Example 1 Divide and write the quotient $\dfrac{2}{x} \div \dfrac{x-1}{x+2}$ in the simplest form.

Solution

$$\dfrac{2}{x} \div \dfrac{x-1}{x+2} = \dfrac{2}{x} \cdot \dfrac{x+2}{x-1} = \dfrac{2(x+2)}{x(x-1)} = \dfrac{2(x+2)}{x(x-1)}$$

Example 2 Divide and write the quotient $\dfrac{x^2-4}{x^2-x-6} \div \dfrac{x-2}{x^2-9}$ in the simplest form.

Solution

$$\dfrac{x^2-4}{x^2-x-6} \div \dfrac{x-2}{x^2-9} = \dfrac{x^2-4}{x^2-x-6} \cdot \dfrac{x^2-9}{x-2}$$

$$\dfrac{(x^2-4)(x^2-9)}{(x^2-x-6)(x-2)} = \dfrac{(x-2)(x+2)(x-3)(x+3)}{(x-3)(x+2)(x-2)} = x+3$$

Example 3 Divide and write the quotient $\dfrac{x^2-2x-3}{2x^2-5x-3} \div \dfrac{x^2-1}{2x^2-x-1}$ in the simplest form.

Solution

$$\dfrac{x^2-2x-3}{2x^2-5x-3} \div \dfrac{x^2-1}{2x^2-x-1} =$$

$$\dfrac{x^2-2x-3}{2x^2-5x-3} \cdot \dfrac{2x^2-x-1}{x^2-1} = \dfrac{(x-3)(x+1)(2x+1)(x-1)}{(x-3)(2x+1)(x+1)(x-1)} = 1$$

Simplify each **fractional expression** completely and write in simplest form.

$\dfrac{3xy}{2} \div \dfrac{9x}{8}$	$\dfrac{4y}{3x} \div \dfrac{8y}{6x}$	$\dfrac{x}{2y} \div \dfrac{3xz^2}{6y^2} \cdot \dfrac{z^2}{y}$	$\dfrac{x^2z}{y} \div \dfrac{y^5}{x^4z^3} \div \dfrac{x^6z^4}{2y^6}$
$\dfrac{4x^2}{3y} \div \dfrac{8x^2}{6yz}$	$\dfrac{x^2-4}{x+1} \div \dfrac{x-2}{2x+2}$	$\dfrac{x^2-y^2}{x+y} \div \dfrac{4x-4y}{2}$	$\dfrac{y+1}{x^2} \div \dfrac{y^2+y}{x^3}$
$\dfrac{3x+6}{2} \div \dfrac{6x+12}{4x}$	$\dfrac{2x+6}{2-x} \div \dfrac{6x+18}{3x-6}$	$\dfrac{x^2-1}{x+3} \div \dfrac{x+1}{2x+6}$	$\dfrac{x+2}{x^2+x+1} \div \dfrac{x^2-4}{x^3-1}$
$\dfrac{x^2-1}{x^2-4} \div \dfrac{2x+2}{2x+4} \cdot \dfrac{x-2}{x-1}$	$\dfrac{x^2-16}{x-3} \div \dfrac{x^2+5x+4}{x^2-2x-3}$	$\dfrac{x^2-5x-6}{x^2-1} \div \dfrac{x^2-7x+6}{(x-1)^2}$	$\dfrac{x^2-3x-4}{x^2+x} \div \dfrac{x^2-16}{2x^2+8x}$
$\dfrac{x^2-3x}{x-1} \div \dfrac{x^3-2x^2-3x}{x^2-1}$	$\dfrac{x^2-x}{2x+1} \div \dfrac{3x^2-x-2}{6x^2+7x+2}$	$\dfrac{2y^2-5y-3}{2y^2+5y+2} \div \dfrac{y^2-2y-3}{y^2+3y+2}$	$\dfrac{z^2+7z-8}{z^2+16z+64} \div \dfrac{z-1}{3z+24}$
$\dfrac{x^3-1}{x^2+3x-4} \div \dfrac{x^2+x+1}{2x+8}$	$\dfrac{x^2-3x+2}{x^2+5x-6} \div \dfrac{x-2}{x^2+6x}$	$\dfrac{2x^2+x}{x^2-2x-8} \div \dfrac{4x^2+4x+1}{2x^2-7x-4}$	$\dfrac{2x^2-x-1}{2x^2+11x+5} \div \dfrac{x^2-25}{x-5}$
$\dfrac{(x+2)^2}{2x+3} \div \dfrac{x^3+x-2}{2x^2+x-3}$	$\dfrac{x^2-1}{2x^2+4x-6} \div \dfrac{x+1}{2x+6}$	$\dfrac{x-2}{2} \div \dfrac{x^3+2x^2-4x-8}{x^2+4x+4}$	$\dfrac{x+2}{x-2} \div \dfrac{2x^2+3x-2}{2x^2-9x+4}$

■ ADDITION AND SUBTRACTION OF FRACTIONAL EXPRESSIONS

To add or subtract fractional expressions we use the LCD method.

Example 1 Add and write the sum $\dfrac{2}{3} + \dfrac{1}{6}$ in the simplest form.

Solution

$$\frac{2}{3} + \frac{1}{6} = \frac{2\cdot2}{3\cdot2} + \frac{1}{6} = \frac{4+1}{6} = \frac{5}{6} \qquad\qquad LCD = 6$$

Example 2 Add and write the sum $\dfrac{x+2}{x^2-x-6} + \dfrac{x-2}{x^2-9}$ in the simplest form.

Solution

$$\frac{x+2}{x^2-x-6} + \frac{x-2}{x^2-9} = \frac{x+2}{(x-3)(x+2)} + \frac{x-2}{(x-3)(x+3)} \qquad LCD=(x+2)(x-3)(x+3)$$

$$\frac{(x+2)\cdot(x+3)}{(x-3)(x+2)\cdot(x+3)} + \frac{(x-2)\cdot(x+2)}{(x-3)(x+3)\cdot(x+2)} = \frac{(x+2)(x+3)+(x-2)(x+2)}{(x-3)(x+3)(x+2)}$$

$$\frac{x^2+3x+2x+6+x^2-4}{(x-3)(x+3)(x+2)} = \frac{2x^2+5x+2}{(x-3)(x+3)(x+2)} = \frac{(2x+1)(x+2)}{(x-3)(x+3)(x+2)} = \frac{2x+1}{x^2-9}$$

Example 3 Subtract and write the difference $\dfrac{x+1}{2x^2-5x-3} - \dfrac{x-1}{2x^2-x-1}$ in the simplest form.

Solution

$$\frac{x+1}{2x^2-5x-3} - \frac{x-1}{2x^2-x-1} = \frac{x+1}{(2x+1)(x-3)} - \frac{x-1}{(2x+1)(x-1)}$$

$$LCD=(2x+1)(x-3)(x-1)$$

$$\frac{(x+1)\cdot(x-1)}{(2x+1)(x-3)\cdot(x-1)} - \frac{(x-1)\cdot(x-3)}{(2x+1)(x-1)\cdot(x-3)} = \frac{x^2-1-x^2+4x-3}{(2x+1)(x-3)(x-1)}$$

$$\frac{4x-4}{(2x+1)(x-3)(x-1)} = \frac{4(x-1)}{(2x+1)(x-3)(x-1)} = \frac{4}{(2x+1)(x-3)}$$

Add or subtract. Write each sum or difference in the simplest form.

$\dfrac{x}{2} + \dfrac{x}{8}$	$\dfrac{4y}{3x} - \dfrac{4y}{6x}$	$\dfrac{x}{2y} - \dfrac{3xy}{6y^2} + \dfrac{1}{y}$	$\dfrac{1}{y^4} - \dfrac{2}{y^5} + \dfrac{1}{y^6}$
$\dfrac{4x}{3yz} + \dfrac{8xz}{6yz^2}$	$\dfrac{x+4}{x+1} + \dfrac{x-8}{2x+2}$	$\dfrac{1}{x+y} - \dfrac{1}{x-y}$	$\dfrac{1}{x+y} + \dfrac{1}{x-y}$
$\dfrac{3x+6}{2} + \dfrac{6x^2-12x}{4x}$	$\dfrac{x+6}{2-x} - \dfrac{6x-18}{3x-6}$	$\dfrac{x^2-1}{x+3} + \dfrac{6x+2}{2x+6}$	$\dfrac{x+2}{x^2+x+1} - \dfrac{x^2-1}{x^3-1}$
$\dfrac{x^2-x}{x^2-4} - \dfrac{2x+2}{2x+4} \cdot \dfrac{x}{x-2}$	$\dfrac{x-1}{x-3} - \dfrac{x^2+x}{x^2-2x-3}$	$\dfrac{1}{x^2-1} + \dfrac{1}{(x-1)^2}$	$\dfrac{x^2-3x-4}{x^2+x} + \dfrac{x+4}{x}$
$\dfrac{x^2-3x}{x-1} - \dfrac{x^3-2x^2-3}{x^2-1}$	$\dfrac{x+2}{2x+1} + \dfrac{3x^2-x-2}{6x^2+7x+2}$	$\dfrac{y+1}{2y^2+5y+2} - \dfrac{1}{y^2+3y+2}$	$\dfrac{z^2+7z-8}{z^2+16z+64} + \dfrac{9}{z+8}$
$\dfrac{x^3-1}{x^2+3x-4} - \dfrac{x^2-x-7}{x+4}$	$\dfrac{x^2-3x+2}{x^2+5x-6} + \dfrac{8x}{x^2+6x}$	$\dfrac{x^2+x}{x^2-x-2} + \dfrac{4x-6}{2x^2-7x-6}$	$\dfrac{2x^2-x-1}{2x^2+11x+5} + \dfrac{x-5}{x^2-25}$

1. Find the result of each operation.

$(-2)^2 \cdot (-3)^3$	$(2^2 x^2 y) \cdot (x^4 y^5)$	$\dfrac{2^5}{3^6}\left(\dfrac{9}{2}\right)^3$	$\dfrac{x^4 y^7}{x^2 y^6} \cdot \dfrac{x^5 y^7}{x^6 y^8}$	$\dfrac{x^2 y^3}{x^7} \div \dfrac{x^{-9} y^{-2}}{x^{-3} y^{-5}}$

2. Write each number or expression in scientific notation.

350,000	0.095	326	9
0.0000000005	$456 \cdot 100$	$5 \cdot 0.0008$	$625 \div 0.05$
$(5 \cdot 10^5)(4 \cdot 10^0)(10^{-3})$	$0.5(4 \cdot 10^4)(7 \cdot 10^{-6})$	$\dfrac{(2.8 \cdot 10^{-8})(0.5 \cdot 10^3)}{(8 \cdot 10^6)(2.5 \cdot 10^{-15})}$	$\dfrac{(1.7 \cdot 10^{15})(9 \cdot 10^{-5})}{(2.4 \cdot 10^7)(5 \cdot 10^{-10})}$

3. Classify each polynomial by number of terms, by degree, and as binomial or trinomial.

$x^3 y^6 + 2 y^5 z^4$	$x + 4x^2$	$5x^3 + 3x^6 + 4x^2$	$x + 8x^5 + y^6$	$2y^4 - 2x^4 y^2$

4. Evaluate each polynomial for **x = 2, y = 1, z = −1**.

$x^3 - 2y^2 + z$	$y + x^2 - z^2$	$2x + zy + 3y^2$	$x^2 + 4xy^3 z^2 + 2$	$y^2 - 3x^2 - x$

5. Combine like terms in each expression and write the answer in standard form.

$x^2 - 2x + 2x^2 - 3x$	$(4 + 2x^2 - 5x) + (x - 3x^2)$	$(6 - 2x^3 + x^4) - (2x^3 - x^4)$

6. Multiply or divide the following polynomials. Write the answer in simplest form.

$(x - 3y)(3y + x)$	$(x + 2) \cdot (x^2 - 2x + 4)$	$\dfrac{x^2 - 4y^2}{x + 2y}$

7. Divide a polynomial by using **long division** and **synthetic division**.

$\dfrac{x^2 - 8x + 16}{x - 4}$	$\dfrac{x^2 + 4x - 5}{x - 1}$	$\dfrac{2y^2 - 3y - 5}{y + 1}$

8. Simplify.

$\dfrac{4x^2 - 9y^2}{2x + 3y} - 2x + y$	$\dfrac{x^2 - 4y^2}{x + y}(x + y) - x^2 + 2y^2$	$\dfrac{x^2 - 3xy^2 + 2x}{x} - 2 + y^2$

1. Find the result of each operation.

$(-3)^2 \cdot (-2)^3$	$(3^2 x^4 y) \cdot (x^2 y^{-2})$	$\dfrac{3^5}{2^6} \left(\dfrac{4}{3}\right)^3$	$\dfrac{x^2 y^{-7}}{x^{-2} y^6} \cdot \dfrac{x^{-5} y^7}{x^6 y^{-8}}$	$\dfrac{x^3 y^4}{x^8} \div \dfrac{x^{-8} y^{-3}}{x^{-3} y^{-1}}$

2. Write each number or expression in scientific notation.

10,000	0.000234	26	1.7
0.0000000057	$45 \cdot 1000000$	$0.05 \cdot 0.008$	$1.25 \div 0.05$
$(7 \cdot 10^9)(5 \cdot 10^0)(10^{-7})$	$0.8(5 \cdot 10^3)(7 \cdot 10^{-4})$	$\dfrac{(5.8 \cdot 10^{-7})(0.5 \cdot 10^2)}{(4 \cdot 10^4)(2.5 \cdot 10^{-10})}$	$\dfrac{(1.5 \cdot 10^{10})(8 \cdot 10^{-15})}{(2.5 \cdot 10^8)(4 \cdot 10^{-12})}$

3. Classify each polynomial by number of terms, by degree, and as binomial or trinomial.

$2x^3 y^6 + 2y^5 z^4$	$5 + x + 4x^2$	$3x^3 + x^6 + 4x^9$	$8x + 5x^4 + y^3$	$y^7 - 3x^6 y^2$

4. Evaluate each polynomial for $x = 1$, $y = 2$, $z = -2$.

$x^2 - y^2 + 2z$	$3y + x^2 - z^2$	$x + 2zy + y^2$	$3x^2 + 4xz^2 + 2y$	$x^2 - 3y^2 - z$

5. Combine like terms in each expression and write the answer in standard form.

$2x^2 - 2x + 23x^2 - 4x$	$(1 + 3x^2 - 4x) + (x - x^2)$	$(1 - 2x^3 + 2x^4) - (2x^3 + 2x^4)$

6. Multiply or divide the following polynomials. Write the answer in simplest form.

$(3x - 2y)(y + x)$	$(x - 2) \cdot (x^2 + 2x + 4)$	$\dfrac{2x^2 - 18y^2}{x + 3y}$

7. Divide a polynomial by using **long division** and **synthetic division**.

$\dfrac{x^2 - 4x + 4}{x - 2}$	$\dfrac{x^2 - 4x - 5}{x + 1}$	$\dfrac{2y^2 - 3y - 5}{2y - 5}$

8. Simplify.

$\dfrac{x^2 - 9y^2}{x + 3y} \cdot (x + y)$	$\dfrac{4x^2 - y^2}{2x + y}(x + y) - 2x^2 + y^2$	$\dfrac{x^2 - 3xy^2 + 2x}{x^2 + xy}(x + y) + 3y^2$

Chapter 3. Power Roots

3.1.   Power Roots and Their Properties
3.2.   Power Roots and Fractional Powers
3.3.   Operations with Power Roots
3.4.   Pythagorean Theorem and Distance Formula

Chapter 3   Classwork
Chapter 3   Homework

$$\sqrt[n]{a} = b \;\;\rightarrow\;\; b^n = a$$

$$\sqrt[n]{a} \cdot \sqrt[n]{b} = \sqrt[n]{ab}$$

$$\frac{\sqrt[n]{a}}{\sqrt[n]{b}} = \sqrt[n]{\frac{a}{b}}$$

$$\sqrt[n]{a^n} = \begin{cases} a & \text{if } n = 2k+1 \\ |a| & \text{if } n = 2k \end{cases}$$

$$\sqrt[n]{a^m} = a^{\frac{m}{n}}$$

3.1. Power Roots and Their Properties

- **DEFINITIONS**
- **PROPERTIES OF POWER ROOTS**

■ DEFINITIONS

We will use the following general definition of **n**[th] **power root**, and two special definitions:

IMPORTANT DEFINITIONS

General Definition ▶

A power root of degree **n** (or **n**[th] power root) of a real number a is:

degree   power root symbol **(or radical symbol)**
        radicand

$$\sqrt[n]{a} = b, \text{ if } b^n = a, \text{ and } \begin{cases} b \text{ is a real number, if } n \geq 3 \text{ odd integer} \\ b \geq 0 \text{ real number, if } a \geq 0, \text{ and } n \geq 2 \text{ even integer} \\ b \text{ is a complex number, if } a < 0, \text{ and } n \geq 2 \text{ even integer} \end{cases}$$

power root

❶ If **n** is ≥ 3 odd integer (for example, **n = 3**, or **n = 5**, or **n = 7**, and so on), then $\sqrt[n]{a}$ is a real number **b**.
❷ If **n** is ≥ 2 even integer (for example, **n = 2**, or **n = 4**, or **n = 6**, and so on) and $a \geq 0$, then $\sqrt[n]{a}$ is a **nonnegative** real number **b** ($b \geq 0$).
❸ If **n** is ≥ 2 even integer (for example, **n = 2**, or **n = 4**, or **n = 6**, and so on) and $a < 0$, then $\sqrt[n]{a}$ does not exist as a real number: $\sqrt[n]{a}$ is a complex number.

We read $\sqrt[n]{a}$ as the "**n**[th] power root of a".

Special Definitions ▶

A **square root** of a real nonnegative number a ($a \geq 0$) is $\sqrt[2]{a} \equiv \sqrt{a}$ (second power root of a).
A **cube root** of a real number a is $\sqrt[3]{a}$ (third power root of a).

Examples

Evaluate the expressions: 1. $\sqrt{4}$;   2. $\sqrt[3]{27}$;   3. $\sqrt[5]{-32}$;   4. $\sqrt[4]{625}$;   5. $\sqrt[3]{-27x^3} \cdot \sqrt[4]{16}$
Solution

1. $\sqrt{4} = 2$;   2. $\sqrt[3]{27} = 3$;   3. $\sqrt[5]{-32} = -2$;   4. $\sqrt[4]{625} = 5$;   5. $\sqrt[3]{-27x^3} \cdot \sqrt[4]{16} = -6x$

■ PROPERTIES OF POWER ROOTS

We will use the following properties of roots.

Property ❶ $\sqrt[n]{a} \cdot \sqrt[n]{b} = \sqrt[n]{ab}$ and $\sqrt[n]{ab} = \sqrt[n]{a} \cdot \sqrt[n]{b}$ (if $\sqrt[n]{a}$ and $\sqrt[n]{b}$ exist)

Property ❷ $\dfrac{\sqrt[n]{a}}{\sqrt[n]{b}} = \sqrt[n]{\dfrac{a}{b}}$ and $\sqrt[n]{\dfrac{a}{b}} = \dfrac{\sqrt[n]{a}}{\sqrt[n]{b}}$ (if $\sqrt[n]{a}$ and $\sqrt[n]{b}$ exist)

Property ❸ $\sqrt[n]{a^n} = \begin{cases} a & \text{if } n = 2k + 1 \text{ (odd number)} \\ |a| & \text{if } n = 2k \text{ (even number)} \end{cases}$

Examples

EXAMPLE 1: Evaluate the expressions: **1.** $\sqrt{3} \cdot \sqrt{27}$; **2.** $\dfrac{\sqrt[3]{16}}{\sqrt[3]{2}}$; **3.** $\sqrt[4]{x^4y^4}$; **4.** $\sqrt[5]{-x^5}$

Solution

1. $\sqrt{3} \cdot \sqrt{27} = \sqrt{3 \cdot 27} = \sqrt{81} = 9$

2. $\dfrac{\sqrt[3]{16}}{\sqrt[3]{2}} = \sqrt[3]{\dfrac{16}{2}} = \sqrt[3]{8} = 2$

3. $\sqrt[4]{x^4y^4} = \sqrt[4]{x^4} \cdot \sqrt[4]{y^4} = |x| \cdot |y| = |xy|$

4. $\sqrt[5]{-x^5} = \sqrt[5]{(-x)^5} = -x$

EXAMPLE 2: Simplify the expression: $\sqrt[3]{-27x^3} \cdot \sqrt[4]{16}$

Solution

$$\sqrt[3]{-27x^3} \cdot \sqrt[4]{16} = \sqrt[3]{(-3x)^3} \cdot \sqrt[4]{2^4} = -3x \cdot 2 = -6x$$

EXAMPLE 3: Simplify the expression: $\sqrt[3]{16a^4} + \sqrt[4]{81} - 2a\sqrt[3]{2a} - 3$

Solution

$$\sqrt[3]{16a^4} + \sqrt[4]{81} - 2a\sqrt[3]{2a} - 3 = \sqrt[3]{(2a)^3 \cdot 2a} + \sqrt[4]{3^4} - 2a\sqrt[3]{2a} - 3 =$$

$$= 2a\sqrt[3]{2a} + 3 - 2a\sqrt[3]{2a} - 3 = 0$$

Use properties of power roots to simplify the expressions.

$\sqrt{49}$	$\sqrt{121}$	$\sqrt[3]{-64}$	$\sqrt[4]{81}$	$\sqrt[5]{32}$
$\sqrt{-4}$	$-\sqrt{49}$	$-\sqrt[5]{-32}$	$\dfrac{-\sqrt{27}}{\sqrt{3}}$	$\sqrt{2}\cdot\sqrt{2}$
$\sqrt{2}\cdot\sqrt{-2}$	$\sqrt{2}\cdot\sqrt{18}$	$\sqrt{3x}\cdot\sqrt{27x}$	$\dfrac{\sqrt{18}}{\sqrt{2}}$	$\dfrac{\sqrt{12x^3}}{\sqrt{3x}}$
$\sqrt[3]{-125x^3}$	$\sqrt[4]{81x^4}$	$\sqrt[4]{16x^3}\cdot\sqrt[4]{x}$	$\sqrt[3]{-8x}\cdot\sqrt[3]{x^2y^3}$	$\sqrt[3]{2x}\cdot\sqrt[3]{3x}\cdot\sqrt[3]{36x}$
$\sqrt{\dfrac{5^2}{x^{-2}}}$	$\dfrac{\sqrt{4y^2}}{\sqrt[3]{-x^{-3}}}$	$\dfrac{\sqrt{x^2}\cdot\sqrt[4]{y^{-4}}}{\sqrt[3]{z^{-3}}}$	$\sqrt[12]{\left(\left(\dfrac{x^{-2}}{2y}\right)^2\right)^3}$	$\dfrac{\sqrt{x^2}+\sqrt{9}}{x+3},\ x>0$
$\sqrt{\dfrac{8\sqrt{x^2+4}}{\sqrt[3]{x}+2}},\ x>0$	$\sqrt{\dfrac{y^2(2xy^{-3})^2}{4x^2y^{-4}z^{-8}}}$	$\sqrt{\sqrt{\dfrac{81x^4}{16}}}$	$\sqrt{\sqrt[3]{\dfrac{64x^6}{y^{12}}}}$	$\sqrt[12]{\left(\left(\left(\dfrac{2x^2}{y^{-2}}\right)^2\right)^3\right)^2}$

3.2. Power Roots and Fractional Powers

■ DEFINITIONS

We will also use the following general definition of **rational** $\dfrac{m}{n}$ **power**.

> **Definition ▶**
>
> A fractional $\dfrac{m}{n}$ power of a real number a is the nth power root of a^m.
>
> base   fractional exponent
>
> $$a^{m/n} = \sqrt[n]{a^m}$$
>
> rational power
>
> ❶ If n is ≥ 3 odd integer (for example, $n = 3$, or $n = 5$, or $n = 7$, and so on), then $\sqrt[n]{a^m}$ is a real number **b**.
>
> ❷ If n is ≥ 2 even integer (for example, $n = 2$, or $n = 4$, or $n = 6$, and so on) and $a^m \geq 0$, then $\sqrt[n]{a^m}$ is a nonnegative real number.
>
> ❸ If n is ≥ 2 even integer (for example, $n = 2$, or $n = 4$, or $n = 6$, and so on) and $a^m < 0$, then $\sqrt[n]{a^m}$ does not exist as a real number: $\sqrt[n]{a^m}$ is a complex number.

Examples

EXAMPLE 1: Evaluate the expressions: 1. $4^{1/2}$; 2. $(-8)^{1/3}$; 3. $(-16)^{1/4}$; 4. $(-27)^{2/3}$

Solution

1. $4^{1/2} = \sqrt{4} = 2$
2. $(-8)^{1/3} = \sqrt[3]{-8} = -2$
3. $(-16)^{1/4} = \sqrt[4]{-16}$ does not exist as a real number
4. $(-27)^{2/3} = (\sqrt[3]{-27})^2 = (-3)^2 = 9$

EXAMPLE 2: Simplify the expression: $\sqrt[3]{-27x^3} \cdot \sqrt[4]{16}$

Solution

$$\sqrt[3]{-27x^3} \cdot \sqrt[4]{16} = \sqrt[3]{(-3x)^3} \cdot \sqrt[4]{2^4} = (-3x)^{3/3} \cdot 2^{4/4} = -3x \cdot 2 = -6x$$

■ PROPERTIES OF FRACTIONAL POWERS

$\mathbf{W}$e will use the following properties of fractional/rational powers (**a, b ∈ R and m, n ∈ N**).

PROPERTY OF FRACTIONAL POWERS	EXAMPLE
1. $\sqrt[n]{\sqrt[m]{a^k}} = \sqrt[nm]{a^k} = a^{k/nm}$	1. $\sqrt[3]{\sqrt{64}} = \sqrt[6]{2^6} = 2^{6/6} = 2$
2. $a^{m/n} = (a^{1/n})^m = (\sqrt[n]{a})^m = (a^m)^{1/n} = \sqrt[n]{a^m}$	2. $8^{2/3} = (\sqrt[3]{8})^2 = 2^2 = 4$
3. $a^{\frac{m}{p}} \cdot a^{\frac{k}{j}} = a^{\frac{m}{p}+\frac{k}{j}}$	3. $3^{\frac{1}{3}} \cdot 3^{\frac{2}{3}} = 3^{\frac{1}{3}+\frac{2}{3}} = 3^{\frac{3}{3}} = 3$
4. $a^{\frac{m}{p}} / a^{\frac{k}{j}} = a^{\frac{m}{p}-\frac{k}{j}}$	4. $16^{\frac{1}{2}} / 16^{\frac{1}{4}} = 16^{\frac{1}{2}-\frac{1}{4}} = 16^{\frac{1}{4}} = 2$
5. $(a^{\frac{m}{p}})^{\frac{k}{j}} = a^{\frac{mk}{pj}}$	5. $(125^{\frac{2}{3}})^{\frac{1}{2}} = 125^{\frac{2}{6}} = 125^{\frac{1}{3}} = 5$
6. $(ab)^{\frac{m}{n}} = a^{\frac{m}{n}} \cdot b^{\frac{m}{n}}$	6. $(4x^4)^{\frac{1}{2}} = (2)^{\frac{2}{2}} x^{\frac{4}{2}} = 2x^2$
7. $(a/b)^{\frac{m}{n}} = a^{\frac{m}{n}} / b^{\frac{m}{n}}$	7. $(9/4)^{\frac{1}{2}} = 9^{\frac{1}{2}} / 4^{\frac{1}{2}} = \dfrac{\sqrt{9}}{\sqrt{4}} = \dfrac{3}{2}$
8. All **other properties** of negative integer powers also work for the fractional/rational powers	8. $(1/8)^{-\frac{2}{3}} = (8)^{\frac{2}{3}} = (2^3)^{\frac{2}{3}} = 4$

Examples

EXAMPLE 1: Simplify the expression: $\sqrt[3]{-2xy^2} \cdot \sqrt[3]{4x^5y}$

Solution

$$\sqrt[3]{-2xy^2} \cdot \sqrt[3]{4x^5y} = \sqrt[3]{-8x^6y^3} = (-8x^6y^3)^{1/3}$$

$$(-2^3)^{1/3}(x^6)^{1/3}(y^3)^{1/3} = -2x^2y$$

EXAMPLE 3: Simplify the expression: $\sqrt[3]{16a^4} + \sqrt[4]{81} - 2a\sqrt[3]{2a} - 3$

Solution

$$\sqrt[3]{16a^4} + \sqrt[4]{81} - 2a\sqrt[3]{2a} - 3 = \sqrt[3]{2^3 \cdot 2 \cdot a \cdot a^3} + \sqrt[4]{3^4} - 2a\sqrt[3]{2a} - 3$$

$$(2^3 \cdot 2 \cdot a \cdot a^3)^{1/3} + 3^{4/4} - 2a(2a)^{1/3} - 3 = 2a(2a)^{1/3} + 3 - 2a(2a)^{1/3} - 3 = 0$$

Use the definitions and properties of fractional powers to simplify the expressions.

$(16)^{1/2}$	$\left(\dfrac{1}{25}\right)^{1/2}$	$(-27)^{1/3}$	$\left(\dfrac{9}{64}\right)^{-1/2}$	$(0.01)^{-3/2}$
$(9)^{1/2}$	$(4x^2)^{1/2}$	$(-27y^6)^{1/3}$	$(-8y^3)^{2/3}$	$(64x^{-2}y^4)^{-3/2}$
$\left(\dfrac{49}{25}\right)^{1/2}$	$\left(\dfrac{-27x^6}{8y^3}\right)^{2/3}$	$\left(\dfrac{32x^{15}}{y^5}\right)^{1/5}$	$\left(\dfrac{x^4}{16y^8}\right)^{1/4}$	$\left(\dfrac{x^9}{125y^3x^6}\right)^{-4/3}$
$(-8x^6)^{1/3}$	$\sqrt[4]{x^3}\cdot(x)^{1/4}$	$(x)^{1/3}\cdot\sqrt[3]{x^2}$	$\sqrt[5]{x^3}\cdot(x^2y^5)^{1/5}$	$\sqrt[3]{x}\cdot(3x)^{1/3}\cdot\sqrt[3]{9x}$
$\left(\dfrac{25x^2}{4y^{-2}}\right)^{1/2}$	$\left(\left(\dfrac{8x^3}{y^{-3}}\right)^2\right)^{1/6}$	$\dfrac{(x)^{1/3}}{\sqrt[3]{x^{-2}}}$	$\dfrac{(x^4)^{1/4}\cdot(-y^4)^{1/4}}{\sqrt[3]{z^{-9}}}$	$\dfrac{(x^3)^{1/3}+(4)^{1/2}}{x+2}$
$\left(\dfrac{x^{3/2}\cdot y^{-2}}{\sqrt[4]{x^{-2}}}\right)^{1/2}$	$\left(\dfrac{y^2(2xy^{-3})^2}{4x^2y^{-4}z^{-8}}\right)^{1/2}$	$\sqrt{\left(\dfrac{16x^4}{625}\right)^{1/2}}$	$\sqrt[3]{\left(\dfrac{x^6}{y^{-6}}\right)^{1/2}}$	$\left(\left(\left(\dfrac{x}{y^{-1}}\right)^3\right)^2\right)^{1/6}$

3.3. Operations with Power Roots

- **BASIC 4 OPERATIONS**
- **RATIONALIZING**

■ BASIC 4 OPERATIONS

In this section we will study addition, subtraction, multiplication, and division of expressions with power roots and fractional powers. To add, subtract, multiply, or divide any two expressions with power roots and rational powers we will use the definitions and properties of power roots and fractional/rational powers.

Examples

EXAMPLE 1: Evaluate the expression $\sqrt{32} + \sqrt{2}$

Solution

$$\sqrt{32} + \sqrt{2} = \sqrt{16 \cdot 2} + \sqrt{2} = \sqrt{16} \cdot \sqrt{2} + \sqrt{2} = 4 \cdot \sqrt{2} + \sqrt{2} = 5\sqrt{2}$$

EXAMPLE 2: Simplify the expression $\sqrt[3]{-8x} - \sqrt[3]{x}$

Solution

$$\sqrt[3]{-8x} - \sqrt[3]{x} = \sqrt[3]{-8} \cdot \sqrt[3]{x} - \sqrt[3]{x} = -2\sqrt[3]{x} - \sqrt[3]{x} = -3\sqrt[3]{x}$$

EXAMPLE 3: Evaluate the expression $\sqrt{2x} \cdot \sqrt{8x}$

Solution

$$\sqrt{2x} \cdot \sqrt{8x} = \sqrt{2x \cdot 8x} = \sqrt{16x^2} = 4x$$

EXAMPLE 4: Simplify the expression $\dfrac{\sqrt[4]{32x^5}}{(2x)^{1/4}}, \ x > 0$

Solution

$$\frac{\sqrt[4]{32x^5}}{(2x)^{1/4}} = \frac{\sqrt[4]{32x^5}}{\sqrt[4]{2x}} = \sqrt[4]{\frac{32x^5}{2x}} = \sqrt[4]{16x^4} = 2|x| = 2x \ \text{(because } x > 0)$$

■ RATIONALIZING

We will use the definitions and properties of power roots and rational/fractional powers as well as multiplying with conjugates and the identity formula $a^2 - b^2 = (a - b)(a + b)$ to rationalize (turn into a rational number) the denominators or numerators of fractional expressions with power roots or fractional powers.

Examples

EXAMPLE 1: Rationalize the denominator of expression $\dfrac{1}{\sqrt{2}}$

Solution

$$\frac{1}{\sqrt{2}} = \frac{1 \cdot \sqrt{2}}{\sqrt{2} \cdot \sqrt{2}} = \frac{\sqrt{2}}{\sqrt{4}} = \frac{\sqrt{2}}{2}$$

EXAMPLE 2: Rationalize the denominator of expression $\dfrac{1}{\sqrt[3]{4}}$

Solution

$$\frac{1}{\sqrt[3]{4}} = \frac{1}{\sqrt[3]{2^2}} = \frac{1 \cdot \sqrt[3]{2}}{\sqrt[3]{2^2} \cdot \sqrt[3]{2}} = \frac{\sqrt[3]{2}}{\sqrt[3]{2^3}} = \frac{\sqrt[3]{2}}{2}$$

EXAMPLE 3: Rationalize the denominator of expression $\dfrac{\sqrt{2}}{\sqrt{3} + \sqrt{2}}$

Solution

$$\frac{\sqrt{2}}{\sqrt{3} + \sqrt{2}} = \frac{\sqrt{2} \cdot (\sqrt{3} - \sqrt{2})}{(\sqrt{3} + \sqrt{2}) \cdot (\sqrt{3} - \sqrt{2})} = \frac{\sqrt{2} \cdot \sqrt{3} - \sqrt{2} \cdot \sqrt{2}}{(\sqrt{3})^2 - (\sqrt{2})^2} = \frac{\sqrt{6} - \sqrt{4}}{3 - 2} = \frac{\sqrt{6} - 2}{1} = \sqrt{6} - 2$$

EXAMPLE 4: Rationalize the numerator of expression $\dfrac{\sqrt{2}}{\sqrt{3} + \sqrt{2}}$

Solution

$$\frac{\sqrt{2}}{\sqrt{3} + \sqrt{2}} = \frac{\sqrt{2} \cdot \sqrt{2}}{(\sqrt{3} + \sqrt{2}) \cdot \sqrt{2}} = \frac{\sqrt{4}}{\sqrt{3} \cdot \sqrt{2} + \sqrt{2} \cdot \sqrt{2}} = \frac{2}{\sqrt{6} + \sqrt{4}} = \frac{2}{\sqrt{6} + 2}$$

EXAMPLE 5: Rationalize the numerator of expression $\dfrac{\sqrt{3} + \sqrt{5}}{\sqrt{5}}$

Solution

$$\frac{\sqrt{3} + \sqrt{5}}{\sqrt{5}} = \frac{(\sqrt{3} + \sqrt{5}) \cdot (\sqrt{5} - \sqrt{3})}{\sqrt{5} \cdot (\sqrt{5} - \sqrt{3})} = \frac{(\sqrt{5})^2 - (\sqrt{3})^2}{\sqrt{5} \cdot \sqrt{5} - \sqrt{5} \cdot \sqrt{3}} = \frac{5 - 3}{\sqrt{25} - \sqrt{15}} = \frac{2}{5 - \sqrt{15}}$$

Simplify

$\sqrt{27} + \sqrt{3}$	$\sqrt{8} - 2\sqrt{2}$	$\sqrt{4x} - \sqrt{x}$	$\sqrt{50x^2} - \sqrt{2}\,\lvert x\rvert$	$\sqrt{32} - \sqrt{2} + \sqrt{8}$
$\sqrt{18y} + \sqrt{2y}$	$\sqrt[3]{8x^2} - (x)^{2/3}$	$\sqrt{9x^5} - x^2\sqrt{x}$	$\sqrt[6]{27} + \sqrt{3} - 3^{1/2}$	$\sqrt{x^3} - x\sqrt{x} + (x)^{3/2}$
$\left(\dfrac{49}{5}\right)^{1/2} \cdot \dfrac{\sqrt{5}}{7}$	$\left(\dfrac{x^3}{y}\right)^{2/3} \cdot \dfrac{\sqrt[3]{y^2}}{x^2}$	$\dfrac{\sqrt[6]{x^3}}{\sqrt{y}} \div \dfrac{\sqrt{x}}{2y^{1/2}}$	$\sqrt{x}\,(\sqrt{x} + (x)^{1/2})$	$\dfrac{x\sqrt[6]{x^3} + y^{1/2}}{\sqrt{y} + x^{3/2}}$

Rationalize the denominator

$\dfrac{1}{\sqrt{7}}$	$\dfrac{1}{\sqrt[5]{x^3}}$	$\dfrac{1}{\sqrt{5} - 2}$	$\dfrac{2}{\sqrt{5} + \sqrt{3}}$	$\dfrac{x - y}{\sqrt{x} - \sqrt{y}}$

Rationalize the numerator

$\dfrac{\sqrt{27}}{9}$	$\dfrac{\sqrt[5]{x^3}}{2}$	$\dfrac{\sqrt{3} - \sqrt{2}}{\sqrt{3}}$	$\dfrac{\sqrt{2} - 1}{\sqrt{2}}$	$\dfrac{\sqrt{x} - \sqrt{y}}{\sqrt{xy}}$

3.4. Pythagorean Theorem and Distance Formula

- **PYTHAGOREAN THEOREM**
- **DISTANCE FORMULA**

■ PYTHAGOREAN THEOREM

In this section we will study relation between **Pythagorean Theorem** for right triangles and power roots.

PYTHAGOREAN THEOREM AND RELATED ROOT FORMULAS

Pythagorean Theorem ▶ For a right triangle $c^2 = a^2 + b^2$, where **a** and **b** are measures of legs, and c is the measure of hypotenuse of a triangle.

Related Root Formulas ▶

❶ $c = \sqrt{a^2 + b^2}$

❷ $a = \sqrt{c^2 - b^2}$

❸ $b = \sqrt{c^2 - a^2}$

Pythagorean Triplets ▶ A triplet of positive integers **x, y, z** is a *Pythagorean Triplet* if and only if the square of biggest integer is equal to sum of squares of other two integers.
For example, the triplet of integers **5, 4, 3** is a Pythagorean Triplet because $5^2 = 4^2 + 3^2$.

Examples

EXAMPLE 1: Find hypotenuse of right triangle with legs **6**in and **8**in.

Solution

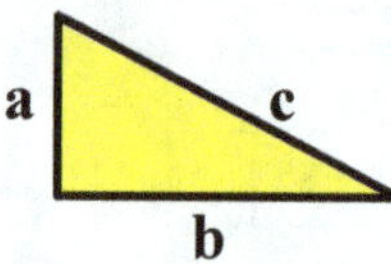

$$c = \sqrt{a^2 + b^2} = \sqrt{6^2 + 8^2} = \sqrt{36 + 64} = \sqrt{100} = 10$$

$c = 10$in

EXAMPLE 2: Find the length of leg of the right triangle if the length of the other leg is **12f** and the length of hypotenuse is **13f.**

Solution

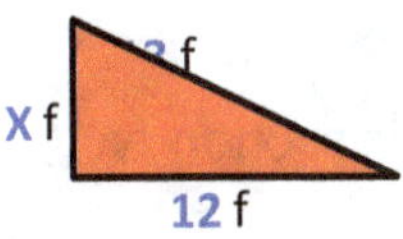

$$a = \sqrt{c^2 - b^2} = \sqrt{13^2 - 12^2} = \sqrt{196 - 144} = \sqrt{25} = 5$$

$$a = 5f$$

EXAMPLE 3: Find the length of legs of the right isosceles triangle if the length of the hypotenuse is **8$\sqrt{2}$ m.**

Solution

$$a^2 + b^2 = c^2$$

$$y^2 + y^2 = \left(8\sqrt{2}\right)^2$$

$$2y^2 = 128$$

$$y^2 = 64$$

$$y = 8 \text{ m}$$

EXAMPLE 4: Write 5 Pythagorean Triplets.

Solution

$$3, 4, 5 ; \quad 30, 40, 50 ; \quad 6, 8, 10 ; \quad 60, 80, 100 ; \quad 5, 12, 13$$

$$3^2 + 4^2 = 5^2 \qquad \leftarrow \qquad 9 + 16 = 25$$

$$30^2 + 40^2 = 50^2 \qquad \leftarrow \qquad 900 + 1600 = 2500$$

$$6^2 + 8^2 = 10^2 \qquad \leftarrow \qquad 36 + 64 = 100$$

$$60^2 + 80^2 = 100^2 \qquad \leftarrow \qquad 3600 + 6400 = 10000$$

$$5^2 + 12^2 = 13^2 \qquad \leftarrow \qquad 25 + 144 = 169$$

EXAMPLE 5: Find the length of longest diagonal of cube with side **5.**

Solution

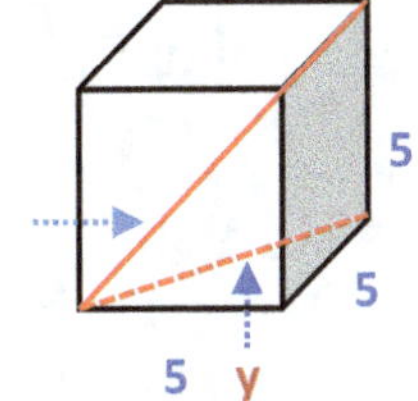

$$y = \sqrt{5^2 + 5^2} = \sqrt{50} = 5\sqrt{2}$$

$$x = \sqrt{50 + 5^2} = \sqrt{75} = 5\sqrt{3}$$

$$x = 5\sqrt{3}$$

■ DISTANCE FORMULA

In this section we will study relation between **Distance Formula** of 2 points and power roots.

Distance Formula ▶ The distance between two points (x_1, y_1) and (x_2, y_2) of the coordinate plane (**xy**-plane) is given by formula

$$d = \sqrt{(x_1 - x_2)^2 + (y_1 - y_2)^2}$$

Examples

EXAMPLE 1: Find the distance between points $(1, 3)$ and $(4, 7)$.

Solution

$$d = \sqrt{(1 - 4)^2 + (3 - 7)^2} = \sqrt{(-3)^2 + (-4)^2} = \sqrt{9 + 16} = \sqrt{25} = 5$$

$$d = 5$$

EXAMPLE 2: Find the distance between points $(-4, 0)$ and $(0, 3)$.

Solution

$$d = \sqrt{(-4 - 0)^2 + (0 - 3)^2} = \sqrt{(-4)^2 + (-3)^2} = \sqrt{16 + 9} = \sqrt{25} = 5$$

$$d = 5$$

EXAMPLE 3: Find the perimeter of triangle with vertices $A(-1, 0)$, $B(-4, 4)$ and $C(-5, 3)$.

Solution

$$d_{AB} = \sqrt{(-1 + 4)^2 + (0 - 4)^2} = \sqrt{(3)^2 + (-4)^2} = \sqrt{9 + 16} = \sqrt{25} = 5$$

$$d_{BC} = \sqrt{(-4 + 5)^2 + (4 - 3)^2} = \sqrt{(1)^2 + (1)^2} = \sqrt{1 + 1} = \sqrt{2}$$

$$d_{CA} = \sqrt{(-5 + 1)^2 + (3 - 0)^2} = \sqrt{(-4)^2 + (3)^2} = \sqrt{16 + 9} = \sqrt{25} = 5$$

$$P_{ABC} = 5 + \sqrt{2} + 5 = 10 + \sqrt{2}$$

Find the third side

$a = 3, b = 5$ $c =$	$a = 1, b = 4$ $c =$	$a = 6, b = 9$ $c =$	$a = 7, b = 7$ $c =$
$c = 4, a = 3$ $b =$	$c = 6, a = 4$ $b =$	$c = 13, a = 5$ $b =$	$c = 13, a = 12$ $b =$
$c = 7, b = 4$ $a =$	$c = 10, b = 6$ $a =$	$c = 9, b = 7$ $a =$	$c = 10, b = 8$ $a =$

Is the triplet of numbers Pythagorean Triplet?

1, 2, 3	12, 9, 15	4, 5, 7	12, 16, 20

Find the distance between two points.

$A(1,2)$ $B(1,5)$ $d =$	$A(3,4)$ $B(5,6)$ $d =$	$A(3,1)$ $B(6,5)$ $d =$	$A(5,2)$ $B(9,2)$ $d =$
$A(-1,2)$ $B(3,5)$ $d =$	$A(1,-4)$ $B(5,-1)$ $d =$	$A(8,-2)$ $B(2,6)$ $d =$	$A(0,0)$ $B(3,4)$ $d =$

Simplify

$\sqrt{8} + \sqrt{18}$	$\sqrt{12} - 2\sqrt{3}$	$\sqrt{x} - \sqrt{9x}$	$\sqrt{100x^2} - 5|x|$	$\sqrt{24} - \sqrt{6} + \sqrt{54}$
$\sqrt{9y} + \sqrt{y}$	$\sqrt[3]{x^2} - (x)^{2/3}$	$\sqrt{4x^3} - x\sqrt{x}$	$\sqrt[6]{8} + \sqrt{2} - 2^{1/2}$	$3\sqrt{x^3} - 5(x)^{3/2}$
$\left(\dfrac{64}{3}\right)^{1/2} \cdot \dfrac{\sqrt{3}}{8}$	$\left(\dfrac{x^3}{y}\right)^{1/3} \cdot \dfrac{\sqrt[3]{y}}{x^2}$	$\dfrac{\sqrt[3]{x^2}}{\sqrt{y}} \div \dfrac{\sqrt{y}}{2x^{1/3}}$	$(x)^{1/2}\left(\sqrt{x} + 1\right)$	$\dfrac{x\sqrt[6]{x^3} + \sqrt[3]{y}}{y^{1/3} + x^{3/2}}$

Rationalize the denominator

$\dfrac{1}{\sqrt{5}}$	$\dfrac{1}{\sqrt[5]{x^2}}$	$\dfrac{1}{\sqrt{6} - 2}$	$\dfrac{2}{\sqrt{7} + \sqrt{3}}$	$\dfrac{x - y}{\sqrt{x} + \sqrt{y}}$

Rationalize the numerator

$\dfrac{\sqrt{8}}{2}$	$\dfrac{\sqrt[4]{x^3}}{3}$	$\dfrac{\sqrt{5} - \sqrt{3}}{\sqrt{3}}$	$\dfrac{\sqrt{3} - 1}{\sqrt{3}}$	$\dfrac{\sqrt{x} + \sqrt{y}}{\sqrt{y}}$

Simplify

$\sqrt{8} + \sqrt{50}$	$\sqrt{27} - 2\sqrt{3}$	$\sqrt{4x} - \sqrt{9x}$	$\sqrt{x^2} - |x|$	$\sqrt{18} - \sqrt{2} + \sqrt{50}$
$\sqrt{4y} + \sqrt{9y}$	$\sqrt[3]{8x^2} - (x)^{2/3}$	$\sqrt{9x^3} - x\sqrt{4x}$	$\sqrt[6]{27} - 3^{1/2}$	$2\sqrt{x^3} - (x)^{3/2}$
$\left(\dfrac{25}{3}\right)^{1/2} \cdot \dfrac{\sqrt{3}}{5}$	$\left(\dfrac{x^4}{y}\right)^{1/4} \cdot \dfrac{\sqrt[4]{y}}{x}$	$\dfrac{\sqrt[5]{x}}{\sqrt{x}} \div \dfrac{\sqrt{x}}{x^{4/5}}$	$(y)^{1/2} \left(\sqrt{y} + 1\right)$	$\dfrac{\sqrt[6]{x^3} + \sqrt[3]{y}}{y^{1/3} + x^{1/3}}$

Rationalize the denominator

$\dfrac{1}{\sqrt{6}}$	$\dfrac{1}{\sqrt[6]{x^4}}$	$\dfrac{1}{\sqrt{3} - 1}$	$\dfrac{2}{\sqrt{5} + \sqrt{2}}$	$\dfrac{\sqrt{x}}{\sqrt{x} + \sqrt{y}}$

Rationalize the numerator

$\dfrac{\sqrt{2}}{3}$	$\dfrac{\sqrt[4]{x}}{3}$	$\dfrac{\sqrt{7} - \sqrt{5}}{\sqrt{5}}$	$\dfrac{\sqrt{5} - 1}{\sqrt{5}}$	$\dfrac{\sqrt{x} - \sqrt{y}}{\sqrt{x} + \sqrt{y}}$

1. In the set $\{-2, -4.3, -1.77, -\frac{1}{2}, 0, \sqrt{2}, 2, \pi, 4\frac{1}{4}, 8, 11.\overline{3}, 12.5\}$ list all of the:

- Natural Numbers
- Whole Numbers
- Integers
- EVEN Numbers
- Odd numbers
- Prime Numbers
- Rational Numbers
- Irrational Numbers
- Real Numbers

2. Evaluate the numerical expression

$$-5 + 4 \cdot (-3) \cdot (-1)$$

3. Let $A = \{2, 3, 4, 6\}$, $B = \{-1, 2, 3, 5, 7\}$. Find each of the following.

a) $A \cup B$

b) $A \cap B$

c) $A \cap A$

d) $(B \cup B) \cap A$

e) $(A \cap B) \cup A$

f) $\{x/\ x \in A,\ x \notin B\}$

g) $\{x/\ x \in B,\ x \geq 3\}$

4. Evaluate the numerical expression.

$$\frac{|4 - 5| + |-(3 - 9) - 9 \div 3|}{|1 - 2| + |-4 \div 2 + (5 - 6)|}$$

5. Evaluate the variable expression for $x = 1$, $y = 3$.

$$\frac{x + 3y}{2x + y} + 2$$

6. Use PEMDAS + L$\rightarrow$R to evaluate the expression.

$$(-3) \cdot \{[-(-1)^2 + 3] + 1\} - 12 \div (-4) + 5$$

7. Find the result of operations.

$$\frac{x^2 y^3}{x^7} \div \frac{x^{-9} y^{-2}}{x^{-3} y^{-5}} \cdot (x\,y^5)$$

8. Write expression in scientific notation form.

$$5(4 \cdot 10^{14})(9 \cdot 10^{-9})$$

9. Classify the polynomial by number of terms, by degree, and as binomial or trinomial.

$$x^3 + 3x^6 - 5x^4$$

10. Evaluate the polynomial for $x = 1$, $y = 1$, $z = -1$.

$$2x^3 + xy^3 z^2 + 2z + y - 3x$$

11. Combine like terms in each expression and write the answer in standard form.

$$(x^4 + x - 3x^2 + x^3) - (2x^3 + x^4 - 2x^2 + x)$$

12. Multiply the following polynomials. Write the answer in simplest form.

$$(x + 2) \cdot (x^2 - 2x + 4)$$

13. Divide a polynomial by using **long division** and **synthetic division**.

$$\frac{y^2 + 2y - 3}{x + 3}$$

14. Simplify.

$$\frac{x^2 - 4y^2}{x + 2y} - x + y$$

15. Factor

$$x^5 - x^3$$

16. Factor

$$x^3 - x^2 - 20x$$

17. Factor

$$2x^2 - 4x - 7$$

18. Simplify

$$\sqrt{27} - \sqrt{12} + 4\sqrt{3}$$

19. Simplify

$$\left(\frac{8x^3}{y^2}\right)^{1/3} \cdot \frac{\sqrt[3]{y^2}}{2x^2}$$

20. Simplify

$$\frac{x\sqrt[3]{x^2} + \sqrt[3]{y}}{y^{1/3} + x^{5/2}}$$

21. Rationalize the denominator

$$\frac{\sqrt{5}}{\sqrt{5} + 2}$$

22. Rationalize the numerator

$$\frac{\sqrt{7} - \sqrt{5}}{2}$$

PART 2

EQUATIONS, INEQUALITIES, AND GRAPHS

Chapter 4. Equations and Inequalities

4.1. Linear Equations and Inequalities
4.2. Absolute Value Equations and Inequalities
4.3. Quadratic Equations and Inequalities
4.4. Polynomial Equations and Inequalities
4.5. Fractional Equations and Inequalities
4.6. Radical Equations and Inequalities

Chapter 4 Classwork
Chapter 4 Homework

Chapter 5. Graphs of Lines and Inequalities

5.1. Systems of Coordinates
5.2. Equation of a Line
5.3. Graphs of Lines
5.4. Parallel, Coincident, Perpendicular, and Intersecting Lines
5.5. Linear Inequalities on the Coordinate Plane

Chapter 5 Classwork
Chapter 5 Homework

Part 2. Pre-Test

Chapter 4. Equations and Inequalities

4.1. Linear Equations and Inequalities
4.2. Absolute Value Equations and Inequalities
4.3. Quadratic Equations and Inequalities
4.4. Polynomial Equations and Inequalities
4.5. Fractional Equations and Inequalities
4.6. Radical Equations and Inequalities

Chapter 4   Classwork
Chapter 4   Homework

$$Ax + By = C$$

$$ax + b = 0, \qquad a \neq 0$$

$$ax + b > 0,\, ax + b < 0,\, ax + b \geq 0,\, ax + b \leq 0,\, a \neq 0$$

$$|E(x)| > k,\, |E(x)| < k,\, |E(x)| \geq k,\, |E(x)| \leq k,\, k > 0$$

$$|E(x)| = m, \qquad m \in R$$

$$Ax + By < C$$

4.1. Equations and Inequalities

- **LINEAR EQUATIONS**
- **SOLVING EQUATIONS USING FACTORING**
- **LINEAR INEQUALITIES**

■ LINEAR EQUATIONS

We will use the following definitions.

IMPORTANT DEFINITIONS

Definition ▶ *A linear equation* is an equation $E_1 = E_2$ where E_1 or E_2 or both E_1 and E_2 are linear expressions with one or more variables.

Definition ▶ *A linear equation with one variable* is an equation $E_1 = E_2$ where E_1 or E_2 or both E_1 and E_2 are linear expressions with one variable.

Definition ▶ *A Linear Equation with One Variable* is any equation that can be presented in the form $ax + b = 0$, where $a, b \in \mathbf{R}$ and $a \neq 0$.

Examples The equations $x + 2 = 4$, $5(x - 2) = 4x + 2$, $2x + y = x - 4$, $5x + y = 1$ are linear equations.

The equations $x + 2 = 4$ and $5(x - 2) = 4x + 2$, $2x + y = x - 4$ are linear equations with one variable.

The equations $2x + y = x - 4$, $5x + y = 1$ are linear equations with two variables.

Below we will solve several types of linear equations with one variable.

Example 1 **Solve** the equation $5 + 4(y - 1) + 2y = 2(y + 4) - 7$

SOLUTION ▶

$$5 + 4(y - 1) + 2y = 2(y + 4) - 7$$

$$5 + 4y - 4 + 2y = 2y + 8 - 7$$

$$6y + 1 = 2y + 1 \implies 6y - 2y = 1 - 1$$

$$4y = 0 \implies 4y/4 = 0/4 \implies y = 0$$

 Solve the equation.

$$\frac{x}{4} - \frac{x}{2} = -1$$

LCD for denominators **4** and **2** is **4**. We will multiply both sides of the equation by **4** to simplify the equation.

$$\frac{x}{4} - \frac{x}{2} = -1 \Rightarrow 4 \cdot \left(\frac{x}{4} - \frac{x}{2}\right) = 4 \cdot (-1) \Rightarrow 4 \cdot \frac{x}{4} - 4 \cdot \frac{x}{2} = -4 \Rightarrow$$

$$x - 2x = -4 \Rightarrow -x = -4 \Rightarrow x = 4$$

 Solve the equation.

$$0.5 + 0.5y = 0.25y + 4.5$$

$$0.5 + 0.5y = 0.25y + 4.5 \Rightarrow -0.25y + 0.5y = 4.5 - 0.5 \Rightarrow 0.25y = 4 \Rightarrow$$
$$0.25y/0.25 = 4/0.25 \Rightarrow y = 16$$

 Solve the proportion.

$$\frac{2x+3}{5} = \frac{3x+2}{7}$$

We will use the cross-multiplication property of proportions

$$\frac{2x+3}{5} = \frac{3x+2}{7}$$

$$7(2x+3) = 5(3x+2)$$

$$14x+21 = 15x+10$$

$$21 - 10 = 15x - 14x$$

$$11 = x \Rightarrow x = 11$$

Example 1 5 more than **3** times a number is **26.** Find the number

$$5 + 3x = 26$$

SOLUTION ▶

$$5 + 3x = 26 \Rightarrow 3x = 26 - 5 \Rightarrow 3x = 21 \Rightarrow 3x/3 = 21/3 \Rightarrow x = 7$$

Example 2 A mobile phone that regularly costs **$200** was on sale for **$150.** What was the percentage of discount?

SOLUTION ▶

Let **x** be the decimal equivalent of the percent discount. The discount was **$200 · x**, and the equation for the sale cost will be:
$$\$200 - \$200 \cdot x = \$150 \Rightarrow \$200 - \$150 = \$200 \cdot x \Rightarrow \$50 = \$200 \cdot x \Rightarrow$$
$$\$50/\$200 = \$200 \cdot x/\$200 \Rightarrow 1/4 = x \Rightarrow x = 1/4 = 0.25 \Rightarrow x = 25\%$$

If the decimal equivalent of the percent discount is **x = 0.25,** then the percent of discount is **25%.**

Example 3 Marina has **$10** in coins. She has twice as many quarters as dimes and four times as many nickels as quarters. How many of each coin does she have?

SOLUTION ▶

Let x be the number of dimes. So, the number of quarters will be **2x**, and the number of nickels will be **4(2x) = 8x.** The equation of the total amount in terms of **x** will be the following:
$$\$0.10 \cdot x + \$0.25 \cdot (2x) + \$0.05 \cdot (8x) = \$10 \Rightarrow \$0.10 \cdot x + \$0.50 \cdot x + \$0.40 \cdot x = \$10$$
$$\Rightarrow \$1 \cdot x = \$10 \Rightarrow x = 10 \Rightarrow 2x = 20 \Rightarrow 8x = 80$$
So, the number of **dimes** is **10,** the number of **quarters** is **20,** and the number of **nickels** is **80.**

  Find three consecutive numbers, the sum of which is **39**.

Let x be the first number. The second number then will be $x + 1$, and the third number will be $x + 2$. The equation for the sum of the three consecutive numbers will be the following:

$x + (x + 1) + (x + 2) = 39 \Rightarrow x + x + 1 + x + 2 = 39 \Rightarrow 3x + 3 = 39 \Rightarrow$

$3x = 39 - 3 \Rightarrow 3x = 36 \Rightarrow 3x/3 = 36/3 \Rightarrow$

$x = 12 \Rightarrow x + 1 = 13 \Rightarrow x + 2 = 14$

So, the three consecutive numbers are **12**, **13**, and **14**.

  Bob is **5** years older than Narek, and Narek is **2** years younger than Nick. The sum of their ages is **16**. What are their ages?

Let x be Narek's age. Then Bob's age will be $x + 5$, and Nick's age will be $x + 2$. The equation for the sum of their ages will be the following:

$x + (x + 5) + (x + 2) = 16 \Rightarrow x + x + 5 + x + 2 = 16 \Rightarrow 3x + 7 = 16 \Rightarrow$

$3x = 16 - 7 \Rightarrow 3x = 9 \Rightarrow 3x/3 = 9/3 \Rightarrow$

$x = 3 \Rightarrow x + 5 = 8 \Rightarrow x + 2 = 5$

So, Narek's age is **3**, Bob's age is **8**, and Nick's age is **5**.

  Mary deposited **$10,000**, splitting that money into two accounts with **5%** and **7%** annual interests. If at the end of the year the total interest earned was **$620**, how much was originally deposited in each account?

Let x represent the amount originally deposited in the **5%** account. So, **10,000 – x** will be the amount originally deposited in the **7%** account. The equation for total yearly interest will be:

$0.05x + 0.07(10{,}000 - x) = 620 \Rightarrow$

$[0.05x + 0.07(10{,}000 - x)] \cdot 100 = 620 \cdot 100 \Rightarrow$

$5x + 7(10{,}000 - x)] = 62{,}000 \quad 5x + 70{,}000 - 7x = 62{,}000 \Rightarrow$

$70{,}000 - 2x = 62{,}000 \Rightarrow 70{,}000 - 62{,}000 = 2x \Rightarrow$

$8{,}000 = 2x \Rightarrow 8{,}000/2 = 2x/2 \Rightarrow x = 4{,}000 \Rightarrow 10{,}000 - x = 6{,}000$

So, the amount originally deposited in the **5%** account is **$4,000** and the amount originally deposited in the **7%** account is **$6,000**.

■ SOLVING EQUATIONS USING FACTORING

Several solving methods of linear equations can help us solve a set of polynomial equations in which polynomials are factorable polynomials.

We will use different methods of factoring of polynomials to solve factorable quadratic or other types of polynomial equations.

Example 1 **Solve** the equation $x^2 + 2x = 0$

SOLUTION ▶ To solve this equation we will use the "factor out method".

$$x^2 + 2x = 0$$

$$x(x + 2) = 0$$
$$\Downarrow$$
$$x = 0 \text{ or } x + 2 = 0$$
$$\Downarrow$$
$$x = 0 \text{ or } x = -2$$

Example 2 **Solve** the equation $4x^3 - 12x^2 = 0$

SOLUTION ▶ To solve this equation we will use the "factor out method".

$$4x^3 - 12x^2 = 0$$

$$4x^2(x - 3) = 0$$
$$\Downarrow$$
$$x^2 = 0 \text{ or } x - 3 = 0$$
$$\Downarrow$$
$$x = 0 \quad \text{or} \quad x = 3$$

Example 3 **Solve** the equation $x^2 + 5x + 6 = 0$

SOLUTION ▶ To solve this equation we will use "branch factoring".

$$x^2 + 5x + 6 = 0$$

$$(x + 2)(x + 3) = 0$$
$$\Downarrow$$
$$x = -2 \text{ or } x = -3$$

Example 4 **Solve** the equation $x^2 - 5x + 6 = 0$

SOLUTION ▶ To solve this equation we will use "branch factoring".

$$x^2 - 5x + 6 = 0$$
$$\overset{-2 \quad -3}{}$$
$$(x - 2)(x - 3) = 0$$
$$\Downarrow$$
$$x = 2 \quad \text{or} \quad x = -3$$

Example 5 **Solve** the equation $2x^2 - 3x - 5 = 0$

SOLUTION ▶ To solve this equation we will use "big X factoring".

$$\overset{-10}{2x^2 + 3x - 5 = 0}$$

$$\frac{5}{2x} = 5/2x \quad \overset{3}{\times}_{-10} \quad -2/2x = \frac{-1}{x}$$

$$(2x + 5)(x - 1) = 0$$
$$\Downarrow$$
$$2x + 5 = 0 \quad \text{or} \quad x - 1 = 0$$
$$\Downarrow$$
$$x = -\frac{5}{2} \quad \text{or} \quad x = 1$$

Example 6 **Solve** the equation $12x^3 - 27x = 0$

SOLUTION ▶ To solve this equation we will use the "identity formulas" method of factoring.

$$12x^3 - 27x = 0$$
$$3x(4x^2 - 9) = 0$$
$$3x[(2x)^2 - 3^2] = 0$$
$$3x[(2x)^2 - 3^2] = 0$$
$$3x(2x - 3)(2x + 3) = 0$$
$$x = 0 \quad \text{or} \quad 2x - 3 = 0 \quad \text{or} \quad 2x + 3 = 0$$
$$\Downarrow$$
$$x = 0, \quad x = 3/2, \quad x = -3/2$$

■ LINEAR INEQUALITIES

We will use the following definitions.

> **Definition** ▶ *A simple linear inequality* is an inequality $E_1 < E_2$ (or $E_1 > E_2$, or $E_1 \leq E_2$, or $E_1 \geq E_2$) where E_1 or E_2 or both E_1 and E_2 are linear expressions with one or more variables.
>
> **Definition** ▶ *A simple linear inequality with one variable* is an inequality $E_1 < E_2$ (or $E_1 > E_2$, or $E_1 \leq E_2$, or $E_1 \geq E_2$) where E_1 or E_2 or both E_1 and E_2 are linear expressions with only one variable.
>
> **Definition** ▶ *A simple linear inequality with one variable* is any inequality that can be presented in the form $ax + b < 0$, or $ax + b \leq 0$, or $ax + b > 0$, or $ax + b \geq 0$, where $a, b \in \mathbf{R}$ and $a \neq 0$.

Examples The inequalities $x + 5 > 4$, $2x + 3y < 6$, $3x - 2 \leq 4x + 2$, $x + y \geq -4$ are simple linear inequalities with one or two variables.

Below we will solve several types of simple and compound linear inequalities with one variable.

SIMPLE LINEAR INEQUALITIES WITH ONE VARIABLE

Example 1 Solve and present the solutions by formula, graph, set, and interval notations.

$$6(x - 1) + 3x \geq 4(x + 2) + 6$$

SOLUTION ▶

$$6(x - 1) + 3x \geq 4(x + 2) + 6$$

$$6x - 6 + 3x \geq 4x + 8 + 6$$

$$-4x + 9x \geq 14 + 6$$

$$5x \geq 20$$

$$x \geq 4$$

INEQUALITY	FORMULA NOTATION OF SOLUTIONS	GRAPH NOTATION OF SOLUTIONS	SET NOTATION OF SOLUTIONS	INTERVAL NOTATION OF SOLUTIONS
$6(x - 1) + 3x \geq 4(x + 2) + 6$	$x \geq 4$	$\xleftarrow{\qquad} 4 \xrightarrow{\qquad} \infty$	$\{\, x / x \geq 4 \,\}$	$x \in [4, \infty)$

Example 2 Solve: $6x - 8 \leq 2(x + 2)$

SOLUTION ▶

$$6x - 8 \leq 2x + 4$$

$$6x - 2x \leq 8 + 4$$
$$4x \leq 12$$
$$4x/4 \leq 12/4$$

$x \leq 3$; $-\infty$ 3   ; $\{x/\, x \leq 3\}$; $x \in (-\infty, 3]$

Example 3 Solve: $0.1x + 0.3 < 0.5$

SOLUTION ▶

$$0.1x + 0.3 < 0.5$$
$$0.1x < 0.5 - 0.3$$
$$0.1x < 0.2$$
$$0.1x/0.1 < 0.2/0.1$$

$x < 2$; $-\infty$ 2   ; $\{x/\, x < 2\}$; $x \in (-\infty, 2)$

Example 4 Solve: $\dfrac{x}{3} - \dfrac{x}{5} > \dfrac{4}{15}$

SOLUTION ▶

LCD for denominators **3** and **5** is **15**. We will multiply both sides of the inequality by **15** to simplify the inequality.

$$\frac{15x}{3} - \frac{15x}{5} > \frac{15 \cdot 4}{15}$$
$$5x - 3x > 4$$
$$5x - 3x > 4$$
$$2x > 4$$
$$2x/2 > 4/2$$

$x > 2$;   2 ∞ ; $\{x/\, x > 2\}$; $x \in (2, \infty)$

COMPOUND LINEAR INEQUALITIES WITH ONE VARIABLE

We will use the following definitions for **compound** (**combined, complex**) linear inequalities with one variable.

Definition ▶ *Compound (or Combined, or Complex) inequalities* are two or more simple inequalities considered together.

Definition ▶ A *Compound (or Combined, or Complex) inequality* which is formed by joining two simple inequalities with the word **and** (or with the symbol $\cap$) or one which can be represented as a **double inequality** is called a *Conjunction*.

For example, the inequalities $x > -3$ **and** $x < 2$; $x \le 5 \cap x > -5$; $3z-1 < 5$ **and** $2z+3 \ge 5$; $y-1 < 5 \cap y \ge 2$; $2 < x < 7$; $x - 6 < 5x-2 \le 3x +8$ are *Conjunctions*.

Definition ▶ Any inequality like $P(x) < Q(x) < R(x)$; or $P(x) \le Q(x) < R(x)$; or $P(x) < Q(x) \le R(x)$; or $P(x) \le Q(x) \le R(x)$, where $P(x)$, $Q(x)$, and $R(x)$ are expressions, is called *Double Inequality*.

Every **double inequality** is a **conjunction**, but there are many **conjunctions** which are not double inequalities.

A **conjunction** is true only if both inequalities are true.

For example, any double inequality like $b < x < a$ can be represented as a conjunction "$x < a$ **and** $x > b$". But, for example, the conjunction $x < 3$ **and** $x < 5$ cannot be represented as a double inequality.

Definition ▶ A *Compound (or Combined, or Complex) inequality* formed by joining two inequalities with the word **or** (or with the symbol $\cup$) is called a *Disjunction.*

A **disjunction** is true if either of the inequalities is true.

For example, the inequalities $x < -2$ **or** $x > 2$; $y \le 5$ **or** $y > 9$; $2z-1 < 5$ **or** $2z+3 \ge 5$; $4 < 3t-2$ **or** $t \le - 3$ are *Disjunctions.*

Examples The inequalities $x + 5 > 4$ **and** $2x + 3 < 6$, $3x - 2 \le 4x + 2$ **or** $x + 5 \ge - 4$, and $3x - 2 \le 4x < - 4$ are compound linear inequalities with one variable.

To solve any compound inequalities either with the word **and** (or with the symbol $\cap$) or with the word **or** (or with the symbol $\cup$) we will work with the constituent inequalities as separate simple inequalities and then we will take the intersection ($\cap$) of solution sets (in case of inequalities with the word **and** (or with the symbol $\cap$) or union ($\cup$) of solution sets (in case of inequalities with the word **or**, or with the symbol $\cup$).

 Solve $3z - 1 < 5$ and $2z + 3 \geq -5$ **Conjunction**

SOLUTION ▶

$$
\begin{array}{lcl}
3z-1 < 5 & \text{and} & 2z+3 \geq -5 \\
+1 \quad +1 & & -3 \quad -3 \\
3z < 6 & \text{and} & 2z \geq -8 \\
3z/3 < 6/3 & \text{and} & 2z/2 \geq -8/2 \\
z < 2 & \text{and} & z \geq -4
\end{array}
$$

$z < 2 \cap z \geq -4$; $\longleftarrow\!\!\!\!\!\!\!\!\!\!\longrightarrow$; $\{\, z/-4 \leq z < 2 \,\}$; $z \in [-4, 2)$
$\qquad\qquad\qquad\quad -4 \qquad 2$

 Solve $x-1 > 5$ or $x+3 \leq 5$ **Disjunction**

SOLUTION ▶

$$
\begin{array}{lcl}
x-1 > 5 & \text{or} & x+3 \leq 5 \\
+1 \quad +1 & & -3 \quad -3 \\
x > 6 & \text{or} & x \leq 2
\end{array}
$$

$x > 6 \cup x \leq 2$; $\longleftarrow\!\!\!\!\!\!\!\!\!\!\longrightarrow$; $\{\, x/x \leq 2 \text{ or } x > 6 \,\}$; $x \in (-\infty, 2] \cup (6, \infty)$
$\qquad\qquad\qquad\quad 2 \qquad 6$

 Solve $2x-3 \geq 1 \cap x+2 \leq 9$ **Conjunction**

SOLUTION ▶

$$
\begin{array}{lcl}
2x-3 \geq 1 & \cap & x+2 \leq 9 \\
+3 \quad +3 & & -2 \quad -2 \\
2x \geq 4 & \cap & x \leq 7 \\
x \geq 2 & \cap & x \leq 7
\end{array}
$$

$2 \leq x \leq 7$; $\longleftarrow\!\!\!\!\!\!\!\!\!\!\longrightarrow$; $\{\, x / 2 \leq x \leq 7 \,\}$; $x \in [2, 7]$
$\qquad\qquad\quad 2 \qquad 7$

 Solve $x-2 < 1 \cup x-3 \geq 2$ **Disjunction**

SOLUTION ▶

$$
\begin{array}{lcl}
x-2 < 1 & \cup & x-3 \geq 2 \\
+2 \quad +2 & & +3 \quad +3 \\
x < 3 & \cup & x \geq 5
\end{array}
$$

$x < 3 \cup x \geq 5$; $\longleftarrow\!\!\!\!\!\!\!\!\!\!\longrightarrow$; $\{\, x/x < 3 \text{ or } x \geq 5 \,\}$; $x \in (-\infty, 3] \cup (5, \infty)$
$\qquad\qquad\qquad 3 \qquad 5$

To solve the double inequalities we can work with both inequalities with the same operations at the same time or we can work with inequalities as two separate inequalities and then take the intersection of solution sets.

Example 5 Solve $2 < 2x - 4 \leq 8$ **Double Inequality = Conjunction**

SOLUTION ▶

$$2 < 2x - 4 \leq 8$$
$$+4 \qquad +4 \quad +4$$
$$6 < \quad 2x \quad \leq 12$$
$$6/2 < \quad 2x/2 \quad \leq 12/2$$
$$3 < \quad x \quad \leq 6$$

$$3 < x \leq 6 \; ; \quad \overset{(\quad]}{\underset{3 \qquad 6}{\longleftrightarrow}} \quad ; \quad \{x / 3 < x \leq 6\}; \quad x \in (3, 6]$$

Example 6 Solve $5 < 3x - 1 \leq x + 7$ **Double Inequality**

SOLUTION ▶

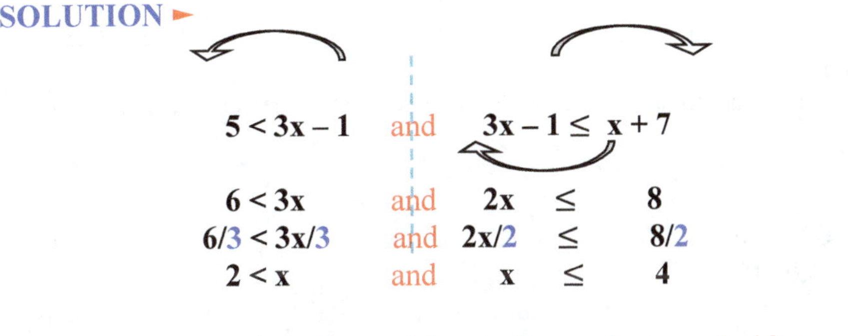

$$5 < 3x - 1 \quad \text{and} \quad 3x - 1 \leq x + 7$$

$$6 < 3x \qquad \text{and} \quad 2x \quad \leq \quad 8$$
$$6/3 < 3x/3 \qquad \text{and} \quad 2x/2 \quad \leq \quad 8/2$$
$$2 < x \qquad \text{and} \quad x \quad \leq \quad 4$$

$$2 < x \leq 4 \; ; \quad \overset{(\quad]}{\underset{2 \qquad 4}{\longleftrightarrow}} \quad ; \quad \{x / 2 < x \leq 4\}; \quad x \in (2, 4]$$

Example 7 Solve $9 - x < 3x + 1 \leq 2x + 5$ **Double Inequality**

SOLUTION ▶

$$9 - x < 3x + 1 \quad \text{and} \quad 3x + 1 \leq 2x + 5$$

$$8/4 < 4x/4 \qquad \text{and} \quad x \quad \leq \quad 4$$
$$2 < x \qquad \text{and} \quad x \quad \leq \quad 4$$

$$2 < x \leq 4 \; ; \quad \overset{(\quad]}{\underset{2 \qquad 4}{\longleftrightarrow}} \quad ; \quad \{x / 2 < x \leq 4\}; \quad x \in (2, 4]$$

Example 1 2 more than **5** times a positive integer is **less than 8.** Find the positive integer.

SOLUTION▶ Let **n** is a given positive integer. So, we can write the inequality:

$$2 + 5n < 8$$
$$\underline{-2 \qquad\quad -2}$$
$$5n < 6$$
$$5n/5 < 6/5$$
$$n < 6/5 = 1.2$$
$$n = 1 \quad\longleftarrow\quad \text{There is only one positive integer less than } 1.2. \text{ That is } 1.$$

Example 2 A company, which produces markers, has yearly overhead costs of $64,000. Each marker costs $0.2 to produce and sells at $0.6. How many markers have to be sold to make a profit?

SOLUTION▶ Let **x** is the number of markers sold yearly. The production cost is **64,000 + 0.2x** and revenue is **0.6x**.
To make a profit means to have revenue that is greater than the cost.

$$0.6x > 64,000 + 0.2x$$
$$\underline{-0.2x \qquad\qquad -0.2x}$$
$$0.4x > 64,000$$
$$0.4x/0.4 > 64,000/0.4$$
$$x > 160,000 \quad\longleftarrow\quad \text{The company has to sell more than } 160,000 \text{ markers per year in order to make a profit.}$$

Example 3 The sum of four consecutive even integers is less than 100. What are the greatest possible values for these integers?

SOLUTION▶ Let the four consecutive even integers be **x, x + 2, x + 4, x + 6.**

$$x + (x + 2) + (x + 4) + (x + 6) < 100$$
$$4x + 12 < 100$$
$$4x < 88$$
$$x < 22$$
$$x = 20, x + 2 = 22, x + 4 = 24, x + 6 = 26 \qquad \text{The greatest even integer} < 22 \text{ is } 20.$$

David's scores are **83, 85, 89** and **93**. What is the lowest score that David needs to achieve an average of at least **87**?

SOLUTION ▶ Let **x** is David's score for the last quiz. So, we can write the inequality:

$$(83+85+89+93+x)/5 \geq 87$$
$$(350+x)/5 \geq 87$$
$$5 \cdot (350+x)/5 \geq 5 \cdot 87$$
$$350+x \geq 435$$
$$-350 \quad\quad -350$$
$$x \geq 85 \implies x_{min} = 85 \quad \text{So, the lowest grade is } 85.$$

Example 5

Mary and George work at the mobile phone company 5 days per week. George works 2 more hours than Mary each day, and together they work no less than 80 hours per week. What is the least number of hours per day that each works?

SOLUTION ▶ Let **x** be the number of hours Mary works per day. The number of hours George works per day will be **x + 2.** So, we can write the following inequality:

$$5(x + x + 2) \geq 80$$
$$5(2x + 2) \geq 80$$
$$10x + 10 \geq 80$$
$$-10 \quad -10$$
$$10x \geq 70$$
$$10x/10 \geq 70/10$$
$$x \geq 7 \implies x_{min} = 7$$

The least number of Mary's work hours is **7**, and the least number of George's work hours is **9**.

Example 6

The width of a rectangle is **10**in. The perimeter is at least **120**in. What is the minimum possible size of the length?

SOLUTION ▶

$$20 + 2x \geq 120$$
$$-20 \quad\quad\quad -20$$
$$2x/2 \geq 100/2$$
$$x \geq 50 \implies \text{The minimum possible size of length is } 50\text{in.}$$

Solve the linear equations with one variable.

1. $x + 2 = 6$	2. $x - 7 = 4$
3. $3x = 15$	4. $\frac{3}{4}x = 9$
5. $3x + 2 = 5$	6. $4x - 3 = 9$
7. $2(x - 5) + x = 2$	8. $\frac{x}{4} = \frac{x}{3} - 1$
9. $5x - 2(x - 5) = x + 2$	10. $x - 7(x - 1) + 4 = 4(x + 2) + 3$
11. $\frac{x}{4} - \frac{x+1}{2} = -\frac{x-2}{4} - 1$	12. $\frac{x+2}{4} = \frac{x-1}{3}$
13. $x^2 - 3x = 0$	14. $x^3 - 4x^2 = 0$
15. $x^2 - 3x - 10 = 0$	16. $4x^3 - 4x^2 - 24x = 0$
17. $4x^2 - 5x - 9 = 0$	18. $6x^2 + x - 1 = 0$
19. $x^2 - 16 = 0$	20. $4x^3 - 36x = 0$

Solve the linear inequalities. Present the solutions set in graph and interval notations.

1. $2x < 4$	2. $x - 1 \leq 7$
3. $\frac{3x}{4} \geq 6$	4. $4(x + 1) - 4 > 8$
5. $2x - 6 < 4x + 2$	6. $-3x + 4 \leq 10$
7. $\frac{x}{2} \leq -\frac{x}{4} + 1$	8. $-x + 2.5 \geq 7.5$
9. $3(x + 2) - 3 < 4(x - 4) + x + 1$	10. $-(x - 5) + 2x \geq 4(x + 2) - 2x + 1$
11. $x - 6 < 4$ and $-2x \leq 4$	12. $3x + 5(x - 1) + 2 \leq 7(x + 2) + 5$
13. $\frac{x}{4} < \frac{x+1}{2} \leq -\frac{x-5}{4}$	14. $x + 0.5 \leq 1.5$ or $3(x + 1) > 9$

4.2. Absolute Value Equations and Inequalities

- **ABSOLUTE VALUE EQUATIONS**
- **ABSOLUTE VALUE INEQUALITIES**

■ ABSOLUTE VALUE EQUATIONS

We will use the following definitions.

> **Definition** ▶ *An equation with absolute value* is an equation $E_1 = E_2$ where E_1 or E_2 or both E_1 and E_2 are expressions with absolute value.
>
> **Definition** ▶ *A linear equation with absolute value* is an equation $E_1 = E_2$ where E_1 or E_2 or both E_1 and E_2 are linear expressions with absolute value.
>
> **Definition** ▶ *Absolute value linear equation with one variable* is an equation $E_1 = E_2$ where E_1 or E_2 or both E_1 and E_2 are single-variable linear absolute value expressions.

Below we study standard and nonstandard absolute value linear equations with one variable.

STANDARD LINEAR EQUATIONS WITH ABSOLUTE VALUE

We will solve different examples from the following six types of standard linear absolute value equations with one variable:

1) $|E(x)| = -k,$ k is a positive real number ($k \in R, \, k > 0$) and
2) $|E(x)| = 0,$ $E(x)$ is a linear expression with one variable
3) $|E(x)| = k,$
4) $|E(x)| = -|P(x)|,$ $P(x) \neq 0$ and $E(x)$ and $P(x)$ are linear expressions
5) $|E(x)| = |P(x)|,$
6) $|E(x)| = Q(x),$ $Q(x) \neq 0$ and $E(x)$ and $Q(x)$ are linear expressions

Examples $|E(x)| = -k,$ where k is a positive real number and $E(x)$ is a linear expression.

1. $|x - 4| = -2$

     $\boxed{\text{No solution}}$

2. $-|2x - 1| = 4$

     $-|2x - 1| \, / {-1} = 4 / {-1}$

     $|2x - 1| = -4$

     $\boxed{\text{No solution}}$

3. $6 + |1 - x| = 2$

   $-6 \qquad\quad -6$

     $|1 - x| = -4$

     $\boxed{\text{No solution}}$

Examples $|E(x)| = 0$, where $E(x)$ is a linear expression with one variable.

1. $|x - 5| = 0$
$x - 5 = 0$
$+5 \quad +5$
$\boxed{x = 5}$

2. $-|3x - 6| = 0$
$|3x - 6| = 0$
$3x = 6$
$3x/3 = 6/3$
$\boxed{x = 2}$

3. $7 + |1 - x| = 7$
$-7 \qquad -7$
$|1 - x| = 0$
$1 - x = 0$
$\boxed{x = 1}$

Examples $|E(x)| = k$, where k is a positive real number and $E(x)$ is a linear expression.

1. $|x - 5| = 2$
$x - 5 = -2 \quad \text{or} \quad x - 5 = 2$
$+5 \quad +5 \qquad\qquad +5 \quad +5$
$x = 3 \qquad\qquad\qquad x = 7$
$\boxed{x = 3, 7}$

2. $4 - |2x - 6| = 0$
$4 = 0 + |2x - 6|$
$|2x - 6| = 4$
$2x - 6 = -4 \qquad \text{or} \qquad 2x - 6 = 4$
$+6 \quad +6 \qquad\qquad\qquad +6 \quad +6$
$2x = 2 \qquad\qquad\qquad\qquad 2x = 10$
$2x/2 = 2/2 \qquad\qquad\qquad 2x/2 = 10/2$
$x = 1 \qquad\qquad\qquad\qquad x = 5$
$\boxed{x = 1, 5}$

3. $|x - 5| = 2$
$x - 5 = -2 \quad \text{or} \quad x - 5 = 2$
$+5 \quad +5 \qquad\qquad +5 \quad +5$
$x = 3 \qquad\qquad\qquad x = 7$
$\boxed{x = 3, 7}$

4. $4 - |2x - 6| = 0$
$4 = 0 + |2x - 6|$
$|2x - 6| = 4$
$2x - 6 = -4 \qquad \text{or} \qquad 2x - 6 = 4$
$+6 \quad +6 \qquad\qquad\qquad +6 \quad +6$
$2x = 2 \qquad\qquad\qquad\qquad 2x = 10$
$2x/2 = 2/2 \qquad\qquad\qquad 2x/2 = 10/2$
$x = 1 \qquad\qquad\qquad\qquad x = 5$
$\boxed{x = 1, 5}$

Examples $|E(x)| = -|P(x)|$, where $P(x) \neq 0$, $E(x)$ and $P(x)$ are linear expressions.

1. $|4x - 2| = -|3 - x|, \ x \neq 3$
$-|3 - x| < 0$
$\Downarrow$
$|4x - 2| = \text{negative number}$ False
$\Downarrow$
$\boxed{\textbf{No solution}}$

2. $|5x - 2| = -|x - 1|, \ x \neq 1$
$-|x - 1| < 0$
$\Downarrow$
$|5x - 2| = \ < 0 \ \text{negative number}$ False
$\Downarrow$
$\boxed{\textbf{No solution}}$

 $|E(x)| = |P(x)|$, where $P(x) \neq 0$, $E(x)$ and $P(x)$ are linear expressions.

1. $|x - 6| = |4 - x|$

$$x - 6 = 4 - x \quad \text{or} \quad x - 6 = -(4 - x)$$
$$x + 6 \;\; +6 + x \qquad \qquad x - 6 = -4 + x$$
$$2x \;\; = 10 \qquad\qquad -x \qquad\qquad -x$$
$$2x/2 \;\; = 10/2 \qquad\qquad -6 = -4 \quad \text{False}$$
$$x = 5 \qquad\qquad\qquad \textbf{no solution}$$

$$\boxed{x = 5}$$

2. $|2x - 8| = |x - 1|$

$$2x - 8 = x - 1 \quad \text{or} \quad 2x - 8 = -(x - 1)$$
$$-x + 8 \;\; -x + 8 \qquad\qquad 2x - 8 = -x + 1$$
$$x = 7 \qquad\qquad\qquad x + 8 \qquad x + 8$$
$$\qquad\qquad\qquad\qquad 3x/3 = 9/3$$
$$x = 7 \qquad\qquad\qquad x = 3$$

$$\boxed{x = 3, 7}$$

 $|E(x)| = Q(x)$, where $Q(x) \neq 0$, $E(x)$ and $Q(x)$ are linear expressions.

1. $|2x - 6| = x - 2$ We have restriction on x: $x - 2 \geq 0$ or $x \geq 2$

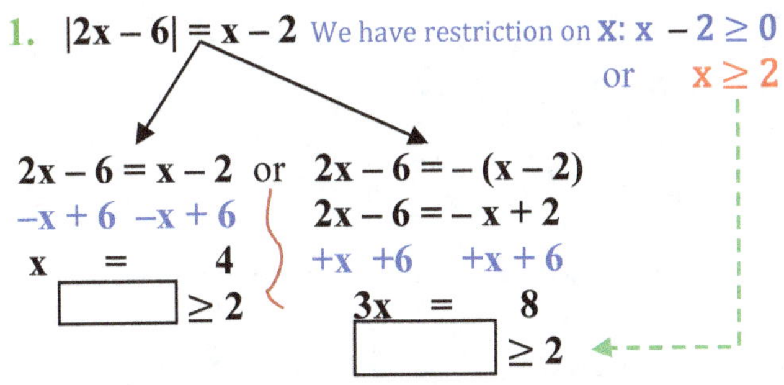

$$2x - 6 = x - 2 \quad \text{or} \quad 2x - 6 = -(x - 2)$$
$$-x + 6 \;\; -x + 6 \qquad\qquad 2x - 6 = -x + 2$$
$$x = 4 \qquad\qquad\qquad +x \;\; +6 \quad +x + 6$$
$$\boxed{} \geq 2 \qquad\qquad 3x = 8$$
$$\boxed{} \geq 2$$

Both $x = 4$ and $x = 8/3$ are the solutions because both are greater than 2 ($4 \geq 2$ and $8/3 \geq 2$).

2. $|3x - 2| = 2x - 2 \rightarrow 2x - 2 \geq 0 \rightarrow x \geq 1$

$$3x - 2 = 2x - 2 \quad \text{or} \quad 3x - 2 = -(2x - 2)$$
$$\qquad\qquad\qquad\qquad 3x - 2 = -2x + 2$$
$$x = 0 \qquad\qquad 2x + 2 \qquad 2x + 2$$
$$\qquad\qquad\qquad\qquad 5x = 4$$
$$x = 0 < 1 \qquad\qquad x = 4/5 < 1$$

$$\boxed{\text{No solution}}$$

Both $x = 0$ and $x = 4/5$ are not the solutions because both are less than 1 ($0 < 1$ and $4/5 < 1$).

3. $|x - 6| = -x \;\; \rightarrow -x \geq 0 \rightarrow x \leq 0$

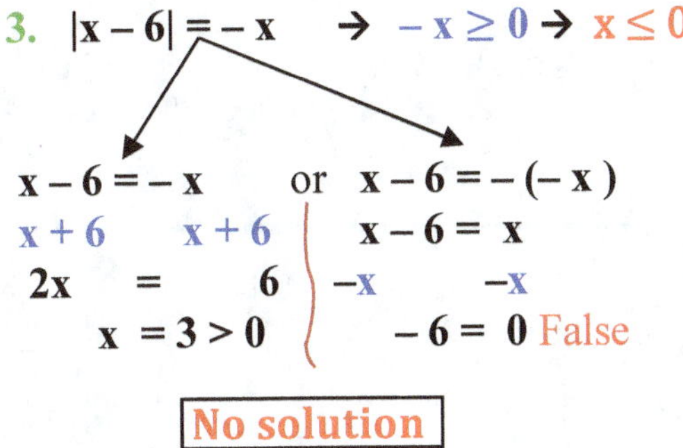

$$x - 6 = -x \quad \text{or} \quad x - 6 = -(-x)$$
$$x + 6 \quad x + 6 \qquad\qquad x - 6 = x$$
$$2x = 6 \qquad\qquad -x \qquad -x$$
$$x = 3 > 0 \qquad\qquad -6 = 0 \quad \text{False}$$

$$\boxed{\text{No solution}}$$

$x = 3$ is not a solution because $3 > 0$.

4. $|2x - 2| = x + 1 \;\; \rightarrow x + 1 \geq 0 \rightarrow x \geq -1$

$$2x - 2 = x + 1 \quad \text{or} \quad 2x - 2 = -x - 1$$
$$\qquad\qquad\qquad\qquad 3x = 1$$
$$x = 3 \qquad\qquad\qquad x = 1/3$$
$$x = 3 \geq -1 \qquad\qquad x = 1/3 \geq -1$$

$$\boxed{x = 3, \;\; x = 1/3}$$

Both $x = 3$ and $x = 1/3$ are the solutions because both a greater than -1 ($3 \geq -1$ and $1/3 \geq -1$).

NONSTANDARD LINEAR EQUATIONS WITH ABSOLUTE VALUE

We will also study several kinds of nonstandard linear equations with absolute value.

Example 1 Solve the linear equation $|x + 3| + |x - 2| = 4$.

SOLUTION▶ We will use the definition of absolute value: $|x| = \begin{cases} x & \text{if } x \geq 0 \\ -x & \text{if } x < 0 \end{cases}$

Zero-point for term $|x + 3|$ is $x = -3$ and zero-point for term $|x - 2|$ is $x = 2$: $|x + 3| + |x - 2| = 4$.
We will divide the interval $(-\infty; \infty)$ of possible values of x by those zero-points into 3 intervals:

1) $x < -3$ (or $x \in (-\infty; -3)$); 2) $-3 \leq x < 2$ (or $x \in [-3; 2)$); 3) $x \geq 2$ (or $x \in [2, \infty)$).

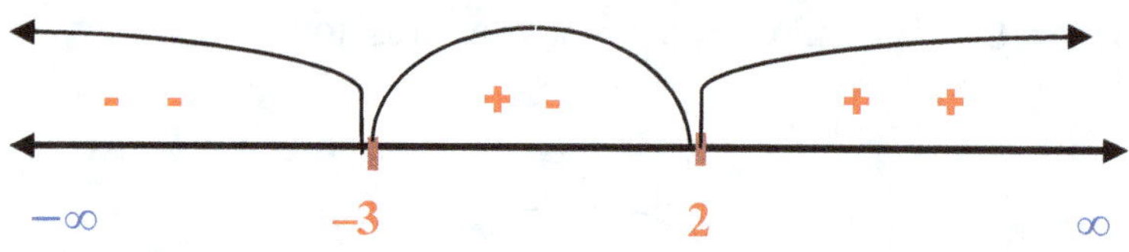

1) Let $x < -3$ (or $x \in (-\infty; -3)$)

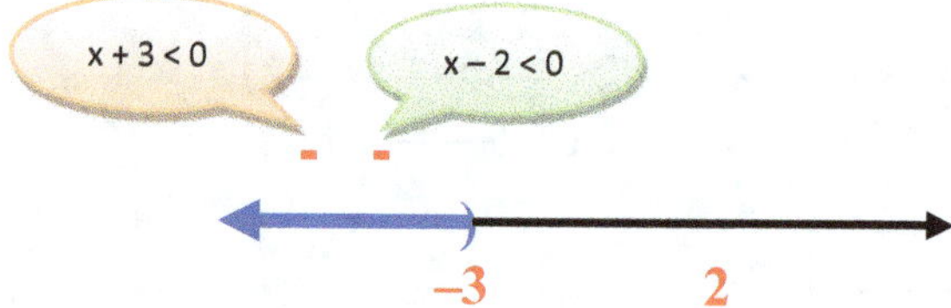

From $x < -3 \rightarrow x + 3 < 0$ and $x - 2 < 0$

So, from the definition of absolute value we can write:

$$|x + 3| + |x - 2| = 4$$
$$-(x + 3) - (x - 2) = 4$$
$$-x - 3 - x + 2 = 4$$
$$-2x = 5$$
$$x = -5/2 = -2.5 \longrightarrow \text{No Solution} \quad (-2.5 > -3, \text{ so, } -2.5 \text{ is not in the interval } x < -3)$$

2) Let $-3 \leq x < 2$ (or $x \in [-3; 2)$)

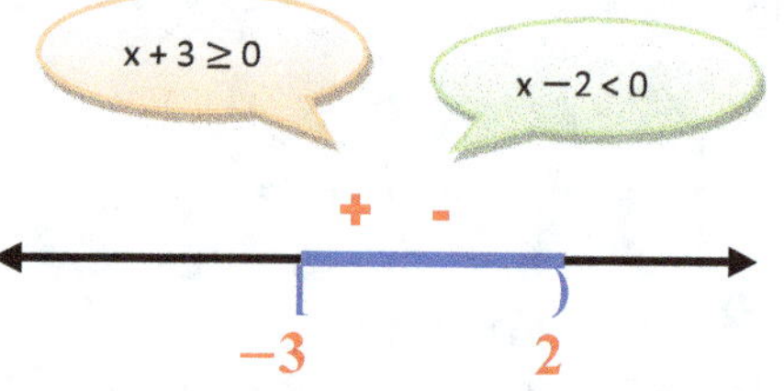

From $-3 \leq x < 2 \rightarrow x + 3 \geq 0$ and $x - 2 < 0$

So, from the definition of absolute value we can write:

$$|x + 3| + |x - 2| = 4$$
$$(x + 3) - (x - 2) = 4$$
$$x + 3 - x + 2 = 4$$
$$5 = 4 \longrightarrow \text{No Solution} \quad (5 = 4 \text{ is a false statement, so we have no solution in this interval}).$$

3) Let $x \geq 2$ (or $x \in [2, \infty)$)

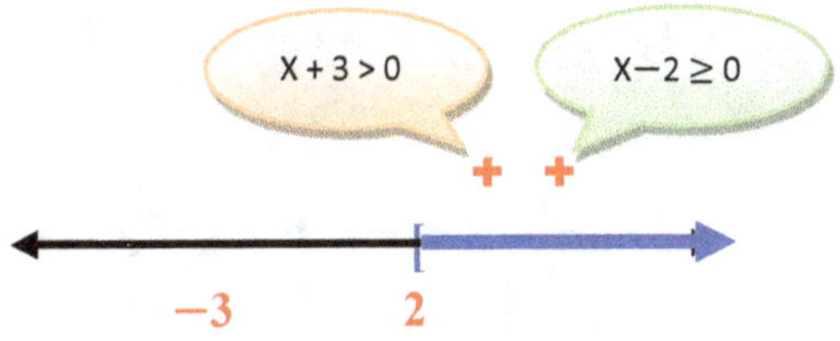

From $x \geq 2 \rightarrow x + 3 > 0$ and $x - 2 \geq 0$

So, from the definition of absolute value we can write:

$$|x + 3| + |x - 2| = 4$$
$$(x + 3) + (x - 2) = 4$$
$$2x + 1 = 4$$
$$x = 3/2 = 1.5 \longrightarrow \text{No Solution} \quad (1.5 < 2, \text{ so, } 1.5 \text{ is not in the interval } x \geq 2).$$

Answer: If we have no solution in these 3 intervals then there's no solution in the real number set.

Example 2   Solve the linear equation $|x + 3| + |x - 2| = 6$.

SOLUTION ► We will use the definition of absolute value.
Zero-point for term $|x + 3|$ is $x = -3$ and zero-point for term $|x - 2|$ is $x = 2$: $|x + 3| + |x - 2| = 4$.
We will divide the whole interval $(-\infty; \infty)$ of possible values of x by those zero-points into three intervals:

1) $x < -3$ (or $x \in (-\infty; -3)$);   **2)** $-3 \leq x < 2$ (or $x \in [-3; 2)$);   **3)** $x \geq 2$ (or $x \in [2, \infty)$).

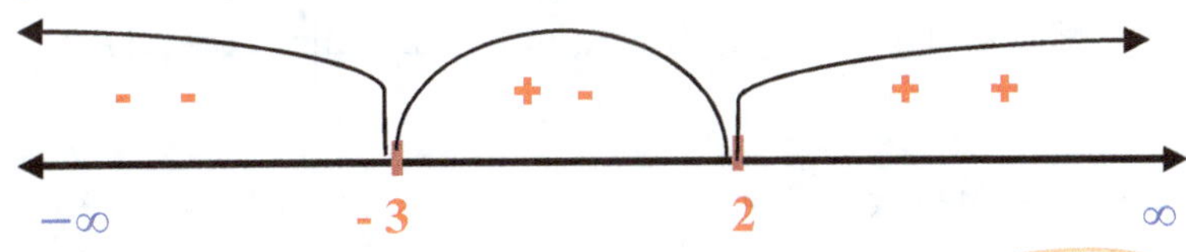

1) Let $x < -3$ (or $x \in (-\infty; -3)$).

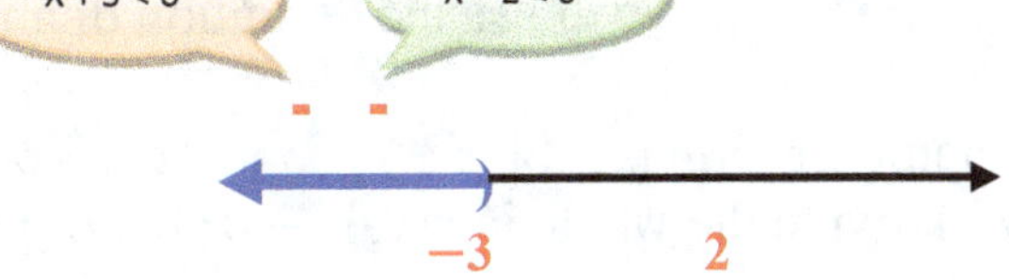

From $x < -3 \rightarrow x + 3 < 0$ and $x - 2 < 0$

So, from the definition of absolute value we can write:
$$|x + 3| + |x - 2| = 6$$
$$-(x + 3) - (x - 2) = 6$$
$$-x - 3 - x + 2 = 6$$
$$-2x - 1 = 6$$
$$+1 \quad +1$$
$$-2x = 7$$
$$-2x/-2 = 7/-2$$
$$x = -7/2 = -3.5 \longrightarrow x = -3.5 \text{ is a solution} \quad (-3.5 < -3)$$

2) Let $-3 \leq x < 2$ (or $x \in [-3; 2)$).

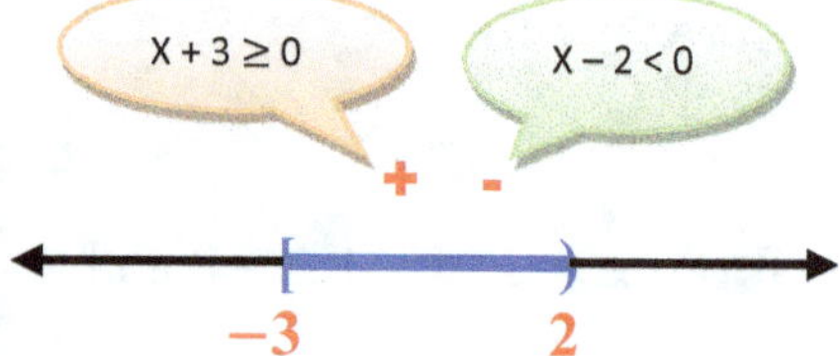

From $-3 \leq x < 2$ → $x + 3 \geq 0$ and $x - 2 < 0$

So, from the definition of absolute value we can write:
$$|x + 3| + |x - 2| = 6$$
$$(x + 3) - (x - 2) = 6$$
$$x + 3 - x + 2 = 6$$
$$5 = 6 \longrightarrow \text{No Solution} \quad (5 = 4 \text{ is a false statement, so we do not have any solution in this interval).}$$

3) Let $x \geq 2$ (or $x \in [2, \infty)$);

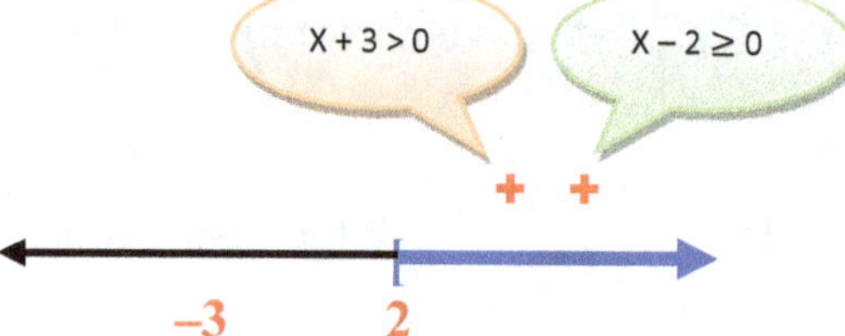

From $x \geq 2$ → $x + 3 > 0$ and $x - 2 \geq 0$

So, from the definition of absolute value we can write:
$$|x + 3| + |x - 2| = 6$$
$$(x + 3) + (x - 2) = 6$$
$$2x + 1 = 6$$
$$x = 5/2 = 2.5 \longrightarrow x = 2.5 \quad (2.5 > 2, \text{ so } 2.5 \text{ is in the interval } x \geq 2).$$

Answer: We have only two solutions $x = -3.5$ and $x = 2.5$ for the equation $|x + 3| + |x - 2| = 6$.

Example 3 Solve the linear equation $|x + 3| + |x - 2| = 5$.

SOLUTION ► We will use the definition of absolute value.

Zero-point for term $|x + 3|$ is $x = -3$ and zero-point for term $|x - 2|$ is $x = 2$: $|x + 3| + |x - 2| = 5$.
We will divide the whole interval $(-\infty; \infty)$ of possible values of x by those zero-points into three intervals:

1) $x < -3$ (or $x \in (-\infty; -3)$); 2) $-3 \leq x < 2$ (or $x \in [-3; 2)$); 3) $x \geq 2$ (or $x \in [2, \infty)$).

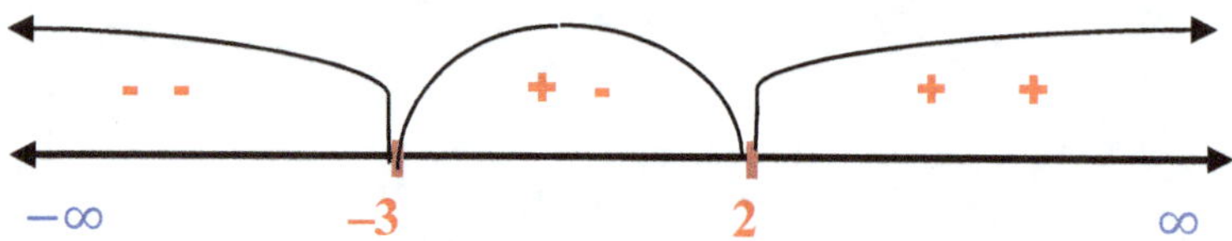

1) Let $x < -3$ (or $x \in (-\infty; -3)$)

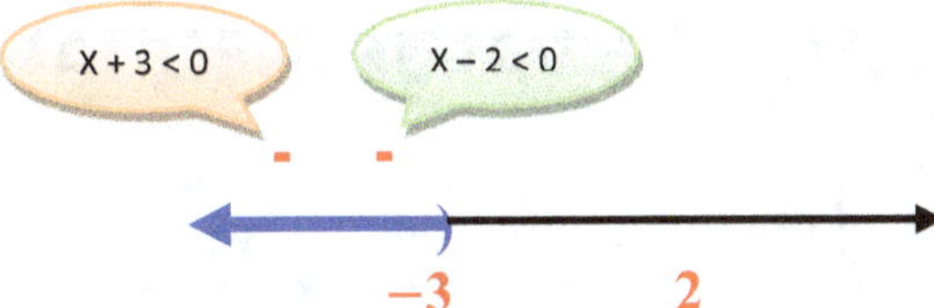

From $x < -3 \rightarrow x + 3 < 0$ and $x - 2 < 0$

So, from the definition of absolute value we can write:
$$|x + 3| + |x - 2| = 5$$
$$-(x + 3) - (x - 2) = 5$$
$$-x - 3 - x + 2 = 5$$
$$-2x - 1 = 5$$
$$+1 \quad +1$$
$$-2x = 6$$
$$-2x/-2 = 6/-2$$
$$x = -3 \rightarrow \quad \text{No Solution} \quad (-3 \text{ is not in the interval } x < -3)$$

2) Let $-3 \le x < 2$ (or $x \in [-3; 2)$).

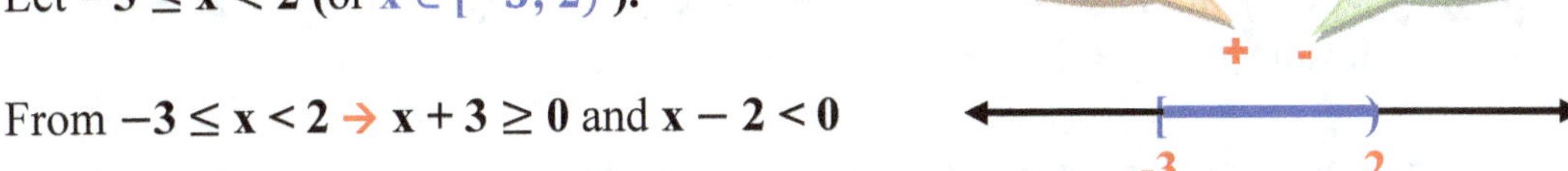

From $-3 \le x < 2 \rightarrow x + 3 \ge 0$ and $x - 2 < 0$

So, from the definition of absolute value we can write:
$$|x + 3| + |x - 2| = 5$$

$$(x + 3) - (x - 2) = 5$$
$$x + 3 - x + 2 = 5$$
$$5 = 5 \rightarrow \quad \text{all } x \in [-3, 2) \quad (5 = 5 \text{ is an identity, so any } x \text{ from this interval is a solution})$$

3) Let $x \ge 2$ (or $x \in [2, \infty)$);

$x \ge 2 \rightarrow x + 3 > 0$ and $x - 2 \ge 0$

So, from the definition of absolute value we can write:

$$|x + 3| + |x - 2| = 5$$
$$(x + 3) + (x - 2) = 5$$
$$2x + 1 = 5$$
$$x = 4/2 = 2 \rightarrow \quad \boxed{x = 2} \quad (2 = 2, \text{ so, } 2 \text{ is in the interval } x \ge 2).$$

Answer: We have infinitely many solutions from the interval $[-3, 2]$ or $x \in [-3, 2]$.

■ ABSOLUTE VALUE INEQUALITY

We will use the following definitions.

> **Definition** ▶ *An absolute value inequality* is an inequality $E_1 < E_2$ (or $E_1 > E_2$, or $E_1 \leq E_2$, or $E_1 \geq E_2$) where E_1 or E_2 or both E_1 and E_2 are expressions with absolute value.
>
> **Definition** ▶ *A linear absolute value inequality* is an inequality $E_1 < E_2$ (or $E_1 > E_2$, or $E_1 \leq E_2$, or $E_1 \geq E_2$) where E_1 or E_2 or both E_1 and E_2 are linear expressions with absolute value.
>
> **Definition** ▶ *Absolute value linear inequality with one variable* is an inequality $E_1 < E_2$ (or $E_1 > E_2$, or $E_1 \leq E_2$, or $E_1 \geq E_2$) where E_1 or E_2 or both E_1 and E_2 are linear expressions with absolute value and with one variable.

Examples The equations $|x + 2| < 3$, $|2x - 1| > x$, $|x - 2y| \leq x - 2$, $|x + 2| + 1 \geq |x - 3|$ are absolute value linear inequalities with one or two variables.

Below we will solve several types of standard and nonstandard absolute value linear inequalities with one variable.

STANDARD LINEAR INEQUALITIES WITH ABSOLUTE VALUE

We will study four standard types of linear inequalities with absolute value.

1) $|E(x)| < \quad k, \quad k > 0, k \in R$ $\qquad$ (or $|E(x)| \leq \quad k$)
2) $|E(x)| > \quad k, \quad k > 0, k \in R$ $\qquad$ (or $|E(x)| \geq \quad k$)
3) $|E(x)| < -k, \quad k > 0, k \in R$ $\qquad$ (or $|E(x)| \leq -k$)
4) $|E(x)| > -k, \quad k > 0, k \in R$ $\qquad$ (or $|E(x)| \geq -k$)

Examples Solve each linear absolute value inequality with one variable.

1. $|x - 5| < 3$
$-3 < x - 5 < 3$
$+5 \qquad +5 \quad +5$
$2 < x < 8$

$2 \qquad 8$

$x \in (2, 8)$

2. $|2x - 3| \leq 5$
$-5 \leq 2x - 3 \leq 5$
$+3 \qquad +3 \quad +3$
$-2 \leq 2x \leq 8$
$-2/2 \leq 2x/2 \leq 8/2$
$-1 \leq x \leq 4$

$-1 \qquad 4$

$x \in (-1, 4)$

3. $\qquad |2 - x| < 4$
$|2 - x| < 4$
$-4 < 2 - x < 4$
$-6 < -x < 2$
$-6/-1 > -x/-1 > 2/-1$
$-2 < x < 6$

$0 \qquad 4$

$x \in (0, 4)$

 Solve each linear absolute value inequality with one variable.

1. $|x - 4| > 1$

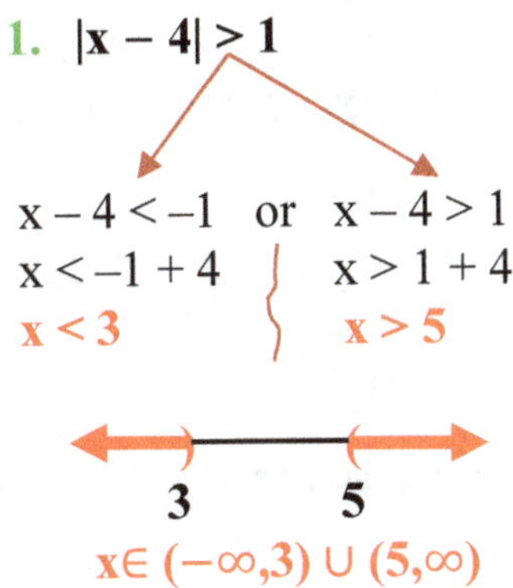

$x - 4 < -1$ or $x - 4 > 1$
$x < -1 + 4$ $x > 1 + 4$
$x < 3$ $x > 5$

$x \in (-\infty, 3) \cup (5, \infty)$

2. $|3x - 1| \geq 8$

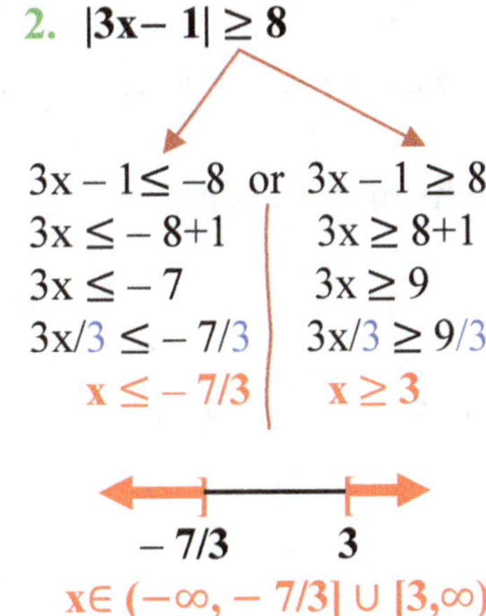

$3x - 1 \leq -8$ or $3x - 1 \geq 8$
$3x \leq -8 + 1$ $3x \geq 8 + 1$
$3x \leq -7$ $3x \geq 9$
$3x/3 \leq -7/3$ $3x/3 \geq 9/3$
$x \leq -7/3$ $x \geq 3$

$x \in (-\infty, -7/3] \cup [3, \infty)$

 Solve each linear absolute value inequality with one variable.

1. $|x - 1| < -1$
$\Downarrow$
No solution or $x \in \emptyset$

2. $|2x - 3| + 3 < 2$
$-3 \quad -3$
$|2x - 3| < -1$
$\Downarrow$
No solution or $x \in \emptyset$

3. $1 - |2x| \geq 5$
$-1 \qquad -1$
$-|2x| \geq 4$
$-|2x|/-1 \geq 4/-1$
$|2x| \leq -4$
$\Downarrow$
No solution or $x \in \emptyset$

All inequalities above have no solutions, because absolute value of any expression never can be less than a negative number for any $x \in \mathbf{R}$.

 Solve each linear absolute value inequality with one variable.

1. $|x + 1| > -1$
$\Downarrow$
All real numbers or $x \in \mathbf{R}$

$x \in (-\infty, \infty)$

2. $|x - 3| + 4 \geq 1$
$-4 \quad -4$
$|x - 3| \geq -3$
$\Downarrow$
All real numbers or $x \in \mathbf{R}$

$x \in (-\infty, \infty)$

3. $-|2x| \leq 4$
$-|2x|/-1 \leq 4/-1$
$|2x| \geq -4$
$\Downarrow$
All real numbers or $x \in \mathbf{R}$

$x \in (-\infty, \infty)$

All inequalities above have infinite number of solutions, because absolute value of any expression always is greater than a negative number for any $x \in \mathbf{R}$.

NONSTANDARD LINEAR INEQUALITIES WITH ABSOLUTE VALUE

We will consider some examples of nonstandard linear inequalities with absolute value.

Example 1 Solve the linear inequality $|2x - 6| \leq 0$.

SOLUTION ▶ $|2x - 6| \leq 0$

$\Downarrow$

$|2x - 6| = 0$ ⟵ Absolute value of an expression is always nonnegative number.

$\Downarrow$

$2x - 6 = 0$

$+6 \quad +6$

$2x = 6$

$2x/2 = 6/2$

$x = 3$ ⟶ Inequality has a unique solution.

Example 2 Solve the linear inequality $|3x - 5| \geq 0$.

SOLUTION ▶ $|3x - 5| \geq 0$

$\Downarrow$

All real numbers; ⟷ ; $x \in (-\infty, \infty)$ Absolute value of an expression is always nonnegative number.

Example 3 Solve the linear inequality $|2x - 4| \leq |x + 2|$

SOLUTION ▶ We will use the definition of absolute value.
Zero-point for term $|2x - 4|$ is $x = 2$ and zero-point for term $|x + 2|$ is $x = -2$.
We will divide the interval $(-\infty; \infty)$ of possible values of x by those zero-points into **3** intervals:

1) $x < -2$ (or $x \in (-\infty; -2)$); 2) $-2 \leq x < 2$ (or $x \in [-2; 2)$); 3) $x \geq 2$ (or $x \in [2, \infty)$).

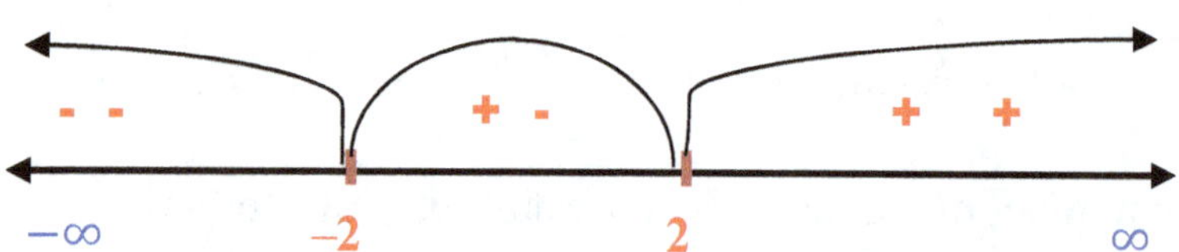

1) Let **x < −2** (or **x ∈ (−∞ ; −2)**).

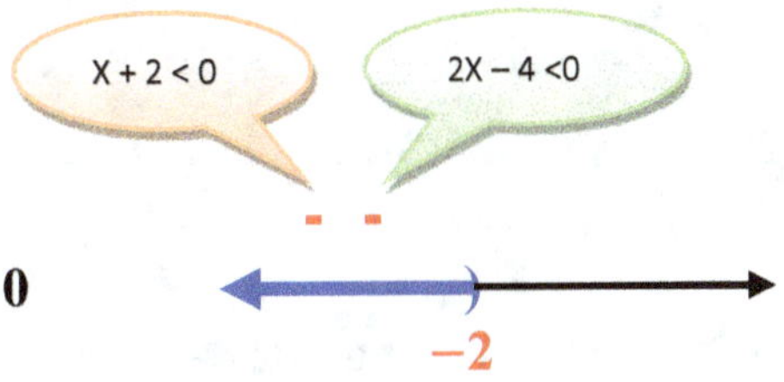

From **x < − 2 → x + 2 < 0** and **2x − 4 < 0**

So, from the definition of absolute value we can write:

$$|2x − 4| \le |x+2|$$
$$\Updownarrow$$

$$− (2x − 4) \le − (x + 2)$$
$$−2x + 4 \le − x − 2$$
$$+ x − 4 \qquad + x − 4$$
$$−x \le − 6$$
$$−x/−1 \ge −6/−1 \quad \longleftarrow \text{Division by negative number will change the direction of inequality sign.}$$
$$x \ge 6$$
$$\Downarrow$$

No solution   ⟵ $\{x/x < −2\} \cap \{x/x \ge 6\} = \emptyset$

2) Let **− 2 ≤ x < 2** (or **x ∈ [−2; 2)**).

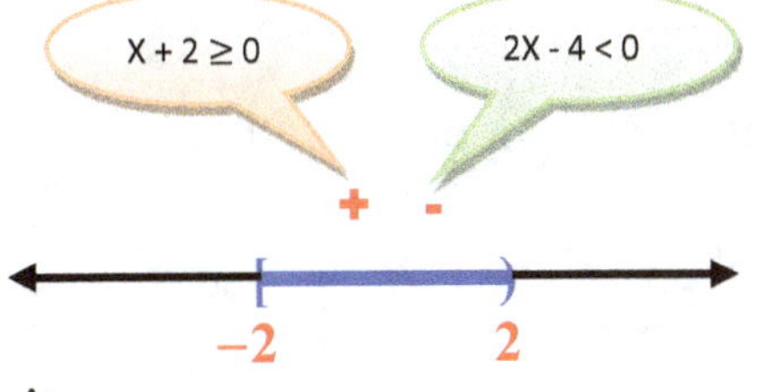

From **−2 ≤ x < 2 → x + 2 ≥ 0** and **2x − 4 < 0**

So, from the definition of absolute value we can write:

$$|2x − 4| \le |x + 2|$$
$$\Updownarrow$$

$$− (2x − 4) \le x + 2$$
$$− 2x + 4 \le x + 2$$
$$−x \;\; −4 \quad −x \; −4$$
$$−3x \le −2$$
$$− 3x/−3 \ge − 2/− 3$$
$$x \ge 2/3$$
$$\Downarrow$$

2/3 ≤ x < 2   ⟵ $\{x / −2 \le x < 2\} \cap \{x / x \ge 2/3\} = \{x / 2/3 \le x < 2\}$

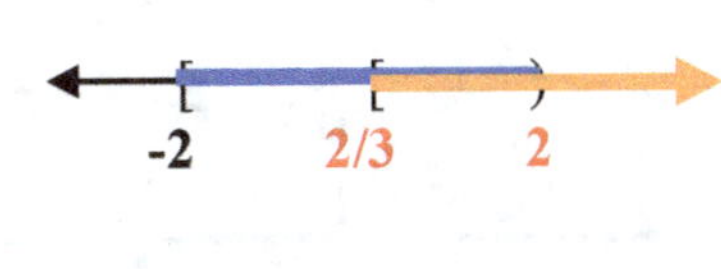

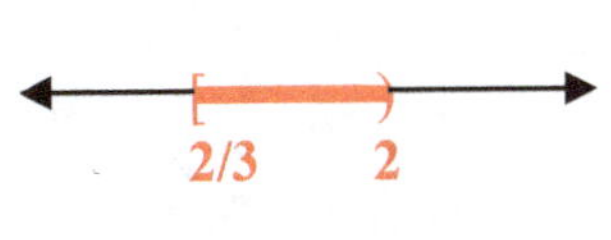

x ∈ [2/3, 2)

3) Let $x \geq 2$ (or $x \in [2, \infty)$);

From $x \geq 2 \rightarrow x + 2 > 0$ and $2x - 4 \geq 0$

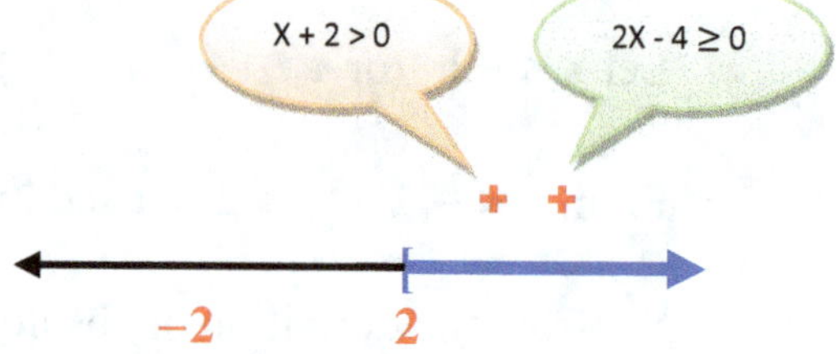

So, from the definition of absolute value we can write:

$$|2x - 4| \leq |x + 2|$$
$$\Updownarrow$$
$$2x - 4 \leq x + 2$$
$$-x + 4 \quad -x + 4$$
$$x \leq 6$$
$$\Downarrow$$
$$2 \leq x \leq 6 \quad \longleftarrow \quad \{x \,/\, 2 \leq x\} \cap \{x \,/\, x \leq 6\} = \{x \,/\, 2 \leq x \leq 6\}$$

$$x \in [2, 6]$$

So, the total solution set will be the union of two sub solution sets: $x \in [2/3, 2) \cup [2, 6] = [2/3, 6]$.

Answer: $2/3 \leq x \leq 6$; $\longleftarrow$; $x \in [2/3, 6]$

Example 4 Solve the linear inequality $|x + 3| + |x - 2| < 4$.

SOLUTION ▶ We will use the definition of absolute value.

Zero-point for term $|x + 3|$ is $x = -3$ and zero-point for term $|x - 2|$ is $x = 2$: $|x + 3| + |x - 2| < 4$. We will divide the interval $(-\infty; \infty)$ of possible values of x by those zero-points into 3 intervals:

1) $x < -3$ (or $x \in (-\infty; -3)$); **2)** $-3 \leq x < 2$ (or $x \in [-3; 2)$); **3)** $x \geq 2$ (or $x \in [2, \infty)$).

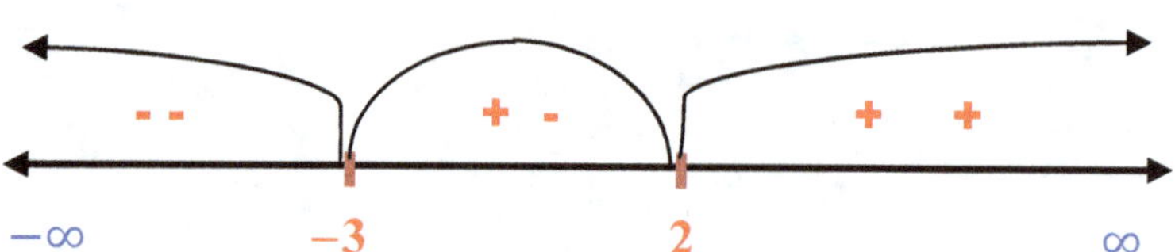

1) Let $x < -3$ (or $x \in (-\infty\,;\,-3)$).

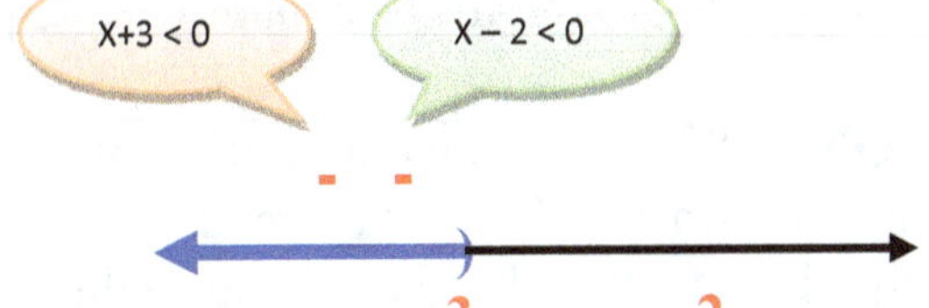

From $x < -3 \rightarrow x + 3 < 0$ and $x - 2 < 0$

So, from the definition of absolute value we can write:
$$|x + 3| + |x - 2| < 4$$
$$\Updownarrow$$
$$-(x + 3) - (x - 2) < 4$$
$$-x - 3 \; -x + 2 < 4$$
$$-2x - 1 < 4$$
$$+1 \; +1$$
$$-2x < 5$$

$-2x/-2 > 5/-2$ ← Division by negative number will change the direction of inequality sign.

$x > -2.5 \rightarrow$ **No Solution** ← $(-\infty; -3) \cap (2.5, \infty) = \emptyset$

2) Let $-3 \le x < 2$ (or $x \in [-3; 2)$).

From $-3 \le x < 2 \rightarrow x + 3 \ge 0$ and $x - 2 < 0$.

From the definition of absolute value we can write:
$$|x + 3| + |x - 2| < 4$$
$$\Updownarrow$$
$$(x + 3) - (x - 2) < 4$$
$$x + 3 - x + 2 < 4$$
$$5 < 4 \rightarrow \textbf{No Solution}$$
← $5 < 4$ is a false statement.

3) Let $x \ge 2$ (or $x \in [2, \infty)$);

From $x \ge 2 \rightarrow x + 3 > 0$ and $x - 2 \ge 0$

So, from the definition of absolute value we can write:
$$|x + 3| + |x - 2| < 4$$
$$\Updownarrow$$
$$(x + 3) + (x - 2) < 4$$
$$2x + 1 < 4$$
$$x < 3/2 = 1.5$$
$$x < 1.5 \rightarrow \textbf{No Solution} \quad \text{(because } (-\infty; 1.5) \cap [2, \infty) = \emptyset)$$

Answer: $x \in \emptyset$. We have **no solution** in all real number set if we have no solutions in all 3 intervals.

 Solve the linear equation $|x + 3| + |x - 2| \geq 4$.

SOLUTION► We will use the definition of absolute value.
Zero-point for term $|x + 3|$ is $x = -3$ and zero-point for term $|x - 2|$ is $x = 2$: $|x + 3| + |x - 2| < 4$.
We will divide the interval $(-\infty; \infty)$ of possible values of x by those zero-points into 3 intervals:

1) $x < -3$ (or $x \in (-\infty; -3)$); 2) $-3 \leq x < 2$ (or $x \in [-3; 2)$); 3) $x \geq 2$ (or $x \in [2, \infty))$.

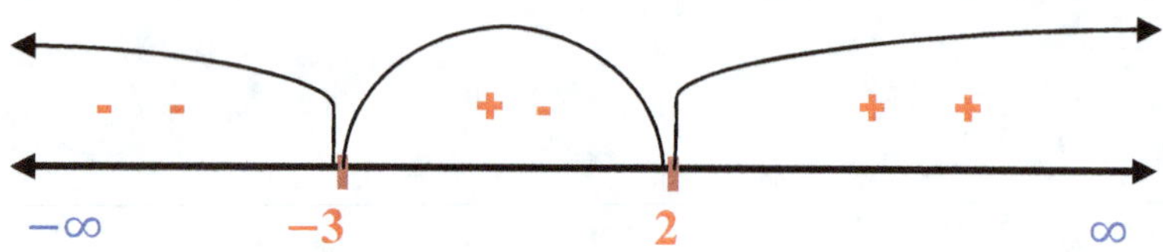

1) Let $x < -3$ (or $x \in (-\infty; -3)$).

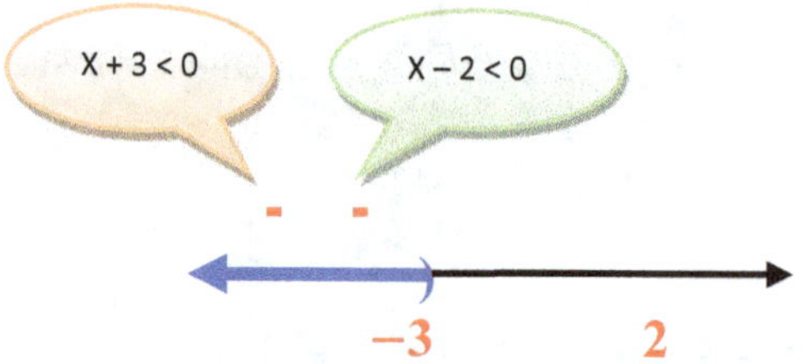

From $x < -3 \rightarrow x + 3 < 0$ and $x - 2 < 0$

From the definition of absolute value we can write:

$$|x + 3| + |x - 2| \geq 4$$
$$\Updownarrow$$
$$-(x + 3) - (x - 2) \geq 4$$
$$-x - 3 - x + 2 \geq 4$$
$$-2x - 1 \geq 4$$

$$-2x \geq 5$$
$$-2x/-2 \leq 5/-2 \leftarrow \text{Division by negative number will change the direction of the inequality sign.}$$
$$x \leq -5/2 = -2.5$$
$$x \leq -2.5$$
$$\Downarrow$$

$x < -3 \longleftarrow \{x/x \leq -2.5\} \cap \{x/x < -3\} = \{x/x < -3\}$ or interval $(-\infty; -3))$

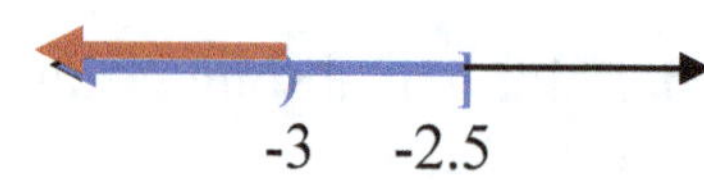

$;\ x \in (-\infty; -3)$
-3

2) Let $-3 \leq x < 2$ (or $x \in [-3; 2)$).

From $-3 \leq x < 2 \rightarrow x + 3 \geq 0$ and $x - 2 < 0$

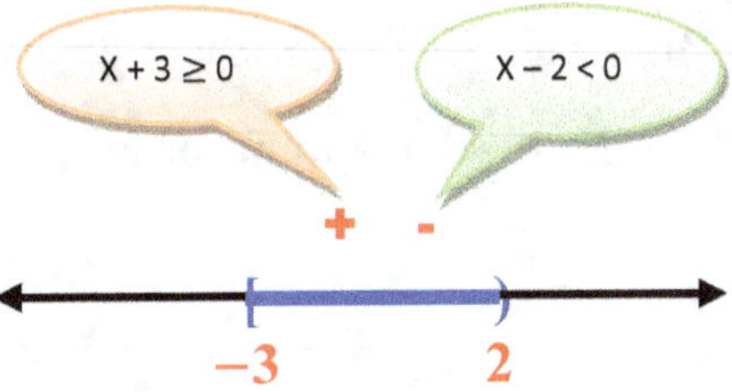

So, from the definition of absolute value we can write:
$|x + 3| + |x - 2| \geq 4$
⇕
$(x + 3) - (x - 2) \geq 4$
$x + 3 - x + 2 \geq 4$
$5 \geq 4$ True statement. So, all the numbers from set $-3 \leq x < 2$ are solutions.
⇓
$-3 \leq x < 2$

; $x \in [-3; 2)$

3) Let $x \geq 2$ (or $x \in [2, \infty)$);

From $x \geq 2 \rightarrow x + 3 > 0$ and $x - 2 \geq 0$

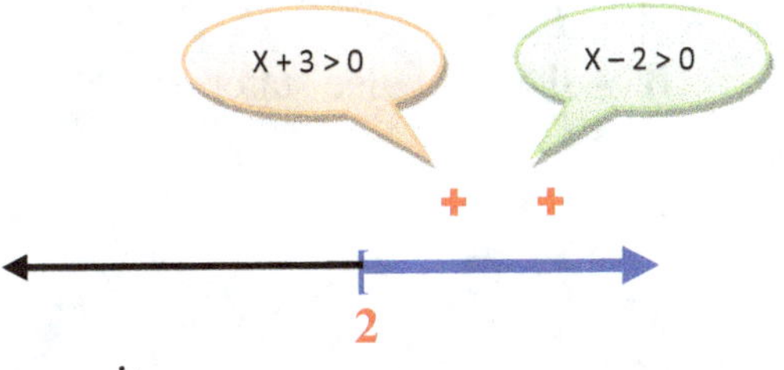

So, from the definition of absolute value we can write:

$|x + 3| + |x - 2| \geq 4$
⇕
$(x + 3) + (x - 2) \geq 4$
$2x + 1 \geq 4$
$x \geq 3/2 = 1.5$
⇓

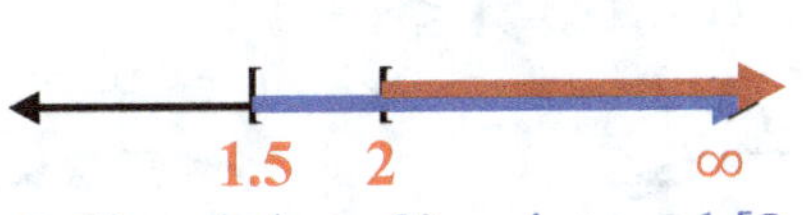

$x \geq 2$ (because $\{x/x \geq 1.5\} \cap \{x/x \geq 2\} = \{x/x \geq 2\}$ or interval $[2, \infty)$)

2

$x \in [2; \infty)$

So, solution set is **all real number set**, because $x \in (-\infty; -3) \cup [-3; 2) \cup [2; \infty) = R$

; $x \in (-\infty; \infty)$

- ∞   ∞

 Solve the linear inequality $|2x - 4| \leq x + 2$

SOLUTION ▶ Zero-point for term $x + 2$ is $x = -2$.

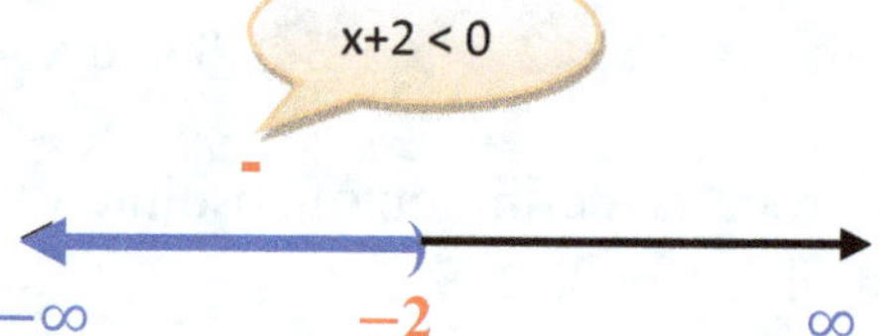

1) Let $x < -2$ ($x \in (-\infty; -2)$) or $x + 2 < 0$.

$$|2x - 4| \leq x + 2 < 0$$
$$|2x - 4| < 0 \quad \longleftarrow \quad \text{Absolute value of an expression always is nonnegative number.}$$
$$\Downarrow$$

No solution

2) Let $x = -2$.

$$|2 \cdot (-2) - 4| \leq -2 + 2$$
$$|-8| \leq 0$$
$$8 \leq 0 \quad \longleftarrow \quad \text{False statement}$$
$$\Downarrow$$

No solution

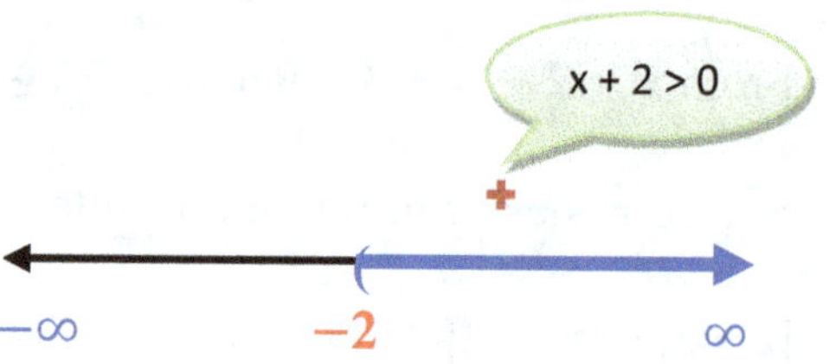

3) Let $x > -2$ ($x \in (-2; \infty)$) or $x + 2 > 0$.

$$|2x - 4| \leq x + 2$$
$$-(x + 2) \leq 2x - 4 \leq x + 2$$
$$-(x + 2) \leq 2x - 4 \quad \cap \quad 2x - 4 \leq x + 2$$
$$-x - 2 \leq 2x - 4 \quad \cap \quad 2x - 4 \leq x + 2$$

$$2 \leq 3x \quad \cap \quad x \leq 6$$
$$2/3 \leq x \quad \cap \quad x \leq 6$$

$$\Downarrow$$

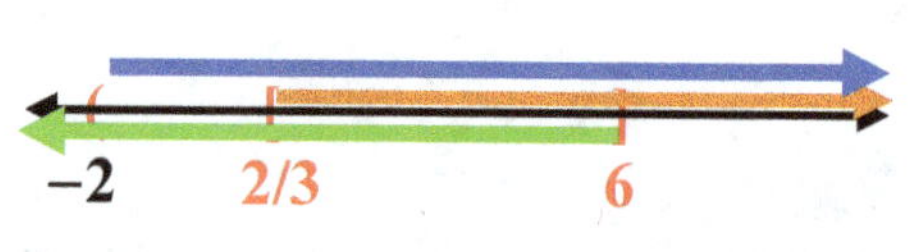

Answer: $2/3 \leq x \leq 6$; $\longleftarrow$ ⊢━━━⊣ $\longrightarrow$; $x \in [\,2/3, 6\,]$; $\{x \,/\, 2/3 \leq x \leq 6\}$
2/3   6

Solve the absolute value linear equations with one variable.

1. $\lvert x + 4 \rvert = 3$	2. $\lvert 2x - 1 \rvert + 2 = 1$
3. $\lvert -x \rvert = -5$	4. $\lvert 6 - 2x \rvert = 6$
5. $2\lvert x + 2 \rvert + 1 = 5$	6. $4 - \lvert x - 2 \rvert = 3$
7. $\lvert \lvert x + 1 \rvert + 2 \rvert = 3$	8. $\lvert x + 2 \rvert = x$
9. $\lvert \lvert x + 4 \rvert = 3 \lvert x + 1 \rvert$	10. $\lvert x \rvert + \lvert x - 1 \rvert = 1$
11. $\lvert 4x - 1 \rvert = -3 + 6x$	12. At what values of **c** equation $\lvert \lvert x - 2 \rvert + c \rvert = 3$ has a unique solution? Find the unique solution.

Solve the absolute value linear inequalities with one variable.

1. $\lvert 2x - 1 \rvert - x < 4$	2. $\lvert x + 1 \rvert + \lvert x \rvert \geq 1$
3. $\lvert 2x - 1 \rvert < 4$	4. $\lvert x + 1 \rvert \geq 1$
5. $\lvert x - 1 \rvert + 5 \leq 3$	6. $2\lvert x - 1 \rvert \leq 6$
7. $\lvert 2x - 1 \rvert - x < 4$	8. $\lvert x + 1 \rvert + \lvert x \rvert \geq 1$
9. $\lvert \lvert x - 1 \rvert - 5 \rvert \leq 3$	10. At what values of **c** does the equation $\lvert \lvert x - 2 \rvert + c \rvert \leq 3$ have a unique solution? Find the unique solution.

4.3. Quadratic Equations and Inequalities

■ QUADRATIC EQUATIONS
■ QUADRATIC INEQUALITIES

■ QUADRATIC EQUATIONS

We will use the following definitions.

Definition ▶ *A quadratic equation* is an equation $E_1 = E_2$ where E_1 or E_2 or both E_1 and E_2 are quadratic expressions with one or more variables.

Definition ▶ *A quadratic equation with one variable* is an equation $E_1 = E_2$ where E_1 or E_2 or both E_1 and E_2 are quadratic expressions with one variable.

Definition ▶ *A quadratic equation with one variable* is any equation that can be presented in the form $ax^2 + bx + c = 0,$ where $a, b, c \in \mathbf{R}$ and $a \neq 0.$
This is a **standard form** of a general quadratic equations with one variable, where **x** represents a variable, and the constants **a**, **b**, and **c** are respectively called the quadratic (or leading) coefficient, the linear (or middle) coefficient and the free constant term.

Examples The equations $x^2 - 81 = 0$, $y^2 = 4x$, and $x^2 + y^2 = 4$ are quadratic equations.
The equations $x^2 - 4x - 5 = 0$, $y^2 - 2y = 0$, $4x^2 + 4x + 1 = 0$, and $x^2 - 16 = 0$ are quadratic equations with one variable.

SOLVING QUADRATIC EQUATIONS WITH ONE VARIABLE

There are several methods to solve quadratic equations with one variable.

METHOD 1. SOLVING QUADRATIC EQUATIONS BY QUDRATIC FORMULA

Each quadratic equation $\mathbf{ax^2 + bx + c = 0}$ has **two solutions** (or **two roots**): two different or two same roots (i.e., one). We can find that solutions by **quadratic formula:**

QUADRATIC ORMULA ▶ Solutions of quadratic equation $\mathbf{ax^2 + bx + c = 0}$ $(\mathbf{a \neq 0})$ are:

$$x = \frac{-b \mp \sqrt{b^2 - 4ac}}{2a} \quad \text{or} \quad x_1 = \frac{-b - \sqrt{b^2 - 4ac}}{2a} \quad \& \quad x_2 = \frac{-b + \sqrt{b^2 - 4ac}}{2a}$$

Discriminant

The quantity $D \equiv b^2 - 4ac$ is called the *Discriminant* of the equation $ax^2 + bx + c = 0$ $(a \neq 0)$. The discriminant $D = b^2 - 4ac$ determines the number and the types of solutions of the standard quadratic equation $ax^2 + bx + c = 0$, $a \neq 0$ (see table below):

Table The Discriminant $D = b^2 - 4ac$ and the kinds of solutions of $ax^2 + bx + c = 0$ $(a \neq 0)$

Discriminant $D = b^2 - 4ac$	Kinds of Solutions of the Equation $ax^2 + bx + c = 0$ $(a \neq 0)$	Graph of $y = ax^2 + bx + c$
$D = b^2 - 4ac > 0$	Two different real solutions: $$x_1 = \frac{-b - \sqrt{b^2 - 4ac}}{2a}$$ $$x_2 = \frac{-b + \sqrt{b^2 - 4ac}}{2a}$$	$a > 0 \Rightarrow x_1 < x_2$ $\qquad$ $a < 0 \Rightarrow x_2 < x_1$
$D = b^2 - 4ac = 0$	One real solution (or 2 same solutions): $$x_1 = x_2 = -\frac{b}{2a}$$	$x_1 = x_2$
$D = b^2 - 4ac < 0$	No real solutions. Two complex solutions: $$x_1 = \frac{-b - i\sqrt{4ac - b^2}}{2a}$$ $$x_2 = \frac{-b + i\sqrt{4ac - b^2}}{2a}$$ where $i \equiv \sqrt{-1}$ (or $i^2 \equiv -1$). i is called *imaginary* (or *non-visible*) *unit*.	

Examples 1 Determine the **number** and **types** of solutions of the equation $x^2 - 4x - 5 = 0$.

SOLUTION ▶

| $a = 1; b = -4; c = -5$ $D = b^2 - 4ac =$ $= (-4)^2 - 4 \cdot 1 \cdot (-5) =$ $= 16 + 20 = \mathbf{36 > 0}$ | Two different real solutions: $$x_1 = \frac{4 - \sqrt{36}}{2} = -1 \quad \text{and} \quad x_2 = \frac{4 + \sqrt{36}}{2} = 5$$ | $a = 1 > 0$ |

SOLUTION ▶

$a = 1; b = -6; c = 9$ $D = b^2 - 4ac =$ $= (-6)^2 - 4 \cdot 1 \cdot (9) =$ $= 36 - 36 = 0$	One real solution (or 2 same solutions): $x_1 = x_2 = \dfrac{6}{2} = 3$	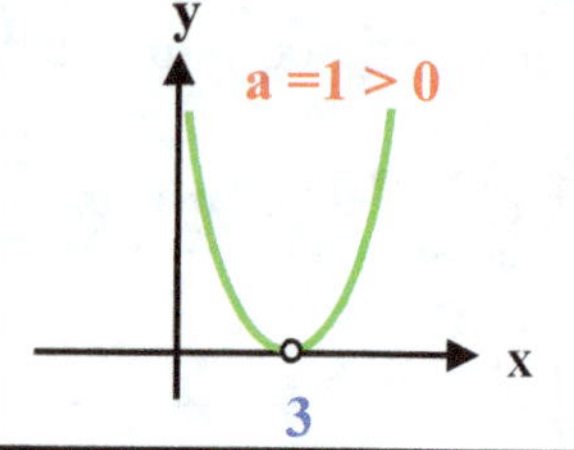

Examples 3 Determine the **number** and **types** of solutions of the equation $x^2 - 2x + 2 = 0$.

SOLUTION ▶

$a = 1; b = -2; c = 2$ $D = b^2 - 4ac =$ $= (-2)^2 - 4 \cdot 1 \cdot (2) =$ $= 4 - 8 = -4 < 0$	No real solutions. Two complex solutions: $x_1 = \dfrac{2 - i\sqrt{4}}{2} = 1 - i$ and $x_2 = \dfrac{2 + i\sqrt{4}}{2} = 1 + i$	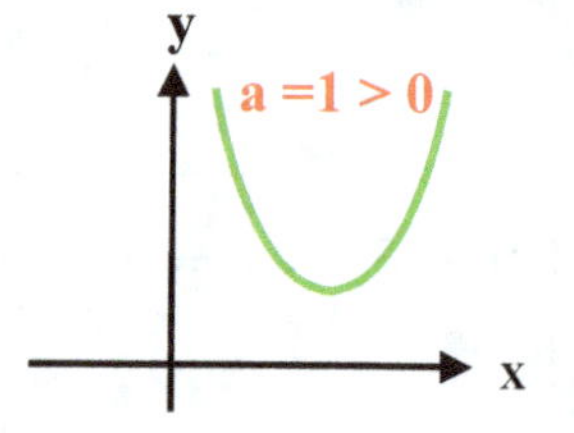

Solutions of any quadratic equation can be presented in different forms.
On a number line we can show only **real solutions**, but not complex (or non-visible) solutions.

EQUATION	FORMULA NOTATION OF THE SOLUTION	GRAPH NOTATION OF THE SOLUTION	SET NOTATION OF THE SOLUTION
$ax^2 + bx + c = 0$ $a \neq 0$	**1.** $D > 0$ $x_{1,2} = \dfrac{-b \pm \sqrt{b^2 - 4ac}}{2a}$	if $a > 0$ $x_1 = \dfrac{-b - \sqrt{b^2 - 4ac}}{2a}$ $x_2 = \dfrac{-b + \sqrt{b^2 - 4ac}}{2a}$ if $a < 0$ $x_2 = \dfrac{-b + \sqrt{b^2 - 4ac}}{2a}$ $x_1 = \dfrac{-b - \sqrt{b^2 - 4ac}}{2a}$	$\left\{ x = \dfrac{-b \pm \sqrt{b^2 - 4ac}}{2a} \right\}$
	2. $D = 0$ $x_1 = x_2 = -\dfrac{b}{2a}$	$x_1 = x_2 = -\dfrac{b}{2a}$	$\left\{ x = -\dfrac{b}{2a} \right\}$
	3. $D < 0$ $x_{1,2} = \dfrac{-b \pm i\sqrt{4ac - b^2}}{2a}$		$\left\{ x = \dfrac{-b \pm i\sqrt{4ac - b^2}}{2a} \right\}$

Below we will solve some quadratic equations by **quadratic formula method**.

Example 4 Solve $x^2 - 3x + 2 = 0$ using the quadratic formula.

SOLUTION ▶ We will use the quadratic formula with $a = 1$, $b = -3$, $c = 2$, and $D = 1$.

$$x = \frac{-b \pm \sqrt{b^2 - 4ac}}{2a} = \frac{-(-3) \pm \sqrt{(-3)^2 - 4 \cdot 1 \cdot 2}}{2 \cdot 1} = \frac{3 \pm \sqrt{1}}{2} = \frac{3 \pm 1}{2} = 2, 1$$

$$\Downarrow$$

$$x = 2, 1$$

Example 5 Solve $2x^2 + 10x + 12 = 0$ using the quadratic formula.

SOLUTION ▶ We will use the quadratic formula with $a = 2$, $b = 10$, $c = 12$, and $D = 4$.

$$x = \frac{-b \pm \sqrt{b^2 - 4ac}}{2a} = \frac{-10 \pm \sqrt{10^2 - 4 \cdot 2 \cdot 12}}{2 \cdot 2} = \frac{-10 \pm \sqrt{4}}{4} = \frac{-10 \pm 2}{4} = -2, -3$$

$$\Downarrow$$

$$x = -2, -3$$

Example 6 Solve $x^2 - 2x + 1 = 0$ using the quadratic formula.

SOLUTION ▶ We will use the quadratic formula with $a = 1$, $b = -2$, $c = 1$, and $D = 0$.

$$x = \frac{-b \pm \sqrt{b^2 - 4ac}}{2a} = \frac{-(-2) \pm \sqrt{(2)^2 - 4 \cdot 1 \cdot 1}}{2 \cdot 1} = \frac{2 \pm \sqrt{0}}{2} = \frac{2}{2} = 1, 1$$

$$\Downarrow$$

$$x = 1, 1$$

Example 7 Solve $x^2 + x + 2 = 0$ using the quadratic formula.

SOLUTION ▶ We will use the quadratic formula with $a = 1$, $b = 1$, $c = 2$, and $D = -7$.

$$x = \frac{-b \pm \sqrt{b^2 - 4ac}}{2a} = \frac{-1 \pm \sqrt{1^2 - 4 \cdot 1 \cdot 2}}{2 \cdot 1} = \frac{-1 \pm i\sqrt{7}}{2}$$

$$\Downarrow$$

$$x_1 = -\frac{1}{2} + i\frac{\sqrt{7}}{2}, \quad x_2 = -\frac{1}{2} - i\frac{\sqrt{7}}{2}$$

Two Important Theorems

We will prove **2 important theorems**.

Theorem 1 ▶ Sum of solutions of a standard quadratic equation with one variable $ax^2+bx+c = 0$, $a\neq 0$, is equal to the coefficient of the middle term with a negative sign divided by the coefficient of the leading term: $x_1 + x_2 = -b/a$.

Proof: $x_1 + x_2 = \dfrac{-b - \sqrt{b^2 - 4ac}}{2a} + \dfrac{-b + \sqrt{b^2 - 4ac}}{2a} = \dfrac{-b - \sqrt{b^2-4ac} - b + \sqrt{b^2-4ac}}{2a} = \dfrac{-2b}{2a} = -b/a$

Theorem 2 ▶ The product of solutions of a quadratic equation with one variable $ax^2+bx+c = 0$, $a\neq 0$, is equal to the third term (c) divided by the coefficient of the first term: $x_1 x_2 = c/a$.

Proof: $x_1 \cdot x_2 = \left(\dfrac{-b - \sqrt{b^2 - 4ac}}{2a}\right)\left(\dfrac{-b + \sqrt{b^2 - 4ac}}{2a}\right) = \dfrac{b^2 - (b^2 - 4ac)}{4a^2} = \dfrac{4ac}{4a^2} = c/a$

Example 1 For the equation $x^2 - 2x - 3 = 0$, show that $x_1 + x_2 = 2$ and $x_1 \cdot x_2 = -3$.

SOLUTION ▶

$$x = \frac{2\pm\sqrt{(-2)^2 - 4\cdot 1\cdot(-3)}}{2\cdot 1} = \frac{2\pm\sqrt{16}}{2} = \frac{2\pm 4}{2} = -1, 3$$

$x_1 = -1,\ x_2 = 3 \Rightarrow x_1 + x_2 = 2$ and $x_1 \cdot x_2 = -3$

Example 2 For the equation $2x^2 - 2x - 12 = 0$, show that $x_1 + x_2 = 2/2$ and $x_1 \cdot x_2 = -12/2$.

SOLUTION ▶

$$x = \frac{2\pm\sqrt{(-2)^2 - 4\cdot 2\cdot(-12)}}{2\cdot 2} = \frac{2\pm\sqrt{100}}{4} = \frac{2\pm 10}{4} = -2, 3$$

$x_1 = -2,\ x_2 = 3 \Rightarrow x_1 + x_2 = 1 = 2/2$ and $x_1 \cdot x_2 = -6 = -12/2$

We can solve the standard quadratic equation $ax^2 + bx + c = 0$ with one variable by **factoring** if we can find the factors of expression $ax^2 + bx + c$ easily. If we cannot find the factor form of a quadratic equation, we have to solve that equation by quadratic formula or by completing the square method (each method works for every quadratic equation).

Solving by factoring is based on the **"Zero" Property of Multiplication**: $a \cdot 0 = 0$ and $0 \cdot b = 0$. If $E = ax^2 + bx + c = E_1 \cdot E_2 = 0$ then $E_1 = 0$ or $E_2 = 0.$ It means that the solutions of two linear equations $E_1 = 0$ and $E_2 = 0$ are the solutions of quadratic equation $ax^2 + bx + c = 0$.

There are following standard factoring methods: **1) "Branch" factoring; 2) "Big X" factoring; 3) Factoring by "Algebraic Identities"; 4) Factoring by "Factor out" common factors**.

METHOD 2.1. "BRANCH" FACTORING

We will use the **"branch" factoring (Hayk Yegoryan's method)** for the quadratic equations $x^2 + bx + c = 0$ ($a = 1$, $b \neq 0$, $c \neq 0$ and $b, c \in R$), if the expression $x^2 + bx + c$ is an **easily factorable expression**. For example, $x^2 + 3x + 2 = (x + 2)(x + 1)$; $x^2 - 3x - 4 = (x - 4)(x + 1)$; $x^2 - 4x + 3 = (x - 3)(x - 1)$ are **easily factorable expressions**. The expressions $x^2 + 3x - 2$; $x^2 - 3x + 4$; and $x^2 + 4x - 3$ are not easily factorable expressions, but each of these is factorable expression like $(x - x_1)(x - x_2)$, were x_1 and x_2 are solutions of the corresponding quadratic equations. For these equations $x^2 + 3x - 2 = 0$ or $x^2 - 3x + 4 = 0$ or $x^2 + 4x - 3 = 0$ it is not easy to find x_1 and x_2 mentally.

> **Theorem 3** ▶ If we can find two numbers c_1 and c_2 for quadratic equation $x^2 + bx + c = 0$ such that $c_1 \cdot c_2 = c$ and $c_1 + c_2 = b$, then
> 1) we can present the quadratic equation $x^2 + bx + c$ in factoring form
> $$x^2 + bx + c = (x + c_1)(x + c_2) = 0$$
> 2) the solutions of quadratic equation $x^2 + bx + c = 0$ are $x = -c_1$ and $x = -c_2$.

Proof: Let $x^2 + bx + c = 0$, and, let $c = c_1 \cdot c_2$ and $b = c_1 + c_2$. Then we can write:

$$x^2 + bx + c = 0$$

$$\overset{\displaystyle c_1 \cdot\ c_2}{\wedge} \qquad \leftarrow \quad c_1 \cdot c_2 = c\, , \ c_1 + c_2 = b \qquad (c_1 \text{ and } c_2 \text{ are branch factors})$$

$$x^2 + (c_1 + c_2)x + c_1 \cdot c_2 = 0$$
$$x^2 + c_1 x + c_2 x + c_1 \cdot c_2 = 0$$
$$(x^2 + c_1 x) + (c_2 x + c_1 c_2) = 0$$
$$x(x + c_1) + c_2(x + c_1) = 0$$
$$(x + c_1)(x + c_2) = 0 \ \Rightarrow \ x^2 + bx + c = (x + c_1)(x + c_2) \quad \leftarrow \text{Factorable quadratic expression}$$

$$x + c_1 = 0 \quad \text{or} \quad x + c_2 = 0 \qquad \leftarrow \quad \text{"Zero" Property of Multiplication}$$

$$x = -c_1 \quad \text{or} \quad x = -c_2$$

Example 1 Solve $x^2 + 5x + 6 = 0$ by **branch factoring method**.

$$x^2 + 5x + 6 = 0$$

$$2 \cdot 3 \longleftarrow 2 \cdot 3 = 6, \ 2 + 3 = 5 \qquad \text{(2 and 3 are branch factors)}$$

$$(x + 2)(x + 3) = 0$$

$$x + 2 = 0 \ \text{or} \ x + 3 = 0$$

$$x = -2, \ x = -3$$

Example 2 Solve $x^2 + 3x - 4 = 0$ by **branch factoring method**.

$$x^2 + 3x - 4 = 0$$

$$-1 \cdot 4 \longleftarrow -1 \cdot 4 = -4, \ -1 + 4 = 3 \qquad \text{(−1 and 4 are branch factors)}$$

$$(x - 1)(x + 4) = 0$$

$$x = 1, \ x = -4$$

Example 3 Solve $x^2 - 3x - 10 = 0$ by **branch factoring method**.

$$x^2 - 3x - 10 = 0$$

$$-5 \cdot 2 \longleftarrow -5 \cdot 2 = -10, \ -5 + 2 = -3 \qquad \text{(−5 and 2 are branch factors)}$$

$$(x - 5)(x + 2) = 0$$

$$x = 5, \ x = -2$$

Example 4 Solve $x^2 - 8x + 15 = 0$ by **branch factoring method**.

$$x^2 - 8x + 15 = 0$$

$$-5 \cdot -3 \longleftarrow -5 \cdot (-3) = 15, \ -5 - 3 = -8 \qquad \text{(−5 and −3 are branch factors)}$$

$$(x - 5)(x - 3) = 0$$

$$x = 5, \ x = 3$$

METHOD 2.2. "BIG X" FACTORING

$\mathbf{W}$e will use the **"Big X" factoring** method to solve the standard quadratic equation
$ax^2 + bx + c = 0$ ($a \neq 1, a \neq 0, b \neq 0, c \neq 0$ and $a, b, c \in R$), if the expression $ax^2 + bx + c$ is a factorable expression.

Theorem 4 ▶ If we can find two numbers c_1 and c_2 for quadratic equation $ax^2 + bx + c = 0$,
with $a \neq 0, 1$ such that $c_1 \cdot c_2 = ac$ and $c_1 + c_2 = b$,

then

1) $ax^2 + bx + c = \dfrac{1}{a} \cdot (ax + c_1) \cdot (ax + c_2) = 0$

2) $x_1 = - c_1/a$ and $x_2 = - c_2/a$ are solutions of equation $ax^2 + bx + c = 0$.

Proof: Let $ax^2 + bx + c = 0$, and, there are c_1 and c_2 such that $c_1 \cdot c_2 = ac$ and $c_1 + c_2 = b$.

So we can write:

$ax^2 + bx + c = 0$

$ax^2 + (c_1 + c_2)x + c_1 \cdot c_2 /a = 0$ ⟵ $c = c_1 \cdot c_2 / a$, $b = c_1 + c_2$

$ax^2 + c_1x + c_2x + c_1 \cdot c_2 /a = 0$

$\dfrac{a}{a} \cdot (ax^2 + c_1x + c_2x + c_1 \cdot c_2 /a) = \dfrac{a}{a} \cdot 0$

$\dfrac{1}{a} \cdot (a^2 x^2 + ac_1x + ac_2x + c_1 \cdot c_2) = 0$

$\dfrac{1}{a} \cdot [(a^2 x^2 + ac_1x) + (ac_2x + c_1 c_2)] = 0$

$\dfrac{1}{a} \cdot ax(ax + c_1) + c_2(ax + c_1) = 0$

$\dfrac{1}{a} \cdot (ax + c_1) (ax + c_2) = 0 \Rightarrow ax^2 + bx + c = \dfrac{1}{a} \cdot (ax + c_1) \cdot (ax + c_2)$ ← **Factorable expression**

$ax + c_1 = 0$ or $ax + c_2 = 0$ ⟵   **"Zero" Property of Multiplication**

$x_1 = - c_1/a$ or $x_2 = - c_2/a$

Example 1 Solve $6x^2 + 7x + 2 = 0$ by **big X factoring method**.

SOLUTION ▶

$6x^2 + 7x + 2 = 0$

$$\frac{1}{2x} \leftarrow \frac{3}{6} \leftarrow 3 = c_1 \qquad \overset{3+4}{\underset{12}{\overset{7}{\times}}} \qquad c_2 = 4 \rightarrow \frac{4}{6} \rightarrow \frac{2}{3x} \quad \leftarrow \; c_1 \cdot c_2 = 12, \; c_1 + c_2 = 7 \Rightarrow c_1 = 3, \; c_2 = 4$$

$(2x + 1)(3x + 2) = 0$

$x_1 = -1/2, \; x_2 = -2/3 \quad (x_1 = -c_1/a = -3/6 = -1/2, \quad x_2 = -c_2/a = -4/6 = -2/3)$

Check: $6(-1/2)^2 + 7(-1/2) + 2 = 0$
$6(1/4) - 7/2 + 2 = 0$
$3/2 - 7/2 + 2 = 0$
$-4/2 + 2 = 0$
$0 = 0$ True

Check: $6(-2/3)^2 + 7(-2/3) + 2 = 0$
$6(4/9) - 14/3 + 2 = 0$
$24/9 - 14/3 + 2 = 0$
$-18/9 + 2 = 0$
$0 = 0$ True

Example 2 Solve $2x^2 - x - 1 = 0$ by **big X factoring method**.

SOLUTION ▶

$2x^2 - x - 1 = 0$

$$\frac{1}{2x} \leftarrow \frac{1}{2} \leftarrow 1 \qquad \overset{-2+1}{\underset{-2}{\overset{-1}{\times}}} \qquad -2 \rightarrow \frac{-2}{2} \rightarrow \frac{-1}{1x} \quad \leftarrow \; c_1 \cdot c_2 = -2, \; c_1 + c_2 = 1 \Rightarrow c_1 = -2, \; c_2 = 1$$

$(2x + 1)(x - 1) = 0$

$x_1 = 1, \; x_2 = -1/2$

Check: $2(-1/2)^2 - (-1/2) - 1 = 0$
$2(1/4) + 1/2 - 1 = 0$
$1/2 + 1/2 - 1 = 0$
$1 - 1 = 0$
$0 = 0$ True

Check: $2(1)^2 - 1 - 1 = 0$
$2 - 2 = 0$
$0 = 0$ True

<u>**METHOD 2.3.**</u> **SOLVING BY "ALGEBRAIC IDENTITIES" FORMULAS**

Sometimes we will use the **"Algebraic Identities"** to solve the standard quadratic equations.
There are **7** important algebraic identities shown below.

1. $a^2 - b^2 = (a - b)(a + b)$
2. $(a + b)^2 = a^2 + 2ab + b^2$
3. $(a - b)^2 = a^2 - 2ab + b^2$
4. $a^3 + b^3 = (a + b)(a^2 - ab + b^2)$
5. $a^3 - b^3 = (a - b)(a^2 + ab + b^2)$
6. $(a + b)^3 = a^3 + 3a^2 b + 3a b^2 + b^3$
7. $(a - b)^3 = a^3 - 3a^2 b + 3a b^2 - b^3$

All these formulas work for any $a, b \in R$

Example 1   Solve the following quadratic equations $x^2 - 4 = 0$

SOLUTION▶

$$x^2 - 4 = 0$$
$$x^2 - 2^2 = 0$$
$$(x - 2)(x + 2) = 0$$

$$x_1 = 2, \; x_2 = -2$$

Example 2   Solve the following quadratic equations $4x^2 - 12x + 9 = 0$

SOLUTION▶

$$4x^2 - 12x + 9 = 0$$
$$(2x)^2 - 2(2x)(3) + (3)^2 = 0$$
$$(2x - 3)^2 = 0$$
$$(2x - 3)(2x - 3) = 0$$
$$2x - 3 = 0 \quad \text{or} \quad 2x - 3 = 0$$
$$2x = 3 \quad \text{or} \quad 2x = 3$$
$$2x/2 = 3/2 \quad \text{or} \quad 2x/2 = 3/2$$

$$x_1 = x_2 = 3/2$$

Example 3   Solve the following quadratic equations $x^2 + 4x + 4 = 0$

SOLUTION▶

$$x^2 + 4x + 4 = 0$$
$$(x + 2)^2 = 0$$

$$x_1 = x_2 = -2$$

We will use this factoring method to solve quadratic equations like $ax^2 + bx = 0$, $a \neq 0$. The factoring by "factor out" common factors helps us immediately find the solutions.

Example 1   Solve the following quadratic equations $x^2 - 3x = 0$

SOLUTION▶

$$x^2 - 3x = 0$$
$$x(x - 3) = 0$$
$$x = 0, 3$$

Example 2   Solve the following quadratic equations $4x^2 - 16x = 0$

SOLUTION▶

$$4x^2 - 16x = 0$$
$$4x(x - 4) = 0$$
$$x = 0, 4$$

Example 3   Solve the following quadratic equations $3x^2 = 27x$

SOLUTION▶

$$3x^2 = 27x$$
$$3x^2 - 27x = 0$$
$$3x(x - 9) = 0$$
$$x = 0, 9$$

Example 4   Solve the following quadratic equations $8x^2 = 16x$

SOLUTION▶

$$8x = 16x^2$$
$$16x^2 - 8x = 0$$
$$8x(2x - 1) = 0$$
$$x = 0, 1/2$$

Example 5   Solve the following quadratic equations $ax^2 + bx = 0$, $a \neq 0$

SOLUTION▶

$$ax^2 + bx = 0, \; a \neq 0$$
$$x(ax + b) = 0$$
$$x = 0, -b/a$$

__METHOD 3.__ **SOLVING QUADRATIC EQUATIONS BY COMPLETING THE SQUARE**

The procedure used for completing the square to solve the quadratic equation with one variable $ax^2 + bx + c = 0$, $a \neq 0$ has the following steps:

Step 1: $(ax^2 + bx + c)\,/a = 0\,/a$
            $x^2 + bx/a + c/a = 0$

Step 2: $x^2 + bx/a + c/a - c/a = 0 - c/a$
            $x^2 + bx/a + \qquad\quad = -c/a$

Step 3: $x^2 + bx/a + (b/2a)^2 = -c/a + (b/2a)^2$ ⟵ We use the identity formula $a^2 + 2ab + b^2 = (a+b)^2$
            $(x + b/2a)^2 \qquad\quad = (b^2 - 4ac)\,/4a^2$ ⟵ **Square is completed**

Step 4: $\sqrt{(x + b/2a)^2} = \sqrt{(b^2 - 4ac)\,/4a^2}$

$$|x + b/2a| = \left|\frac{\sqrt{b^2 - 4ac}}{2a}\right|$$

$$x + \frac{b}{2a} = \pm\frac{\sqrt{b^2 - 4ac}}{2a}$$

Step 5: $x + \dfrac{b}{2a} - \dfrac{b}{2a} = -\dfrac{b}{2a} \pm \dfrac{\sqrt{b^2 - 4ac}}{2a}$

Answer: $x = -\dfrac{b}{2a} \pm \dfrac{\sqrt{b^2 - 4ac}}{2a}$

Examples   Solve the following quadratic equations by completing the square.

1. $x^2 - 4x - 5 = 0$
     $x^2 - 4x - 5 + 5 = 0 + 5$
     $x^2 - 4x \qquad = 5$
     $x^2 - 4x + (4/2)^2 = 5 + (4/2)^2$
     $x^2 - 4x + 2^2 = 5 + 2^2$
     $(x - 2)^2 = 9$
     $\sqrt{(x - 2)^2} = \sqrt{9}$
     $|x - 2| = 3$
     $x - 2 = \pm 3$
     $x - 2 + 2 = +2 \pm 3$
         $x = 2 \pm 3$
Answer: $x = 5, -1$

2. $x^2 - 6x = 16$
     $x^2 - 6x + 9 = 16 + 9$
     $(x - 3)^2 = 25$
     $\sqrt{(x - 3)^2} = \sqrt{25}$
     $|x - 3| = 5$
     $x - 3 = \pm 5$
     $x - 3 + 3 = +3 \pm 5$
         $x = 3 \pm 5$
Answer: $x = 8, -2$

3. $x^2 - 77 = 4x$
     $x^2 - 4x = 77$
     $x^2 - 4x + 4 = 77 + 4$
     $(x - 2)^2 = 81$
     $\sqrt{(x - 2)^2} = \sqrt{81}$
     $|x - 2| = 9$
     $x - 2 = \pm 9$
     $x - 2 + 2 = +2 \pm 9$
         $x = 2 \pm 9$
Answer: $x = 11, -7$

QUDRATIC - LIKE EQUATIONS

We will use the following definition.

> **Definition** ▶ The equation $ay^{2n} + by^n + c = 0$, where $a, b, c, n \in R$, and $a \neq 0$ is called *Quadratic-Like Equation with one variable*.

Examples　The equations $x^4 + 2x^2 + 4 = 0$, $y^4 - 13y^2 + 36 = 0$;　$2x^6 + 3x^3 - 2 = 0$, $z - 5z^{1/2} + 4 = 0$;　$w^8 - 1 = 0$ are quadratic-like equations.

Any quadratic-like equation after being rewritten into a quadratic equation form can be solved as a quadratic equation. The general quadratic-like equation $ay^{2n} + by^n + c = 0$ can be rewritten as a quadratic equation $ax^2 + bx + c = 0$, where $x \equiv y^n$. It means, first of all we have to solve the quadratic equation $ax^2 + bx + c = 0$, and then the equation $y^n = x$.

Examples　Solve each of the following quadratic-like equations with one variable:
$y^4 - 13y^2 + 36 = 0$, $2q^6 + 3q^3 - 2 = 0$, $z - 5z^{1/2} + 4 = 0$, and $w^8 - 1 = 0$.

SOLUTIONS▶

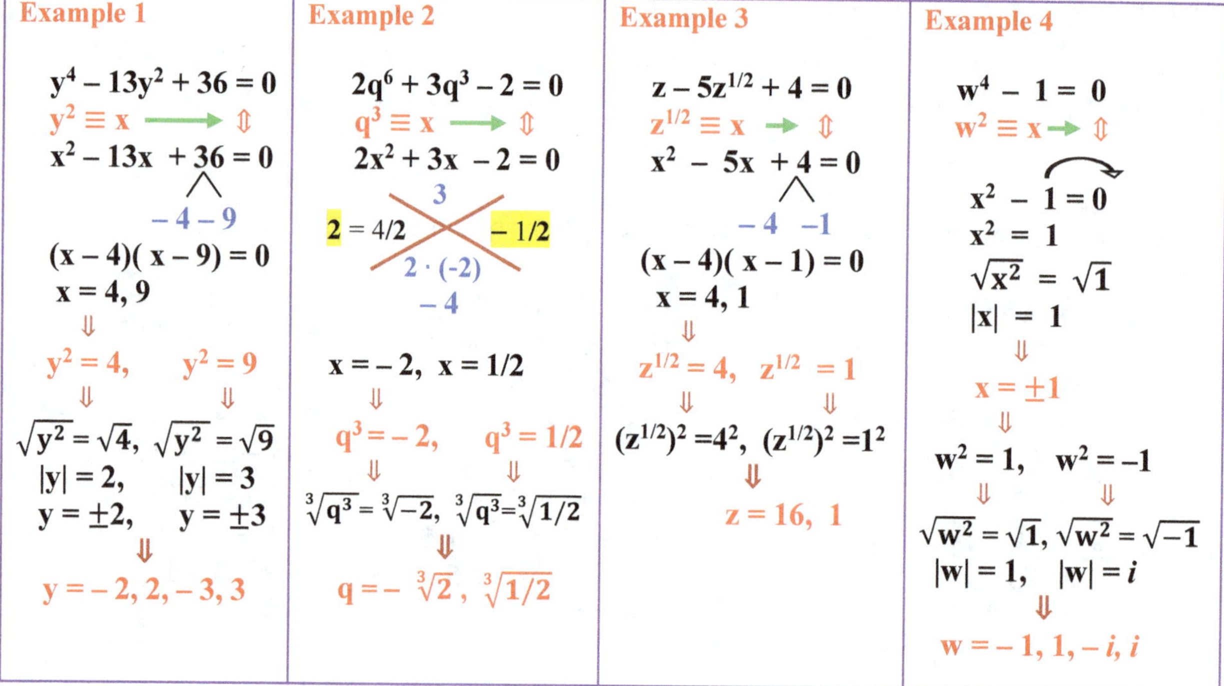

Example 1 The product of **two consecutive positive** numbers is **132**. Find the numbers.

SOLUTION ▶

Let **x** represents the first number and **x + 1** represents the second number. So, we can write:

$$x(x + 1) = 132$$

$$x^2 + x = 132$$
$$x^2 + x - 132 = 0$$

$$-11 \quad 12$$
$$(x - 11)(x + 12) = 0$$
$$\Downarrow$$

$$x = 11, -12$$ ⟵ The solution −12 does not work, because numbers are positive.
$$\Downarrow$$

$$x = 11, x + 1 = 12$$ ➡ The consecutive positive numbers are 11 and 12.

Example 2 The sum of two positive integers is **22** and their product is **40**. Find the numbers.

SOLUTION ▶

Let **x** represents the first number and **22 − x** represents the second number. So, we can write:

$$x(22 - x) = 40$$

$$22x - x^2 = 40$$

$$0 = x^2 - 22x + 40$$

$$-2 \quad -20$$
$$(x - 2)(x - 20) = 0$$
$$\Downarrow$$

$$x = 2, 20$$ ➡ The positive integers are 2 and 20.

<table><tr><td>Example 3</td><td>Ani has coins. She has twice as many quarters as dimes and four times as many nickels as quarters. The product of the numbers of dimes and quarters is 10 more than the number of nickels. How many of each coin does she have?</td></tr></table>

SOLUTION ▶

Let **x** be the number of dimes. So, the number of quarters will be **2x**, and the number of nickels will be **4(2x) = 8x.** The relating equation in terms of **x** will be the following:

$$x(2x) = 8x + 10$$

$$2x^2 = 8x + 10$$

$$2x^2 - 8x - 10 = 0$$
$$(2x^2 - 8x - 10)/2 = 0/2$$
$$x^2 - 4x - 5 = 0$$

$$-5 \quad 1$$
$$(x - 5)(x + 1) = 0$$

$$x = 5, -1 \quad \longleftarrow \text{The solution } -1 \text{ does not work, because number of dimes is positive.}$$

$$x = 5, \ 2x = 10, \ 8x = 40$$

The number of dimes is **5,** the number of quarters is **10,** and the number of nickels is **40.**

<table><tr><td>Example 4</td><td>The area of the square shown below is 36m. Find the length of the side.</td></tr></table>

The area of the square shown below is **36m²**. Find the length of the side.

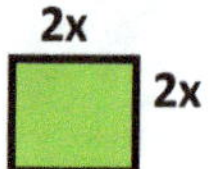

SOLUTION ▶

$$A = (2x)^2 = 36$$
$$\sqrt{(2x)^2} = \sqrt{36}$$
$$|2x| = 6$$
$$2x = \pm 6$$
$$x = 3, -3 \quad \longleftarrow \text{The solution } -3 \text{ does not work, because length of the side is positive.}$$
$$x = 3$$

COMPLEX NUMBERS

IMAGINARY NUMBERS

We will use the following definitions for **imaginary unit** and **imaginary numbers**.

> **Definition** ▶ The *imaginary unit* is denoted by symbol i, and $i = \sqrt{-1}$ or $i^2 = -1$.
>
> **Definition** ▶ An *imaginary number* is a number in the form bi, where $b \in R$ and $i = \sqrt{-1}$.

Examples The numbers i, $-i$, $-2i$, $3i$, $1.2i$, $-5.7i$, $2i/5$, and $-3i/7$ are imaginary numbers.

COMPLEX NUMBERS

Real numbers and imaginary numbers make **complex numbers**.

> **Definition** ▶ A *complex number* is a number in the form $a + bi$, were $a, b \in R$ and $i = \sqrt{-1}$. The number a is the **real part** of the complex number $a + bi$ and bi is the **imaginary part** of the complex number $a + bi$.
>
> **Definition** ▶ A *conjugate* of complex number $c = a + bi$ is the complex number $\bar{c} = a - bi$.
>
> **Definition** ▶ The *complex number set* C is a set of all **real** and **imaginary** numbers together.

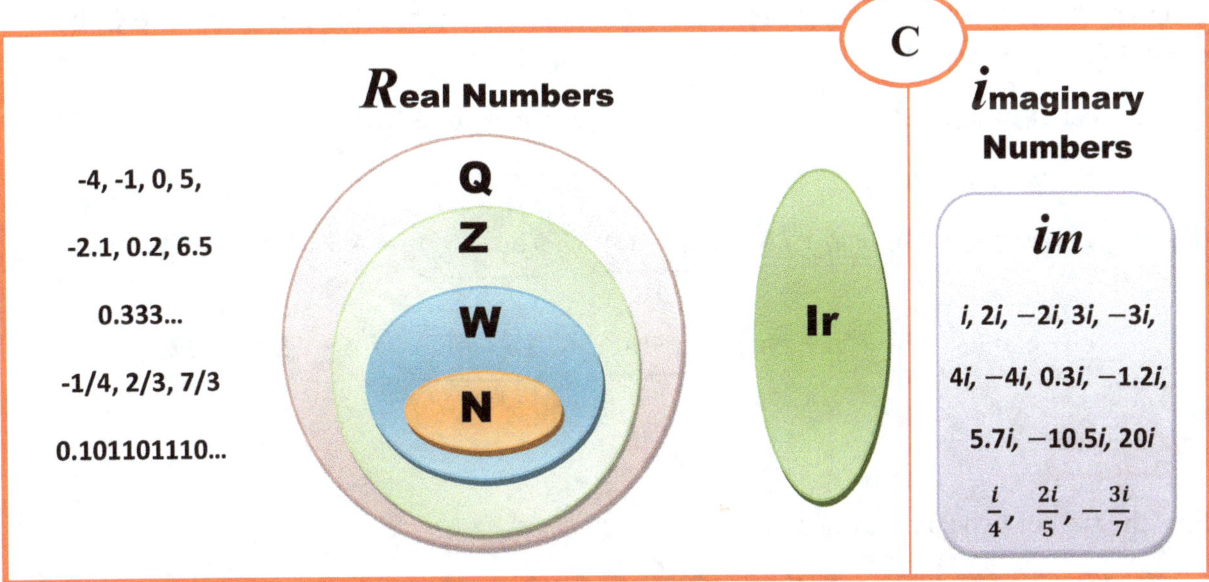

The diagram of complex number set C

Examples The numbers $3 = 3 + 0 \cdot i$, $2i = 0 + 2i$, $1 - 2i$, $5 + 3i$ are complex numbers.

$\mathbf{W}$e will use the following definition of **equality of two complex numbers**.

> **Definition** ▶ Two complex numbers $c_1 = a_1 + i\,b_1$ and $c_2 = a_2 + i\,b_2$ are equal if and only if their real parts are equal ($a_1 = a_2$) and their imaginary parts are equal ($b_1 = b_2$).

Examples

1. Two complex numbers $x + 6i$ and $4 - 3yi$ are equal if and only if $x = 4$ and $y = -2$.
2. Two complex numbers yi and $x + 5i$ are equal if and only if $x = 0$ and $y = 5$.
3. Two complex numbers xi and yi are equal if and only if $x = y$.
4. Two complex numbers x and $y + zi$ are equal if and only if $x = y, z = 0$.

ALGEBRA OF COMPLEX NUMBERS

$\mathbf{C}$omplex numbers can be added, subtracted, multiplied, and divided. The set of complex numbers is **closed under the basic algebraic operations** (**closure property**).
It means any algebraic combination of any two complex numbers is a complex number.

Important Note: *The set of complex numbers is an algebraically closed set.*

■ **ADDITION OF COMPLEX NUMBERS**

$\mathbf{T}$he set of complex numbers is **closed under the addition.** It means if $c_1 \in C$ and $c_2 \in C$ then $c_1 + c_2 \in C$ (and $c_2 + c_1 \in C$).

Proof: Let $a_1, b_1, a_2, b_2 \in R$, $c_1 = a_1 + ib_1 \in C$ and $c_2 = a_2 + ib_2 \in C$.

So, we can write

$$c_1 + c_2 = a_1 + i\,b_1 + a_2 + i\,b_2 = \underbrace{(a_1 + a_2)}_{A} + i\,\underbrace{(b_1 + b_2)}_{B} = A + i\,B = c_3 \in C$$

$$c_2 + c_1 = a_2 + i\,b_2 + a_1 + i\,b_1 = \underbrace{(a_1 + a_2)}_{A} + i\,\underbrace{(b_1 + b_2)}_{B} = A + i\,B = c_3 \in C$$

Examples

Find the sum of two complex numbers:

1. $(3 - 2i) + (5 + 4i) = (3 + 5) + (-2 + 4)i = 8 + 2i$
2. $5 + (2 - 4i) = (5 + 2) - 4i = 7 - 4i$
3. $(4 - 3i) + 8i = 4 + (-3 + 8)i = 4 + 5i.$

The set of complex numbers is **closed under the subtraction.** It means if $c_1 \in C$ and $c_2 \in C$ then $c_1 - c_2 \in C$ (and $c_2 - c_1 \in C$).

Proof: Let $a_1, b_1, a_2, b_2 \in R$, $c_1 = a_1 + i\, b_1 \in C$ and $c_2 = a_2 + i\, b_2 \in C$.

So, we can write

$$c_1 - c_2 = (a_1 + i\, b_1) - (a_2 + i\, b_2) = \underbrace{(a_1 - a_2)}_{A} + i\, \underbrace{(b_1 - b_2)}_{B} = A + i\, B = c_3 \in C$$

$$c_2 - c_1 = (a_2 + i\, b_2) - (a_1 + i\, b_1) = \underbrace{(a_2 - a_1)}_{D} + i\, \underbrace{(b_2 - b_1)}_{E} = D + i\, E = c_4 \in C$$

Examples

Find the difference of two complex numbers:

1. $(3 - 2i) - (5 + 4i) = (3 - 5) + (-2 - 4)i = -2 - 6i$
2. $5 - (2 - 4i) = (5 - 2) + 4i = 3 + 4i$
3. $(4 - 3i) - 8i = 4 + (-3 - 8)i = 4 - 11i.$

■ MULTIPLICATION OF COMPLEX NUMBERS

The set of complex numbers is **closed under the multiplication.** It means if $c_1 \in C$ and $c_2 \in C$ then $c_1 c_2 \in C$ (and $c_2 c_1 \in C$).

Proof: Let $a_1, b_1, a_2, b_2 \in R$, $c_1 = a_1 + ib_1 \in C$ and $c_2 = a_2 + ib_2 \in C$.

So, we can write

$$c_1 \cdot c_2 = (a_1 + i\, b_1) \cdot (a_2 + i\, b_2) = a_1 a_2 + i\, a_1 b_2 + i\, b_1 a_2 + i^{2}{}^{-1} b_1 b_2 =$$
$$= \underbrace{(a_1 a_2 - b_1 b_2)}_{A} + i\, \underbrace{(a_1 b_2 + b_1 a_2)}_{B} = c_2 \cdot c_1 = A + i\, B = c_3 \in C$$

Examples

Find the product of two complex numbers:

1. $(3 - 2i) \cdot (5 + 4i) = 3 \cdot 5 + 3 \cdot 4i - 2 \cdot 5i - 2i \cdot 4i = 15 + 12i - 10i - 8i^{2} = 23 + 2i$
2. $5 \cdot (2 - 4i) = 5 \cdot 2 - 5 \cdot 4i = 10 - 20i$
3. $(4 - 3i) \cdot 8i = 4 \cdot 8i - 3i \cdot 8i = 32i - 24i^{2} = 24 + 32i.$

POWERS OF *i*

We can calculate any power of *i* with natural exponent.

$$i^1 = i$$
$$i^2 = -1$$
$$i^3 = i \cdot i^2 = -i$$
$$i^4 = i^2 \cdot i^2 = (-1) \cdot (-1) = 1$$
$$i^5 = i^4 \cdot i = 1 \cdot i = i$$
$$i^6 = i^4 \cdot i^2 = 1 \cdot (-1) = -1$$
$$i^7 = i^4 \cdot i^3 = 1 \cdot (-i) = -i$$
$$i^8 = i^4 \cdot i^4 = 1 \cdot 1 = 1$$

...

$$i^{4n} = (i^4)^n = 1^n = 1$$
$$i^{4n+1} = i^{4n} \cdot i = 1 \cdot i = i$$
$$i^{4n+2} = i^{4n} \cdot i^2 = 1 \cdot (-1) = -1$$
$$i^{4n+3} = i^{4n} \cdot i^3 = 1 \cdot (-i) = -i$$

...

Find the powers of *i* with natural exponents.

1. $i^{2000} = (i^4)^{500} = 1^{500} = 1$
2. $i^{2009} = i^{2008+1} = i^{2008} \cdot i = (i^4)^{502} \cdot i = 1^{502} \cdot i = 1 \cdot i = i$
3. $i^{2010} = i^{2008+2} = i^{2008} \cdot i^2 = (i^4)^{502} \cdot i^2 = 1^{502} \cdot i^2 = 1 \cdot (-1) = -1$
4. $i^{67} = i^{64+3} = i^{64} \cdot i^3 = (i^4)^{16} \cdot i^3 = 1^{16} \cdot i^3 = 1 \cdot i^3 = 1 \cdot (-i) = -i$

We can also calculate powers of *i* with **0** and negative integer exponents.

$$i^0 = (i^4)^0 = 1^0 = 1$$
$$i^{-1} = \frac{1}{i} = \frac{1}{i} \cdot \frac{i}{i} = i/i^2 = i/(-1) = -i$$
$$i^{-2} = 1/i^2 = 1/(-1) = -1$$
$$i^{-3} = 1/i^3 = 1 \cdot i/i^3 \cdot i = i/i^4 = i/1 = i$$
$$i^{-4} = 1/i^4 = 1/1 = 1$$

...

$$i^{-4n} = 1/(i^4)^n = 1/1^n = 1$$
$$i^{-4n-1} = 1/(i^4)^n \cdot i^{-1} = 1 \cdot i^{-1} = 1 \cdot (-i) = -i$$
$$i^{-4n-2} = 1/(i^4)^n \cdot i^{-2} = 1 \cdot (-1) = -1$$
$$i^{-4n-3} = 1/(i^4)^n \cdot i^{-3} = 1 \cdot i = i$$

...

$\mathbf{T}$he set of complex numbers is **closed under the division.** It means if $c_1 \in C$ and $c_2 \in C$ then $c_1 / c_2 \in C$ (and $c_2 / c_1 \in C$).

Proof: Let $a_1, b_1, a_2, b_2 \in R,\ \ c_1 = a_1 + i\,b_1 \in C$ and $c_2 = a_2 + i\,b_2 \in C.$

So, we can write

$\blacktriangleright c_1 / c_2 = (a_1 + i\,b_1) / (a_2 + i\,b_2) = (a_1 + i\,b_1)(a_2 - i\,b_2) / (a_2 + i\,b_2)(a_2 - i\,b_2) =$

$(a_1 a_2 - i\,a_1 b_2 + i\,b_1 a_2 - \overset{-1}{i^2}\,b_1 b_2) / (a_2 + i\,b_2)(a_2 - i\,b_2) =$

$\underbrace{(a_1 a_2 + b_1 b_2)/(a_2{}^2 + b_2{}^2)}_{A} + i\,\underbrace{(b_1 a_2 - a_1 b_2)/(a_2{}^2 + b_2{}^2)}_{B} = A + i\,B = c_3 \in C$

$\blacktriangleright c_2 / c_1 = (a_2 + i\,b_2) / (a_1 + i\,b_1) = (a_2 + i\,b_2)(a_1 - i\,b_1) / (a_1 + i\,b_1)(a_1 - i\,b_1) =$

$(a_2 a_1 - i\,a_2 b_1 + i\,b_2 a_1 - \overset{-1}{i^2}\,b_2 b_1) / (a_1 + i\,b_1)(a_1 - i\,b_1) =$

$\underbrace{(a_2 a_1 + b_2 b_1)/(a_1{}^2 + b_1{}^2)}_{D} + i\,\underbrace{(b_2 a_1 - a_2 b_1)/(a_1{}^2 + b_1{}^2)}_{E} = D + i\,E = c_4 \in C$

Examples

Find the quotient of two complex numbers:

1. $\dfrac{3 - 2i}{5 + 4i} = \dfrac{3 - 2i}{5 + 4i} \cdot \dfrac{5 - 4i}{5 - 4i} = \dfrac{(3 - 2i)(5 - 4i)}{(5 + 4i)(5 - 4i)} = \dfrac{15 - 12i - 10i + 8i^2}{5^2 - (4i)^2} = \dfrac{15 - 12i - 10i + 8i^2}{25 + 16} = \dfrac{7}{41} - \dfrac{22}{41}\,i$

2. $\dfrac{3}{1 + 2i} = \dfrac{3}{1 + 2i} \cdot \dfrac{1 - 2i}{1 - 2i} = \dfrac{3(1 - 2i)}{(1 + 2i)(1 - 2i)} = \dfrac{3 - 6i}{1^2 - (2i)^2} = \dfrac{3 - 6i}{5} = \dfrac{3}{5} - \dfrac{6}{5}\,i$

3. $\dfrac{2i}{1 - i} = \dfrac{2i}{1 - i} \cdot \dfrac{1 + i}{1 + i} = \dfrac{2i(1 + i)}{(1 - i)(1 + i)} = \dfrac{2i + 2i^2}{1^2 - (i)^2} = \dfrac{2i - 2}{2} = -1 + i$

4. $\dfrac{1 + 2i}{i} = \dfrac{1 + 2i}{i} \cdot \dfrac{i}{i} = \dfrac{i + 2i^2}{(i)^2} = \dfrac{i - 2}{-1} = 2 - i$

THE COMPLEX PLANE

We can write the complex number **c = a + b*i*** also in **ordered pare notation c = < a, b >.**
So, we can graph the complex numbers on a plane in a manner similar to the graph of usual
ordered pare of numbers on **xy-plane**.

A complex number can be showed by the **point** or the **position radius -vector** in
two-dimensional **Cartesian** (or orthogonal) coordinate system called the **complex plane**.

The **x-axis ($\mathcal{R}$)** is called the **Real axis**, and **y-axis ($\mathcal{I}$)** is called the **Imaginary axis.**

There is a one-to-one correspondence between the complex numbers and the points on a $\mathcal{RI}$
plane.

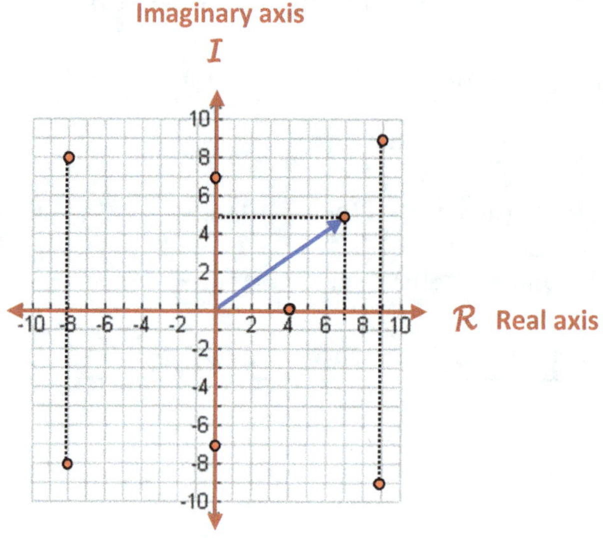

A complex number **7 + 5*i*** is plotted as a point (red) and as a position radius-vector (blue) on a plane.
Complex numbers **7*i* = < 0, 7 >; - 7*i* = < 0, -7 >; - 8 + 8*i* = < - 8, 8 >; - 8 - 8*i* = < - 8, - 8 >;**
9 + 9*i* = < 9, 9 >;   9 - 9*i* = < 9, - 9 >; and **4 = < 4, 0 >;** are plotted as points on the $\mathcal{RI}$ plane.

In the Figure **1.4.2.1**, shown above, we can see that the corresponding points of the conjugate
complex numbers are symmetric points with respect to the real axis $\mathcal{R}$ of the plane $\mathcal{RI}$.
The corresponding points of the conjugate complex numbers **7*i* = < 0, 7 >** and **−7*i* = < 0,−7 >**
are symmetric points with respect to the real axis $\mathcal{R}$. The corresponding points of the conjugate
complex numbers **−8 + 8*i* = < −8, 8 >** and **−8 −8*i* = < − 8, −8 >,** as well as the conjugate
complex numbers **9 + 9*i* = < 9, 9 >** and **9 − 9*i* = < 9, − 9 >** also are symmetric points with
respect to the real axis $\mathcal{R}$. The conjugate number of the complex number **4 = < 4, 0 >** is 4, so
the corresponding points are the same.

ABSOLUTE VALUE OF COMPLEX NUMBERS

$\mathbf{W}$e will use the following definition of **absolute value of a complex numbers**

Definition ▶ *Absolute value* of a complex number $\mathbf{a + bi}$ is a real number $|a + bi| = \sqrt{a^2 + b^2}$.

---

1. Absolute value of a complex number $3 + 4i$ is $|3 + 4i| = \sqrt{3^2 + 4^2} = 5.$
2. Absolute value of a complex number $7i$ is $|0 + 7i| = \sqrt{0^2 + 7^2} = 7.$
3. Absolute value of a complex number $6 - 8i$ is $|6 - 8i| = \sqrt{6^2 + (-8)^2} = 10.$

$\mathbf{A}$bsolute value of a complex number is the length of the corresponding radius-vector.

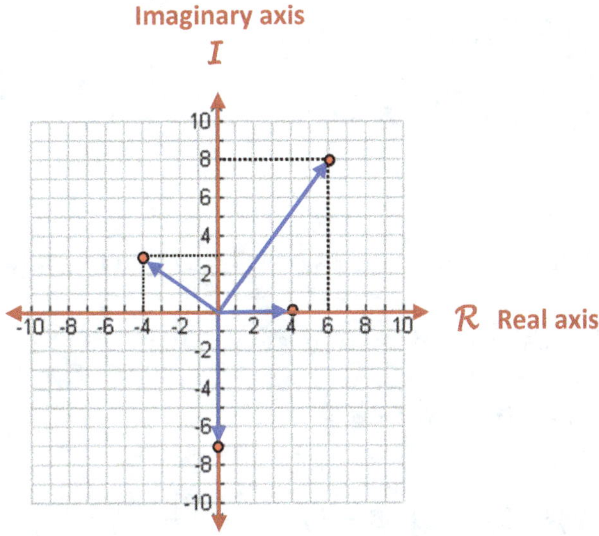

Absolute value of each complex number $6 + 8i$, $-4 + 3i$, $4 + 0i$, and $0 - 6i$
is the length (or magnitude) of the corresponding radius-vector.

$\mathbf{U}$sing Pythagorean theorem we can find the values of the length of radius-vectors of complex numbers $6 + 8i$ and $-4 + 3i$, equal to **10** and **5** respectively. The length of the complex number $4 + 0i$ is **4**, and the length of complex number $0 - 6i$ is **6**.
Definition of absolute value of a complex number gives us the same numbers:

$$|6 + 8i| = \sqrt{6^2 + 8^2} = \sqrt{36 + 64} = \sqrt{100} = 10$$
$$|-4 + 3i| = \sqrt{(-4)^2 + 3^2} = \sqrt{16 + 9} = \sqrt{25} = 5$$
$$|4 + 0i| = \sqrt{4^2 + 0^2} = \sqrt{16} = 4$$
$$|-6 + 0i| = \sqrt{6^2 + 0^2} = \sqrt{36} = 6$$

$\textbf{W}$e will use definition of absolute value of complex numbers to solve some additional examples.

Example 1 Simplify each expression.

  a) $\sqrt{-4}$

  b) $\sqrt{-24}$

  c) $\sqrt{-5}$

SOLUTION ▶ We will use the definition of imaginary unit $i = \sqrt{-1}$ or $i^2 = -1$, and the definition of absolute value of complex numbers.

  a) $\sqrt{-4}\ = \sqrt{(i)^2 \cdot 4} = 2i$

  b) $\sqrt{-24} = \sqrt{(i)^2 \cdot 4 \cdot 6} = 2\sqrt{6}i$

  c) $\sqrt{-5}\ = \sqrt{(i)^2 \cdot 5} = \sqrt{5}i$

Example 2 Simplify the complex expression $[(3 + 2i) + (1-i)]\,[(1 + i) - 1/(1-i)]$.

SOLUTION ▶ We will use the definition of imaginary unit $i = \sqrt{-1}$ or $i^2 = -1$.

$$[\,(3 + 2i) + (1 - i)\,]\,[(1 + i) - \frac{1}{1-i}\,] =$$

$$[\,(3 + 1) + (2 - 1)\,i\,]\,[(1 + i) - \frac{1}{1-i} \cdot \frac{1+i}{1+i}\,] =$$

$$[\,4 + i\,]\,[1 + i - \frac{1+i}{2}\,] =$$

$$[\,4 + i\,]\,[\,\frac{1+i}{2}\,] =$$

$$\frac{4 + 4i + i + i^2}{2} =$$

$$\frac{3}{2} + \frac{5}{2}\,i$$

 Find the following powers of i.

  a) i^{39}
  b) i^{141}
  c) i^{1624}
  d) i^{2999}

SOLUTION ▶ We will use the definition of imaginary unit $i = \sqrt{-1}$ or $i^2 = -1$.

  a) $i^{39} = i^{36+3} = i^{36} \cdot i^3 = (i^4)^9 \cdot i^3 = 1^9 \cdot i^3 = i^3 = i^{2+1} = i^2 \cdot i = -i$
  b) $i^{141} = i^{140+1} = i^{140} \cdot i = (i^4)^{35} \cdot i = 1^{35} \cdot i = i$
  c) $i^{1624} = (i^4)^{406} = 1^{406} = 1$
  d) $i^{2998} = i^{2996+2} = i^{2996} \cdot i^2 = (i^4)^{749} \cdot i^2 = 1^{749} \cdot i^2 = i^2 = -1$

 Show the complex numbers below on the **complex plane** by the points and by the position radius-vectors.

  a) $6 + 8i$
  b) $-4 + 6i$
  c) $-8 - 8i$
  d) $6 - 4i$
  e) 7
  f) $-6i$

SOLUTION ▶

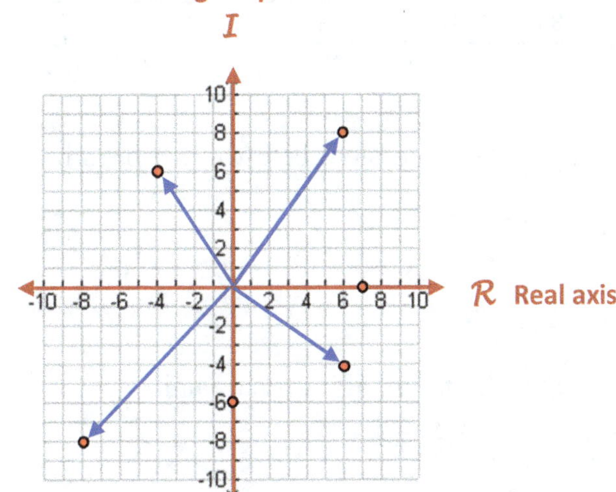

Example 5 Find the absolute value of the complex numbers

a) $|8 + 6i|$
b) $|-4 + 3i|$
c) $|-30 - 40i|$
d) $|6 - 8i|$
e) $|3|$
f) $|-9i|$
g) $\left|\dfrac{i+2}{i-2}\right|$

SOLUTION ▶ We will use the definition of imaginary unit $i = \sqrt{-1}$ or $i^2 = -1$, and the definition of absolute value of complex numbers.

a) $|8 + 6i| = \sqrt{8^2 + 6^2} = \sqrt{100} = 10$

b) $|-4 + 3i| = \sqrt{(-4)^2 + (3)^2} = \sqrt{25} = 5$

c) $|-30 - 40i| = \sqrt{(-30)^2 + (-40)^2} = \sqrt{2500} = 50$

d) $|6 - 8i| = \sqrt{6^2 + (-8)^2} = \sqrt{100} = 10$

e) $|3| = |3 + 0i| = \sqrt{3^2 + 0^2} = \sqrt{9} = 3$

f) $|-9i| = |0 + 9i| = \sqrt{0^2 + 9^2} = \sqrt{81} = 9$

g) $\left|\dfrac{i+2}{i-2}\right| = \left|\dfrac{i+2}{i-2} \cdot \dfrac{i+2}{i+2}\right| = \left|\dfrac{(i+2)(i+2)}{i^2 - 2^2}\right| = \left|\dfrac{i^2 + 4i + 2^2}{-5}\right| = \left|\dfrac{3}{-5} + \dfrac{4i}{-5}\right|$

$\left|-\dfrac{3}{5} - \dfrac{4i}{5}\right| = \sqrt{\left(-\dfrac{3}{5}\right)^2 + \left(-\dfrac{4}{5}\right)^2} = \sqrt{\dfrac{25}{25}} = 1$

Example 6 Write the complex expressions below in $\mathbf{a} + \mathbf{b}i$ form.

a) $\dfrac{1}{i^3}$

b) $\dfrac{1}{i^6}$

SOLUTION ▶ We will use the definition of imaginary unit $i = \sqrt{-1}$ or $i^2 = -1$.

a) $\dfrac{1}{i^3} = \dfrac{1}{i^3} \cdot \dfrac{i}{i} = \dfrac{i}{i^4} = \dfrac{i}{1} = i$

b) $\dfrac{1}{i^6} = \dfrac{1}{i^6} \cdot \dfrac{i^2}{i^2} = \dfrac{i^2}{i^8} = \dfrac{i^2}{(i^4)^2} = \dfrac{i^2}{1} = -1$

■ QUADRATIC INEQUALITIES

$\mathbf{W}$e will use the following definitions.

Definition ▶ *A quadratic inequality* is an inequality $E_1 < E_2$ (or $E_1 > E_2$, or $E_1 \leq E_2$, or $E_1 \geq E_2$) where E_1 or E_2 or both E_1 and E_2 are quadratic expressions with one or more variables.

Definition ▶ *A quadratic inequality with two variables* is an inequality $E_1 < E_2$ (or $E_1 > E_2$, or $E_1 \leq E_2$, or $E_1 \geq E_2$) where E_1 or E_2 or both E_1 and E_2 are quadratic expressions with two variables.

Definition ▶ *A quadratic inequality with two variables in standard form* is any inequality that can be presented
in the form $Ax^2 + Bxy + Cy^2 + Dx + Ey + F < 0$, or
in the form $Ax^2 + Bxy + Cy^2 + Dx + Ey + F > 0$, or
in the form $Ax^2 + Bxy + Cy^2 + Dx + Ey + F \leq 0$, or
in the form $Ax^2 + Bxy + Cy^2 + Dx + Ey + F \geq 0$,
where $A, B, C, D, E, F \in \mathbf{R}$ and at least $A \neq 0$, or $B \neq 0$, or $C \neq 0$.
If $A = 0, B = 0,$ and $C = 0,$ the inequality become a linear inequality.

Definition ▶ *A quadratic inequality with one variable* is an inequality $E_1 < E_2$ (or $E_1 > E_2$, or $E_1 \leq E_2$, or $E_1 \geq E_2$) where E_1 or E_2 or both E_1 and E_2 are quadratic expressions with only one variable.

Definition ▶ *A quadratic inequality with one variable in standard form* is any inequality that can be presented in the form $ax^2 + bx + c > 0$, or in the form $ax^2 + bx + c < 0$, or in the form $ax^2 + bx + c \geq 0$, or in the form $ax^2 + bx + c \leq 0$, where $a, b, c \in \mathbf{R}$ and $a \neq 0$.

In each of these inequalities **x** represents a variable and the constants **a**, **b**, and **c** are respectively called, the quadratic (or leading) coefficient, the linear (or middle) coefficient, and the free constant.

Examples The inequalities $x^2 + y^2 + 2z^2 - xy - 3xz + 6xz - 5x - 2y + z - 1 < 0,\ y^2 + 1 \leq 0$, $x^2 - 81 > 0$, and $x^2 + 3x + 1 \geq 0$ are quadratic inequalities.

The inequalities $5x^2 - 4xy + 4 - y + 3x < 0,\ y^2 - 4x > 0,\ x^2 + y^2 - 4 \leq 0,$ and $x^2/4 + y^2/9 - 1 \geq 0$ are quadratic inequalities with two variables.

The inequalities $x^2 - 4x - 5 > 0,\ y^2 - 2y < 0,\ 4x^2 + 4x + 1 \geq 0,$ and $x^2 - 16 \leq 0$ are quadratic inequalities with one variable.

SOLVING QUADRATIC INEQUALITIES WITH ONE VARIABLE

METHOD 1. SOLVING BY QUDRATIC FORMULA

We will solve all possible 8 general quadratic inequalities with one variable: $ax^2 + bx + c > 0$, $a > 0$; $ax^2 + bx + c < 0$, $a > 0$; $ax^2 + bx + c \geq 0$, $a > 0$; $ax^2 + bx + c \leq 0$, $a > 0$; and $ax^2 + bx + c > 0$, $a < 0$; $ax^2 + bx + c < 0$, $a < 0$; $ax^2 + bx + c \geq 0$, $a < 0$; $ax^2 + bx + c \leq 0$, $a < 0$.

1. Let we have the inequality $ax^2 + bx + c > 0$, with $a > 0$. To solve this quadratic inequality by quadratic formula, we will use **equivalent transformations** and **properties of inequalities**.

$$ax^2 + bx + c > 0, \quad a > 0$$

$$ax^2/a + bx/a + c/a > 0/a$$

$$x^2 + \frac{b}{a}x + \frac{c}{a} > 0$$

$$x^2 + \frac{b}{a}x > -\frac{c}{a}$$

$$x^2 + \frac{b}{a}x + \left(\frac{b}{2a}\right)^2 > -\frac{c}{a} + \left(\frac{b}{2a}\right)^2$$

$$\left(x + \frac{b}{2a}\right)^2 > \frac{b^2 - 4ac}{4a^2}$$

1)   If $D = b^2 - 4ac < 0$, then

x is any real number

2)   If $D = b^2 - 4ac = 0$, then

x is any real number $x \neq -\dfrac{b}{2a}$

3)   If $D = b^2 - 4ac > 0$, then

$$\sqrt{\left(x + \frac{b}{2a}\right)^2} > \sqrt{\frac{b^2 - 4ac}{4a^2}}$$

$$\left|x + \frac{b}{2a}\right| > \left|\frac{\sqrt{b^2 - 4ac}}{2a}\right|$$

$$x + \frac{b}{2a} < -\frac{\sqrt{b^2 - 4ac}}{2a} \quad \text{or} \quad x + \frac{b}{2a} > \frac{\sqrt{b^2 - 4ac}}{2a}$$

$$x < x_1 = -\frac{b}{2a} - \frac{\sqrt{b^2 - 4ac}}{2a} \quad \text{or} \quad x > x_2 = -\frac{b}{2a} + \frac{\sqrt{b^2 - 4ac}}{2a} \quad \text{*)}$$

---

*) $x_1 = -\dfrac{b}{2a} - \dfrac{\sqrt{b^2 - 4ac}}{2a}$ and $x_2 = -\dfrac{b}{2a} + \dfrac{\sqrt{b^2 - 4ac}}{2a}$ are small and big solutions of equation $ax^2 + bx + c = 0$, $a > 0$.

We can represent the solution set of this inequality by the **formula**, **graph**, **set**, and **interval notations**.

INEQUALITY	FORMULA NOTATION OF SOLUTIONS	GRAPH NOTATION OF SOLUTIONS	SET NOTATION OF SOLUTIONS	INTERVAL NOTATION OF SOLUTIONS
$ax^2 + bx + c > 0$ $a > 0$	1) If $b^2 - 4ac < 0$, then x is any real number		$\{x/\, x \in R\}$	$x \in (-\infty, \infty)$
	2) If $b^2 - 4ac = 0$, then $x \neq -\dfrac{b}{2a}$ and $x \in R$		$\{x/\, x \in R - \{-\dfrac{b}{2a}\}\}$	$x \in (-\infty, -\dfrac{b}{2a}) \cup (-\dfrac{b}{2a}, \infty)$
	3) If $b^2 - 4ac > 0$, then $x < x_1 = -\dfrac{b}{2a} - \dfrac{\sqrt{b^2-4ac}}{2a}$ *or* $x > x_2 = -\dfrac{b}{2a} + \dfrac{\sqrt{b^2-4ac}}{2a}$	$x_1 = -\dfrac{b}{2a} - \dfrac{\sqrt{b^2-4ac}}{2a}$ $-\dfrac{b}{2a} + \dfrac{\sqrt{b^2-4ac}}{2a} = x_2$	$\{x/x < -\dfrac{b}{2a} - \dfrac{\sqrt{b^2-4ac}}{2a}\}$ *or* $\{x/x > -\dfrac{b}{2a} + \dfrac{\sqrt{b^2-4ac}}{2a}\}$	$x \in (-\infty, x_1) \cup (x_2, \infty)$

Example Solve the inequality $x^2 - 4x - 5 > 0$.

SOLUTION ▶ We will use the quadratic formula method with $a = 1 > 0$, $b = -4$, and $c = -5$.

$$D = b^2 - 4ac = (-4)^2 - 4 \cdot 1 \cdot (-5) = 16 + 20 = 36 > 0.$$

$$x < x_1 = -\frac{b}{2a} - \frac{\sqrt{b^2-4ac}}{2a} \quad \text{or} \quad x > x_2 = -\frac{b}{2a} + \frac{\sqrt{b^2-4ac}}{2a}$$

$$x < -\frac{-4}{2} - \frac{\sqrt{36}}{2} \quad \text{or} \quad x > -\frac{-4}{2} + \frac{\sqrt{36}}{2}$$

$$x < 2 - 3 \quad \text{or} \quad x > 2 + 3$$

$$x < -1 \quad \text{or} \quad x > 5 \qquad x < -1 \quad \text{or} \quad x > 5$$

$$\{x/\, x < -1 \text{ or } x > 5\} \qquad x \in (-\infty, -1) \cup (5, \infty)$$

2. Let we have the inequality $ax^2 + bx + c < 0$, with $a > 0$.

We can solve this inequality by quadratic formula and represent the solution set of this inequality by the **formula**, **graph**, **set**, and **interval notations**.

INEQUALITY	FORMULA NOTATION OF SOLUTIONS	GRAPH NOTATION OF SOLUTIONS	SET NOTATION OF SOLUTIONS	INTERVAL NOTATION OF SOLUTIONS
$ax^2 + bx + c < 0$ $a > 0$	1) If $b^2 - 4ac < 0$, then No solution 2) If $b^2 - 4ac = 0$, then No solution 3) If $b^2 - 4ac > 0$, then $x > x_1 = -\dfrac{b}{2a} - \dfrac{\sqrt{b^2 - 4ac}}{2a}$ ***and*** $x < x_2 = -\dfrac{b}{2a} + \dfrac{\sqrt{b^2 - 4ac}}{2a}$		$\{x/\, x \in \varnothing\}$ $\{x/\, x \in \varnothing\}$ $\{x/x > -\dfrac{b}{2a} - \dfrac{\sqrt{b^2 - 4ac}}{2a}\}$ ***and*** $\{x/x < -\dfrac{b}{2a} + \dfrac{\sqrt{b^2 - 4ac}}{2a}\}$	$x \in \varnothing$ $x \in \varnothing$ $x \in (x_1, x_2)$ $\Updownarrow$ $x \in (-\dfrac{b}{2a} - \dfrac{\sqrt{b^2 - 4ac}}{2a}, -\dfrac{b}{2a} + \dfrac{\sqrt{b^2 - 4ac}}{2a})$

Example   Solve the inequality $x^2 - 5x - 6 < 0$.

SOLUTION▸   We will use the quadratic formula method.

$$D = b^2 - 4ac = (-5)^2 - 4 \cdot 1 \cdot (-6) = 25 + 24 = 49 > 0.$$

$$-\frac{b}{2a} - \frac{\sqrt{b^2 - 4ac}}{2a} = x_1 < x < x_2 = -\frac{b}{2a} + \frac{\sqrt{b^2 - 4ac}}{2a}$$

$$-\frac{-5}{2} - \frac{\sqrt{49}}{2} < x < -\frac{-5}{2} + \frac{\sqrt{49}}{2}$$

$$\frac{5}{2} - \frac{7}{2} = x_1 < x < x_2 = \frac{5}{2} + \frac{7}{2}$$

$$-1 < x < 6 \qquad\qquad -1 < x < 6$$

$$\{x/{-1} < x < 6\} \qquad\qquad x \in (-1, 6)$$

3. Let we have the inequality $ax^2 + bx + c \geq 0$, with $a > 0$.

We can solve this inequality by quadratic formula and represent the solution set of this inequality by the **formula**, **graph**, **set**, and **interval notations**.

INEQUALITY	FORMULA NOTATION OF SOLUTIONS	GRAPH NOTATION OF SOLUTIONS	SET NOTATION OF SOLUTIONS	INTERVAL NOTATION OF SOLUTIONS
$ax^2 + bx + c \geq 0$ $a > 0$	1) If $b^2 - 4ac < 0$, then x is any real number		$\{x / x \in R\}$	$x \in (-\infty, \infty)$
	2) If $b^2 - 4ac = 0$, then x is any real number		$\{x / x \in R\}$	$x \in (-\infty, \infty)$
	3) If $b^2 - 4ac > 0$, then $x \leq x_1 = -\dfrac{b}{2a} - \dfrac{\sqrt{b^2-4ac}}{2a}$ **or** $x \geq x_2 = -\dfrac{b}{2a} + \dfrac{\sqrt{b^2-4ac}}{2a}$		$\{x/x \leq -\dfrac{b}{2a} - \dfrac{\sqrt{b^2-4ac}}{2a}\}$ **or** $\{x/x \geq -\dfrac{b}{2a} + \dfrac{\sqrt{b^2-4ac}}{2a}\}$	$x \in (-\infty, x_1] \cup [x_2, \infty)$

Example Solve the inequality $x^2 + 5x - 6 \geq 0$.

SOLUTION ▶ We will use the quadratic formula method with $a = 1 > 0$, $b = 5$, and $c = -6$.

$$D = b^2 - 4ac = (5)^2 - 4 \cdot 1 \cdot (-6) = 25 + 24 = 49 > 0.$$

$$x \leq x_1 = -\frac{b}{2a} - \frac{\sqrt{b^2-4ac}}{2a} \quad \text{or} \quad x \geq x_2 = -\frac{b}{2a} + \frac{\sqrt{b^2-4ac}}{2a}$$

$$x \leq -\frac{5}{2} - \frac{\sqrt{49}}{2} \quad \text{or} \quad x \geq -\frac{5}{2} + \frac{\sqrt{49}}{2}$$

$$x \leq \frac{-5-7}{2} \quad \text{or} \quad x \geq \frac{-5+7}{2}$$

$$x \leq -6 \quad \text{or} \quad x \geq 1 \qquad\qquad x \leq -6 \quad \text{or} \quad x \geq 1$$

$$\{x / x \leq -6 \text{ or } x \geq 1\} \qquad\qquad x \in (-\infty, -6] \cup [1, \infty)$$

4. Let we have the inequality $ax^2 + bx + c \leq 0$, with $a > 0$.

We can solve this inequality by quadratic formula and represent the solution set of this inequality by the **formula**, **graph**, **set**, and **interval notations**.

INEQUALITY	FORMULA NOTATION OF SOLUTIONS	GRAPH NOTATION OF SOLUTIONS	SET NOTATION OF SOLUTIONS	INTERVAL NOTATION OF SOLUTIONS
$ax^2 + bx + c \leq 0$ $a > 0$	1) If $b^2 - 4ac < 0$, then No solution 2) If $b^2 - 4ac = 0$, then $x = -\dfrac{b}{2a}$ 3) If $b^2 - 4ac > 0$, then $x \geq x_1 = -\dfrac{b}{2a} - \dfrac{\sqrt{b^2-4ac}}{2a}$ *and* $x \leq x_2 = -\dfrac{b}{2a} + \dfrac{\sqrt{b^2-4ac}}{2a}$		$\{x/x \in \varnothing\}$ $\{x/x = -\dfrac{b}{2a}\}$ $\{x/x_1 \leq x \leq x_2\}$	$x \in \varnothing$ $x = -\dfrac{b}{2a}$ $x \in [x_1, x_2]$

Example   Solve the inequality $x^2 - 6x + 8 \leq 0$.

SOLUTION ▶   We will use the quadratic formula method:

$$D = b^2 - 4ac = (-6)^2 - 4 \cdot 1 \cdot 8 = 36 - 32 = 4 > 0.$$

$$-\frac{b}{2a} - \frac{\sqrt{b^2-4ac}}{2a} \leq x \leq -\frac{b}{2a} + \frac{\sqrt{b^2-4ac}}{2a}$$

$$-\frac{-6}{2} - \frac{\sqrt{4}}{2} \leq x \leq -\frac{-6}{2} + \frac{\sqrt{4}}{2}$$

$$\frac{6}{2} - \frac{2}{2} = x_1 \leq x \leq x_2 = \frac{6}{2} + \frac{2}{2}$$

$$2 \leq x \leq 4 \qquad\qquad 2 \leq x \leq 4$$

$$\{x/2 \leq x \leq 4\} \qquad\qquad x \in [2, 4]$$

5. Let we have the inequality $ax^2 + bx + c > 0$, with $a < 0$.

We can solve this inequality by quadratic formula and represent the solution set of this inequality by the formula, graph, set, and interval notations.

INEQUALITY	FORMULA NOTATION OF SOLUTIONS	GRAPH NOTATION OF SOLUTIONS	SET NOTATION OF SOLUTIONS	INTERVAL NOTATION OF SOLUTIONS
$ax^2 + bx + c > 0$ $a < 0$	1) If $b^2 - 4ac < 0$, then No solution 2) If $b^2 - 4ac = 0$, then No solution 3) If $b^2 - 4ac > 0$, then $$x > x_1 = -\frac{b}{2a} + \frac{\sqrt{b^2-4ac}}{2a}$$ **and** $$x < x_2 = -\frac{b}{2a} - \frac{\sqrt{b^2-4ac}}{2a}$$		$\{x/\, x \in \emptyset\}$ $\{x/\, x \in \emptyset\}$ $\{x/\, x_1 < x < x_2\}$	$x \in \emptyset$ $x \in \emptyset$ $x \in (x_1, x_2)$

Example Solve the inequality $-x^2 - x + 2 > 0$ $(a = -1, b = -1, c = 2)$.

SOLUTION▶ We will use the quadratic formula method:

$$D = b^2 - 4ac = (-1)^2 - 4 \cdot (-1) \cdot 2 = 9 > 0.$$
$$\Downarrow$$
$$-\frac{b}{2a} + \frac{\sqrt{b^2-4ac}}{2a} = x_1 < x < x_2 = -\frac{b}{2a} - \frac{\sqrt{b^2-4ac}}{2a}$$
$$\Downarrow$$
$$-\frac{-1}{2(-1)} + \frac{\sqrt{9}}{2(-1)} = x_1 < x < x_2 = -\frac{-1}{2(-1)} - \frac{\sqrt{9}}{2(-1)}$$
$$\Downarrow$$
$$-\frac{1}{2} - \frac{3}{2} = x_1 < x < x_2 = -\frac{1}{2} + \frac{3}{2}$$
$$\Downarrow$$
$$-2 < x < 1$$

$$\{x/ -2 < x < 1\} \qquad x \in (-2, 1)$$

6. Let we have the inequality $ax^2 + bx + c < 0$, with $a < 0$.

We can solve this inequality by quadratic formula and represent the solution set of this inequality by the **formula**, **graph**, **set**, and **interval notations**.

INEQUALITY	FORMULA NOTATION OF SOLUTIONS	GRAPH NOTATION OF SOLUTIONS	SET NOTATION OF SOLUTIONS	INTERVAL NOTATION OF SOLUTIONS
$ax^2 + bx + c < 0$ $a < 0$	1) If $b^2 - 4ac < 0$, then All real numbers 2) If $b^2 - 4ac = 0$, then All real numbers except 0 3) If $b^2 - 4ac > 0$, then $$x < x_1 = -\frac{b}{2a} + \frac{\sqrt{b^2-4ac}}{2a}$$ *or* $$x > x_2 = -\frac{b}{2a} - \frac{\sqrt{b^2-4ac}}{2a}$$		$\{x/\ x \in R\}$ $\{x/x \in R - \{-\frac{b}{2a}\}\}$ $\{x/\ x < x_1 \text{ or } x > x_2\}$	$x \in R$ $(-\infty, -\frac{b}{2a}) \cup (-\frac{b}{2a}, \infty)$ $x \in (-\infty, x_1) \cup (x_2, \infty)$

Example Solve the inequality $-x^2 + 2x + 3 < 0$ $(a = -1, b = 2, c = 3)$.

SOLUTION▶ We will use the quadratic formula method:

$$D = b^2 - 4ac = (2)^2 - 4 \cdot (-1) \cdot 3 = 16 > 0.$$
$$\Downarrow$$
$$x < x_1 = -\frac{b}{2a} + \frac{\sqrt{b^2-4ac}}{2a} \quad \text{or} \quad x > x_2 = -\frac{b}{2a} - \frac{\sqrt{b^2-4ac}}{2a}$$
$$\Downarrow$$
$$x < -\frac{2}{2(-1)} + \frac{\sqrt{16}}{2(-1)} \quad \text{or} \quad x > -\frac{2}{2(-1)} - \frac{\sqrt{16}}{2(-1)}$$
$$\Downarrow$$
$$x < \frac{2}{2} - \frac{4}{2} \quad \text{or} \quad x > \frac{2}{2} + \frac{4}{2}$$
$$\Downarrow$$
$$x < -1 \quad \text{or} \quad x > 3$$

$$\{x/x < -1 \text{ or } x > 3\}$$

$$x \in (-\infty, -1) \cup (3, \infty)$$

7. Let we have the inequality $ax^2 + bx + c \geq 0$, with $a < 0$.

We can solve this inequality by quadratic formula and represent the solution set of this inequality by the **formula**, **graph**, **set**, and **interval notations**.

INEQUALITY	FORMULA NOTATION OF SOLUTIONS	GRAPH NOTATION OF SOLUTIONS	SET NOTATION OF SOLUTIONS	INTERVAL NOTATION OF SOLUTIONS
$ax^2 + bx + c \geq 0$ $a < 0$	1) If $b^2 - 4ac < 0$, then No solution 2) If $b^2 - 4ac = 0$, then $x = -\dfrac{b}{2a}$ 3) If $b^2 - 4ac > 0$, then $x \geq x_1 = -\dfrac{b}{2a} + \dfrac{\sqrt{b^2 - 4ac}}{2a}$ *and* $x \leq x_2 = -\dfrac{b}{2a} - \dfrac{\sqrt{b^2 - 4ac}}{2a}$		$\{x/x \in \emptyset\}$ $\{x/x = -\dfrac{b}{2a}\}$ $\{x/x_1 \leq x \leq x_2\}$	$x \in \emptyset$ $x = -\dfrac{b}{2a}$ $x \in [x_1, x_2]$

Example Solve the inequality $-x^2 + 5x - 6 \geq 0$ ($a = -1$, $b = 5$, $c = -6$).

SOLUTION ▶ We will use the quadratic formula method:

$$D = b^2 - 4ac = (5)^2 - 4 \cdot (-1) \cdot (-6) = 1 > 0.$$

$$\Downarrow$$

$$-\frac{b}{2a} + \frac{\sqrt{b^2 - 4ac}}{2a} = x_1 \leq x \leq x_2 = -\frac{b}{2a} - \frac{\sqrt{b^2 - 4ac}}{2a}$$

$$\Downarrow$$

$$-\frac{5}{2(-1)} + \frac{\sqrt{1}}{2(-1)} = x_1 \leq x \leq x_2 = -\frac{5}{2(-1)} - \frac{\sqrt{1}}{2(-1)}$$

$$\Downarrow$$

$$\frac{5}{2} - \frac{1}{2} = x_1 \leq x \leq x_2 = \frac{5}{2} + \frac{1}{2}$$

$$\Downarrow$$

$$2 \leq x \leq 3$$

$$\{x/\, 2 \leq x \leq 3\} \qquad\qquad x \in [2, 3]$$

8. Let we have the inequality $ax^2 + bx + c \leq 0$, with $a < 0$.

We can solve this inequality by quadratic formula and represent the solution set of this inequality by the **formula**, **graph**, **set**, and **interval notations**.

INEQUALITY	FORMULA NOTATION OF SOLUTIONS	GRAPH NOTATION OF SOLUTIONS	SET NOTATION OF SOLUTIONS	INTERVAL NOTATION OF SOLUTIONS
$ax^2 + bx + c \leq 0$ $a < 0$	**1)** If $b^2 - 4ac < 0$, then All real numbers **2)** If $b^2 - 4ac = 0$, then All real numbers **3)** If $b^2 - 4ac > 0$, then $x \leq x_1 = -\dfrac{b}{2a} + \dfrac{\sqrt{b^2-4ac}}{2a}$ *or* $x \geq x_2 = -\dfrac{b}{2a} - \dfrac{\sqrt{b^2-4ac}}{2a}$		$\{x / x \in R\}$ $\{x / x \in R\}$ $\{x/x \leq x_1 \text{ or } x \geq x_2\}$	$x \in R$ $x \in (-\infty, \infty)$ $x \in (-\infty, x_1] \cup [x_2, \infty)$

Example Solve the inequality $-x^2 + 2x + 8 < 0$ $(a = -1, b = 2, c = 8)$.

SOLUTION▶ We will use the quadratic formula method:

$$D = b^2 - 4ac = (2)^2 - 4 \cdot (-1) \cdot 8 = 36 > 0.$$

$$\Downarrow$$

$$x \leq x_1 = -\frac{b}{2a} + \frac{\sqrt{b^2-4ac}}{2a} \quad \text{or} \quad x \geq x_2 = -\frac{b}{2a} - \frac{\sqrt{b^2-4ac}}{2a}$$

$$\Downarrow$$

$$x \leq -\frac{2}{2(-1)} + \frac{\sqrt{36}}{2(-1)} \quad \text{or} \quad x \geq -\frac{2}{2(-1)} - \frac{\sqrt{36}}{2(-1)}$$

$$\Downarrow$$

$$x \leq \frac{2}{2} - \frac{6}{2} \quad \text{or} \quad x \geq \frac{2}{2} + \frac{6}{2}$$

$$\Downarrow$$

$$x \leq -2 \quad \text{or} \quad x \geq 4$$

$$\{x/x \leq -2 \text{ or } x \geq 4\}$$

$$x \in (-\infty, -2] \cup [4, \infty)$$

<u>METHOD 2.</u> **SOLVING QUDRATIC INEQUALITIES BY FACTORING**

We will solve the standard quadratic inequalities $ax^2 + bx + c > 0$, $ax^2 + bx + c < 0$, $ax^2 + bx + c \geq 0$, and $ax^2 + bx + c \leq 0$ with one variable by **factoring method** if we can find the factor form of a quadratic expression $ax^2 + bx + c$ by easy way. If we cannot find the easy factor form of a quadratic expression, then we will solve given inequality by quadratic formula method. Quadratic formula method works for every quadratic inequality.
Factoring method usually we use for only easy factorable quadratic expression $ax^2 + bx + c$.
Solving by factoring method is based on **the following properties**.

1. If $E = ax^2 + bx + c = E_1 \cdot E_2 > 0$ then $E_1 > 0$ & $E_2 > 0$, or $E_1 < 0$ & $E_2 < 0$.
2. If $E = ax^2 + bx + c = E_1 \cdot E_2 \geq 0$ then $E_1 \geq 0$ & $E_2 \geq 0$, or $E_1 \leq 0$ & $E_2 \leq 0$.
3. If $E = ax^2 + bx + c = E_1 \cdot E_2 < 0$ then $E_1 > 0$ & $E_2 < 0$, or $E_1 < 0$ & $E_2 > 0$.
4. If $E = ax^2 + bx + c = E_1 \cdot E_2 \leq 0$ then $E_1 \geq 0$ & $E_2 \leq 0$, or $E_1 \leq 0$ & $E_2 \geq 0$.

We will study only four standard factoring methods: **1) "Branch" factoring; 2) "Big X" factoring; 3) Factoring by "Algebraic Identities"; 4) Factoring by "Factor out" common factors**.
Types and numbers of solutions of equation $ax^2 + bx + c = 0$ will defined the types of solution sets of quadratic inequalities $ax^2+bx+c > 0$, $ax^2+bx+c < 0$, $ax^2+bx+c \geq 0$, and $ax^2+bx+c \leq 0$.

METHOD 2.1. "BRANCH" FACTORING

We will use the **"branch" factoring** to solve the quadratic inequalities like $x^2 + bx + c > 0$, $x^2 + bx + c \geq 0$, $x^2 + bx + c < 0$, $x^2 + bx + c \leq 0$, or like $-x^2 + bx + c > 0$, $-x^2 + bx + c \geq 0$, $-x^2 + bx + c < 0$, $-x^2 + bx + c \leq 0$ with $b \neq 0$, $c \neq 0$ and $b, c \in \mathbf{R}$, if we can factor quadratic expression $x^2 + bx + c$, or quadratic expression $-x^2 + bx + c$, **by branch factoring method**.

Example 1 Solve the quadratic inequality $x^2 - 5x + 6 > 0$ by "branch" factoring method.

SOLUTION▶

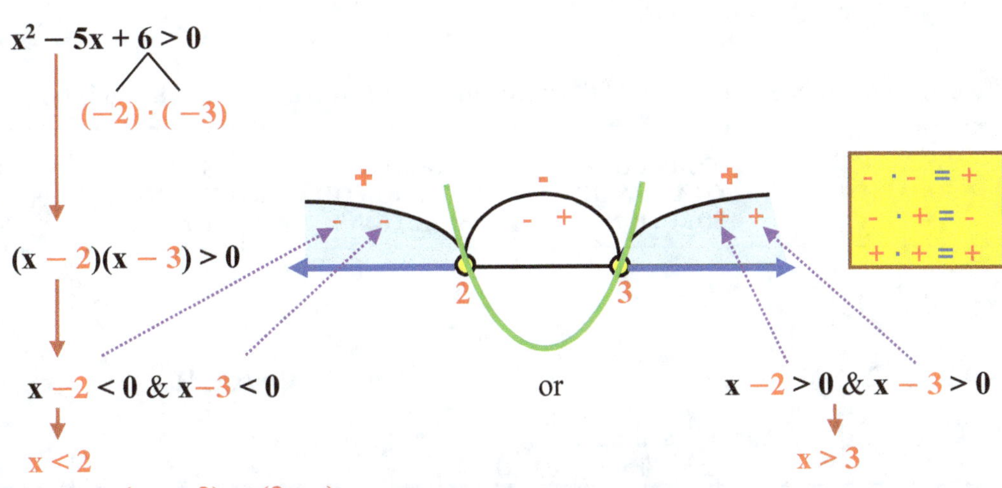

Answer: $x \in (-\infty, 2) \cup (3, \infty)$

We can represent the solution set of this inequality by **formula**, **graph**, **set**, and **interval notations**.

INEQUALITY	FORMULA NOTATION OF SOLUTIONS	GRAPH NOTATION OF SOLUTIONS	SET NOTATION OF SOLUTIONS	INTERVAL NOTATION OF SOLUTIONS
$x^2 - 5x + 6 > 0$ $\updownarrow$ $(x-2)(x-3) > 0$ $D > 0$	$x < 2$ or $x > 3$	$a = 1 > 0$ 2 3	$\{x/x < 2 \text{ or } x > 3\}$	$x \in (-\infty, 2) \cup (3, \infty)$

Example 2 Solve the quadratic inequality $x^2 - 4x + 4 > 0$ by **branch factoring method**.

SOLUTION▶

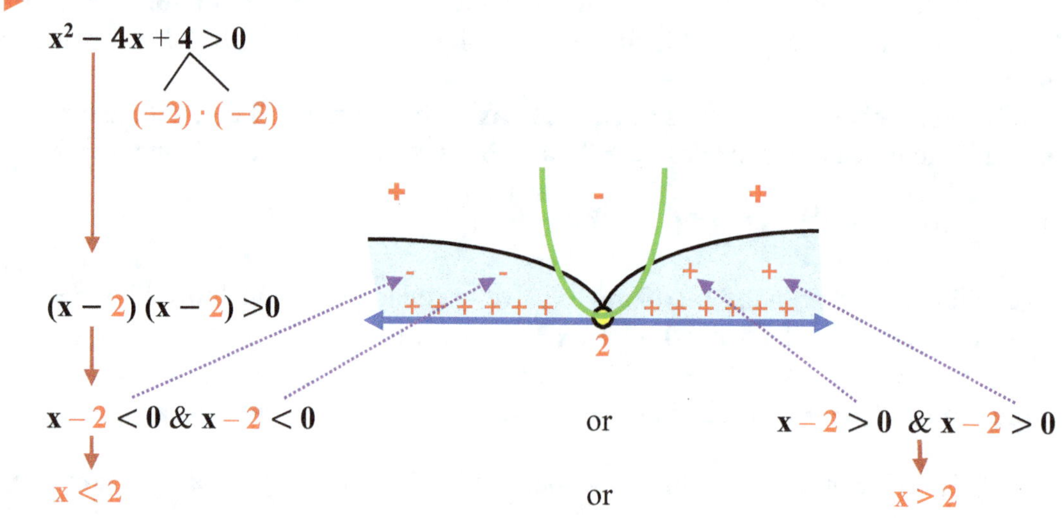

$$x^2 - 4x + 4 > 0$$
$$(-2) \cdot (-2)$$

$$(x - 2)(x - 2) > 0$$

$$x - 2 < 0 \ \& \ x - 2 < 0 \qquad \text{or} \qquad x - 2 > 0 \ \& \ x - 2 > 0$$

$$x < 2 \qquad\qquad \text{or} \qquad\qquad x > 2$$

Answer: $x \in (-\infty, 2) \cup (2, \infty)$

We can represent the solution set of this inequality by **formula**, **graph**, **set**, and **interval notations**.

INEQUALITY	FORMULA NOTATION OF SOLUTIONS	GRAPH NOTATION OF SOLUTIONS	SET NOTATION OF SOLUTIONS	INTERVAL NOTATION OF SOLUTIONS
$x^2 - 4x + 4 > 0$ $\updownarrow$ $(x-2)(x-2) > 0$ $D = 0$	$x < 2$ or $x > 2$ All real numbers $x \neq 2$ If $x = 2 \Rightarrow 0 > 0$ False	$a = 1 > 0$ 2	$\{\, x/x \in \mathbf{R} - \{2\} \,\}$	$x \in (-\infty, 2) \cup (2, \infty)$

Example 3 Solve the quadratic inequality $x^2 + 2x + 2 > 0$.

SOLUTION ▶

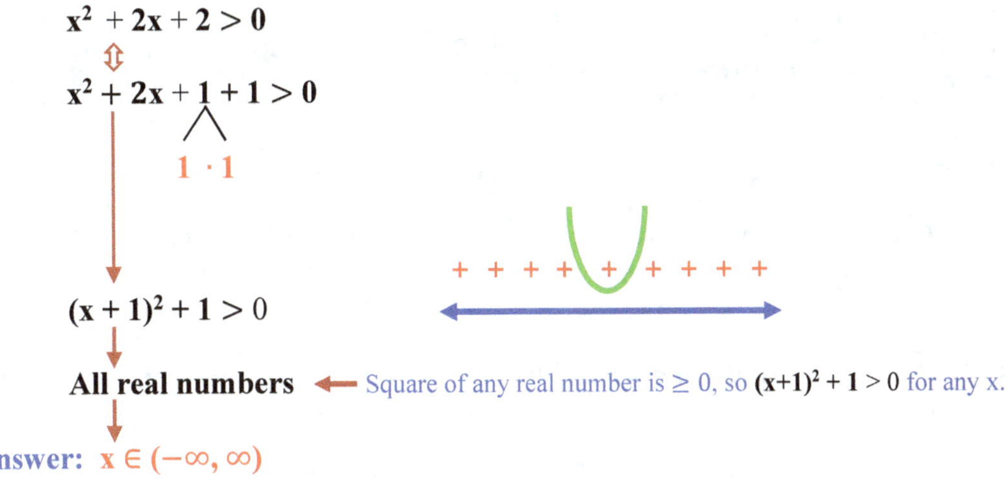

$$x^2 + 2x + 2 > 0$$

$$\updownarrow$$

$$x^2 + 2x + 1 + 1 > 0$$

$$\overset{\wedge}{1 \cdot 1}$$

$$(x + 1)^2 + 1 > 0$$

All real numbers ⟵ Square of any real number is ≥ 0, so $(x+1)^2 + 1 > 0$ for any x.

Answer: $x \in (-\infty, \infty)$

We can represent the solution set of this inequality by **formula**, **graph**, **set**, and **interval notations**.

INEQUALITY	FORMULA NOTATION OF SOLUTIONS	GRAPH NOTATION OF SOLUTIONS	SET NOTATION OF SOLUTIONS	INTERVAL NOTATION OF SOLUTIONS
$x^2 + 2x + 2 > 0$ $D < 0$	$x \in R$ **All real numbers**		$\{\, x/x \in R \,\}$	$x \in (-\infty, \infty)$

Example 4 Solve the quadratic inequality $x^2 + 2x - 3 \geq 0$ by **branch factoring method**.

SOLUTION ▶

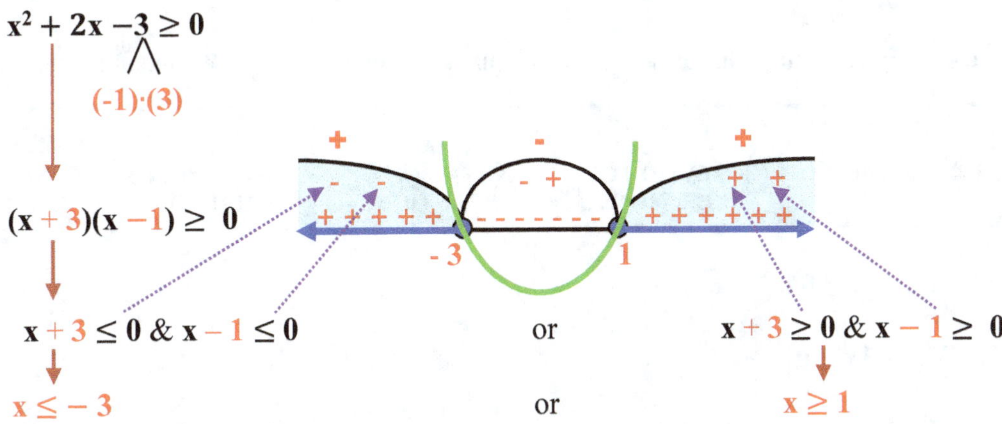

$$x^2 + 2x - 3 \geq 0$$

$$\overset{\wedge}{(-1) \cdot (3)}$$

$$(x + 3)(x - 1) \geq 0$$

$$x + 3 \leq 0 \ \& \ x - 1 \leq 0 \qquad \text{or} \qquad x + 3 \geq 0 \ \& \ x - 1 \geq 0$$

$$x \leq -3 \qquad\qquad\qquad \text{or} \qquad\qquad\qquad x \geq 1$$

Answer: $x \in (-\infty, -3] \cup [1, \infty)$

$\mathbf{W}$e can represent the solution set of this inequality by **formula, graph, set,** and **interval notations**.

INEQUALITY	FORMULA NOTATION OF SOLUTIONS	GRAPH NOTATION OF SOLUTIONS	SET NOTATION OF SOLUTIONS	INTERVAL NOTATION OF SOLUTIONS
$x^2 + 2x - 3 \geq 0$ $\updownarrow$ $(x+3)(x-1) \geq 0$ $D > 0$	$x \leq -3$ or $x \geq 1$	-3 1	$\{x/x \leq -3,\ x \geq 1\}$	$x \in (-\infty, -3] \cup [1, \infty)$

Example 5   Solve the quadratic inequality $x^2 - 6x + 9 \geq 0$ by **branch factoring method**.

SOLUTION▶

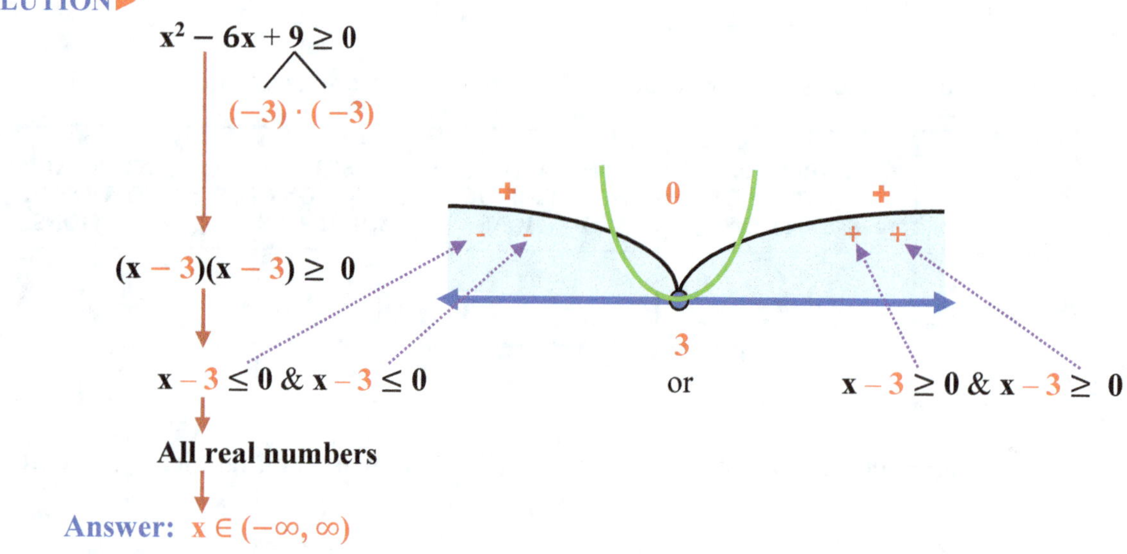

$$x^2 - 6x + 9 \geq 0$$

$$(-3) \cdot (-3)$$

$$(x - 3)(x - 3) \geq 0$$

$$x - 3 \leq 0\ \&\ x - 3 \leq 0 \qquad\qquad \text{or} \qquad\qquad x - 3 \geq 0\ \&\ x - 3 \geq 0$$

All real numbers

Answer: $x \in (-\infty, \infty)$

$\mathbf{W}$e can represent the solution set of this inequality by **formula, graph, set,** and **interval notations**.

INEQUALITY	FORMULA NOTATION OF SOLUTIONS	GRAPH NOTATION OF SOLUTIONS	SET NOTATION OF SOLUTIONS	INTERVAL NOTATION OF SOLUTIONS
$x^2 - 6x + 9 \geq 0$ $\updownarrow$ $(x-3)(x-3) \geq 0$ $D = 0$	$x \in R$ **All real umbers**	3	$\{x/x \in R\}$	$x \in (-\infty, \infty)$

Example 6 Solve the quadratic inequality $-x^2 - 4x + 12 \geq 0$ by **branch factoring method**.

SOLUTION ▶

$$(-x^2 - 4x + 12)(-1) \geq 0(-1)$$

$$\updownarrow$$

$$x^2 + 4x - 12 \leq 0$$

$$(6) \cdot (-2)$$

$$(x + 6)(x - 2) \leq 0$$

$$-6 \leq x \leq 2$$

Answer: $x \in [-6, 2]$

Example 7 Solve the quadratic inequality $-x^2 - 3x + 18 < 0$ by **branch factoring method**.

SOLUTION ▶

$$(-x^2 - 3x + 18)(-1) < 0(-1)$$

$$\updownarrow$$

$$x^2 + 3x - 18 > 0$$

$$(6) \cdot (-3)$$

$$(x + 6)(x - 3) > 0$$

$$x < -6 \text{ or } x > 3$$

Answer: $x \in (-\infty, -6] \cup [3, \infty)$

Example 8 Solve the quadratic inequality $-x^2 - 3x + 40 \leq 0$ by **branch factoring method**.

SOLUTION ▶

$$(-x^2 - 3x + 40)(-1) \leq 0(-1)$$

$$\updownarrow$$

$$x^2 + 3x - 40 \geq 0$$

$$(8) \cdot (-5)$$

$$(x + 8)(x - 5) \geq 0$$

$$x \leq -8 \text{ or } x \geq 5$$

Answer: $x \in (-\infty, -8] \cup [5, \infty)$

<u>**METHOD 2.2. "BIG X" FACTORING**</u>

$\mathbf{W}$e will use the **"Big X" factoring** method to solve the standard quadratic inequalities, like
$\mathbf{ax^2 + bx + c > 0, \ ax^2 + bx + c \geq 0, \ ax^2 + bx + c < 0, \ ax^2 + bx + c \leq 0}$, or like $\mathbf{-ax^2 + bx + c > 0,}$
$\mathbf{-ax^2 + bx + c \geq 0, \ -ax^2 + bx + c < 0, \ -ax^2 + bx + c \leq 0}$ with $\mathbf{a \neq 0, \ a \neq 1, \ b \neq 0, \ c \neq 0}$ and
$\mathbf{a, b, c \in R}$, if we can factor quadratic expression $\mathbf{ax^2 + bx + c}$, or quadratic expression
$\mathbf{-ax^2 + bx + c}$ **by big X factoring method**.

Example 1 Solve the quadratic inequality $\mathbf{2x^2 + 5x - 7 > 0}$ by **big X factoring method**.

SOLUTION ▶

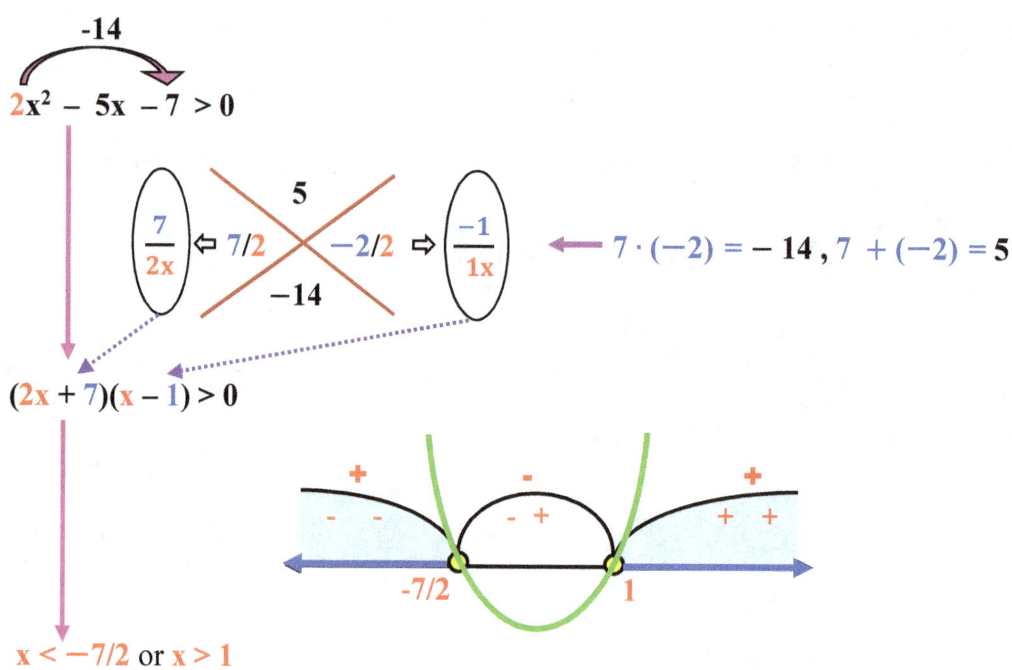

Answer: $x \in (-\infty, -7/2) \cup (1, \infty)$

$\mathbf{W}$e can represent the solution set of this inequality by **formula**, **graph**, **set**, and **interval notations**.

INEQUALITY	FORMULA NOTATION OF SOLUTIONS	GRAPH NOTATION OF SOLUTIONS	SET NOTATION OF SOLUTIONS	INTERVAL NOTATION OF SOLUTIONS
$2x^2 + 5x - 7 > 0$ $\Updownarrow$ $(2x + 7)(x - 1) > 0$ $D > 0$	$x < -7/2 \ , \ x > 3$	-7/2 1	$\{x / x < -7/2 \text{ or } x > 3\}$	$x \in (-\infty, -\frac{7}{2}) \cup (1, \infty)$

We will solve the inequality with same expression, but with another inequality sign.

Example 2 Solve the quadratic inequality $2x^2 + 5x - 7 \leq 0$ by **branch factoring method**.

SOLUTION ▶

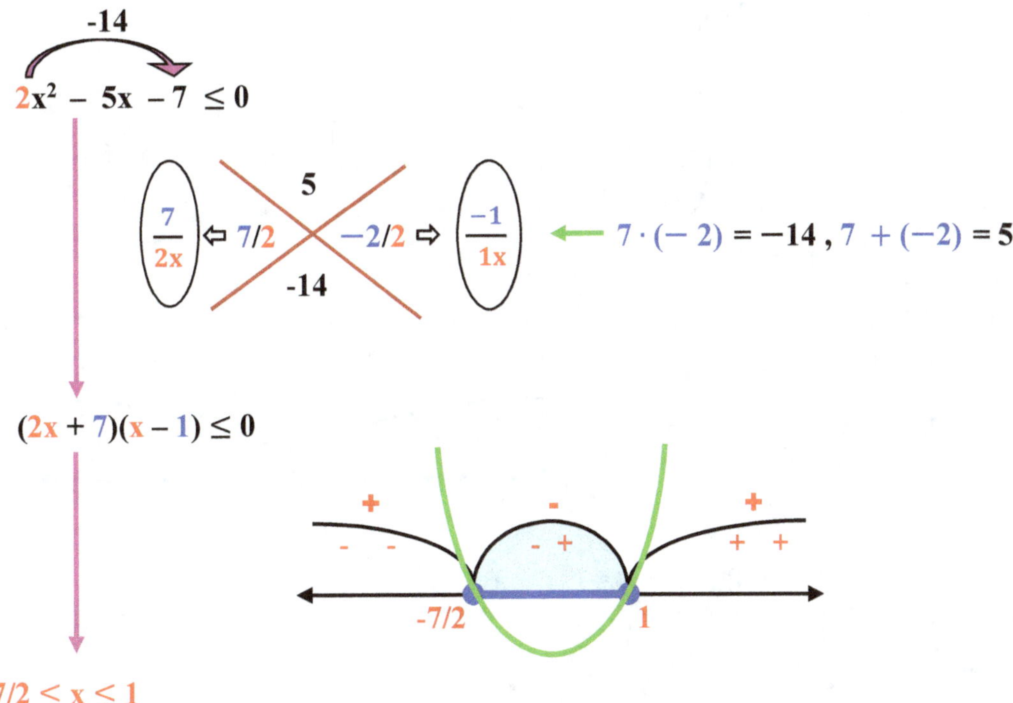

Answer: $x \in [-7/2 , 1]$

We can represent the solution set of this inequality by **formula**, **graph**, **set**, and **interval notations**.

INEQUALITY	FORMULA NOTATION OF SOLUTIONS	GRAPH NOTATION OF SOLUTIONS	SET NOTATION OF SOLUTIONS	INTERVAL NOTATION OF SOLUTIONS
$2x^2 + 5x - 7 \leq 0$ $\Updownarrow$ $(2x + 7)(x - 1) \leq 0$ $D > 0$	$-7/2 \leq x \leq 1$	-7/2 1	$\{x \,/- 7/2 \leq x \leq 1\}$	$x \in \left[-\frac{7}{2}, 1\right]$

 Solve the quadratic inequality $4x^2 - 12x + 9 > 0$ by **branch factoring method**.

SOLUTION ▶

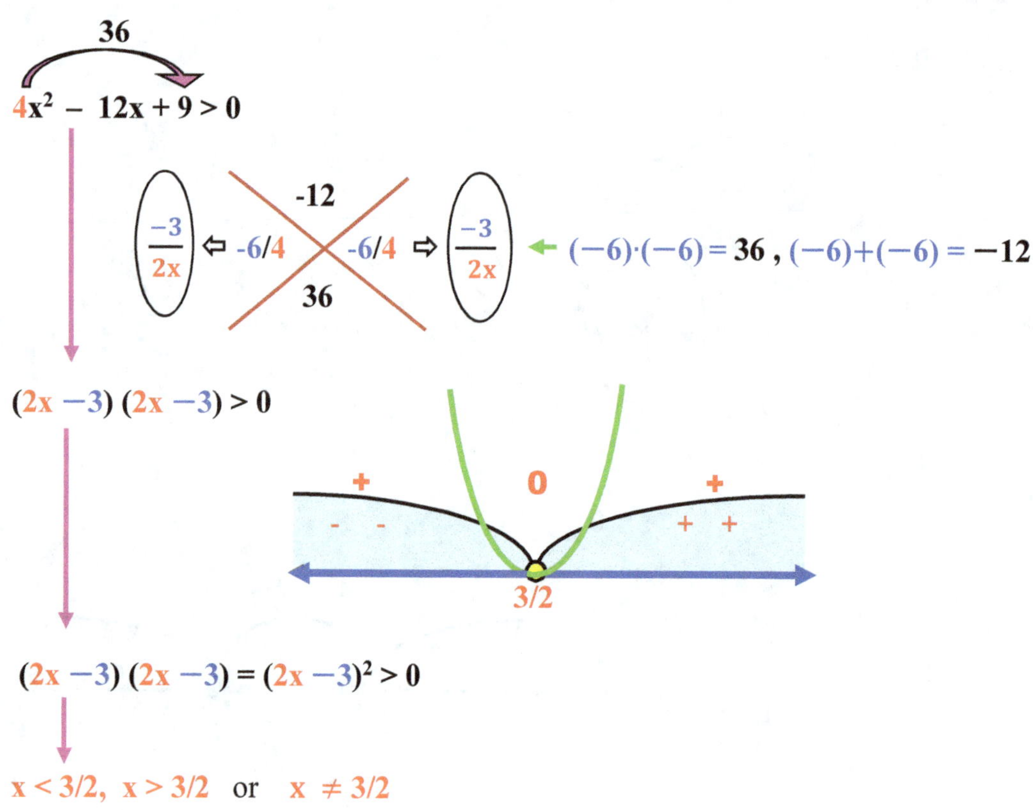

We can represent the solution set of this inequality by **formula**, **graph**, **set**, and **interval notations**.

INEQUALITY	FORMULA NOTATION OF SOLUTIONS	GRAPH NOTATION OF SOLUTIONS	SET NOTATION OF SOLUTIONS	INTERVAL NOTATION OF SOLUTIONS
$4x^2 - 12x + 9 > 0$ ⇕ $(2x-3)(2x-3) > 0$ $D = 0$	$x < 3/2$ or $x > 3/2$ $x \neq 3/2$	3/2	$\{x/\ x < 3/2 \text{ or } x > 3/2 \}$ $\{x/\ x \in R - \{3/2\}\}$	$x \in (-\infty, \frac{3}{2}) \cup (\frac{3}{2}, \infty)$

Example 4 Solve the quadratic inequality $4x^2 - 12x + 9 \leq 0$ by **branch factoring method**.

SOLUTION ▶

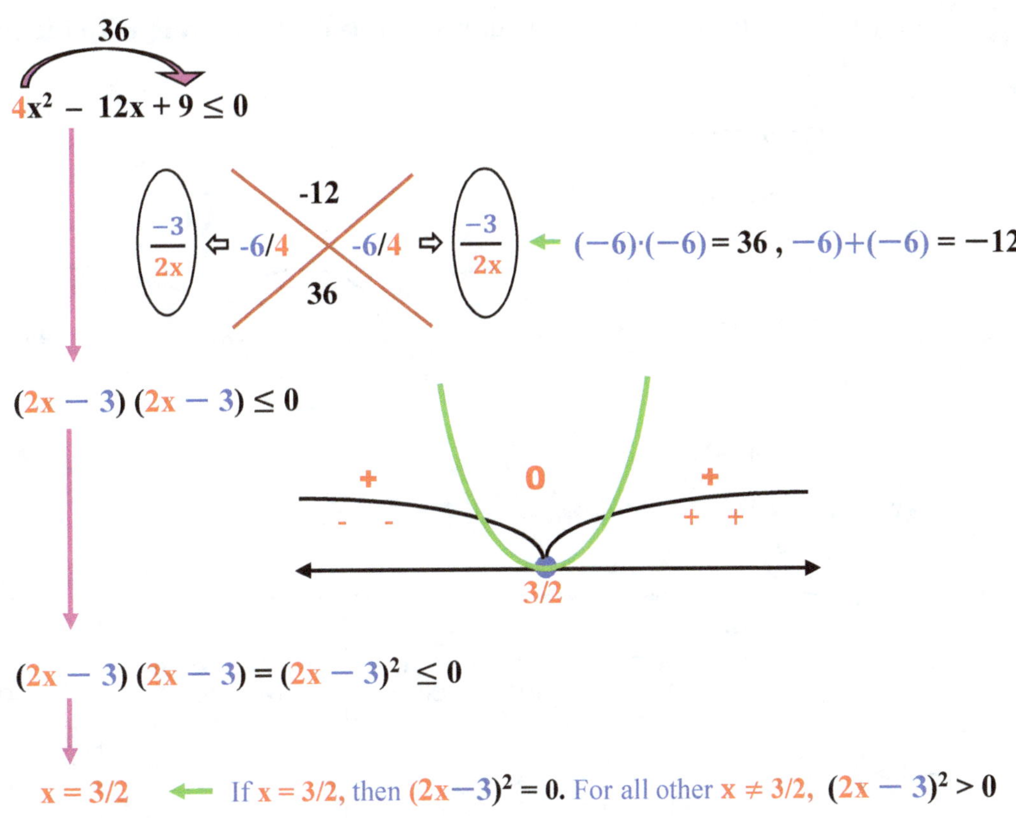

Answer: $x = 3/2$

$\mathbf{W}$e can represent the solution set of this inequality by **formula**, **graph**, **set**, and **interval notations**.

INEQUALITY	FORMULA NOTATION OF SOLUTIONS	GRAPH NOTATION OF SOLUTIONS	SET NOTATION OF SOLUTIONS	INTERVAL NOTATION OF SOLUTIONS
$4x^2 - 12x + 9 \leq 0$ ⇕ $(2x - 3)^2 > 0$ $D = 0$	$x = 3/2$	3/2	$\{x/\ x = 3/2\}$	$x = 3/2$

<u>**METHOD 2.3.**</u> **FACTORING BY "ALGEBRAIC IDENTITIES" FORMULAS**

Sometimes we will use the "Algebraic Identities" to solve the standard quadratic inequalities.

Example 1 Solve the quadratic inequality $x^2 - 4 > 0$ by **factoring with identities formulas**

SOLUTION ▶

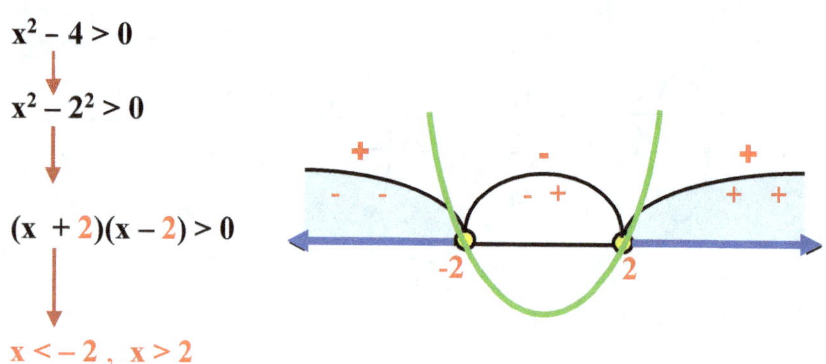

$$x^2 - 4 > 0$$

$$x^2 - 2^2 > 0$$

$$(x + 2)(x - 2) > 0$$

$$x < -2 , \ x > 2$$

Answer: $x \in (-\infty, -2) \cup (2, \infty)$

INEQUALITY	FORMULA NOTATION OF SOLUTIONS	GRAPH NOTATION OF SOLUTIONS	SET NOTATION OF SOLUTIONS	INTERVAL NOTATION OF SOLUTIONS
$x^2 - 4 > 0$ ⇕ $(x+2)(x-2)>0$ $D > 0$	$x <-2 , \ x > 2$		$\{x/x < -2, \ x > 2\}$	$x \in (-\infty, -2) \cup (2, \infty)$

Example 2 Solve the quadratic inequality $x^2 - 9 < 0$ by **factoring with identities formulas**

SOLUTION ▶

$$x^2 - 9 < 0$$

$$x^2 - 3^2 < 0$$

$$(x + 3)(x - 3) < 0$$

$$-3 < x < 3$$

Answer: $x \in (-3, 3)$

Example 3 Solve the quadratic inequality $x^2 - 10x + 25 \geq 0$

SOLUTION ▶

$$x^2 - 10x + 25 \geq 0$$

$$\downarrow$$

$$(x - 5)^2 \geq 0$$

$$\downarrow$$

$$(x - 5)(x - 5) \geq 0$$

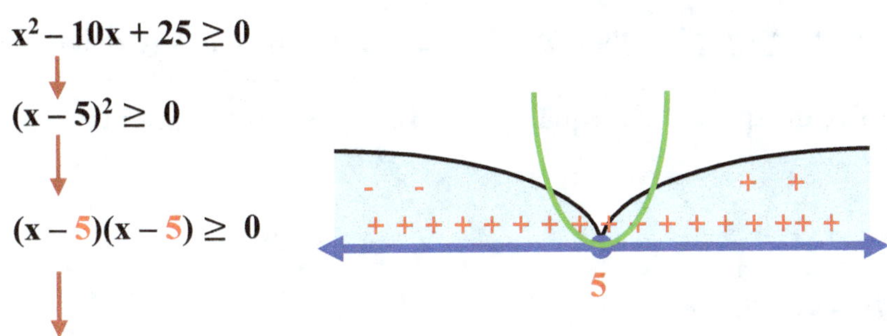

$$\downarrow$$

All real numbers ⟵ Square of any number is positive number or zero.

$$x \in R ; \qquad \longleftrightarrow \qquad ; \; \{\, x \,/\, x \in R \,\}$$

Answer: $x \in (-\infty, \infty)$

Example 4 Solve the quadratic inequality $4x^2 - 25 \leq 0$

SOLUTION ▶

$$4x^2 - 25 \leq 0$$

$$\downarrow$$

$$(2x)^2 - 5^2 \leq 0$$

$$\downarrow$$

$$(2x + 5)(2x - 5) \leq 0$$

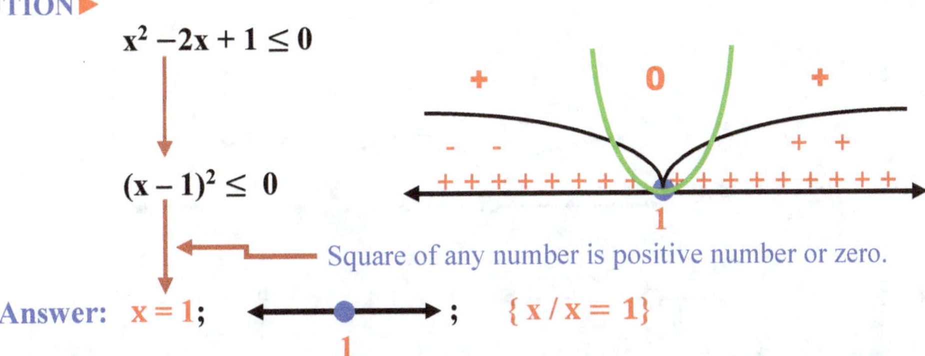

$$\downarrow$$

Answer: $-5/2 \leq x \leq 5/2; \qquad ; \; \{\, x \,/\, -5/2 \leq x \leq 5/2 \,\}; \; x \in (-5/2, 5/2)$

Example 5 Solve the quadratic inequality $x^2 - 2x + 1 \leq 0$

SOLUTION ▶

$$x^2 - 2x + 1 \leq 0$$

$$\downarrow$$

$$(x - 1)^2 \leq 0$$

$$\downarrow$$

⟵ Square of any number is positive number or zero.

Answer: $x = 1; \qquad ; \; \{\, x \,/\, x = 1 \,\}$

<u>**METHOD 2.4.**</u> **FACTORING BY "FACTOR - OUT"**

We will use the **"Factor - out"** method to solve some kinds of standard quadratic inequalities.

Example 1 Solve the quadratic inequality $x^2 - 4x > 0$ by **factor-out method**

SOLUTION▶

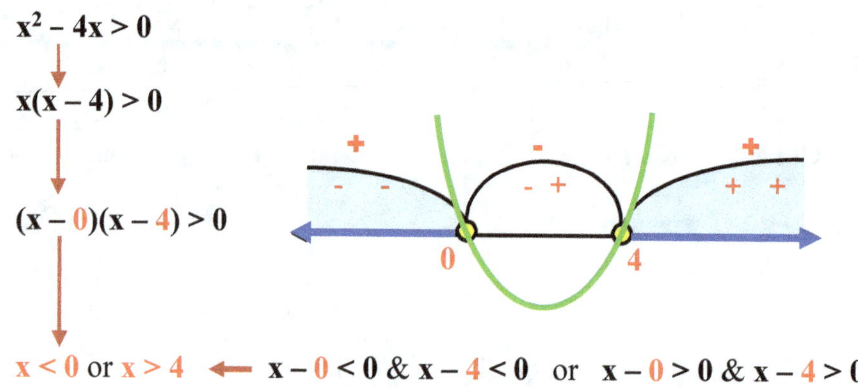

$$x^2 - 4x > 0$$

$$x(x - 4) > 0$$

$$(x - 0)(x - 4) > 0$$

$$x < 0 \text{ or } x > 4 \quad \longleftarrow \quad x - 0 < 0 \ \& \ x - 4 < 0 \quad \text{or} \quad x - 0 > 0 \ \& \ x - 4 > 0$$

Answer: $x \in (-\infty, 0) \cup (4, \infty)$

INEQUALITY	FORMULA NOTATION OF SOLUTIONS	GRAPH NOTATION OF SOLUTIONS	SET NOTATION OF SOLUTIONS	INTERVAL NOTATION OF SOLUTIONS
$x^2 - 4x > 0$ $\Updownarrow$ $(x - 0)(x - 4) > 0$ $D > 0$	$x < 0$, $x > 4$	0 4	$\{x/x < 0 \text{ or } x > 4\}$	$x \in (-\infty, 0) \cup (4, \infty)$

Example 2 Solve the quadratic inequality $2x^2 - 6x < 0$

SOLUTION▶

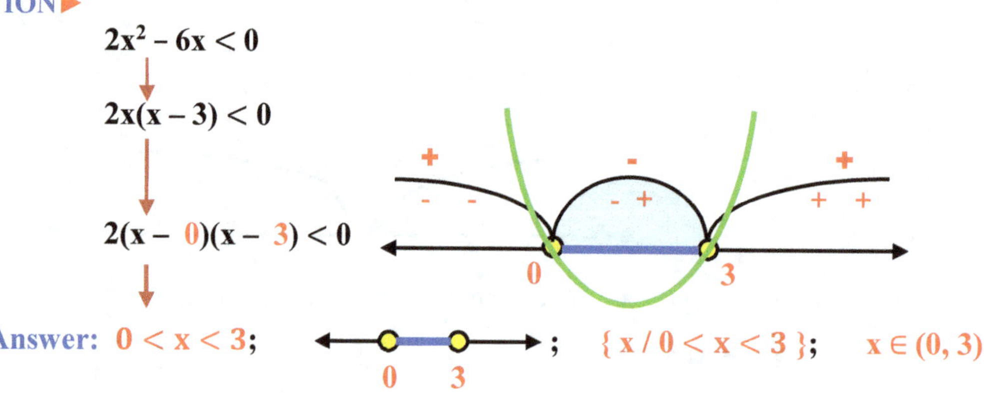

$$2x^2 - 6x < 0$$

$$2x(x - 3) < 0$$

$$2(x - 0)(x - 3) < 0$$

Answer: $0 < x < 3$; $\quad$; $\{x / 0 < x < 3\}$; $\quad x \in (0, 3)$

 Solve the quadratic inequality $2x^2 - 8x + 8 > 0$

SOLUTION ▶

$$2x^2 - 8x + 8 > 0$$

$$2(x - 2)^2 > 0$$

$$2(x - 2)(x - 2) > 0$$

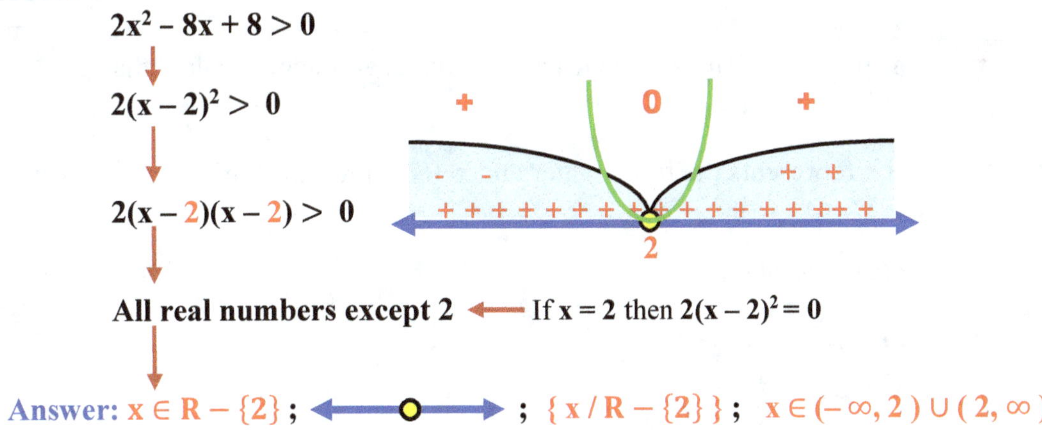

All real numbers except 2 ◀— If $x = 2$ then $2(x - 2)^2 = 0$

Answer: $x \in R - \{2\}$; ⟵ ● ⟶ ; $\{x / R - \{2\}\}$; $x \in (-\infty, 2) \cup (2, \infty)$

Example 4 Solve the quadratic inequality $3x^2 - 75 \leq 0$

SOLUTION ▶

$$3x^2 - 75 \leq 0$$

$$3[(x)^2 - 5^2] \leq 0$$

$$3(x + 5)(x - 5) \leq 0$$

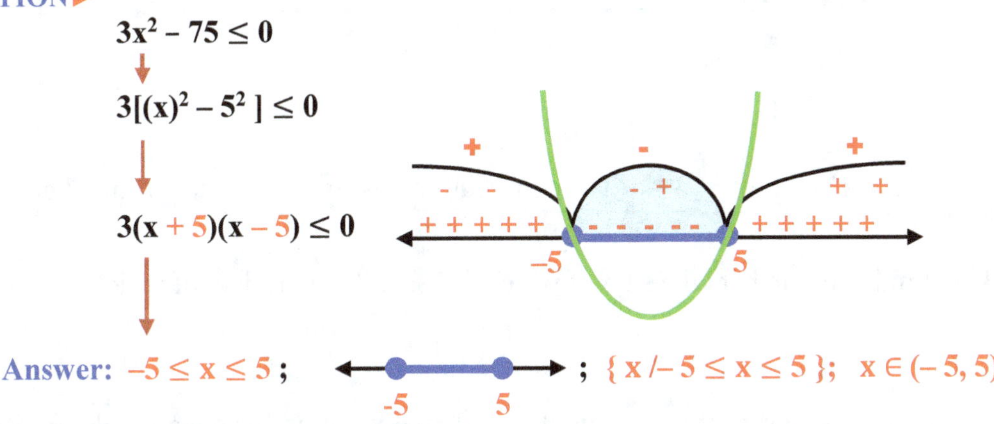

Answer: $-5 \leq x \leq 5$; ⟵ ●——● ⟶ ; $\{x / -5 \leq x \leq 5\}$; $x \in (-5, 5)$
-5   5

Example 5 Solve the quadratic inequality $-4x^2 + 8x - 4 \geq 0$

SOLUTION ▶

$$-4x^2 + 8x - 4 \geq 0$$

$$-4(x - 1)^2 \geq 0$$

$$4(x - 1)^2 \leq 0 \quad \longleftarrow \text{ Square of any number is positive number or zero.}$$

$$x = 1 ; \quad \longleftarrow ● \longrightarrow ; \quad \{x / x = 1\}$$
1

Example 1 The product of **two consecutive positive even** integers is less than **24**. Find the integers.

SOLUTION▶ Let **x** represents the first integer and **x + 2** represents the second integer. So, we can write:

$$x(x+2) < 24$$

$$x^2 + 2x < 24$$

$$x^2 + 2x - 24 < 0$$

$$-4 \quad 6$$

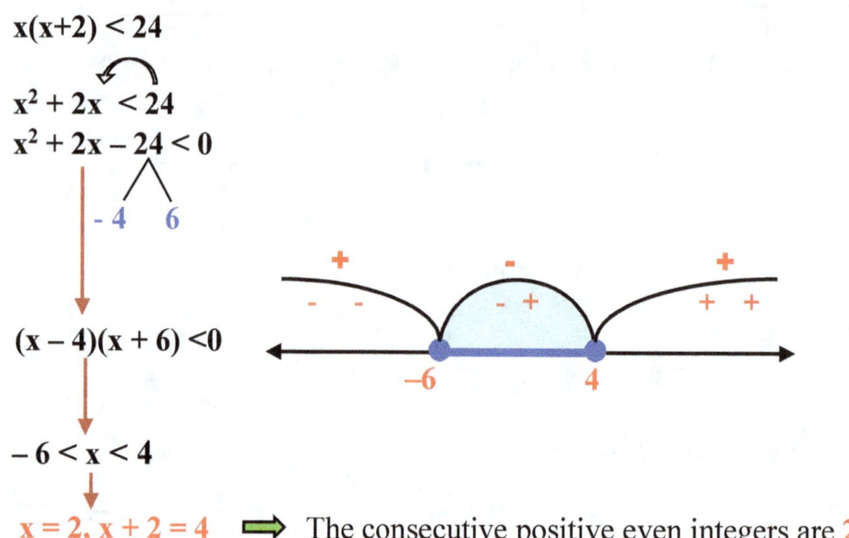

$$(x - 4)(x + 6) < 0$$

$$-6 < x < 4$$

$$x = 2, \; x + 2 = 4 \implies \text{The consecutive positive even integers are 2 and 4.}$$

Example 2 Find two numbers the sum of which is **18**, and the product of which is ≥ 81.

SOLUTION▶

Let **x** represents the first number and **18 − x** represents the second number. So, we can write:

$$x(18 - x) \geq 81$$

$$18x - x^2 \geq 81$$

$$0 \geq x^2 - 18x + 81$$

$$(x - 9)^2 \leq 0$$

$$\Downarrow$$

$$x = 9 \text{ and } 18 - x = 9 \implies \text{The numbers are 9 and 9.}$$

 Marty has coins. He has twice as many quarters as dimes and three times as many nickels as quarters. The product of numbers of dimes and quarters is less than the difference of **20** and the number of nickels. How many of each coin does he have?

SOLUTION ▶

Let x be the number of dimes. So, the number of quarters will be $2x$, and the number of nickels will be $3(2x) = 6x$. The relation inequality in terms of x will be the following: $x(2x) < 20 - 6x$.

$$2x^2 < 20 - 6x$$

$$2x^2 + 6x - 20 < 0$$
$$(2x^2 + 6x - 20)/2 < 0/2$$
$$x^2 + 3x - 10 < 0$$
$$-2 \cdot 5$$

$$(x - 2)(x + 5) < 0$$

$$-5 < x < 2$$

$$x = 1, 2x = 2, 8x = 8$$

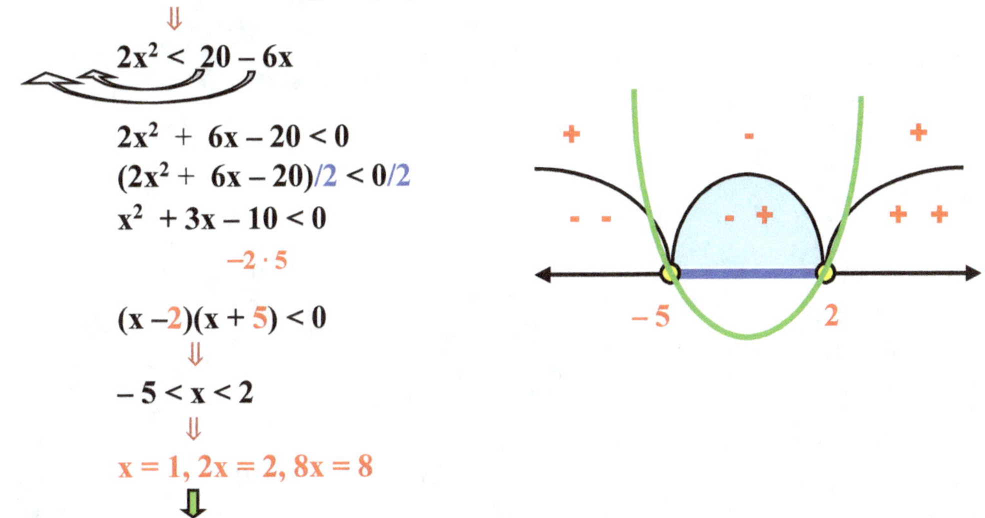

The number of dimes is **1**, the number of quarters is **2**, and the number of nickels is **6**.

 For the given right triangle write and solve an inequality for x.

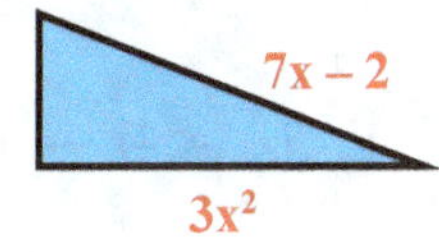

SOLUTION ▶

$$3x^2 < 7x - 2$$
$$3x^2 - 7x + 2 < 0$$
$$(3x - 1)(x - 2) < 0$$

$$-1/3x = -1/3 \quad -6/3 = -2/1x$$

$$1/3 < x < 2; \qquad ; \quad \{x \,/\, \tfrac{1}{3} < x < 2\}; \quad x \in (\tfrac{1}{3}, 2)$$

Solve each quadratic equation by using **Quadratic Formula**.

1. $x^2 - 3x - 4 = 0$	2. $x^2 + 6x - 7 = 0$
3. $x^2 + 2x + 1 = 0$	4. $x^2 - x - 1 = 0$
5. $x^2 + 2x + 2 = 0$	6. $x^2 + 2x - 1 = 0$
7. $x^2 - 4x - 5 = 0$	8. $x^2 - 4x + 3 = 0$
9. $2x^2 - 3x - 5 = 0$	10. $3x^2 - 3x - 6 = 0$
11. $5x^2 - 4x - 1 = 0$	12. $x^2 + x = 12$
13. $-x^2 + 5x - 6 = 0$	14. $x^2 = 8 - 2x$

Solve each quadratic equation by **"Factor Out"** method.

1. $x^2 - 4x = 0$	2. $x^2 + 2x = 0$
3. $x^2 + 4x = 0$	4. $x^2 - 2x = 0$
5. $9x^2 + 3x = 0$	6. $4x^2 - 5x = 0$
7. $x^2 - 4x = 0$	8. $x^2 - 9x = 0$
9. $2x^2 - 4x = 0$	10. $3x^2 - 12x = 0$
11. $5x^2 - 25x = 0$	12. $x^2 = 8x$
13. $-x^2 + 5x = 0$	14. $x^2 = -2x$

Solve each quadratic inequality by "**Factor Out**" method.

1. $x^2 - 2x < 0$	2. $x^2 + x > 0$
3. $x^2 + 5x \geq 0$	4. $x^2 - 4x \leq 0$
5. $12x^2 + 3x > 0$	6. $2x^2 - 5x < 0$
7. $x^2 - x \geq 0$	8. $3x^2 - 6x \leq 0$
9. $2x^2 - x < 0$	10. $3x^2 - 18x > 0$
11. $5x^2 - 15x \geq 0$	12. $x^2 \leq 8x$
13. $-x^2 + 6x > 0$	14. $x^2 < -11x$

Solve each quadratic equation by using **Branch Factoring**.

1. $x^2 - 3x - 10 = 0$

   x =

2. $x^2 + 6x - 7 = 0$

   x =

3. $x^2 + 7x + 10 = 0$

   x =

4. $x^2 - x - 12 = 0$

   x =

5. $x^2 + 3x - 10 = 0$

   x =

6. $x^2 + 6x - 40 = 0$

   x =

7. $x^2 - 8x - 15 = 0$

   x =

8. $x^2 + 2x - 15 = 0$

   x =

9. $x^2 + 9x + 20 = 0$

   x =

10. $x^2 + 8x + 7 = 0$

    x =

11. $x^2 - x - 20 = 0$

    x =

12. $x^2 + x - 12 = 0$

    x =

Solve each quadratic equation by using **Branch Factoring**.

13. $x^2 + 3x - 4 = 0$

$x =$

14. $x^2 - 6x - 7 = 0$

$x =$

15. $x^2 - 7x - 18 = 0$

$x =$

16. $x^2 + 12x + 11 = 0$

$x =$

17. $x^2 + 5x - 14 = 0$

$x =$

18. $x^2 + 6x - 27 = 0$

$x =$

19. $x^2 + 8x + 16 = 0$

$x =$

20. $x^2 + 2x - 24 = 0$

$x =$

21. $x^2 - 10x + 25 = 0$

$x =$

22. $x^2 + 4x - 45 = 0$

$x =$

23. $x^2 - 9x + 14 = 0$

$x =$

24. $x^2 - x - 56 = 0$

$x =$

25. $x^2 + 3x - 40 = 0$

$x =$

26. $x^2 + 2x - 63 = 0$

$x =$

27. $x^2 - x - 30 = 0$

$x =$

28. $x^2 - 16x + 64 = 0$

$x =$

Solve each quadratic inequality by using **Branch Factoring**.

1. $x^2 - 3x - 18 > 0$	**2.** $x^2 + 6x - 16 < 0$
3. $x^2 + 5x + 6 \geq 0$	**4.** $x^2 - x - 12 \leq 0$
5. $x^2 + 3x - 28 > 0$	**6.** $x^2 + 6x - 7 < 0$
7. $x^2 - 8x - 9 \geq 0$	**8.** $x^2 + 2x - 15 \leq 0$
9. $x^2 + 9x - 10 > 0$	**10.** $x^2 + 8x - 9 < 0$
11. $x^2 - x - 30 \geq 0$	**12.** $x^2 + x - 30 \leq 0$

Solve each quadratic equation by using **Big X Factoring**.

1. $2x^2 + 5x + 3 = 0$  x =	**2.** $2x^2 + x - 3 = 0$  x =
3. $3x^2 - x - 4 = 0$ 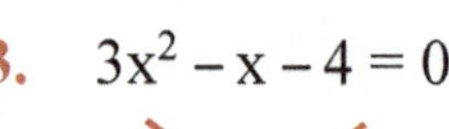 x =	**4.** $4x^2 - 2x - 6 = 0$  x =
5. $7x^2 + 9x + 2 = 0$  x =	**6.** $6x^2 + 8x + 2 = 0$ 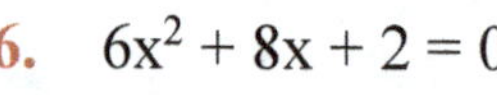 x =
7. $5x^2 - 2x - 3 = 0$ 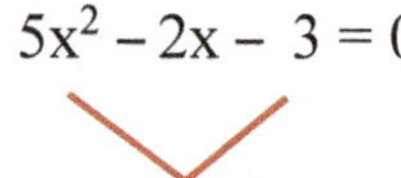 x =	**8.** $3x^2 + 8x + 5 = 0$ 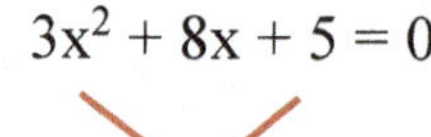 x =
9. $9x^2 - 6x - 3 = 0$ 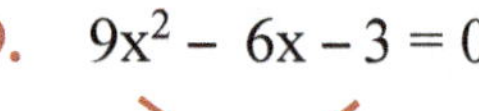 x =	**10.** $5x^2 + 9x + 4 = 0$ 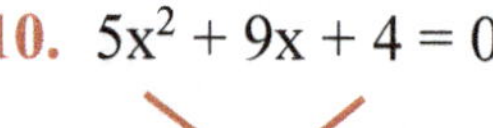 x =

Solve each quadratic equation by using **Big X Factoring**.

11. $7x^2 + 47x - 14 = 0$ 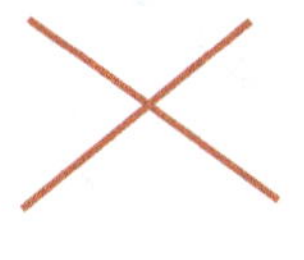 x =	12. $10x^2 + 11x + 1 = 0$ 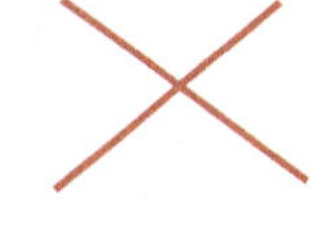 x =
13. $11x^2 - 54x - 5 = 0$ 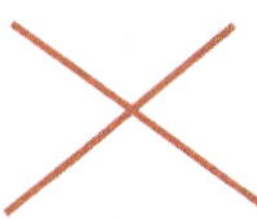 x =	14. $8x^2 - 4x - 4 = 0$  x =
15. $2x^2 + 9x + 7 = 0$ 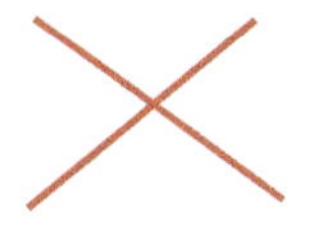 x =	16. $7x^2 - 3x - 4 = 0$  x =
17. $9x^2 + 3x - 6 = 0$ 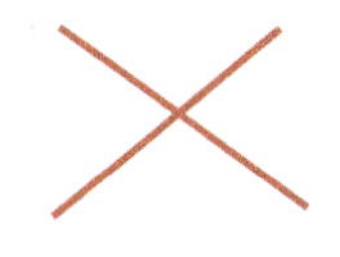 x =	18. $-4x^2 - 2x + 6 = 0$  x =
19. $7x^2 + 5x - 2 = 0$  x =	20. $6x^2 + 5x - 1 = 0$  x =
21. $20x^2 + 22x + 2 = 0$ 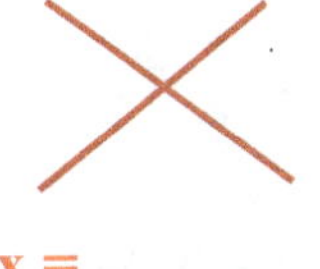 x =	22. $4x^2 + 6x - 10 = 0$  x =
23. $12x^2 + 19x + 5 = 0$ 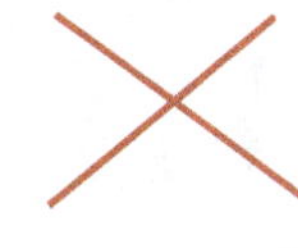 x =	24. $15x^2 - 37x - 8 = 0$ 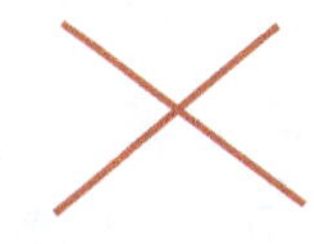 x =

Solve each quadratic inequality by using **Big X Factoring**.

1. $2x^2 + 5x + 3 > 0$ 	**2.** $3x^2 + x - 2 < 0$ 
3. $4x^2 - x - 3 \geq 0$ 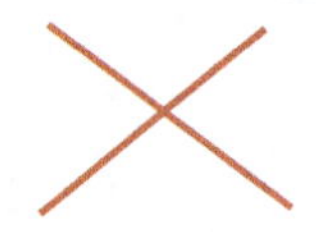	**4.** $6x^2 - 2x - 4 \leq 0$ 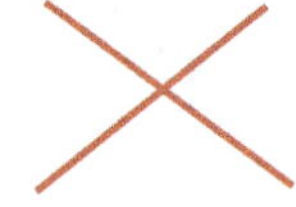
5. $2x^2 + 9x + 7 > 0$ 	**6.** $6x^2 + 8x + 2 < 0$ 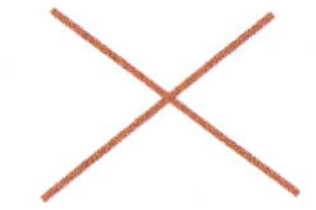
7. $3x^2 - 2x - 5 \geq 0$ 	**8.** $5x^2 + 8x + 3 \leq 0$ 
9. $3x^2 - 6x - 9 > 0$ 	**10.** $4x^2 + 9x + 5 < 0$ 

1. $x^2 - 121 = 0$

$\Downarrow$

$x^2 - 11^2 = 0$
$(x - 11)(x + 11) = 0$

$\Downarrow$

$x = -11, 11$

2. $9x^2 - 24x + 16 = 0$

$\Downarrow$

$(3x)^2 - 2(3x)(4) + (4)^2 = 0$
$(3x - 4)^2 = 0$
$(3x - 4)(3x - 4) = 0$

$3x - 4 = 0$ or $3x - 4 = 0$
$3x = 4$ or $3x = 4$
$3x/3 = 4/3$ or $3x/3 = 4/3$

$\Downarrow$

$x = 4/3$

3. $x^2 + 12x + 36 = 0$

$\Downarrow$

$(x + 6)^2 = 0$

$\Downarrow$

$x = -6$

1. $a^2 - b^2 = (a - b)(a + b)$
2. $(a - b)^2 = a^2 - 2ab + b^2$
3. $(a + b)^2 = a^2 + 2ab + b^2$

Use **Identity Formulas** to factor and solve each quadratic equation.

1. $x^2 - 49 = 0$	**2.** $x^2 + 10x + 25 = 0$
3. $4x^2 - 81 = 0$	**4.** $8x^2 - 24x + 18 = 0$
5. $9x^2 + 25 = 0$	**6.** $12x^2 - 27 = 0$
7. $16x^4 - 81 = 0$	**8.** $25x^2 + 30x + 9 = 0$

Solve each quadratic equation by factoring with **Identity Formulas**.

9. $x^2 - 4 = 0$	10. $x^2 + 4 = 0$
11. $x^2 - 0.36 = 0$	12. $(x + 2)^2 + 12(x + 2) + 36 = 0$
13. $25x^2 - 64 = 0$	14. $100x^4 + 25x^2 = 0$
15. $(x - 1)^4 - 16 = 0$	16. $49x - 4x^3 = 0$
17. $8x^2 - 98x^4 = 0$	18. $\dfrac{1}{16} - \dfrac{1}{2}x^2 + x^4 = 0$
19. $y^4 - 625 = 0$	20. $5x^6 - 125x^4 = 0$
21. $81x^4 - a^4 = 0$ (solve for x)	22. $27x^4 - 48b^4 = 0$ (solve for x)

Solve each quadratic inequality by factoring with **Identity Formulas**.

1. $x^2 - 1 > 0$	2. $x^2 + 4 < 0$
3. $x^2 - 0.4 \geq 0$	4. $(x + 1)^2 - 36 \leq 0$
5. $x^2 - 64 > 0$	6. $100x^2 - 25 < 0$
7. $x^4 - 16 \geq 0$	8. $4x - x^3 \leq 0$
9. $x^2 - 4x^4 > 0$	10. $-\dfrac{1}{4} + x^2 < 0$
11. $y^2 - 121 \geq 0$	12. $5x^2 - 125 \leq 0$
13. $x^2 - a^4 > 0$ (solve for x)	14. $4x^2 - 9b^4 < 0$ (solve for x)

Solve each quadratic equation by factoring with **Identity Formulas**.

15. $x^2 - 2x + 1 = 0$	**16.** $x^2 + 10x + 25 = 0$
17. $4x^2 - 28x + 49 = 0$	**18.** $9x^2 + 48x + 64 = 0$
19. $x^2 + 20x + 100 = 0$	**20.** $x^2 - 24x + 144 = 0$
21. $2x^2 + 20x + 200 = 0$	**22.** $-4x^2 + 44x = 121$
23. $7x^2 - 28x + 28 = 0$	**24.** $9x^2 + 6x = -1$
25. $121x^2 + 22x + 1 = 0$	**26.** $49x^2 + 9 = 42x$
27. $3x^2 + 18x + 27 = 0$	**28.** $50x^2 - 20x + 8 = 0$

Solve each quadratic equation by **Completing the Square** method.

1. $x^2 - 3x - 4 = 0$	2. $x^2 + 5x - 6 = 0$
3. $x^2 + 4x + 4 = 0$	4. $x^2 - 2x - 5 = 0$
5. $x^2 + 2x - 5 = 0$	6. $x^2 + 2x - 3 = 0$
7. $x^2 - 4x + 9 = 0$	8. $x^2 - 3x + 5 = 0$
9. $2x^2 - 4x - 8 = 0$	10. $3x^2 - 3x - 9 = 0$
11. $x^2 - 4x - 4 = 0$	12. $x^2 + 2x = 8$
13. $-x^2 + 5x - 6 = 0$	14. $x^2 = 7 - 2x$

Solve each quadratic inequality by any method.

1. $x^2 + 3x + 2 < 0$	**2.** $x^2 + 6x \geq 7$
3. $2x^2 + x - 3 > 0$	**4.** $x^2 + 4 < 0$
5. $x^2 + 5 > 0$	**6.** $3x + 2\sqrt{x} - 5 \leq 0$
7. $x^2 + 5x < 0$	**8.** $x^2 - 9 \geq -5$
9. $3x^2 + 2x - 5 \geq 0$	**10.** $x^2 + 2x + 1 \leq 0$
11. $x^2 + 5x > 0$	**12.** $4x^2 - 9 < 0$

Solve each quadratic equation and inequality by any method.

13. $x^2 + 2x + 2 = 0$	14. $x^2 + 7x = 8$
15. $2x^2 + 5x - 3 = 0$	16. $x^2 + 4 = 0$
17. $\frac{1}{2}x^2 - \frac{1}{3}x - \frac{1}{6} = 0$	18. $3x + 2\sqrt{x} - 5 = 0$

Simplify.

1. $(1 + \sqrt{-3})(1 - \sqrt{-3})$	2. $[(3 + 2i) - (2 + 3i)](1 + i)$
3. $\dfrac{(3 + 2i)(2 + 3i)}{1 + i}$	4. $i^{124} \cdot i^{215} \cdot (1 - i)^2$
5. $\dfrac{(3 - i)(1 + 2i)}{1 + i}$	6. $\dfrac{i^{99}}{i^{27}} \cdot i^{42} \cdot \left\lvert\dfrac{2+i}{2-i}\right\rvert$

Solve each quadratic inequality with one variable.

1. $x^2 + 11x > 12$	2. $-2x^2 + 5x + 3 < 0$
3. $x^2 + 2x + 1 \le 0$	4. Find the value of **k** so that the given inequality will have one real solution. $x^2 - 4x + 4k \le 0$

4.4. Polynomial Equations and Inequalities

- **POLYNOMIAL EQUATIONS**
- **POLYNOMIAL INEQUALITIES**

■ POLYNOMIAL EQUATIONS

We will use the following definitions.

Definition ▶ *A polynomial equation* is an equation $E_1 = E_2$ where E_1 or E_2 or both E_1 and E_2 are polynomial expressions with one or more variables.

Definition ▶ *A polynomial equation with one variable* is an equation $E_1 = E_2$ where E_1 or E_2 or both E_1 and E_2 are polynomial expressions with one variable.

Definition ▶ *A quadratic equation with one variable* is any equation that can be presented in the form $ax^2 + bx + c = 0$, where $a, b, c \in \mathbf{R}$ and $a \neq 0$.
This is a **standard form** of a general quadratic equations with one variable, where **x** represents a variable, and the constants **a**, **b**, and **c** are respectively called the quadratic (or leading) coefficient, the linear (or middle) coefficient and the free constant term.

Definition ▶ *A single-variable Polynomial Equation of degree n* is an equation in form

$$a_n x^n + a_{n-1} x^{n-1} + a_{n-2} x^{n-2} + \ldots + a_1 x + a_0 = 0$$

where $\mathbf{a_n}, \mathbf{a_{n-1}}, \mathbf{a_{n-2}}, \ldots, \mathbf{a_1}, \mathbf{a_0}$ are any complex/real numbers with $\mathbf{a_n} \neq \mathbf{0}$; $\mathbf{n} \in \mathbf{N}$.

Definition ▶ *A single-variable Polynomial Equation of degree 1* is called an equation in form

$$a_1 x + a_0 = 0$$

where $\mathbf{a_1}, \mathbf{a_0}$ complex/real numbers; and $\mathbf{a_1} \neq \mathbf{0}$ (or $\mathbf{ax + b = 0}$ and $\mathbf{a} \neq \mathbf{0}$).

Definition ▶ *A single-variable Polynomial Equation of degree 2* is called an equation in form

$$a_2 x^2 + a_1 x + a_0 = 0$$

where $\mathbf{a_2}, \mathbf{a_1}, \mathbf{a_0} \in \mathbf{C}$ (or $\mathbf{R}$); and $\mathbf{a_2} \neq \mathbf{0}$ (or $\mathbf{ax^2 + bx + c = 0}$ where $\mathbf{a} \neq \mathbf{0}$).

Examples Equations $x + 2 = 0$, $x^2 - 8x + 7 = 0$, $x^3 + 2x^2 - x - 2 = 0$, $x^4 - 5x^2 + 4 = 0$, $4x^5 = 8$, and $x^8 - 1 = 0$ are polynomial equations with one variable.

We have already studied polynomial equations of degree **1** and **2**. In this section we will study single-variable polynomial equations of degree $\mathbf{n} \geq \mathbf{3}$.

We will use the following important theorems to solve single-variable polynomial equations.

▶ **Every single-variable Polynomial Equation of degree n ≥ 1 has exactly n solutions** (or **roots**). Each solution/root has to be counted up to its **multiplicity.**

The Factor Theorem

▶ The factor theorem states that a polynomial $P(x)$ has a factor $(x - k)$ if and only if $P(k) = 0$.

or $P(x) = (x - k) Q(x)$ if and only if $P(k) = 0$

The Complete Factorization Theorem

▶ If $P(x) = a_n x^n + a_{n-1} x^{n-1} + a_{n-2} x^{n-2} + \ldots + a_1 x + a_0$ is a single-variable Polynomial Equation of degree **n** ($a_n \neq 0$ and $n \geq 1$), then there are complex/real numbers $k_1, k_2, \ldots, k_n$ such that

$$P(x) = a_n(x - k_1)(x - k_2)(x - k_3) \ldots (x - k_n)$$

The Conjugate Pairs Theorem

▶ If $P(x) = a_n x^n + a_{n-1} x^{n-1} + a_{n-2} x^{n-2} + \ldots + a_1 x + a_0 = 0$ is a single-variable Polynomial Equation of degree **n** ($a_n \neq 0$ and $n \geq 1$) with real coefficients ($a_n, a_{n-1}, \ldots, a_1, a_0 \in R$) and the complex number $c = a + i b$ is a root of the equation $P(x) = 0$, then complex conjugate number $\bar{c} = a - i b$ is also a root of equation $P(x) = 0$.

SOLVING POLYNOMIAL EQUATIONS

We will consider some polynomial equations with one variable and with degree $n \geq 3$.

Example 1 Solve the polynomial equation $x^5 - 32 = 0$.

SOLUTION ▶ To solve this polynomial equation of degree **5** we will use addition property of equality

$$x^5 - 32 = 0$$
$$+32 \quad +32$$
$$x^5 = 32$$
$$x = \sqrt[5]{32}$$
$$\Downarrow$$
$$x = 2$$

Example 2 Solve the polynomial equation $x^3 - 4x = 0$.

SOLUTION ▶ To solve this polynomial equation of degree **3** we will use the "factor out" method

$$x^3 - 4x = 0$$
$$x(x^2 - 4) = 0$$
$$x(x - 2)(x + 2) = 0$$
$$x = 0, \ x - 2 = 0, \ x + 2 = 0$$
$$\Downarrow$$
$$x = 0, \ 2, \ -2$$

Example 3 Solve the polynomial equation $x^3 - 16x = 0$.

SOLUTION ▶ To solve this polynomial equation of degree **3** we will use the combination of factoring by "factor out" and factoring by "identity formulas" methods

$$x^3 - 16x = 0$$
$$x(x^2 - 16) = 0$$
$$x(x^2 - 4^2) = 0$$
$$x(x - 4)(x + 4) = 0$$
$$x = 0, \ x - 4 = 0, \ x + 4 = 0$$
$$\Downarrow$$
$$x = 0, \ 4, \ -4$$

Example 4 Solve the polynomial equation $x^3 - 2x^2 - 9x + 18 = 0$.

SOLUTION ▶ To solve this polynomial equation we will use the combination of factoring by "groups" and factoring by "identity formulas" methods

$$x^3 - 2x^2 - 9x + 18 = 0$$
$$x^2(x - 2) - 9(x - 2) = 0$$
$$(x - 2)(x^2 - 9) = 0$$
$$(x - 2)(x^2 - 3^2) = 0$$
$$(x - 2)(x - 3)(x + 3) = 0$$
$$x - 2 = 0, \ x - 3 = 0, \ x + 3 = 0$$
$$\Downarrow$$
$$x = 2, \ x = 3, \ x = -3$$

Example 5   Solve the equation $x^4 - 1 = 0$.

SOLUTION ▶ To solve this polynomial equation of degree **4** we will use the factoring by "identity formulas" method

$$x^4 - 1 = 0$$
$$(x^2)^2 - 1^2 = 0$$
$$(x^2 - 1)(x^2 + 1) = 0$$
$$[(x)^2 - 1^2][(x)^2 - i^2] = 0$$
$$(x - 1)(x + 1)(x - i)(x + i) = 0$$
$$x - 1 = 0, \quad x + 1 = 0, \quad x - i = 0, \quad x + i = 0$$
$$\Downarrow$$
$$x = 1, \quad x = -1, \quad x = i, \quad x = -i$$

Example 6   Solve $2x^5 - 10x^3 + 8x = 0$

SOLUTION ▶ To solve this polynomial equation of degree **5** we will use the combination of factoring by "factor out", "miscellaneous factoring", factoring by "identity formulas", and "branch factoring" methods.

$$2x^5 - 10x^3 + 8x = 0$$
$$2x(x^4 - 5x^2 + 4) = 0 \qquad \text{factor out}$$
$$2x(x^4 - 4x^2 - x^2 + 4) = 0 \qquad \text{miscellaneous factoring}$$
$$2x(x^4 - 4x^2 + 4 - x^2) = 0$$
$$2x[(x^2 - 2)^2 - x^2] = 0 \qquad \text{identity formula}$$
$$2x(x^2 - 2 - x)(x^2 - 2 + x) = 0$$
$$2x(x^2 - x - 2)(x^2 + x - 2) = 0 \qquad \text{branch factoring}$$

$$2x(x - 2)(x + 1)(x - 1)(x + 2) = 0$$
$$x = 0, \quad x - 2 = 0, \quad x + 1 = 0, \quad x - 1 = 0, \quad x + 2 = 0$$
$$\Downarrow$$
$$x = 0, \quad x = 2, \quad x = -1, \quad x = 1, \quad x = -2$$

Example 7   Solve the polynomial equation $x^7 + 128 = 0$.

SOLUTION ▶ To solve this polynomial equation we will use addition property of equality

$$x^7 + 128 = 0$$
$$ -128 \quad -128$$
$$x^7 = -128$$
$$x = \sqrt[7]{-128}$$
$$\Downarrow$$
$$x = -2$$

■ POLYNOMIAL INEQUALITIES

We will use the following definitions.

Definition ▶ *A polynomial inequality* is an inequality $E_1 < E_2$ (or $E_1 > E_2$, or $E_1 \leq E_2$, or $E_1 \geq E_2$) where E_1 or E_2 or both E_1 and E_2 are polynomial expressions with one or more variables.

Definition ▶ *A polynomial inequality with two variables* is an inequality $E_1 < E_2$ (or $E_1 > E_2$, or $E_1 \leq E_2$, or $E_1 \geq E_2$) where E_1 or E_2 or both E_1 and E_2 are polynomial expressions with two variables.

Definition ▶ *A polynomial inequality with one variable* is an inequality $E_1 < E_2$ (or $E_1 > E_2$, or $E_1 \leq E_2$, or $E_1 \geq E_2$) where E_1 or E_2 or both E_1 and E_2 are polynomial expressions with only one variable.

Definition ▶ *A polynomial inequality with one variable in standard form* is any inequality that can be presented in one of the following forms:

$$a_n x^n + a_{n-1} x^{n-1} + a_{n-2} x^{n-2} + \ldots + a_1 x + a_0 < 0$$

$$a_n x^n + a_{n-1} x^{n-1} + a_{n-2} x^{n-2} + \ldots + a_1 x + a_0 > 0$$

$$a_n x^n + a_{n-1} x^{n-1} + a_{n-2} x^{n-2} + \ldots + a_1 x + a_0 \leq 0$$

$$a_n x^n + a_{n-1} x^{n-1} + a_{n-2} x^{n-2} + \ldots + a_1 x + a_0 \geq 0$$

where a_n, a_{n-1}, a_{n-2}, ..., a_1, a_0 are any numbers, $a_n \neq 0$; $n \in N$.

In each of these inequalities x is a variable, the constants a_n, a_{n-1}, ..., a_1, a_0 are coefficients, and a_n leading coefficient.

Examples The inequalities $x^5 + y^3 + 2z^2 < 0$, $y^4 + 1 > 0$, $x^4 - 81 \leq 0$, and $x^6 + 3x \geq 0$ are polynomial

The inequalities $x^3 - 4x > 0$, $y^4 - 2y^2 < 0$, $4x^4 + 4x^2 + 1 \geq 0$, and $x^4 - 16 \leq 0$ are quadratic inequalities with one variable in standard form.

In this section we will solve several polynomial inequalities.

Example 1 Solve the polynomial inequality $x^4 + 3x^3 - 4x^2 \geq 0$.

SOLUTION ▶

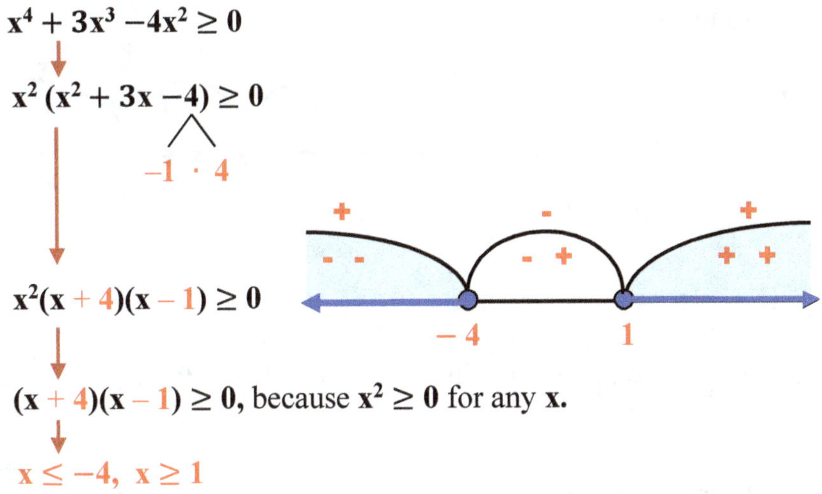

$$x^4 + 3x^3 - 4x^2 \geq 0$$

$$x^2(x^2 + 3x - 4) \geq 0$$

$$-1 \cdot 4$$

$$x^2(x + 4)(x - 1) \geq 0$$

$(x + 4)(x - 1) \geq 0$, because $x^2 \geq 0$ for any x.

$$x \leq -4, \ x \geq 1$$

Answer: $x \in (-\infty, -4] \cup [1, \infty)$

Example 2 Solve the polynomial inequality $x^6 - 4x^5 + 4x^4 \geq 0$.

SOLUTION ▶

$$x^6 - 4x^5 + 4x^4 \geq 0$$

$$x^4(x^2 - 4x + 4) \geq 0$$

$$(-2) \cdot (-2)$$

$$x^4(x - 2)(x - 2) \geq 0$$

$$(x - 2)(x - 2) \geq 0$$

All real numbers

Answer: $x \in (-\infty, \infty)$

Example 3 Solve the polynomial inequality $x^3 - x \le 0$.

SOLUTION ▶

$$x^3 - x \le 0$$
$$x(x^2 - 1) \le 0$$

$$(x + 0)(x + 1)(x - 1) \le 0$$

$$x \le -1, \quad 0 \le x \le 1$$

Answer: $x \in (-\infty, -1] \cup [0, 1]$

Example 4 Solve the polynomial inequality $-x^4 - 2x^2 + 8x^2 > 0$.

SOLUTION ▶

$$-x^2(x^2 + 2x - 8) > 0$$
$$x^2 + 2x - 8 < 0, \quad \text{because } -x^2 \le 0 \text{ for any } x.$$

$$(4) \cdot (-2)$$

$$(x + 4)(x - 2) < 0$$

$$-4 < x < 2$$

Answer: $x \in (-4, 2)$

Example 5 Solve the polynomial inequality $x^6 + 3x^5 - 40x^4 \ge 0$.

SOLUTION ▶

$$x^6 + 3x^5 - 40x^4 \ge 0$$
$$x^4(x^2 + 3x - 40) \ge 0$$
$$x^2 + 3x - 40 \ge 0, \text{ because } x^4 \ge 0 \text{ for any } x.$$

$$(8) \cdot (-5)$$

$$(x + 8)(x - 5) \ge 0$$

$$x \le -8 \text{ or } x \ge 5$$

Answer: $x \in (-\infty, -8] \cup [5, \infty)$

Example 6 Solve the polynomial inequality $4x^4 - 12x^3 + 9x^2 > 0$.

SOLUTION▶

$$4x^4 - 12x^3 + 9x^2 > 0$$
$$x^2(4x^2 - 12x + 9) > 0$$
$4x^2 - 12x + 9 > 0$, because $x^2 \geq 0$ for any $\mathbf{x}$.

$$\left(\frac{-3}{2x}\right) \Leftarrow -6/4 \quad -6/4 \Rightarrow \left(\frac{-3}{2x}\right) \quad \leftarrow (-6)\cdot(-6) = 36 , (-6)+(-6) = -12$$

-12 , 36

$$(2x - 3)(2x - 3) > 0$$

$$(2x - 3)(2x - 3) = (2x - 3)^2 > 0$$

$x < 3/2, \ x > 3/2$ or $x \neq 3/2$

Answer: $x \in (-\infty, 3/2) \cup (3/2, \infty)$

Example 7 Solve the polynomial inequality $x^4 - 16 > 0$.

SOLUTION▶

$$x^4 - 16 > 0$$
$$(x^2 - 4)(x^2 + 4) > 0$$
$(x^2 - 4) > 0$, because $x^2 + 4 > 0$ for any $\mathbf{x}$.

$$(x + 2)(x - 2) > 0$$

$x < -2 , \ x > 2$

Answer: $x \in (-\infty, 2) \cup (2, \infty)$

Solve each polynomial equation with one variable.

1. $x^5 - 25x^3 = 0$	**2.** $x^4 - 81 = 0$
3. $x^5 - 9x^3 = 0$	**4.** $x^4 - 2x^3 - 3x^2 = 0$
5. $x^3 + x^2 + x + 1 = 0$	**6.** Use the Factor Theorem to proof that $x - 1$ is a factor of $x^{2012} - 1$

Solve each polynomial inequality with one variable.

1. $x^6 - 81x^2 < 0$	**2.** $-2x^4 + 5x^3 + 3x^2 < 0$
3. $x^4 + 8x^2 - 9 \geq 0$	**4.** $x^5 - 5x^3 + 4x > 0$
5. $x^4 + 5x^3 - 6x \geq 0$	**6.** $x^6 - x^5 - 6x^4 \leq 0$

4.5. Fractional Equations and Inequalities

- **FRACTIONAL EQUATIONS**
- **FRACTIONAL INEQUALITIES**

■ FRACTIONAL EQUATIONS

We will use the following definitions.

> **Definition** ▶ *A fractional (rational) equation* is an equation $E_1 = E_2$ where E_1 or E_2 or both E_1 and E_2 are fractional expressions with one or more variables.
>
> **Definition** ▶ *A fractional (rational) equation* is any equation that can be rewritten in the form $P/Q = 0$, where P and Q are any expressions, and $Q \neq 0$.
>
> **Definition** ▶ *A single-variable fractional equation* is an equation $E_1 = E_2$ where E_1 or E_2 or both E_1 and E_2 are fractional expressions with one variable.
>
> **Definition** ▶ *A single-variable fractional equation* is any equation that can be represented in the form $P(x)/Q(x) = 0$, where $P(x)$ and $Q(x)$ are any expressions with one variable, and $Q(x) \neq 0$.

Examples Equations $\dfrac{x}{2} + 3 = x - 1$ and $\dfrac{x}{3y} = \dfrac{x}{5y} + 4$ are fractional (rational) equations.

Equations $\dfrac{x}{2} + 3 = x - 1$ and $\dfrac{x+2}{x-1} = \dfrac{x}{3}$ are fractional equations with one variables.

To solve fractional equations we will use the **cross multiplication** and **LCD** methods, as well as the method with **moving all terms of equation to the left or to the right side of equation**.

CROSS MULTIPLICATION METHOD

This method we will use if fractional equation is a **proportion like** the equation $\dfrac{P}{Q} = \dfrac{R}{S}$.

Let we have a **proportion like** the equation with one variable $\dfrac{P}{Q} = \dfrac{R}{S}$.

To solve this equation, first of all, we have to find restrictions for the variable ($Q \neq 0$, $S \neq 0$), then multiply crossly denominators with numerators ($PS = QR$). After that we have to solve the corresponding **polynomial equation PS = QR.** On the final phase we have to compare each solution of equation $PS = QR$ with restrictions $Q \neq 0$, $S \neq 0$ and as the solutions of the initial fractional equation $\dfrac{P}{Q} = \dfrac{R}{S}$ write only those solutions of equivalent polynomial equation $PS = QR$, which are not in conflict with restrictions $Q \neq 0$, $S \neq 0$.

We can also just multiply crossly denominators with numerators (**PS = QR**) and solve the corresponding **polynomial equation PS = QR.** Then we have to check what solutions of polynomial equation will satisfy initial fractional equation. We have to exclude from the solution set of the corresponding **polynomial equation PS = QR** any value that makes denominators of initial fractional equation $\dfrac{P}{Q} = \dfrac{R}{S}$ equal to **0 (Q ≠ 0, S ≠ 0)**.

Example 1 Solve the fractional equation $\dfrac{2}{x+1} = \dfrac{1}{x-1}$

SOLUTION▶

$$\dfrac{2}{x+1} = \dfrac{1}{x-1}$$

Restrictions for x: **x + 1 ≠ 0; x − 1 ≠ 0; or**
x ≠ − 1; x ≠ 1

$$2(x-1) = x+1$$

$$2x - 2 = x + 1$$

comparing:

Possible solution **x = 3**, is not in the conflict with restrictions for x: **x ≠ −1; x ≠ 1 (3 ≠ −1; 3 ≠ 1)**, so x = 3 is a unique solution.

$$2x - x = 1 + 2$$
$$x = 3$$

Answer: **x = 3**

Check: $\dfrac{2}{3+1} = \dfrac{1}{3-1} \Rightarrow \dfrac{2}{4} = \dfrac{1}{2} \Rightarrow \dfrac{1}{2} = \dfrac{1}{2}$ (true statement)

Example 2 Solve the fractional equation $\dfrac{2}{x-2} = \dfrac{x}{x-2}$

SOLUTION▶

$$\dfrac{2}{x-2} = \dfrac{x}{x-2}$$

Restrictions for x: **x − 2 ≠ 0;**
x ≠ 2

$$2x - 4 = x^2 - 2x$$

$$x^2 - 4x + 4 = 0$$
$$(x-2)^2 = 0$$
$$x = 2$$

comparing

Possible solution **x = 2**, is in the conflict with restriction for x: **x ≠ 2**, so we have no any solution.

Answer: **No solution**

Check: $\dfrac{2}{2-2} = \dfrac{2}{2-2} \Rightarrow \dfrac{2}{0} = \dfrac{2}{0} \Rightarrow$ Solution **x = 2** is not a solution for the initial fractional because that makes denominators of the initial fractional equation equal to **0.**

Example 3 Solve the fractional equation $\dfrac{x+1}{2}=\dfrac{4}{x-1}$

SOLUTION ▶

$$\dfrac{x+1}{2}=\dfrac{4}{x-1}$$

Restrictions for x: $x-1\neq 0$;

$$x\neq 1$$

$$x^2-1=8$$
$$x^2=9$$
$$\sqrt{x^2}=\sqrt{9}$$
$$|x|=3$$
$$x=3,\,-3$$

comparing

Possible solutions $x=3$ and $x=-3$ are not in the conflict with restriction for x: $x\neq 1$ ($3\neq 1$; $-3\neq 1$), so we have 2 solutions: $x=3$ and $x=-3$.

Answer: $x=3,\ x=-3$

Check: $\dfrac{3+1}{2}=\dfrac{4}{3-1}\ \Rightarrow\ \dfrac{4}{2}=\dfrac{4}{2}$ (true statement)

$$\dfrac{-3+1}{2}=\dfrac{4}{-3-1}\ \Rightarrow\ \dfrac{-2}{2}=\dfrac{4}{-4}\ \Rightarrow\ -1=-1\ \text{(true statement)}$$

Example 4 Solve the fractional equation $\dfrac{x-2}{3}=\dfrac{x}{x+2}$

SOLUTION ▶

$$\dfrac{x-2}{3}=\dfrac{x}{x+2}$$

Restrictions for x: $x+2\neq 0$;

$$x\neq -2$$

$$x^2-4=3x$$
$$x^2-3x-4=0$$

$$-4\quad 1$$

$$(x-4)(x+1)=0$$
$$x=4,\,-1$$

comparing

Possible solutions $x=4$ and $x=-1$ are not in the conflict with restriction for x: $x\neq -2$ ($4\neq -2$; $-1\neq -2$), so we have 2 solutions: $x=4$ and $x=-1$.

Answer: $x=4,\ x=-1$

Check: $\dfrac{4-2}{3}=\dfrac{4}{4+2}\ \Rightarrow\ \dfrac{2}{3}=\dfrac{2}{3}$ (true statement)

$$\dfrac{-1-2}{3}=\dfrac{-1}{-1+2}\ \Rightarrow\ \dfrac{-3}{3}=\dfrac{-1}{-1}\ \Rightarrow\ -1=-1\ \text{(true statement)}$$

LCD METHOD

To solve another type of fractional (rational) equations, **proportion unlike** equations (or equations like **P/Q + T = R/S; P/Q = R/S + W; P/Q + U = R/S + V**, where **Q ≠ 0, S ≠ 0**, and **P, Q, R, S, T, U, V**, and **W** are real numbers or expressions), we use **LCD method**. This method includes the following steps: finding restrictions for the variable of the given equation, finding LCD, multiplying each term of the equation by LCD, simplifying each term, solving corresponding polynomial equation, comparing solutions of the polynomial equation with restrictions, and selecting some of the solutions of polynomial equation as the solutions of the initial fractional equation, which are not in the conflict with restrictions of the variable.

To solve the initial fractional equation we can also start from finding LCD and multiplying all the terms by LCD. Then we have to check all possible solutions of corresponding polynomial equation and select only the solutions which satisfy the original fractional equation.

Example 1   Solve the fractional equation $\dfrac{2}{x+2} + \dfrac{2}{x-2} = 1 - \dfrac{1}{x^2-4}$

SOLUTION▶

$$\frac{2}{x+2} + \frac{2}{x-2} = 1 - \frac{1}{x^2-4}$$

$(x-2)(x+2)$

$$\frac{2(x+2)(x-2)}{x+2} + \frac{2(x+2)(x-2)}{x-2} = 1(x+2)(x-2) - \frac{1(x+2)(x-2)}{x^2-4}$$

$$2(x-2) + 2(x+2) = (x+2)(x-2) - 1$$
$$2x - 4 + 2x + 4 = x^2 - 4 - 1$$

$$4x = x^2 - 5$$
$$x^2 - 4x - 5 = 0$$

$-5 \quad 1$

$$(x-5)(x+1) = 0$$

comparing

$$x = -1, \ x = 5$$

Answer: $x = -1, \ x = 5$

Restrictions for x:
$x + 2 \neq 0; \ x - 2 \neq 0;$
$x^2 - 4 \neq 0,$ or
$x \neq -2; \ x \neq 2$

$\text{LCD} = (x+2)(x-2)$

Possible solutions $x = -1,$
$x = 5$ are not in the conflict with restriction for x: $x \neq 2, x \neq -2$
$(-1 \neq 2; \ -1 \neq -2; \ 5 \neq 2; \ 5 \neq -2)$
So set of solutions is $\{-1, 5\}$.

Check: $\dfrac{2}{-1+2} + \dfrac{2}{-1-2} = 1 - \dfrac{1}{1-4} \Rightarrow 2 - \dfrac{2}{3} = 1 + \dfrac{1}{3} \Rightarrow 1\dfrac{1}{3} = 1\dfrac{1}{3}$ (true statement)

$\dfrac{2}{5+2} + \dfrac{2}{5-2} = 1 - \dfrac{1}{25-4} \Rightarrow \dfrac{2}{7} + \dfrac{2}{3} = 1 - \dfrac{1}{21} \Rightarrow \dfrac{20}{21} = \dfrac{20}{21}$ (true statement)

  Solve the fractional equation $\dfrac{1}{x-1} + 2 = \dfrac{1}{x-1}$

SOLUTION ▶

$$\frac{1}{x-1} + 2 = \frac{1}{x-1}$$

Restrictions for x: $x \neq 1$

$$\frac{1(x-1)}{x-1} + 2(x-1) = \frac{1(x-1)}{x-1}$$

$LCD = x - 1$

$$1 + 2(x-1) = 1$$
$$1 + 2x - 2 = 1$$
$$2x - 1 = 1$$
$$2x = 2$$
$$x = 1$$

comparing

Possible solution $x = 1$, is in the conflict with restriction for x: $x \neq 1$, so set of solutions is $\varnothing$

Answer: **No solution** or $x \in \varnothing$

  Solve the fractional equation $\dfrac{x}{x-5} - \dfrac{2x}{x^2-25} = \dfrac{4}{x-5}$

SOLUTION ▶

$$\frac{x}{x-5} - \frac{2x}{x^2-25} = \frac{4}{x-5}$$

Restrictions for x: $x - 5 \neq 0$; $x^2 - 25 \neq 0$;
or $x \neq 5$; $x \neq -5$

$$\frac{x(x-5)(x+5)}{x-5} - \frac{2x(x-5)(x+5)}{x^2-25 \;(x+5)(x-5)} = \frac{4(x-5)(x+5)}{x-5}$$

$LCD = (x-5)(x+5)$

$$x(x+5) - 2x = 4(x+5)$$
$$x^2 + 5x - 2x = 4x + 20$$
$$x^2 + 3x = 4x + 20$$

$$x^2 - x - 20 = 0$$

comparing

$$-5 \quad 4$$
$$(x-5)(x+4) = 0$$
$$x = -4, \; x = 5$$

Possible solution $x = 5$, is in the conflict with restriction for x: $x \neq 5$ ($5 = 5$).
Possible solution $x = -4$, is not in the conflict with restriction for x ($-4 \neq 5$) so we have only one solution: $x = -4$.

Answer: $x = -4$

Check: $\dfrac{-4}{-4-5} - \dfrac{2(-4)}{(-4)^2-25} = \dfrac{4}{-4-5} \;\Rightarrow\; \dfrac{4}{9} - \dfrac{-8}{-9} = \dfrac{4}{-9} \;\Rightarrow\; -\dfrac{4}{9} = -\dfrac{4}{9}$ (true statement)

Example 4 Solve the fractional equation $\dfrac{1}{x-1} - \dfrac{1}{x+1} = 1$

SOLUTION ▶

$$\frac{1}{x-1} - \frac{1}{x+1} = 1$$

Restrictions for x: $x - 1 \neq 0$; $x + 1 \neq 0$
or $\quad x \neq 1,\ x \neq -1$

$$\frac{1 \cdot (x-1)(x+1)}{x-1} - \frac{1 \cdot (x-1)(x+1)}{x+1} = 1 \cdot (x-1)(x+1)$$

$LCD = (x-1)(x+1)$

$$(x+1) - (x-1) = (x-1)(x+1)$$
$$x + 1 - x + 1 = x^2 - 1$$

$$2 = x^2 - 1$$
$$2 + 1 = x^2$$
$$x^2 = 3$$
$$\sqrt{x^2} = \sqrt{3}$$
$$|x| = \sqrt{3}$$
$$x = \sqrt{3},\ x = -\sqrt{3}$$

comparing

Possible solutions $x = \sqrt{3}$, $x = -\sqrt{3}$ are not in the conflict with restriction for x: $x \neq -1$ and $x \neq 1$

Answer: $x = \sqrt{3},\ x = -\sqrt{3}$

So $x = \sqrt{3}$, $x = -\sqrt{3}$ are 2 solutions of the initial fractional equation.

Example 5 Solve the fractional equation $\dfrac{1}{x-4} - \dfrac{8}{x^2-16} = \dfrac{1}{x+4}$

SOLUTION ▶

$$\frac{1}{x-4} - \frac{8}{x^2-16} = \frac{1}{x+4}$$

Restrictions for x: $x - 4 \neq 0$, $x + 4 \neq 0$, $x^2 - 16 \neq 0$ or $x \neq 4$, $x \neq -4$

$$\frac{1(x-4)(x+4)}{x-4} - \frac{8(x-4)(x+4)}{x^2-16} = \frac{1(x-4)(x+4)}{x+4}$$

$LCD = (x-4)(x+4)$

$$x + 4 - 8 = x - 4$$
$$x + 4 - 8 - x + 4 = 0$$
$$0 = 0$$
All real numbers

comparing

Two possible solutions $x = 4$, $x = -4$ from real number set **R** are in the conflict with restriction for x: $x \neq 4$, $x \neq -4$.
So, set of solutions is the set of all real numbers except $x = 4$, $x = -4$ or $x \in R - \{-4, 4\}$

Answer: $x \in R$, $x \neq 4$, $x \neq -4$ or $x \in R - \{-4, 4\}$

Example 6 Solve the fractional equation $\dfrac{x}{2} = \dfrac{x^2 - 5x + 6}{4x - 8}$

SOLUTION ▶

This equation is a **proportion like** fractional equation, so, we will cross multiply.

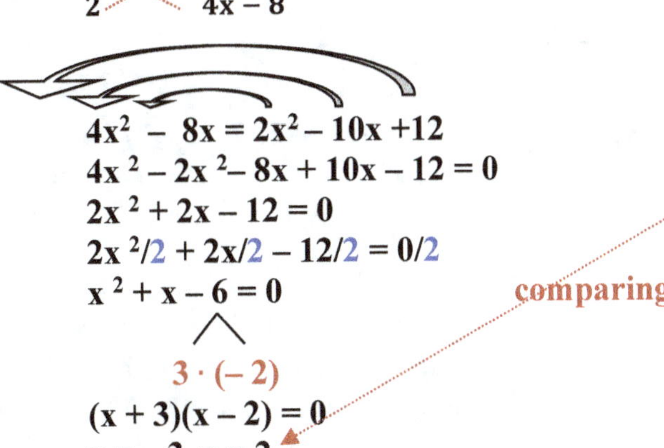

Restrictions for x: $4x - 8 \neq 0$ or

$x \neq 2$

$$4x^2 - 8x = 2x^2 - 10x + 12$$
$$4x^2 - 2x^2 - 8x + 10x - 12 = 0$$
$$2x^2 + 2x - 12 = 0$$
$$2x^2/2 + 2x/2 - 12/2 = 0/2$$
$$x^2 + x - 6 = 0$$

$$3 \cdot (-2)$$
$$(x + 3)(x - 2) = 0$$
$$x = -3, \ x = 2$$

comparing

Possible solution $x = 2$, is in the conflict with restriction for x: $x \neq 2$.
Possible solution $x = -3$, is not in the conflict with restriction for x: $x \neq 2$.
So we have only one solution: $x = -3$.

Answer: $x = -3$

Example 7 Solve the fractional equation $\dfrac{x^2 + 11}{x^2 + 1} = \dfrac{6}{x}$

SOLUTION ▶

This equation is a **proportion like** fractional equation so we will cross multiply.

$$\dfrac{x^2 + 11}{x^2 + 1} = \dfrac{6}{x}$$

Restrictions for x: $x^2 + 1 \neq 0$, $x \neq 0$ or

$x \neq i, \ x \neq -i, \ x \neq 0$

$$x^3 + 11x = 6x^2 + 6$$
$$x^3 - 6x^2 + 11x - 6 = 0$$
$$x^3 - x^2 - 5x^2 + 5x + 6x - 6 = 0$$
$$(x - 1)(x^2 - 5x + 6) = 0$$

$$-2 \cdot (-3)$$
$$(x - 1)(x - 2)(x - 3) = 0$$
$$x = 1, \ x = 2, \ x = 3$$

comparing

$$\begin{array}{r|rrrr}
1 & 1 & -6 & 11 & -6 \\
  & 0 & 1 & -5 & 6 \\
\hline
  & 1 & -5 & 6 & 0
\end{array}$$

Possible solutions $x = 1$, $x = 2$, $x = 3$ are not in the conflict with restriction for x: $x \neq i$, $x \neq -i$, $x \neq 0$
So $x = 1$, $x = 2$, $x = 3$ are the solutions

Answer: $x = 1, \ x = 2, \ x = 3$

Example 8 Solve the fractional equation $\dfrac{x^3 - 2x^2 + 3}{x - 2} - x^2 = \dfrac{4}{x + 2}$

SOLUTION ▶

$$\dfrac{x^3 - 2x^2 + 3}{x - 2} - x^2 = \dfrac{4}{x + 2}$$

Restrictions for x: $x - 2 \neq 0$, $x + 2 \neq 0$ or $x \neq 2$, $x \neq -2$

$$\dfrac{(x^3 - 2x^2 + 3)(x - 2)(x + 2)}{x - 2} - x^2(x - 2)(x + 2) = \dfrac{4(x - 2)(x + 2)}{x + 2}$$

$$(x^3 - 2x^2 + 3)(x + 2) - x^2(x - 2)(x + 2) = 4(x - 2)$$

LCD $= (x - 2)(x + 2)$

$$x^4 - 2x^3 + 3x + 2x^3 - 4x^2 + 6 - x^4 + 4x^2 = 4x - 8$$

$$x^4 - 2x^3 + 3x + 2x^3 - 4x^2 + 6 - x^4 + 4x^2 = 4x - 8$$

$$3x + 6 = 4x - 8$$ **comparing**

$$6 + 8 = 4x - 3x$$

Possible solution $x = 14$ is not in the conflict with restriction for x: $x \neq 2$, $x \neq -2$, so, we have one solution $x = 14$.

$$x = 14$$

Answer: $x = 14$

We also can use division to solve this equation as a proportion.

$$\dfrac{x^3 - 2x^2 + 3}{x - 2} = \dfrac{x^2(x - 2) + 3}{x - 2} = x^2 + \dfrac{3}{x - 2}$$

$$\dfrac{x^3 - 2x^2 + 3}{x - 2} - x^2 = \dfrac{4}{x + 2}$$

$$x^2 + \dfrac{3}{x - 2} - x^2 = \dfrac{4}{x + 2}$$

$$\dfrac{3}{x - 2} = \dfrac{4}{x + 2}$$

$$3(x + 2) = 4(x - 2)$$
$$3x + 6 = 4x - 8$$

Answer: $x = 14$

MOVING ALL TERMS TO THE LEFT OR TO THE RIGHT SIDE OF EQUATION

This method we can use for all the types of fractional equations. To solve any fractional equation we will collect all the terms of equation to the left side (or to the right side) of equation. Then we will use the "zero" property of fraction, which says that **fraction is equal to zero if the numerator of that fraction is zero.**

Example 1 Solve the fractional inequality $\dfrac{2}{x+1} = \dfrac{1}{x-1}$

SOLUTION ▶

$$\dfrac{2}{x+1} = \dfrac{1}{x-1}$$

We will move all the terms of the equation to the left side of equation.

$$\dfrac{2}{x+1} - \dfrac{1}{x-1} = 0$$

$$\dfrac{2 \cdot (x-1)}{(x+1)(x-1)} - \dfrac{1 \cdot (x+1)}{(x-1)(x+1)} = 0$$

LCD $= (x+1)(x-1)$

$$\dfrac{2 \cdot (x-1) - 1 \cdot (x+1)}{(x+1)(x-1)} = 0$$

$$\dfrac{2x - 2 - x - 1}{(x+1)(x-1)} = 0$$

$$\dfrac{x-3}{(x+1)(x-1)} = 0$$

Restrictions for x: $x \neq -1$, $x \neq 1$, because the value $x = -1$ and the value $x = 1$ make fraction undefined

$$\dfrac{x-3}{(x+1)(x-1)} = 0$$

Fraction is equal to zero if the numerator of that fraction is zero.

$$x - 3 = 0$$

Possible solution $x = 3$, is not in the conflict with restrictions for x: $x \neq -1$, $x \neq 1$. **So we have only one solution: $x = 3$.**

Answer: $x = 3$

Example 2 Solve the fractional inequality $\dfrac{3}{x} = \dfrac{1}{x-2}$

SOLUTION ▶

$$\dfrac{3}{x} = \dfrac{1}{x-2}$$

We will move all the terms of the equation to the left side of equation.

$$\dfrac{3}{x} - \dfrac{1}{x-2} = 0$$

$$\dfrac{3 \cdot (x-2)}{x \cdot (x-2)} - \dfrac{1 \cdot x}{(x-2) \cdot x} = 0$$

LCD = $x(x-2)$

$$\dfrac{3 \cdot (x-2) - 1 \cdot x}{x(x-2)} = 0$$

$$\dfrac{3x - 6 - x}{x(x-2)} = 0$$

Restrictions for x: $x \neq 0$, $x \neq 2$, because the value $x = 0$ and the value $x = 2$ make fraction undefined

$$\dfrac{2x - 6}{x(x-2)} = 0$$

$$\dfrac{2(x-3)}{(x+1)(x-1)} = 0$$

Fraction is equal to zero if the numerator of that fraction is zero.

$$x - 3 = 0$$

Possible solution $x = 3$, is not in the conflict with restrictions for x: $x \neq 0$, $x \neq 2$. So we have only one solution: $x = 3$.

Answer: $x = 3$

Example 3 Solve the fractional inequality $\dfrac{2}{x} = \dfrac{1}{3}$

SOLUTION ▶

$$\dfrac{2}{x} = \dfrac{1}{3}$$

We will move all the terms of the equation to the left side of equation.

$$\dfrac{2}{x} - \dfrac{1}{3} = 0$$

LCD = $3x$

$$\dfrac{2 \cdot 3}{x \cdot 3} - \dfrac{1 \cdot x}{3 \cdot x} = 0$$

Restrictions for x: $x \neq 0$, because the value $x = 0$ makes fraction undefined

$$\dfrac{6 - x}{3x} = 0$$

Fraction is equal to zero if the numerator of that fraction is zero.

$$6 - x = 0$$

Possible solution $x = 6$, is not in the conflict with restrictions for x: $x \neq 0$. So we have only one solution: $x = 3$.

Answer: $x = 6$

Example 1 John and Robert together can paint the boat in **6** hours. Alone John needs **5** hours longer to paint the boat than Robert needs. How many hours does John need to paint the boat? How many hours does Robert need to paint the boat?

SOLUTION ▶

Let **t** represent the time in hours John needs to paint the boat. Then **t − 5** represents the time in hours Robert needs to paint the boat.
The equation to paint the boat together is

$$\frac{1}{t} + \frac{1}{t-5} = \frac{1}{6}$$ (John's rate + Robert's rate = together rate)

$$\frac{1 \cdot 6t(t-5)}{t} + \frac{1 \cdot 6t(t-5)}{t-5} = \frac{1 \cdot 6t(t-5)}{6}$$ LCD = 6t(t − 5)

$$6t - 30 + 6t = t^2 - 5t$$

$$12t - 30 = t^2 - 5t$$

$$t^2 - 17t + 30 = 0$$

$$-15 \cdot -2$$

$$(t - 15)(t - 2) = 0$$

$$t = 15, \quad t \neq 2$$ (t = 2 cannot be a solution because t − 5 > 0)

So, the time in hours John needs to paint the boat alone is **15** hours and the time in hours Robert needs to paint the boat alone is **10** hours.

Example 2 Two similar triangles are shown below. Find **x**.

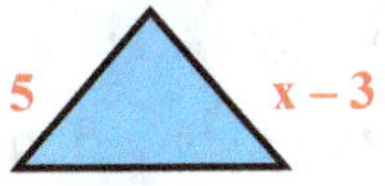

SOLUTION ▶

We can write the following fractional equation from similarity property:

$$\frac{x}{5} = \frac{x+4}{x-3} \implies x(x-3) = 5(x+4) \implies x^2 - 3x = 5x + 20 \implies x^2 - 8x - 20 = 0$$

$$(x-10)(x+2) = 0 \implies x = 10 \quad (x = -2 \text{ cannot be a solution because } x > 0)$$

So, $x = 10$

■ FRACTIONAL INEQUALITIES

We will use the following definitions.

> **Definition** ▶ *A fractional (rational) inequality* is an inequality $E_1 < E_2$ (or $E_1 > E_2$, or $E_1 \le E_2$, or $E_1 \ge E_2$) where E_1 or E_2 or both E_1 and E_2 are fractional expressions with one or more variables.
>
> **Definition** ▶ *A fractional (rational) inequality* is any inequality, which we can rewrite as $P/Q < 0$, $P/Q > 0$, $P/Q \le 0$, or $P/Q \ge 0$, where P and Q are any expressions and $Q \ne 0$.
>
> **Definition** ▶ *A fractional inequality with one variable* is an inequality $E_1 < E_2$ (or $E_1 > E_2$, or $E_1 \le E_2$, or $E_1 \ge E_2$) where E_1 or E_2 or both E_1 and E_2 are fractional expressions with one variables.
>
> **Definition** ▶ *A fractional inequality with one variable* is any inequality, which we can rewrite as $P(x)/Q(x) < 0$, $P(x)/Q(x) > 0$, $P(x)/Q(x) \le 0$, or $P(x)/Q(x) \ge 0$, where $P(x)$ and $Q(x)$ are any expressions with one variable x, and $Q(x) \ne 0$.

Examples Inequalities $\frac{x}{2} + 3 < x - 1$, $\frac{x}{3y} + 6 > \frac{x}{5y} + 4$, and $\frac{x+2}{3x-1} + 1 \le \frac{x}{3x-1} + 2$ are fractional (rational) inequalities.

Inequalities $\frac{x}{2} + 3 < x - 1$, $\frac{1}{x} + 1 > \frac{1}{2x} + 5$, and $\frac{x+2}{3x-1} + 1 \le \frac{x}{3x-1} + 2$ are fractional inequalities with one variable.

We can't use cross multiplication and LCD methods to solve fractional inequalities, because we don't have any information about signs of denominators or LCD, and we know that multiplication by negative expression will change the direction of inequality sign.

For example, the inequality $\frac{x+2}{3x-1} + 1 \le \frac{x}{3x-1} + 2$ we can't solve by LCD method, because LCD $= 3x - 1$, and we don't have any information about the sign of expression $3x - 1$, so we can't multiply both sides of inequality by $3x - 1$.

Some simple fractional inequalities we can solve logically and mentally.

For example, the inequality $\frac{1}{x} < \frac{1}{3}$ we can solve mentally on base of comparing of fractions and we can write that solution is $x > 3$.

To solve fractional inequalities we will use the **"finding total sign"** method.

"FINDING TOTAL SIGN" METHOD

This method we will use for the all types of fractional inequalities.

Example 1 Solve the fractional inequality $\dfrac{2}{x+1} < \dfrac{1}{x-1}$

SOLUTION ▶

$$\frac{2}{x+1} < \frac{1}{x-1}$$

$$\frac{2}{x+1} - \frac{1}{x-1} < 0$$

$$\frac{2\cdot(x-1)}{(x+1)(x-1)} - \frac{1\cdot(x+1)}{(x-1)(x+1)} < 0$$

$$\frac{2\cdot(x-1) - 1\cdot(x+1)}{(x+1)(x-1)} < 0$$

$$\frac{2x-2-x-1}{(x+1)(x-1)} < 0$$

$$\frac{\overset{3}{x-3}}{\underset{-1 \quad 1}{(x+1)(x-1)}} < 0$$

To solve any fractional inequality we have to collect all the terms of the inequality to the left side (or to the right side) of the inequality and find the total sign of the expression on the left side (on the right side). The total sign will help us to find the set of solution.

LCD $= (x+1)(x-1)$

Zero points for the numerator and for the denominator of the expression on the left side are $x = 3$, $x = -1$ and $x = 1$. These points divide the number line (or set of all the possible values of x) into 4 intervals: $(-\infty, -1)$; $(-1, 1)$; $(1, 3)$; $(3, \infty)$. $x \neq -1$, $x \neq 1$, and $x \neq 3$ because the value $x = -1$ and the value $x = 1$ make fraction undefined, and the value $x = 3$ makes fraction equal to **0**.

1. If $x \in (-\infty, -1)$, then $x - 3 < 0$, $x + 1 < 0$ and $x - 1 < 0$. The total sign is $-$ ($-/- \cdot - = -$).
2. If $x \in (-1, 1)$, then $x - 3 < 0$, $x + 1 > 0$, and $x - 1 < 0$. The total sign is $+$ ($-/+\cdot- = +$).
3. If $x \in (1, 3)$, then $x - 3 < 0$, $x + 1 > 0$, and $x - 1 > 0$. The total sign is $-$ ($-/+\cdot+ = -$).
4. If $x \in (3, \infty)$, then $x - 3 > 0$, $x + 1 > 0$, and $x - 1 > 0$. The total sign is $+$ ($+/+\cdot+ = +$).

So, the solution set is union of two intervals

$$x \in (-\infty, -1) \cup (1, 3)$$

where the expression $(x-3)/(x+1)(x-1) < 0$.

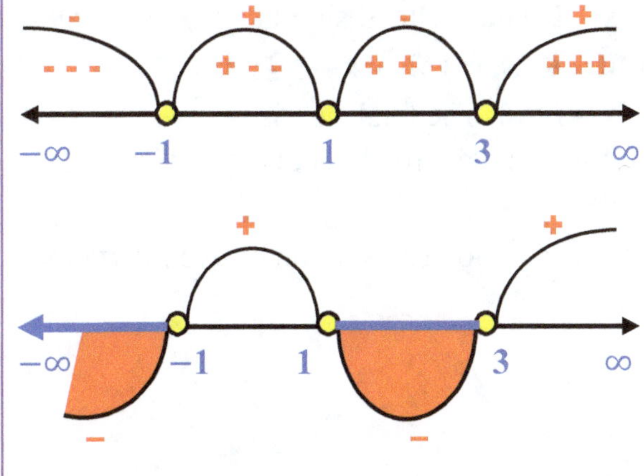

SOLUTION ▶

<table>
<tr><td>

$$\dfrac{2}{x+1} \geq \dfrac{1}{x-1}$$

$$\dfrac{2}{x+1} - \dfrac{1}{x-1} \geq 0$$

$$\dfrac{2\cdot(x-1)}{(x+1)(x-1)} - \dfrac{1\cdot(x+1)}{(x-1)(x+1)} \geq 0$$

$$\dfrac{2\cdot(x-1) - 1\cdot(x+1)}{(x+1)(x-1)} \geq 0$$

$$\dfrac{2x-2-x-1}{(x+1)(x-1)} \geq 0$$

$$\dfrac{\overset{3}{x-3}}{\underset{-1 \quad 1}{(x+1)(x-1)}} \geq 0$$

1. If $x \in (-\infty, -1)$, then $x-3 < 0$, $x+1 < 0$, and $x-1 < 0$. The total sign is **–** (−/−·− = −).
2. If $x \in (-1, 1)$, then $x-3 < 0$, $x+1 > 0$, and $x-1 < 0$. The total sign is **+** (−/+·− = +).
3. If $x \in (1, 3)$, then $x-3 < 0$, $x+1 > 0$, and $x-1 > 0$. The total sign is **–** (−/+·+ = −).
4. If $x \in [3, \infty)$, then $x-3 \geq 0$, $x+1 > 0$, and $x-1 > 0$. The total sign is **+** (+/+·+ = +) if $x > 3$, and expression equals to **0** if $x = 3$.

So, the solution set is union of two intervals

$$x \in (-1, 1) \cup [3, \infty)$$

</td><td>

To solve this fractional inequality we have to collect all the terms of the inequality to the left side of the inequality and find the total sign of the expression on the left side. The total sign will help us to find the set of solution.

$$LCD = (x+1)(x-1)$$

Zero points for the numerator and for the denominator of the expression on the left side are **x = 3**, **x = −1**, and **x = 1**. These points divide the number line (or set of all the possible values of x) into 4 intervals: $(-\infty, -1)$; $(-1, 1)$; $(1, 3)$; $[3, \infty)$.
$x \neq -1$ and $x \neq 1$ because the value $x = -1$ and the value $x = 1$ make fraction undefined. The value $x = 3$ makes fraction equal to **0**, so x can be equal to **3**.

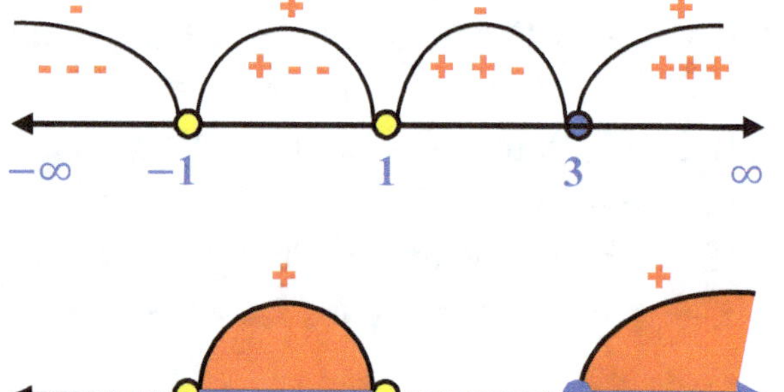

</td></tr>
</table>

<table>
<tr><td>

Example 3   Solve the fractional inequality $\dfrac{2}{x-2} > \dfrac{x}{x-2}$

</td></tr>
</table>

<table>
<tr>
<td>

$$\dfrac{2}{x-2} > \dfrac{x}{x-2}$$

$$\dfrac{2}{x-2} - \dfrac{x}{x-2} > 0$$

$$\dfrac{2-x}{x-2} > 0$$

$$\dfrac{-(x-2)^{\,1}}{x-2} > 0$$

$$-1 > 0 \quad \textbf{False statement}$$

$$\Downarrow$$

No solution   (or set of solutions is ∅ set)

</td>
<td>

To solve this fractional inequality we have to collect all the terms of the inequality to the left side of the inequality.

LCD = x − 2

Restriction for x is **x ≠ 2.**

</td>
</tr>
</table>

<table>
<tr><td>

Example 4   Solve the fractional inequality $\dfrac{2}{x-2} < \dfrac{x}{x-2}$

</td></tr>
</table>

<table>
<tr>
<td>

$$\dfrac{2}{x-2} < \dfrac{x}{x-2}$$

$$\dfrac{2}{x-2} - \dfrac{x}{x-2} < 0$$

$$\dfrac{2-x}{x-2} < 0$$

$$\dfrac{-(x-2)^{\,1}}{x-2} < 0$$

$$-1 < 0 \quad \textbf{True statement}$$

$$\Downarrow$$

All real numbers except x = 2

Or $x \in (-\infty, 2) \cup (2, \infty)$

Or set of solutions is **R − {2}**

</td>
<td>

To solve this fractional inequality we have to collect all the terms of the inequality to the left side of the inequality.

LCD = x − 2

Restriction for x is **x ≠ 2.**

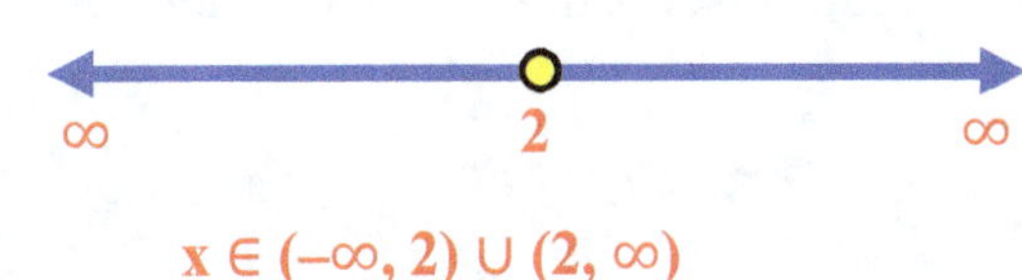

</td>
</tr>
</table>

  Solve the fractional inequality $\dfrac{2}{x+2} + \dfrac{2}{x-2} \leq \dfrac{4}{x^2-4}$

SOLUTION ▶

<table>
<tr><td>

$$\dfrac{2}{x+2} + \dfrac{2}{x-2} \leq \dfrac{4}{x^2-4}$$

$$\dfrac{2}{x+2} + \dfrac{2}{x-2} - \dfrac{4}{x^2-4} \leq 0$$

$$(x+2)(x-2)$$

$$\dfrac{2(x-2)}{(x+2)(x-2)} + \dfrac{2(x+2)}{(x-2)(x+2)} - \dfrac{4}{(x+2)(x-2)} \leq 0$$

$$\dfrac{2(x-2) + 2(x+2) - 4}{(x+2)(x-2)} \leq 0$$

$$\dfrac{2x - 4 + 2x + 4 - 4}{(x+2)(x-2)} \leq 0$$

$$\dfrac{4(x-1)}{(x+2)(x-2)} \leq 0$$

1. If $x \in (-\infty, -2)$, then $x - 1 < 0$, $x + 2 < 0$, and $x - 2 < 0$. The total sign is **–** ($-/-\cdot- = -$).
2. If $x \in (-2, 1)$, then $x - 1 < 0$, $x + 2 > 0$, and $x - 2 < 0$. The total sign is **–** ($-/+\cdot- = +$).
3. If $x \in [1, 2)$, then $x - 1 \geq 0$, $x + 2 > 0$, and $x - 2 < 0$. The total sign is **–** ($+/+\cdot- = -$) or expression $= 0$ if $x = 1$.
4. If $x \in (2, \infty)$, then $x - 1 > 0$, $x + 2 > 0$, and $x - 2 > 0$. The total sign is **+** ($+/+\cdot+ = +$).

So, the solution set is union of two intervals

$$x \in (-\infty, -2) \cup [1, 2)$$

</td><td>

To solve this fractional inequality we have to collect all the terms of the inequality to the left side of the inequality and find the total sign of the expression on the left side. The total sign will help us to find the set of solution.

$$LCD = (x+2)(x-2)$$

Zero points for the numerator and for the denominator of the expression on the left side are $x = 1$, $x = -2$ and $x = 2$. These points divide the number line (or set of all possible values of x) into 4 intervals: $(-\infty, -2)$; $(-2, 1)$; $[1, 2)$; $(2, \infty)$.
$x \neq -2$, $x \neq 2$, because the value $x = -2$ and the value $x = 2$ make fraction undefined. The value $x = 1$ makes fraction equal to **0**, so **x** can be equal to **1**.

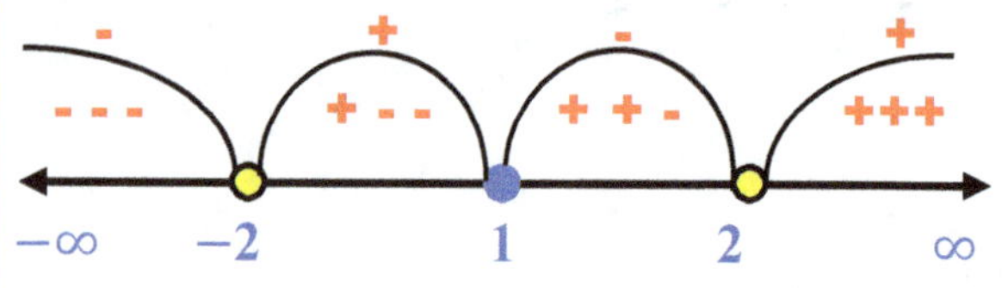

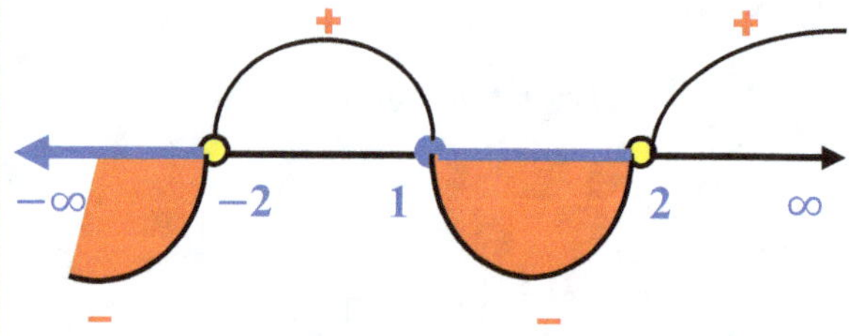

</td></tr>
</table>

 Solve the fractional equation $\dfrac{x}{x+5} < \dfrac{1}{2}$

$$\dfrac{x}{x+5} < \dfrac{1}{2}$$

$$\dfrac{x}{x+5} - \dfrac{1}{2} < 0$$

$$\dfrac{2x}{2(x+5)} - \dfrac{1(x+5)}{2(x+5)} < 0 \qquad\qquad LCD = 2(x+5)$$

$$\dfrac{2x - x - 5}{2(x+5)} < 0$$

$$\dfrac{\overset{5}{x-5}}{2(x+5)} < 0$$

$$-5$$

1. If $x \in (-\infty, -5)$, then $x - 5 < 0$, $x + 5 < 0$,.
The total sign is **+** $(-/- = +)$.

2. If $x \in (-5, 5)$, then $x - 5 < 0$, $x + 5 > 0$.
The total sign is **–** $(-/+ = -)$.

3. If $x \in (5, \infty)$, then $x - 5 > 0$, $x + 5 > 0$.
The total sign is **+** $(+/+ = +)$.

So solution is

$$-5 < x < 5$$

Interval notation of the solution set is

$$x \in (5, 5)$$

where the expression $(x-5)/2(x+5) < 0.$

Zero points for the numerator and for the denominator of the expression on the left side are $x = 5$, and $x = -5$. These points divide the number line into 3 intervals: $(-\infty, -5)$; $(-5, 5)$; $(5, \infty)$. $x \neq -5$ because the value $x = -5$ makes fraction undefined. The value $x = 5$ makes fraction equal to 0, so $x \neq 5$.

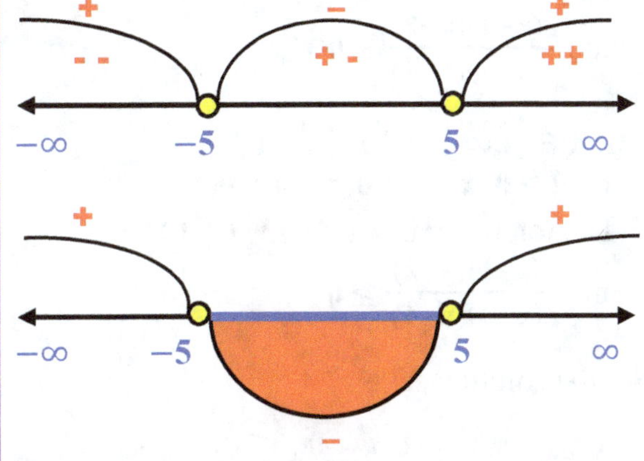

Example 7 Solve the fractional equation $\dfrac{(x-1)(x-2)}{(x-3)(x-4)} \geq 0$

SOLUTION ▶

$$\overset{1\quad 2}{\underset{3\quad 4}{\frac{(x-1)(x-2)}{(x-3)(x-4)}}} \geq 0$$

1. If $x \in (-\infty, 1]$, then $x-1 \leq 0$, $x-2 < 0$, $x-3 < 0$, $x-4 < 0$,
The total sign is **+** $(-\cdot-/-\cdot- = +)$
and $\dfrac{(x-1)(x-2)}{(x-3)(x-4)} \geq 0$

2. If $x \in (1, 2)$, then $x-1 > 0$, $x-2 < 0$, $x-3 < 0$, $x-4 < 0$,
The total sign is **-** $(+\cdot-/-\cdot- = -)$ and $\dfrac{(x-1)(x-2)}{(x-3)(x-4)} < 0$

3. If $x \in [2, 3)$, then $x-1 > 0$, $x-2 \geq 0$, $x-3 < 0$, $x-4 < 0$,
The total sign is **+** $(+\cdot+/-\cdot- = +)$
and $\dfrac{(x-1)(x-2)}{(x-3)(x-4)} \geq 0$

4. If $x \in (3, 4)$, then $x-1 > 0$, $x-2 > 0$, $x-3 > 0$, $x-4 < 0$,
The total sign is **-** $(+\cdot+/+\cdot- = -)$
and $\dfrac{(x-1)(x-2)}{(x-3)(x-4)} < 0$

5. If $x \in (4, \infty)$, then $x-1 > 0$, $x-2 > 0$, $x-3 > 0$, $x-4 > 0$,
The total sign is **+** $(+\cdot+/+\cdot+ = +)$
and $\dfrac{(x-1)(x-2)}{(x-3)(x-4)} \geq 0$

So, the solution set is

$$x \in (-\infty, 1] \cup [2, 3) \cup (4, \infty)$$

Zero points for the numerator and for the denominator of the expression on the left side are **x = 1, x = 2, x = 3,** and **x = 4.** These points divide the number line into **5** intervals: $(-\infty, 1]$; $(1, 2)$; $[2, 3)$; $(3, 4)$; $(4, \infty)$. $x \neq 3$; $x \neq 4$ because the value $x = 3$ and the value $x = 4$ make fraction undefined.

The value **x = 1** and the value **x = 2** make fraction equal to **0,** so **x** can be equal to **1** and **2.**

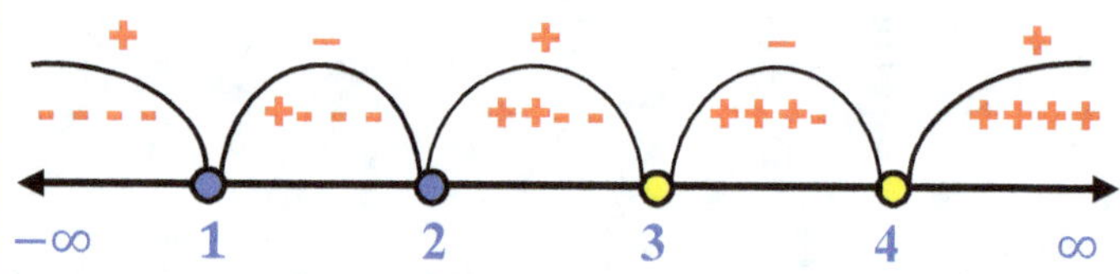

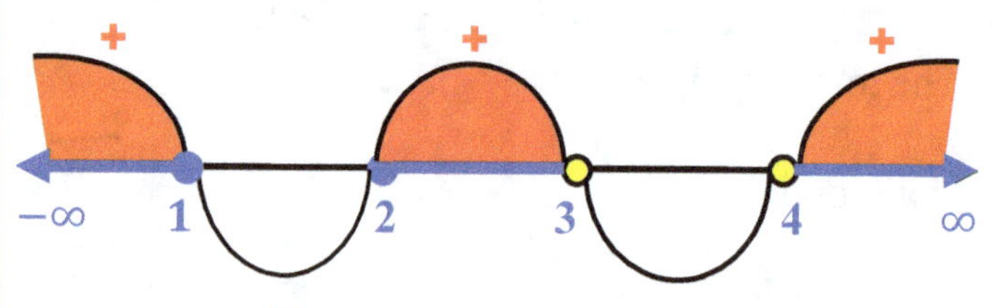

 Solve the fractional inequality $\dfrac{2}{x+2} + \dfrac{2}{x-2} \le 1 - \dfrac{1}{x^2-4}$

SOLUTION ▶

$$\frac{2}{x+2} + \frac{2}{x-2} \le 1 - \frac{1}{x^2-4}$$

$$\frac{2}{x+2} + \frac{2}{x-2} - 1 + \frac{1}{x^2-4} \le 0$$

$(x+2)(x-2)$

$$\frac{2(x-2) + 2(x+1) - 1(x+2)(x-2) + 1}{(x+2)(x-2)} \le 0$$

$$\frac{2x - 4 + 2x + 4 - x^2 + 4 + 1}{(x+2)(x-2)} \le 0$$

$$\frac{-(x^2 - 4x - 5)}{(x+2)(x-2)} \le 0$$

$$\frac{-(x+1)(x-5)}{(x+2)(x-2)} \le 0$$

$$\frac{(x+1)(x-5)}{(x+2)(x-2)} \ge 0$$

1. If $x \in (-\infty, -2)$, then $x + 1 < 0$, $x - 5 < 0$, $x + 2 < 0$ and $x - 2 < 0$.
   The total sign is **+** ($-\cdot-/-\cdot- = +$).
2. If $x \in (-2, -1)$, then $x + 1 < 0$, $x - 5 < 0$, $x + 2 > 0$ and $x - 2 < 0$.
   The total sign is **−** ($-\cdot-/+\cdot- = +$).
3. If $x \in [-1, 2)$, then $x + 1 \ge 0$, $x - 5 < 0$, $x + 2 > 0$ and $x - 2 < 0$.
   The total sign is **+** ($+\cdot-/+\cdot- = +$).
4. If $x \in (2, 5)$, then $x + 1 > 0$, $x - 5 < 0$, $x + 2 > 0$ and $x - 2 > 0$.
   The total sign is **−** ($+\cdot-/+\cdot+ = -$).
5. If $x \in [5, \infty)$, then $x + 1 > 0$, $x - 5 \ge 0$, $x + 2 > 0$ and $x - 2 > 0$.
   The total sign is **+** ($+\cdot+/+\cdot+ = +$).

$$x \in (-\infty, -2) \cup [1, 2) \cup [5, \infty)$$

To solve this fractional inequality we have to collect all the terms of the inequality to the left side of the inequality and find the total sign of the expression on the left side. The total sign will help us to find the set of solution.

$$\text{LCD} = (x+2)(x-2)$$

Zero points for the numerator and for the denominator of expression on the left side are $x = -1$, $x = 5$, $x = -2$, and $x = 2$. These points divide the number line into **5** intervals: $(-\infty, -2)$; $(-2, -1)$; $[-1, 2)$; $(2, 5)$; $[5, \infty)$. $x \ne -2$; $x \ne 2$ because the value $x = -2$ and the value $x = 2$ make fraction undefined. The value $x = -1$ and the value $x = 5$ make fraction equal to **0**, so x can be equal to -1 and 5.

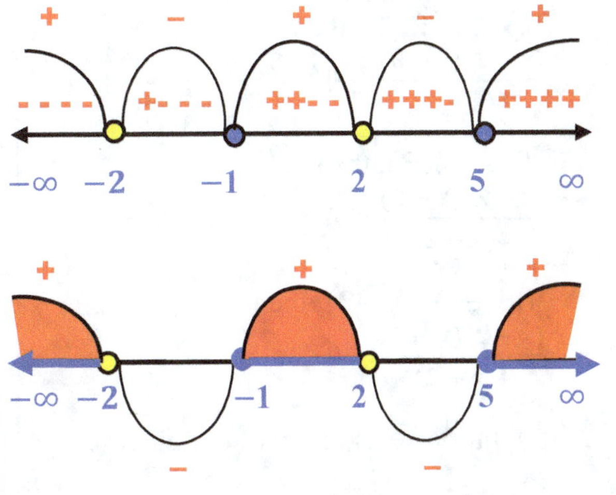

SOLUTION ▶

$$\frac{x}{2} < \frac{x^2 - 5x + 6}{4x - 8}$$

$$\frac{x}{2} - \frac{x^2 - 5x + 6}{4x - 8} < 0$$

$$\frac{x}{2} - \frac{x^2 - 5x + 6}{4(x - 2)} < 0$$

$$\frac{x \cdot 2(x - 2) - (x^2 - 5x + 6)}{4(x - 2)} < 0$$

$$\frac{2x^2 - 4x - x^2 + 5x - 6}{4(x - 2)} < 0$$

$$\frac{x^2 - x - 6}{4(x - 2)} < 0$$

$$\frac{(x + 2)(x - 3)}{4(x - 2)} < 0$$

1. If $x \in (-\infty, -2)$, then
$$\frac{(x + 2)(x - 3)}{4(x - 2)} < 0$$

2. If $x \in (-2, 2)$, then
$$\frac{(x + 2)(x - 3)}{4(x - 2)} > 0$$

3. If $x \in (2, 3)$, then
$$\frac{(x + 2)(x - 3)}{4(x - 2)} < 0$$

4. If $x \in (-\infty, -2)$, then
$$\frac{(x + 2)(x - 3)}{4(x - 2)} < 0$$

So, the solution set is union of two intervals
$$x \in (-\infty, -2) \cup (2, 3)$$

To solve this fractional inequality we have to collect all the terms of the inequality to the left side of the inequality and find the total sign of the expression on the left side.

$$LCD = 4(x - 2)$$

Zero points for the numerator and for the denominator of the expression on the left side are $x = -2$, $x = 3$, and $x = 2$. These points divide the number line into **4** intervals: $(-\infty, -2)$; $(-2, 2)$; $(2, 3)$; $(3, \infty)$. $x \neq 2$ because the value $x = 2$ makes fraction undefined. The value $x = -2$ and the value $x = 3$ make the fraction equal to **0**, so **x** cannot be equal to -2 and **3**.

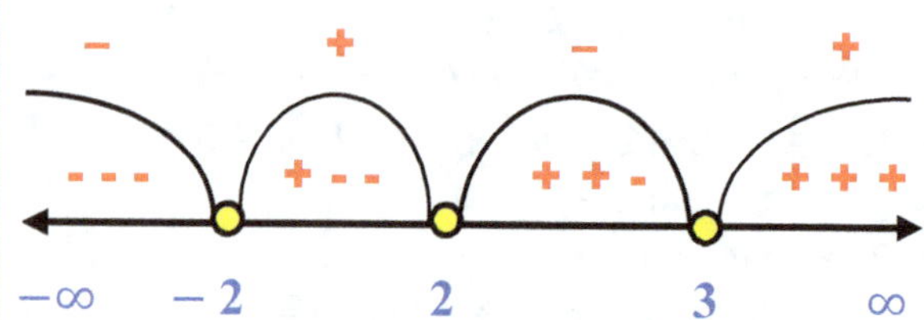

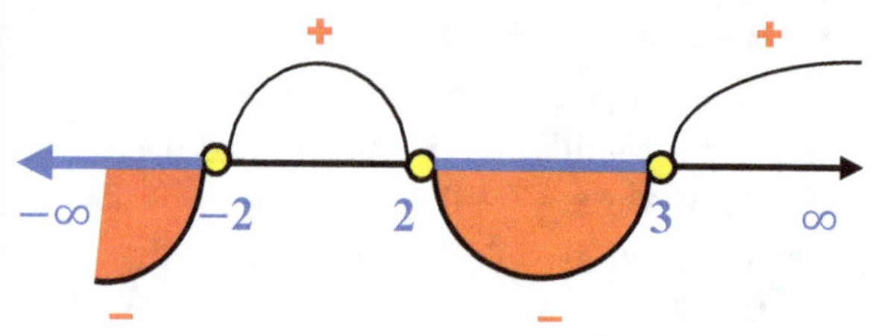

Solve each fractional (rational) equation.

1. $\dfrac{x+1}{x+2} = \dfrac{3}{4}$

2. $\dfrac{5}{y-2} = \dfrac{3}{y-4}$

3. $\dfrac{x+2}{2x+1} = \dfrac{3}{x-2}$

4. $\dfrac{4}{x+1} = \dfrac{2}{x-2} + \dfrac{x-3}{x^2-x-2}$

5. $\dfrac{x^2+1}{x^2-1} - \dfrac{1}{x(x-1)} = \dfrac{1}{x}$

6. $\dfrac{1-\frac{1}{x}}{\frac{1}{x}} + 1 = \dfrac{1}{x}$

Solve each fractional (rational) inequality.

1. $\dfrac{x+1}{x+2} < \dfrac{3}{4}$

2. $\dfrac{5}{y-2} > \dfrac{3}{y-4}$

3. $\dfrac{x+2}{x+1} \geq \dfrac{x}{x-2}$

4. $\dfrac{4}{x+1} \leq \dfrac{2}{x-2} + \dfrac{x-3}{x^2-x-2}$

5. $\dfrac{(x-2)(x-3)}{(x+1)^2(x-1)} > 0$

6. $\dfrac{1-\frac{1}{x}}{\frac{1}{x}} + 1 < \dfrac{1}{x}$

4.6. Radical Equations and Inequalities

- **RADICAL EQUATIONS**
- **RADICAL INEQUALITIES**

■ RADICAL EQUATIONS

We will use the following definitions.

> **Definition** ▶ *A radical equation* is an equation $E_1 = E_2$ where E_1 or E_2 or both E_1 and E_2 are radical expressions with one or more variables.
>
> **Definition** ▶ *A single-variable radical equation* is an equation $E_1 = E_2$ where E_1 or E_2 or both E_1 and E_2 are radical expressions with one variable.

Examples Equations $\sqrt{x-1} + 3 = 5$, $\sqrt{xy} = y$, and $\sqrt{x} = \sqrt{x} - 1$ are radical equations.

Equations $\sqrt{x} = 3$, $\sqrt{1-x} = 2$, $\sqrt{x-1} + 3 = 5$, $\sqrt{x} + \sqrt{x+1} = 2$, $\sqrt{1-2x} = x$, and $\sqrt{x+2} - \sqrt{x-1} = 1$ are radical equations with one variables.

To solve radical equations with one variable we will use the following power and power root properties of expressions.

> **Properties** ▶ If $A(x) = B(x)$, then $A^2 = B^2$, where $A(x)$ and $B(x)$ are any two expressions.
>
> ▶ If we have radical expression $\sqrt{P(x)}$, then $P(x) \geq 0$.
>
> ▶ If we have equation $\sqrt{P(x)} = Q(x)$, then $P(x) \geq 0$ and $Q(x) \geq 0$.

Example If $\sqrt{x-3} = 2x - 4$, then $x - 3 \geq 0$, $2x - 4 \geq 0$, and $(\sqrt{x-3})^2 = (2x-4)^2$.

SOLVING RADICAL EQUATIONS BY SQUARING

The solving of the radical equations by squaring both sides of the equation sometimes can introduce some solutions which do not satisfy the initial radical equation. Such solutions are called **extraneous solutions**. So, when you solve the radical equation by squaring both sides of the equation, you have to check each solution in the original equation. Or to solve a radical equation you have to write all restrictions for the variable before you will start solving and after solving you have to compare each possible solution with restrictions for variable and to select the solutions that are not in the conflict with restrictions for the variable.

<table>
<tr><td>Example 1</td><td>Solve the radical equation $\sqrt{x} = 5$</td></tr>
</table>

SOLUTION ▶

$\sqrt{x} = 5$ Restriction for x: $x \geq 0$.

$(\sqrt{x})^2 = (5)^2$ Square each side.

$x = 25$ **comparing:**   Possible solution $x = 25$, is not in the conflict with restriction for x: $x \geq 0$ $(25 \geq 0)$.

$x = 25$ So $x = 25$ is a unique solution of radical equation.

Check: $\sqrt{x} = 5 \Rightarrow \sqrt{25} = 5 \Rightarrow 5 = 5$ (true statement)

<table>
<tr><td>Example 2</td><td>Solve the radical equation $\sqrt{x + 2} = x$</td></tr>
</table>

SOLUTION ▶

$\sqrt{x + 2} = x$ Restriction for x: $x + 2 \geq 0 \cap x \geq 0 \Rightarrow$

$$x \geq -2 \cap x \geq 0 \Rightarrow x \geq 0.$$

$(\sqrt{x + 2})^2 = (x)^2$ Square each side.

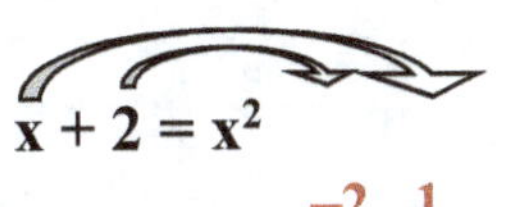

$x + 2 = x^2$

$0 = x^2 - x - 2$
$(x - 2)(x + 1) = 0$
$x = 2, \; x = -1$ **comparing:**

$x = 2$

Possible solution $x = 2$, is not in the conflict with restriction for x: $x \geq 0$ $(2 \geq 0)$.
Possible solution $x = -1$, is in the conflict with restriction for x: $x \geq 0$ $(-1 < 0)$.
So, $x = -1$ is an extraneous solution, and $x = 2$ is a unique solution of radical equation.

Check: $\sqrt{2 + 2} = 2 \Rightarrow \sqrt{4} = 2 \Rightarrow 2 = 2$ (true statement)

<table>
<tr><td>Example 3</td><td>Solve the radical equation $\sqrt{x - 3} = -4$</td></tr>
</table>

SOLUTION ▶

$\sqrt{x - 3} = -4$ If we have equation $\sqrt{P(x)} = Q(x)$, then $P(x) \geq 0$ and $Q(x) \geq 0$.

No solution $(x \in \emptyset)$ But $-4 < 0$. So, equation has **no any solution.**

Example 4 Solve the radical equation $\sqrt{3x+1} + \sqrt{x+4} = \sqrt{9-x}$

SOLUTION▶

$\sqrt{3x+1} + \sqrt{x+4} = \sqrt{9-x}$

$(\sqrt{3x+1} + \sqrt{x+4})^2 = (\sqrt{9-x})^2$

$3x + 1 + x + 4 + 2\sqrt{3x+1}\sqrt{x+4} = 9 - x$

$4x + 5 + 2\sqrt{3x+1}\sqrt{x+4} = 9 - x$

$2\sqrt{3x+1}\sqrt{x+4} = 9 - x - 4x - 5$

$2\sqrt{3x+1}\sqrt{x+4} = 4 - 5x$

$(2\sqrt{3x+1}\sqrt{x+4})^2 = (4 - 5x)^2$

$4(3x+1)(x+4) = (4 - 5x)^2$

$4(3x^2 + 13x + 4) = 16 - 40x + 25x^2$

$12x^2 + 52x + 16 = 16 - 40x + 25x^2$

$13x^2 - 92x = 0$

$x(13x - 92) = 0$

$x = 0,\ x = 92/13$ **comparing:**

⬇

$x = 0$

Restriction for x: $3x+1 \geq 0 \cap x+4 \geq 0 \cap 9-x \geq 0$

$\Rightarrow$ $x \geq -1/3 \cap x \geq -4 \cap x \leq 9$

$\Rightarrow$ $-1/3 \leq x \leq 9.$

Square each side.

$4 - 5x \geq 0 \Rightarrow x \leq 4/5 \Rightarrow$ Final restriction for x will be: $-1/3 \leq x \leq 4/5$

Square each side.

Collect all the terms on the right side of equation.

Possible solution $x = 0$, is not in the conflict with the final restriction for x: $-1/3 \leq x \leq 4/5$. Possible solution $x = 92/13$, is in the conflict with the restriction for x: $-1/3 \leq x \leq 4/5$ ($92/13 > 4/5$). So, $x = 92/13$ is an extraneous solution, and $x = 0$ is a unique solution of the radical equation.

Example 5 Solve the radical equation $3\sqrt{x-1} = 12$

SOLUTION▶

$3\sqrt{x-1} = 12$

$3\sqrt{x-1}/3 = 12/3$

$\sqrt{x-1} = 4$

$(\sqrt{x-1})^2 = (4)^2$

$x - 1 = 16$

$x = 17$ **comparing:**

⬇

$x = 17$

Restriction for x: $x - 1 \geq 0.$

$x \geq 1$

Possible solution $x = 17$, is not in the conflict with restriction for x: $x \geq 1$ ($17 \geq 0$). So, $x = 17$ is a unique solution of the radical equation.

Check: $3\sqrt{17-1} = 12 \Rightarrow 3\sqrt{16} = 12 \Rightarrow 12 = 12$ (true statement)

 Solve the radical equation $\sqrt{4x + 1} = x + 1$

SOLUTION▶

$\sqrt{4x + 1} = x + 1$

Restriction for x: $4x + 1 \geq 0 \cap x + 1 \geq 0 \Rightarrow$

$x \geq -1/4 \cap x \geq -1 \Rightarrow$

$x \geq -1/4.$

$(\sqrt{4x + 1})^2 = (x + 1)^2$

Square each side.

$4x + 1 = x^2 + 2x + 1$

$0 = x^2 + 2x + 1 - 4x - 1$

$0 = x^2 - 2x$

$x(x - 2) = 0$

$x = 0, x = 2$ **comparing:**

$x = 0, \ 2$

Possible solutions $x = 0$ and $x = 2$ are not in the conflict with the restriction for x: $x \geq -1/4.$
$(0 \geq -1/4$ and $2 \geq -1/4).$
So, $x = 0$ and $x = 2$ are solutions of equation.

Check: $\sqrt{4 \cdot 0 + 1} = 0 + 1 \Rightarrow \sqrt{1} = 1 \Rightarrow 1 = 1$ (true statement)

$\sqrt{4 \cdot 2 + 1} = 2 + 1 \Rightarrow \sqrt{9} = 3 \Rightarrow 3 = 3$ (true statement)

Example 7   Solve the radical equation $5\sqrt{x + 1} - 1 = 9$

SOLUTION▶

$5\sqrt{x + 1} - 1 = 9$

Restriction for x: $x + 1 \geq 0.$

$\quad\quad\quad +1 \ +1$

$x \geq -1$

$5\sqrt{x + 1} = 10$

$5\sqrt{x + 1} / 5 = 10 / 5$

$\sqrt{x + 1} = 2$

$(\sqrt{x + 1})^2 = (2)^2$

Square both sides of equation.

$x + 1 = 4$

$x = 3$ **comparing:**

$x = 3$

Possible solution $x = 3$, is not in the conflict with the restriction for x: $x \geq -1 \ (3 \geq -1).$
So $x = 3$ is a unique solution of radical equation.

Check: $5\sqrt{3 + 1} - 1 = 9 \Rightarrow 5\sqrt{4} - 1 = 9 \Rightarrow 5 \cdot 2 - 1 = 9 \Rightarrow 9 = 9$ (true statement)

Example 8 Solve the radical equation $\sqrt{5x-1} = \sqrt{x+3}$

SOLUTION ▶

$\sqrt{5x-1} = \sqrt{x+3}$

Restriction for x: $5x - 1 \geq 0 \cap x + 3 \geq 0 \Rightarrow$
$x \geq 1/5 \cap x \geq -3 \Rightarrow$
$x \geq 1/5.$

$(\sqrt{5x-1})^2 = (\sqrt{x+3})^2$

Square each side.

$5x - 1 = x + 3$

$4x = 4$

$4x/4 = 4/4$

$x = 1$ comparing: Possible solution $x = 1$ is not in the conflict with the restriction for x: $x \geq 1/5$ $(1 \geq 1/\ )$.
So, $x = 1$ is a unique solutions of equation.

$x = 1$

Check: $\sqrt{5 \cdot 1 - 1} = \sqrt{1 + 3} \Rightarrow \sqrt{4} = \sqrt{4}$ (true statement)

Example 9 Solve the radical equation $\sqrt{x+1} - \sqrt{x-1} = 1$

SOLUTION ▶

$\sqrt{x+1} - \sqrt{x-1} = 1$

Restriction for x: $x + 1 \geq 0 \cap x - 1 \geq 0.$
$x \geq 1$

$\sqrt{x+1} - \sqrt{x-1} = 1$
$\sqrt{x+1} = \sqrt{x-1} + 1$
$(\sqrt{x+1})^2 = (\sqrt{x-1} + 1)^2$

Square both sides of equation.

$x + 1 = x - 1 + 2\sqrt{x-1} + 1$
$1 = 2\sqrt{x-1}$
$(1)^2 = (2\sqrt{x-1})^2$

Square both sides of equation.

$1 = 4x - 4$
$5 = 4x$
$x = 5/4$ comparing: Possible solution $x = 5/4$, is not in the conflict with the restriction for x: $x \geq 1$ $(5/4 \geq 1)$.
So $x = 5/4$ is a unique solution of given equation.

$x = 5/4$

Check: $\sqrt{5/4 + 1} - \sqrt{5/4 - 1} = 1 \Rightarrow \sqrt{9/4} - \sqrt{1/4} = 1 \Rightarrow 3/2 - 1/2 = 1 \Rightarrow 1 = 1$ (true)

Example 1 Find the radius of the circle shown below if circumference is **4π** feet.

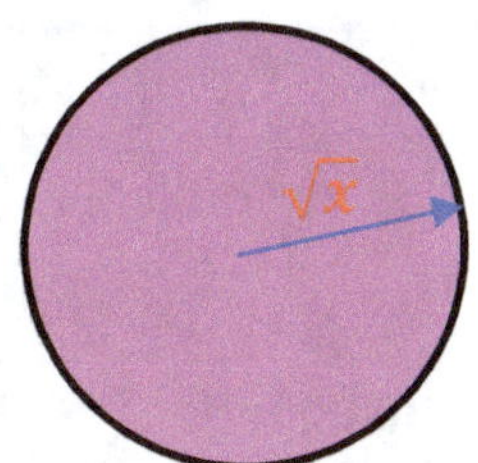

SOLUTION ▶

$$C = 2\pi\sqrt{x} = 4\pi$$
$$2\pi\sqrt{x} \,/\, 2\pi = 4\pi \,/\, 2\pi$$
$$\sqrt{x} = 2$$
$$(\sqrt{x})^2 = (2)^2$$
$$x = 4$$

Example 2 Find the sides of the right triangle shown below if perimeter is **12m**.

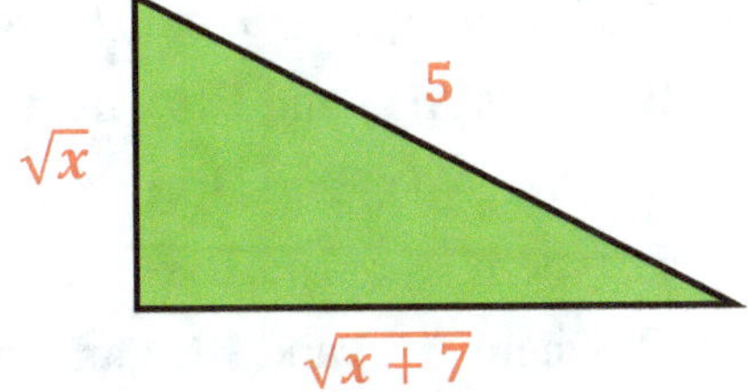

SOLUTION ▶

$$P = \sqrt{x} + \sqrt{x+7} + 5 = 12$$
$$\sqrt{x} + \sqrt{x+7} = 7$$
$$\sqrt{x+7} = 7 - \sqrt{x}$$
$$(\sqrt{x+7})^2 = (7 - \sqrt{x})^2$$
$$x + 7 = 49 - 14\sqrt{x} + x$$

$$14\sqrt{x} = 42$$
$$14\sqrt{x} \,/\, 14 = 42 \,/\, 14$$
$$\sqrt{x} = 3$$
$$(\sqrt{x})^2 = (3)^2$$
$$x = 9$$

■ RADICAL INEQUALITIES

We will use the following definitions.

> **Definition** ▶ *A radical inequality* is an inequality $E_1 < E_2$ (or $E_1 > E_2$, or $E_1 \leq E_2$, or $E_1 \geq E_2$) where E_1 or E_2 or both E_1 and E_2 are radical expressions with one or more variables.
>
> **Definition** ▶ *A single-variable radical inequality* is an inequality $E_1 < E_2$ (or $E_1 > E_2$, or $E_1 \leq E_2$, or $E_1 \geq E_2$) where E_1 or E_2 or both E_1 and E_2 are radical expressions with one variable.

Examples Equations $\sqrt{x} > 3$, $\sqrt{1-x} < 2$, $\sqrt{x-1} + 3 \geq 5$, $\sqrt{x} + \sqrt{x+1} \leq 2$, and $\sqrt{x+2} - \sqrt{x-1} < 1$ are radical inequalities with one variables.

To solve radical inequalities with one variable we will use the following properties.

Properties
▶ If we have radical expression $\sqrt{P(x)}$, then $P(x) \geq 0$ and $\sqrt{P(x)} \geq 0$.

▶ If $\sqrt{A(x)} > \sqrt{B(x)}$, then $A > B$, where $A(x)$ and $B(x)$ are any two expressions.

▶ If $\sqrt{A(x)} < \sqrt{B(x)}$, then $A < B$, where $A(x)$ and $B(x)$ are any two expressions.

▶ If $\sqrt{A(x)} \geq \sqrt{B(x)}$, then $A \geq B$, where $A(x)$ and $B(x)$ are any two expressions.

▶ If $\sqrt{A(x)} \leq \sqrt{B(x)}$, then $A \leq B$, where $A(x)$ and $B(x)$ are any two expressions.

▶ If $\sqrt{P(x)} > Q(x)$, then $\begin{cases} P > Q^2 & \text{if } P \geq 0 \text{ and } Q \geq 0. \\ x \in \{ x \,/\, P(x) \geq 0 \} & \text{if } P \geq 0 \text{ and } Q < 0. \end{cases}$

▶ If $\sqrt{P(x)} \geq Q(x)$, then $\begin{cases} P \geq Q^2 & \text{if } P \geq 0 \text{ and } Q \geq 0. \\ P = Q = 0 & \text{if } P \geq 0 \text{ and } Q \leq 0. \\ x \in \{ x \,/\, P(x) \geq 0 \} & \text{if } P \geq 0 \text{ and } Q < 0. \end{cases}$

▶ If $\sqrt{P(x)} < Q(x)$, then $\begin{cases} P < Q^2 & \text{if } P \geq 0 \text{ and } Q \geq 0. \\ x \in \varnothing & \text{if } P \geq 0 \text{ and } Q < 0. \end{cases}$

▶ If $\sqrt{P(x)} \leq Q(x)$, then $\begin{cases} P < Q^2 & \text{if } P \geq 0 \text{ and } Q \geq 0. \\ P = Q = 0 & \text{if } P \geq 0 \text{ and } Q \leq 0. \\ x \in \varnothing & \text{if } P \geq 0 \text{ and } Q < 0 \end{cases}$

SOLVING RADICAL INEQUALITIES

Example 1 Solve the radical inequality $\sqrt{x-1} > 3$

SOLUTION ▶

$\sqrt{x-1} > 3$ Restriction for x: $x - 1 \geq 0$

$\qquad\qquad\qquad\qquad\qquad\qquad\qquad x \geq 1$

$(\sqrt{x-1})^2 > (3)^2$ Square each side.

$x - 1 > 9$

$x > 10$

$x > 10$ Solution set is $x > 10 \cap x \geq 1$

or in interval notation $x \in (10, \infty)$

Example 2 Solve the radical inequality $\sqrt{x-3} < 1$

SOLUTION ▶

$\sqrt{x-3} < 1$ Restriction for x: $x - 3 \geq 0$

$\qquad\qquad\qquad\qquad\qquad\qquad\qquad x \geq 3$

$(\sqrt{x-3})^2 < (1)^2$ Square each side.

$x - 3 < 1$

$x < 4$

$3 \leq x < 4$ Solution set is $x < 4 \cap x \geq 3$

or in interval notation $x \in [3, 4)$

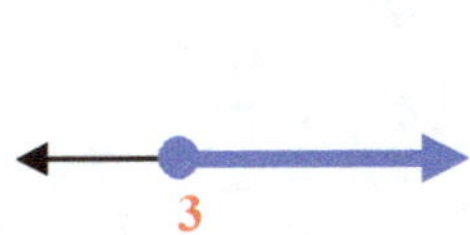

Example 3 Solve the radical inequality $\sqrt{x-3} > -4$

SOLUTION ▶

$\sqrt{x-3} > -4$ Restriction for x: $x - 3 \geq 0$

$\qquad\qquad\qquad\qquad\qquad\qquad\qquad x \geq 3$

$x \geq 3$ Solution set is $x \geq 3$

or in interval notation $x \in [3, \infty)$

 Solve the radical equation $3\sqrt{x-1} > 12$

SOLUTION ▶

$$3\sqrt{x-1} > 12$$

Restriction for x: $x - 1 \geq 0$

$$x \geq 1$$

$$3\sqrt{x-1}\,/\,3 > 12\,/\,3$$

$$\sqrt{x-1} > 4$$

Square each side

$$(\sqrt{x-1})^2 > (4)^2$$

Solution set is $x > 15 \cap x \geq 1$
or in interval notation $x \in (15, \infty)$

$$x - 1 > 16$$

$$x > 15$$

Solution set in interval notation is $x \in (15, \infty)$

$$x > 15$$

 Solve the radical equation $2\sqrt{x-3} \leq 6$

SOLUTION ▶

$$2\sqrt{x-3} \leq 6$$

Restriction for x: $x - 3 \geq 0$

$$x \geq 3$$

$$2\sqrt{x-3}\,/\,2 \leq 6\,/\,2$$

$$\sqrt{x-3} \leq 3$$

Square each side

$$(\sqrt{x-3})^2 \leq (3)^2$$

Solution set is $x \leq 12 \cap x \geq 3$
or $3 \leq x \leq 12$

$$x - 3 \leq 9$$

$$x \leq 12$$

Solution set in interval notation is $x \in [3, 12]$

$$3 \leq x \leq 12$$

 Solve the radical equation $\sqrt{2x-1} \leq -5$

SOLUTION ▶

The nonnegative number $\sqrt{2x-1}$ cannot be ≤ -5.

No solution   or $x \in \emptyset$

Example 7   Solve the radical equation $\sqrt{x+2} < x$

$\sqrt{x+2} < x$ Restriction for x: $x + 2 \geq 0$ $\Rightarrow$
$$x \geq -2.$$

1. If $x \leq 0$, then there is no any solution.

Square root of any expression is a nonnegative number, so it cannot be less than any negative number or zero

$x \in \emptyset$

2. If $x > 0$, then

Square each side.

$$(\sqrt{x+2})^2 > x^2$$

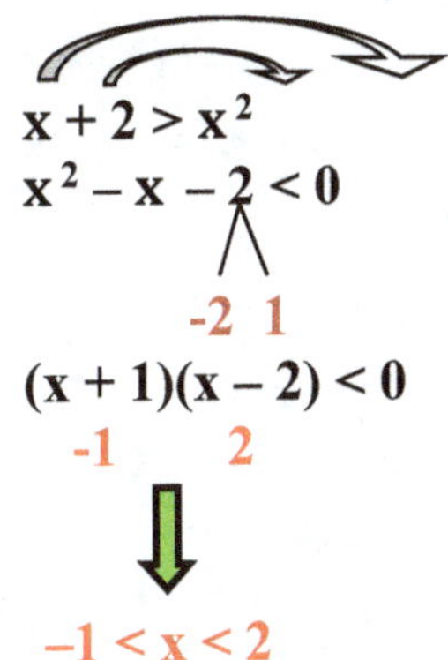

$$x + 2 > x^2$$
$$x^2 - x - 2 < 0$$

Factoring by "branch factoring" method

$$-2 \quad 1$$
$$(x+1)(x-2) < 0$$
$$-1 \qquad 2$$

Zero points of the expression are $x = -1$, and $x = 2$. These points divide the number line into 3 intervals: $(-\infty, -1)$; $(-1, 2)$; $(2, \infty)$.
$x \neq -1$ and $x \neq 2$ because that values make expression equal to 0 and 0 is not < 0.

$$-1 < x < 2$$

In the set notation the solution is

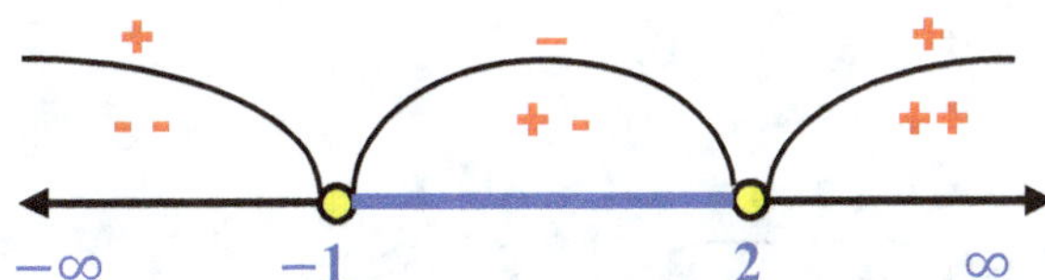

$\{ x \, / -1 < x < 2 \}$

In the interval notation the solution is

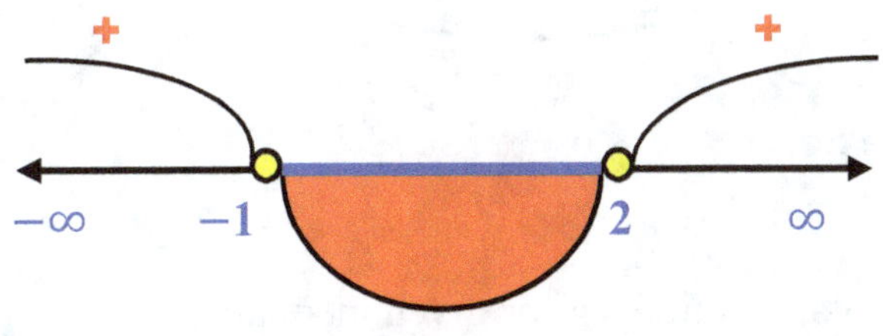

$x \in (-1, 2)$

SOLUTION ▶

$$x + \sqrt{x^2 + x - 6} > -1$$

$$\sqrt{x^2 + x - 6} > -x - 1$$

1. Let $-x - 1 < 0 \Rightarrow x > -1$

Square root of any expression is a nonnegative number, so it always is greater than a negative number. It means that any value of $x > -1$ for which expression $\sqrt{x^2 + x - 6}$ exists (is a real number) is a solution for given inequality. So, the solution is

$$x \geq 2 \quad \text{or} \quad x \in [2, \infty)$$

Restrictions for x: $x^2 + x - 6 \geq 0 \Rightarrow$

$$(x + 3)(x - 2) \geq 0 \Rightarrow$$

$$x \leq -3 \quad \text{or} \quad x \geq 2$$

$\sqrt{x^2 + x - 6} \in \mathbf{R}$ if $x \in (-\infty, -3] \cup [2, \infty)$

Solution set $= \{ x > -1 \} \cap \{ x \leq -3 \text{ or } x \geq 2 \}$

2. Let $-x - 1 = 0$.

$$-x - 1 = 0 \Rightarrow x = -1 \Rightarrow$$

$\sqrt{x^2 + x - 6} = \sqrt{-6}$ is not a real number.

So, $x = -1$ is not a solution.

3. Let $-x - 1 > 0$.

$$-x - 1 > 0 \Rightarrow x < -1$$

$$(\sqrt{x^2 + x - 6})^2 > (-x - 1)^2 \qquad \text{Square each side.}$$

$$x^2 + x - 6 > x^2 + 2x + 1$$

$$-7 > x$$

So, $x < -7$ or $x \in (-\infty, -7)$

We have to compare this possible solution $x < -7$ with $x < -1$ as well as with restrictions for x: $x \leq -3$, $x \geq 2$

Total solution set for given inequality is
$$\{ x \,/\, x < -7 \text{ or } x \geq 2 \}$$
In the interval notation solution set is
$$x \in (-\infty, -7) \cup [2, \infty)$$

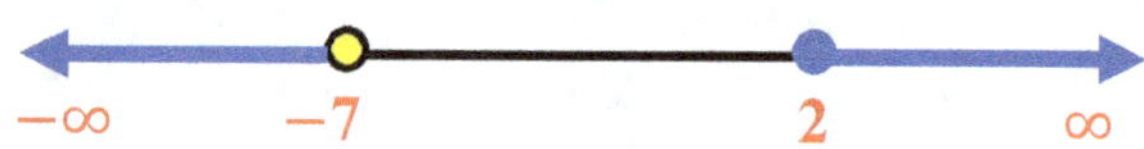

Example 1 The triangle **ABC** is obtuse triangle. Biggest angle is ∡**B**. Find the interval of the possible values for **x**.

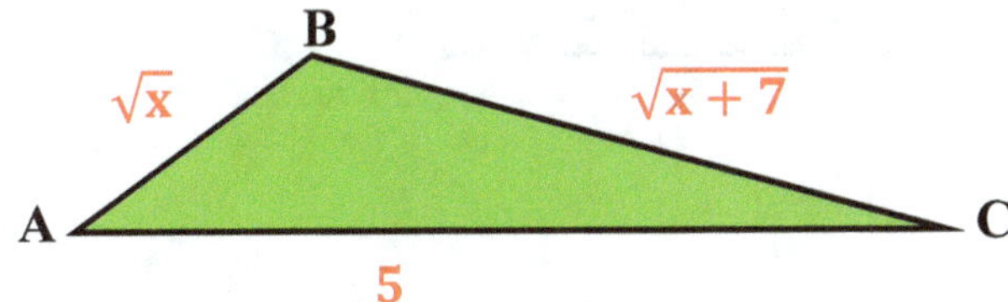

SOLUTION▶

From the inequality properties of the triangle we have the following system of inequalities:

$$\begin{cases} \sqrt{x} < 5 \\ \sqrt{x+7} < 5 \\ \sqrt{x} + \sqrt{x+7} > 5 \end{cases} \Rightarrow \begin{cases} (\sqrt{x})^2 < (5)^2 \\ (\sqrt{x+7})^2 < (5)^2 \\ (\sqrt{x+7})^2 > (5 - \sqrt{x})^2 \end{cases} \Rightarrow$$

$$\begin{cases} x < 25 \\ x + 7 < 25 \\ x + 7 > 25 + x - 10\sqrt{x} \end{cases} \Rightarrow \begin{cases} x < 25 \\ x < 18 \\ 10\sqrt{x} > 18 \end{cases} \Rightarrow$$

$$\begin{cases} x < 25 \\ x < 18 \\ (10\sqrt{x}\,/\,10)^2 > (18/10)^2 \end{cases} \Rightarrow \begin{cases} x < 25 \\ x < 18 \\ x > 81/25 \end{cases} \Rightarrow$$

$$\frac{81}{25} < x < 18$$

Example 2   Triangle **ABC** is a scalene triangle. The biggest angle is **∡B**. The smallest angle is **∡C**. Find the interval of possible values for *x*.

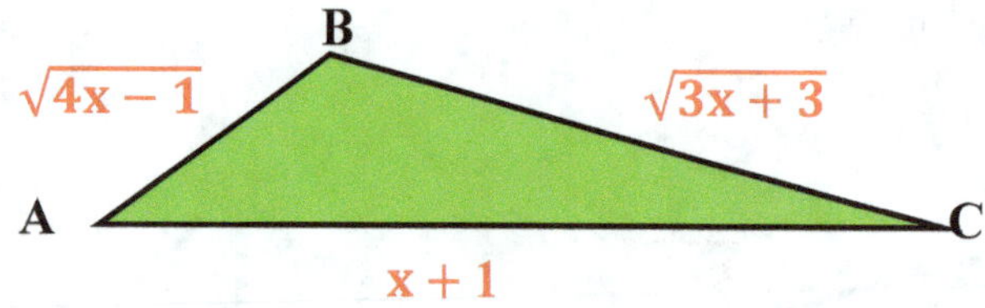

SOLUTION ▶

From inequality properties of the triangle we have the following:

If m∡C < m∡A < m∡B, then the relation between lengths of opposite sides will be

$$\sqrt{4x-1} < \sqrt{3x+3} < x+1$$

⇕

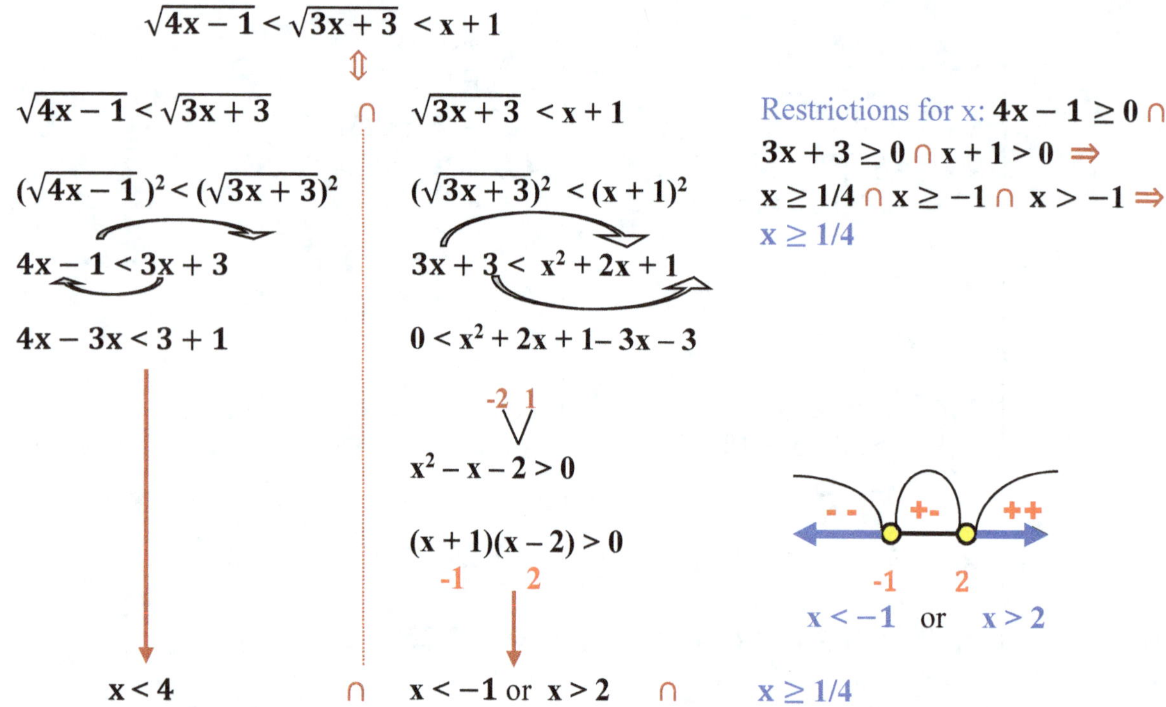

$\sqrt{4x-1} < \sqrt{3x+3}$ ∩ $\sqrt{3x+3} < x+1$ Restrictions for x: $4x-1 \geq 0$ ∩

$(\sqrt{4x-1})^2 < (\sqrt{3x+3})^2$ $(\sqrt{3x+3})^2 < (x+1)^2$ $3x+3 \geq 0$ ∩ $x+1 > 0$ ⇒

$4x-1 < 3x+3$ $3x+3 < x^2+2x+1$ $x \geq 1/4$ ∩ $x \geq -1$ ∩ $x > -1$ ⇒

$4x-3x < 3+1$ $0 < x^2+2x+1-3x-3$ $x \geq 1/4$

$x^2 - x - 2 > 0$

$(x+1)(x-2) > 0$

$x < 4$ ∩ $x < -1$ or $x > 2$ ∩ $x \geq 1/4$

So, the final solution is

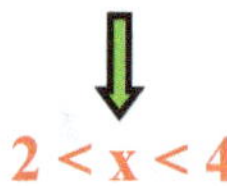

$2 < x < 4$

Solve each radical equation.

1. $\sqrt{-2x} = 4$	**2.** $\sqrt{x-3} = -5$
3. $3\sqrt{5x-1} = 9$	**4.** $\sqrt{15-2x} = x$
5. $\sqrt{x} + \sqrt{x+1} = 1$	**6.** $\sqrt{6-x} - \sqrt{4-x} = \sqrt{x-2}$

Solve each radical inequality.

1. $\sqrt{1-2x} < 4$	**2.** $\sqrt{x-1} > -5$
3. $2\sqrt{5x-1} \geq 4$	**4.** $\sqrt{2x+3} \leq x$
5. $\sqrt{x} + \sqrt{x+1} > 1$	**6.** $\sqrt{2x-1} < \sqrt{x+1} < \sqrt{x^2+2x-1}$

Solve the **linear** equations with one variable.

1. $x + 1 = 4$	2. $x - 3 = 2$	3. $4x = 12$
4. $\frac{3}{5}x = 6$	5. $2x + x - 3 = 3$	6. $\frac{3}{4}(4x + 8) = 6$
7. $2x - 3(x - 1) = x + 1$	8. $4x - 2(x - 1) = 3(x + 2)$	9. $-5(x - 1) + 4x = 3 + x$
10. $\frac{x}{3} - \frac{x-1}{2} = -\frac{x+2}{6} - 2$	11. $\frac{x-3}{5} = \frac{x-2}{6}$	12. $\frac{x}{3} = \frac{x}{2} - 2$

Solve the **linear** inequalities. Present the solutions set in graph and interval notations.

1. $3x > 6$	2. $x + 1 \leq 2$	3. $2x - 1 < 5$
4. $\frac{3x}{5} \geq 9$	5. $2(x + 2) - 4 > 6$	6. $-2(x + 1) + 3x > 2$
7. $\frac{x}{2} < \frac{x+2}{4}$	8. $-3x + 6 \leq 2 - 2(x - 2)$	9. $2(-x + 2.1) \geq 6.4$
10. $x - 2 < 1$ and $-x \leq 2$	11. $x - 3 < 1$ or $-x \leq 5$	12. $\frac{1}{3} < \frac{x-1}{2} \leq \frac{5}{6}$

Solve the **absolute value** linear equations and inequalities with one variable.

1. $|x + 3| = 2$	2. $|x - 1| + 5 = 4$	3. $|2x - 1| + 2 = 0$
4. $3|x + 2| - 1 = 5$	5. $|x + 1| = x$	6. $|x + 2| = x + 2$
7. $2|x + 2| = 4|x - 4|$	8. $|x| = 2x - 1$	9. $||x + 1| + 2| = 3$
10. $|x - 2| < 4$	11. $|x + 1| \geq 2$	12. $|x + 5| \geq -0.005$
13. $|2x - 4| < -4$	14. $|x + 1| - x < 3$	15. $||x + 2| - 3| \geq 1$

Solve the **quadratic** equations and inequalities with one variable.

1. $x^2 - 5x = 0$	2. $4x^2 - 6x = 0$	3. $3x^2 + 7x < 0$
4. $x^2 - 8x - 9 = 0$	5. $x^2 - 6x + 8 = 0$	6. $x^2 + 5x + 4 > 0$
7. $2x^2 - 7x + 6 = 0$	8. $3x^2 - 2x - 5 = 0$	9. $5x^2 + x - 6 \leq 0$
10. $x^2 - 49 = 0$	11. $4x^2 - 100 < 0$	12. $9x^2 - 12x + 4 \geq 0$

Solve the **polynomial** equations and inequalities with one variable.

1. $x^3 - 4x = 0$	2. $2x^4 - 18x^2 = 0$	3. $3x^3 + 7x^2 < 0$
4. $x^4 - 18x^2 + 81 = 0$	5. $x^3 - 6x^2 + 5x = 0$	6. $x^3 + 4x^2 + 3x > 0$
7. $2x^5 - 4x^4 - 6x^3 > 0$	8. $x^4 - 2x^2 + 1 < 0$	9. $5x(x^2 + x - 12) \leq 0$
10. $x^4 - 16 = 0$	11. $16x^4 - 81 \geq 0$	12. $9x^3 - 12x^2 + 4x < 0$

Solve the **fractional** equations and inequalities with one variable.

1. $\dfrac{x+3}{x+5} = \dfrac{2}{3}$	2. $\dfrac{1}{x+5} = \dfrac{x}{6}$	3. $\dfrac{1}{x-2} = \dfrac{1}{x+2} + 1$
4. $\dfrac{1}{x+1} > \dfrac{1}{2}$	5. $\dfrac{1}{x+1} < \dfrac{x}{12}$	6. $\dfrac{1}{x-1} \leq \dfrac{1}{x+1} + 1$
7. $\dfrac{1}{x} = \dfrac{x-2}{3}$	8. $\dfrac{x}{4} \geq \dfrac{1}{x}$	9. $\dfrac{1}{2x+1} > \dfrac{1}{3x-1}$

Solve the **radical** equations and inequalities with one variable.

1. $\sqrt{1-3x} = 6$	2. $\sqrt{x^2-3x} = 2$	3. $\sqrt{2x-1} \leq 3$
3. $\sqrt{4x+5} = x$	4. $\sqrt{7x-4} < -2$	4. $\sqrt{x-1} \geq 2$

Solve the linear equations with one variable.

1. $x + 5 = 2$	**2.** $x - 1 = 6$	**3.** $3x = 15$
4. $\dfrac{1}{25}x = \dfrac{1}{5}$	**5.** $x + 2(x - 3) = 3$	**6.** $\dfrac{1}{2}(2x + 4) = 8 + 2x$
7. $x + 2(x - 2) = 2(x + 1)$	**8.** $3x - (x - 2) = 4(x - 2)$	**9.** $-(x - 1) + x = 2 + x$
10. $\dfrac{x}{3} - \dfrac{x}{2} = -\dfrac{x + 1}{6}$	**11.** $\dfrac{x + 3}{5} = \dfrac{x - 1}{4}$	**12.** $\dfrac{x}{2} = \dfrac{x - 1}{3} + 1$

Solve the linear inequalities. Present the solutions set in graph and interval notations.

13. $2x < 4$	**14.** $x + 3 \leq 3$	**15.** $3x - 4 \geq 2$
16. $\dfrac{3x}{5} < 9$	**17.** $3(x + 2) - 1 > 2x + 4$	**18.** $-3(x + 1) + 2x \leq 1$
19. $\dfrac{x}{3} < \dfrac{x + 1}{4}$	**20.** $-x + 5 \geq -2(x - 2)$	**21.** $3(-x + 2.1) \geq 6.1$
22. $x - 4 < 1$ and $-x \leq 3$	**23.** $x - 1 < 1$ or $-x \leq -6$	**24.** $\dfrac{1}{2} < \dfrac{x + 1}{6} \leq \dfrac{2}{3}$

Solve the absolute value linear equations and inequalities with one variable.

25. $|x + 2| = 1$	**26.** $|2x - 6| + 2 = 3$	**27.** $|x - 1| + 2 = 2$
28. $2|x + 2| - 1 = 3$	**29.** $|x - 1| = x$	**30.** $|x - 2| = x - 2$
31. $3|x + 1| = 6|x - 3|$	**32.** $|2x| = x - 1$	**33.** $||x + 1| + 1| = 2$
34. $|x - 1| < 2$	**35.** $|x + 4| \geq 3$	**36.** $|x + 1| \leq -1$
37. $|x - 1| > -2$	**38.** $|x + 1| - x < 1$	**39.** $||x + 1| - 1| \geq 1$

Solve the **quadratic** equations and inequalities with one variable.

40. $2x^2 - 8x = 0$	**41.** $3x^2 - 6x = 0$	**42.** $x^2 + 5x > 0$
43. $x^2 - 2x - 15 = 0$	**44.** $x^2 - x - 30 = 0$	**45.** $x^2 + 4x + 3 > 0$
46. $2x^2 + 5x - 7 = 0$	**47.** $3x^2 - x - 4 = 0$	**48.** $5x^2 + 2x - 7 \leq 0$
49. $x^2 - 49 < 0$	**50.** $9x^2 - 49 = 0$	**51.** $4x^2 - 12x + 9 \geq 0$

Solve the **polynomial** equations and inequalities with one variable.

52. $4x^3 - x = 0$	**53.** $2x^4 - 50x^2 = 0$	**54.** $x^3 + 3x^2 > 0$
55. $x^4 - 14x^2 + 49 = 0$	**56.** $x^3 - 2x^2 - 8x = 0$	**57.** $x^3 + 5x^2 + 4x < 0$
58. $3x^5 - 6x^4 - 9x^3 = 0$	**59.** $x^4 - 4x^2 + 4 > 0$	**60.** $x(x^2 + x - 6) \leq 0$
61. $x^4 - 25x^2 = 0$	**62.** $49x^3 - 16x \geq 0$	**63.** $4x^3 - 12x^2 + 9x > 0$

Solve the **fractional** equations and inequalities with one variable.

64. $\dfrac{x+1}{x+2} = \dfrac{1}{2}$	**65.** $\dfrac{1}{x+3} = \dfrac{x}{4}$	**66.** $\dfrac{1}{x-1} = \dfrac{1}{x+1} + 1$
67. $\dfrac{1}{x} > \dfrac{1}{x+2}$	**68.** $\dfrac{1}{x-1} \leq \dfrac{x}{6}$	**69.** $\dfrac{1}{x-2} > \dfrac{1}{x+2} + 1$
70. $\dfrac{1}{x} = \dfrac{x-5}{6}$	**71.** $\dfrac{x}{4} \geq \dfrac{1}{x}$	**72.** $\dfrac{1}{x+1} < \dfrac{1}{2x-1}$

Solve the **radical** equations and inequalities with one variable.

73. $\sqrt{1 - 2x} = 5$	**74.** $\sqrt{x^2 - 2x} = \sqrt{3}$	**75.** $\sqrt{x-1} \leq 1$
76. $\sqrt{3 - 2x} = x$	**77.** $\sqrt{x-1} < -3$	**78.** $\sqrt{x^2 - 3x} \geq 2$

Chapter 5. Graphs of Lines and Inequalities

5.1. **System of Coordinates**
5.2. **Equation of a Line**
5.3. **Graphs of Lines**
5.4. **Parallel, Coincident, Perpendicular, and Intersecting Lines**
5.5. **Linear Inequalities on the Coordinate Plane**

Chapter 5   Classwork
Chapter 5   Homework

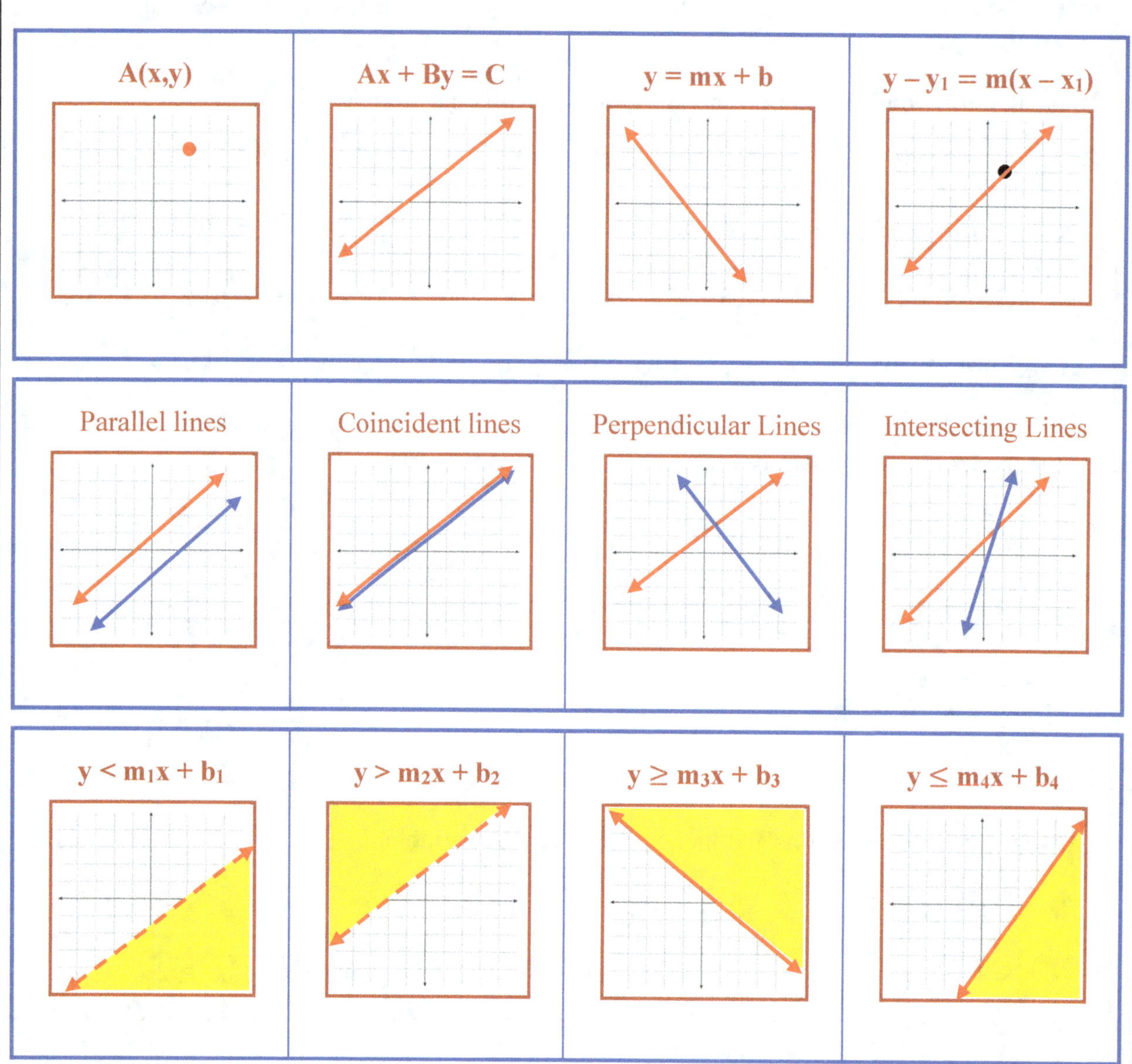

- **THE COORDINATE PLANE**
- **POINTS ON THE COORDINATE PLANE**

THE COORDINATE PLANE

The coordinate plane or Cartesian coordinate system is important in mathematics because it helps us *visualize solution sets* of *linear equations and linear inequalities with two variables* (**x** and **y**). Each solution is an ordered pair of real numbers (**x,y**) or is a point on the coordinate plane (or on the system of coordinates).

We will use the following definitions of coordinate plane/rectangular system of coordinates.

DEFINITION

Definition ▶ *A coordinate plane (or rectangular system of coordinates or Cartesian coordinate system)* is a system of horizontal and vertical number lines on the plane with the same **0**-point/origin.
The horizontal real number line is called the **x-axis**. The vertical real number line with the positive direction up and the negative direction down is called the **y-axis**. The axes divide the plane into 4 quadrants, which are numbered in a counterclockwise direction.

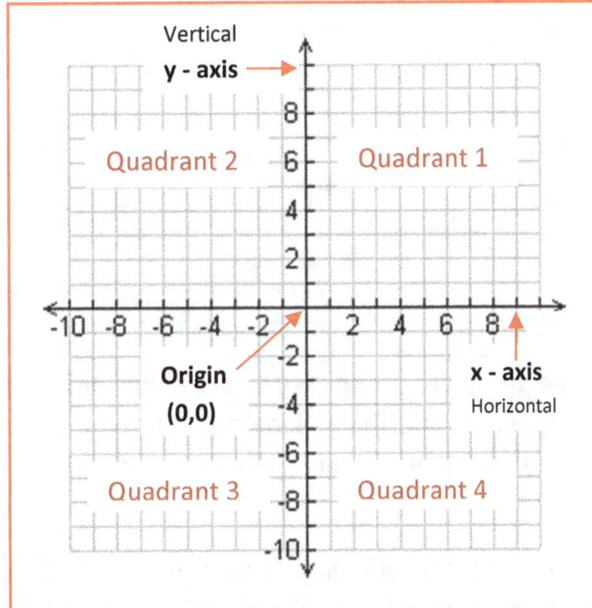

■ POINTS ON THE COORDINATE PLANE

For any point on the coordinate plane we can find a corresponding **ordered pair** of numbers **(x,y)** where the first number **x** (or **x-coordinate** of the given point) corresponds to the projection of the given point on the **x- axis**, and the second number **y** (or **y-coordinate** of the given point) corresponds to the projection of the given point on the **y- axis**.

We will use the following definitions of an **ordered pair of numbers**.

Definition ▶ *An ordered pair of numbers (x,y)* is pair of real numbers enclosed in parentheses and separated by a comma. The number **x** is the first number and the number **y** is the second number. The first number **x** is called the **x-coordinate** and the second number **y** is called the **y-coordinate**.

For the ordered pair **(4, 6)**, the **x-coordinate** is **4**, the **y-coordinate** is **6**.

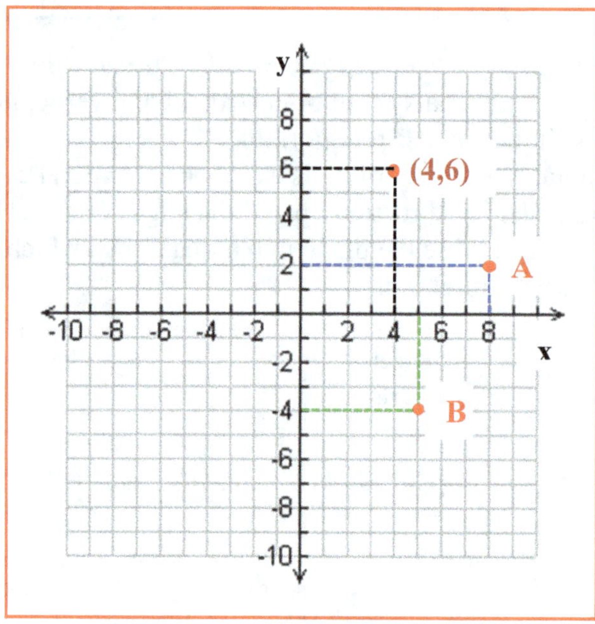

There is **1-1** relation between any point of the coordinate plane and the pair of ordered numbers. This means that for any point on the coordinate plane we can find a corresponding ordered pair of real numbers, and for any ordered pair of real numbers we can find a corresponding point on the coordinate plane.

Examples Corresponding point for ordered pair **(8,2)** is **A** on the figure above.
Corresponding ordered pair for the point **B** is pair of numbers **(5,-4)**.

1. Find the ordered pairs of numbers (or coordinates) associated with the following points

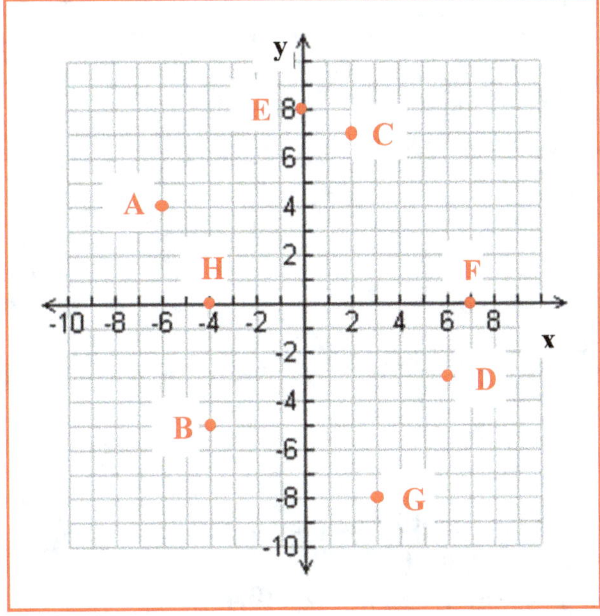

2. Find and label the points on coordinate plane associated with the following pairs of numbers:
(3,2); (-4,0); (0,6); (0,0); (5,-5); (-5,8); (3,9); (6,1); (8,7); (8,-4); (2,0); (7,-7); (-7,7); (2,-8).
Name the quadrant each point is in.

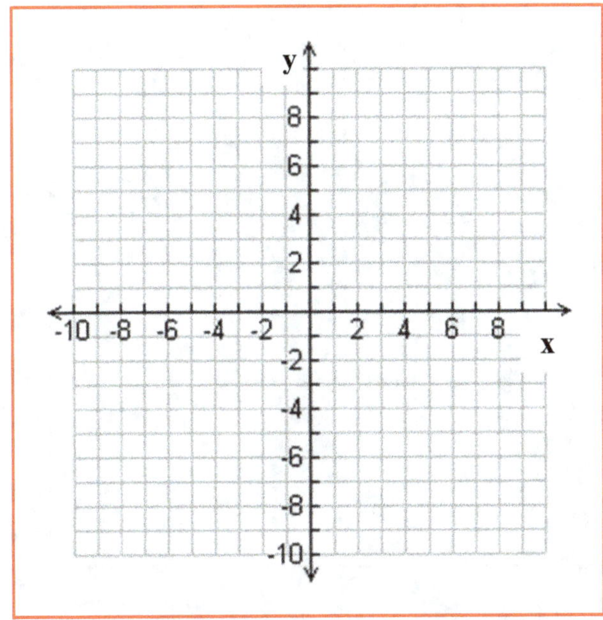

3. Without plotting the point, tell whether it is in Quadrant 1, Quadrant 2, Quadrant 3, Quadrant 4, or on the **x-axis**, or on the **y-axis**.
- **(2,2)**
- **(3,-4)**
- **(-1,5)**
- **(-3,-8)**
- **(0,1)**
- **(-1,0)**
- **(-5,-5)**
- **(6,-6)**
- **(0,4)**
- **(3,0)**
- **(0,-5)**
- **(-1,0)**

4. Draw the following figures on the coordinate plane.
  a) The triangle with the vertices **(0,0)**; **(4,2)**; **(2,4)**.
  b) The square with the vertices **(0,0)**; **(4,0)**; **(4,4)**; **(0,4)**.
  c) The hexagon with the vertices **(-5,6)**; **(5,6)**; **(8,0)**; **(5,-6)**; **(-5,-6)**; **(-8,-0)**.
  d) The line segment with endpoints **(8,9)** and **(-8,-9)**.
  e) The rectangle with the vertices **(-6,-2)**; **(-6,-6)**; **(-2,-6)**; **(-2,-2)**.

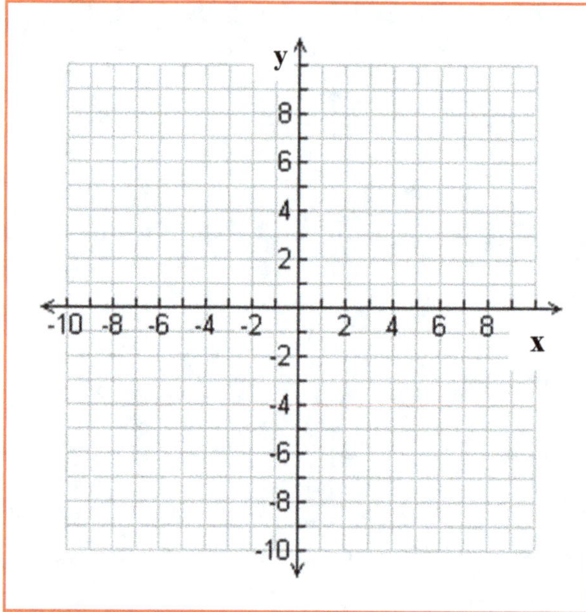

5. Is the graph of the point **(0,0)** lie on the line segment with endpoints **(8,9)** and **(-8,-9)**?

5.2. Equation of a Line

- **DIFFERENT FORMS OF THE EQUATION OF A LINE**
- **THE SLOPE OF A LINE**
- **WRITING EQUATIONS OF LINES**

DIFFERENT FORMS OF THE EQUATION OF A LINE

From the previous section we know that any point on the coordinate plane can be represented by an ordered pair of real numbers (**x,y**), and for any ordered pair of real numbers there is one corresponding point on the coordinate plane.

In this section we will study the relation between lines on the coordinate plane and linear equations with **two** variables **x** and **y**.

There is a one-to-one relation between the lines on the coordinate plane and the linear equations with **two** variables **x** and **y**. For any line on the coordinate plane we can find a corresponding linear equation with **2** variables **x** and **y**. And for any linear equation with **2** variables **x** and **y** we can find a corresponding line on the coordinate plane.

We will use the following three forms of the equation of a line.

DIFFERENT FORMS OF THE EQUATION OF A LINE

Standard form ▶ $Ax + By = C$ where $A, B, C \in \mathbb{R}$

Slope-y-intercept form ▶ $y = mx + b$ where m is the slope and b is y-intercept of the line

Slope-point form ▶ $y - y_1 = m(x - x_1)$ where (x_1, y_1) is the point through of which the line passes

$Ax + By = C$	$y = mx + b$	$y - y_1 = m(x - x_1)$
$y_{int} = \dfrac{C}{B}$ $x_{int} = \dfrac{C}{A}$	b	(x_1, y_1)
Example: $4x + 3y = 12$	Example: $y = 2x - 3$	Example: $y - 3 = 2(x - 1)$

Example 1 Convert the standard form of a line to the slope-y-intercept form.

SOLUTION ▶

$$Ax + By = C$$

$$By = -Ax + C$$

$$\frac{B}{B}y = -\frac{A}{B}x + \frac{C}{B}$$

$$y = -\frac{A}{B}x + \frac{C}{B}$$

$$\mathbf{y = m\,x + b} \qquad \text{where} \quad m = -\frac{A}{B} \text{ and } b = \frac{C}{B}$$

Example 2 Convert the slope-y-intercept form of a line to the standard form.

SOLUTION ▶

$$y = m\,x + b$$

$$-mx + y = b$$

$$\mathbf{Ax + By = C} \qquad \text{where} \quad A = -m, B = 1, \text{ and } C = b$$

Example 3 Show that the slope-point form derives from the slope-y-intercept form.

SOLUTION ▶

Let $\mathbf{y = m\,x + b}$ (1) is the slope-y-intercept form of the equation of the given line

Then $\mathbf{y_1 = mx_1 + b}$ (2) because point (x_1, y_1) lies on the line $y = m\,x + b$

Difference of the equations (1) and (2) give us the slope-point form of the given line

$$\mathbf{y - y_1 = m(x - x_1)}$$

Example 4 Convert the slope-point form of a line to the slope-y-intercept form.

SOLUTION ▶

$$y - y_1 = m(x - x_1)$$

$$y = m(x - x_1) + y_1$$

$$y = mx - mx_1 + y_1$$

$$\mathbf{y = m\,x + b} \qquad \text{were} \quad b = -mx_1 + y_1$$

Example 5   Convert the standard form of a line to the slope-y-intercept form.

SOLUTION ▶

$$2x + 3y = 6$$

$$3y = -2x + 6$$

$$\frac{3}{3}y = -\frac{2}{3}x + \frac{6}{3}$$

$$y = -\frac{2}{3}x + 2$$

$$y = -\frac{2}{3}x + 2 \qquad \text{where} \quad m = -\frac{2}{3} \text{ and } b = 2$$

Example 6   Convert the slope-y-intercept form of a line to standard form.

SOLUTION ▶

$$y = 3x + 4$$

$$-3x + y = 4$$

$$-3x + y = 4 \qquad \text{where} \quad A = -3, B = 1, \text{ and } C = 4$$

Example 7   Write in slope-point form the equation of the line that passes through the point (2,5) with slope 4.

SOLUTION ▶

$$y - y_1 = m(x - x_1) \qquad \text{is the general slope-point form of the equation of the line}$$

$$y - 5 = 4(x - 2) \qquad \text{we plug in } y_1 = 5, m = 4, \text{ and } x_1 = 2 \text{ in general formula}$$

$$y - 5 = 4(x - 2) \qquad \text{is the slope-point form of the equation of the line}$$

Example 8   Convert the slope-point form $y - 5 = 4(x - 2)$ to the slope-y-intercept form.

SOLUTION ▶

$$y - 5 = 4(x - 2) \qquad \text{is the slope-point form of the equation of the line}$$

$$y - 5 = 4(x - 2)$$

$$y = 4x - 8 + 5$$

$$y = 4x - 3 \qquad \text{is the slope-y-intercept form of the equation of the line}$$

Rewrite each equation in slope-y-intercept form.

1. $2x + y = 3$	2. $-x + y = 2$	3. $y - 4x = 5$
4. $3x + 3y = 6$	5. $4x - 2y = 4$	6. $2y + 6 = 4x$
7. $x + 4y = 3$	8. $2x + 5y = 10$	9. $x - y = 4$
10. $2x - 4y = 8$	11. $7x - 2y = -6$	12. $x + \frac{1}{4}y = 1$
13. $x + 2y = 0$	14. $2x - \frac{1}{2}y = \frac{1}{2}$	15. $2x = 3 - y$

Rewrite each equation in standard form.

1. $y = 2x + 3$	2. $y = -x + 1$	3. $y = 5 - 4x$
4. $y = 6x$	5. $-x = 4 + 2y$	6. $3y + 5 = 2x$
7. $y = 3 + 5x$	8. $5y = 10x$	9. $x = -y + 1$
10. $0.2x = -0.4y - 0.8$	11. $7 - 2x = -y$	12. $\frac{1}{4}y = x + 1$
13. $2y = 8 + 3x$	14. $4x = \frac{1}{2} - \frac{1}{2}y$	15. $y = \frac{1}{2} - \frac{1}{2}x$

Write the slope-point form for the equation of the line with the given slope and point.

1. $m = 2,\ (1,4)$	2. $m = -1,\ (0,3)$	3. $m = -2,\ (5,0)$
4. $m = -3,\ (2,-3)$	5. $m = -1,\ (-2,-3)$	6. $m = -0,\ (-1,3)$
7. $m = \frac{1}{2},\ (1,-4)$	8. $m = -\frac{2}{3},\ (-3,2)$	9. $m = \frac{1}{3},\ (-3,0)$

Rewrite each slope-point equation in slope-y-intercept form and then in standard form.

1. $y - 1 = 4(x + 1)$	2. $y - 4 = 3(x - 3)$	3. $y + 2 = 4(x + 1)$
4. $y - 3 = 2(x - 1)$	5. $y - 0.5 = 0.5(x - 1)$	6. $y + 6 = \frac{1}{2}(x - 2)$
7. $y + 1 = \frac{1}{3}(x + 6)$	8. $y - 2 = \frac{3}{4}(x - 8)$	9. $y = 2(x + 2)$
10. $y + 7 = \frac{4}{5}(x + 5)$	11. $y + \frac{1}{2} = \frac{3}{2}(x + 1)$	12. $y - \frac{1}{2} = \frac{1}{4}(x + 2)$

■ THE SLOPE OF A LINE

$\mathbf{W}$e will use the following definitions of the **slope**.

Definition ▶ *The slope of a line* passes through two points $(\mathbf{x_1}, \mathbf{y_1})$ and $(\mathbf{x_2}, \mathbf{y_2})$ is

$$\mathbf{m} = \frac{y_2 - y_1}{x_2 - x_1} = \frac{y_1 - y_2}{x_1 - x_2} = \frac{\text{vertical change}}{\text{horizontal change}} = \frac{\text{rise}}{\text{run}}$$

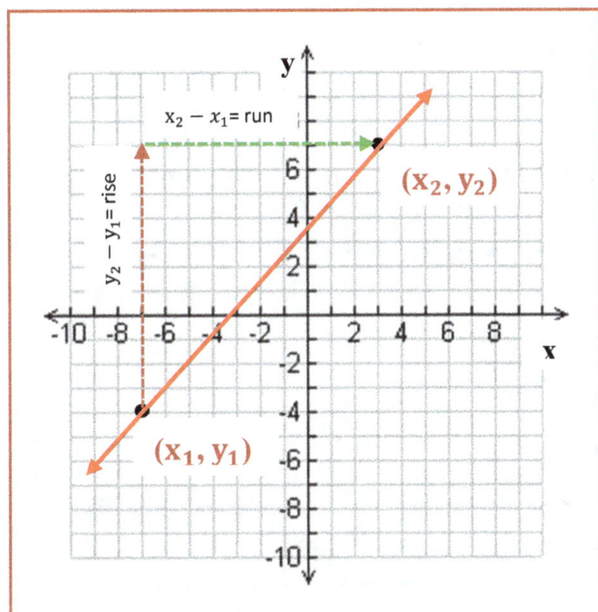

Example: Slope of the line above passes through two points $(-7, -4)$ and $(3, 7)$ is $\mathbf{m} = \dfrac{11}{10}$.

Note: Slope can be **positive**, **negative**, **zero**, or **undefined**.

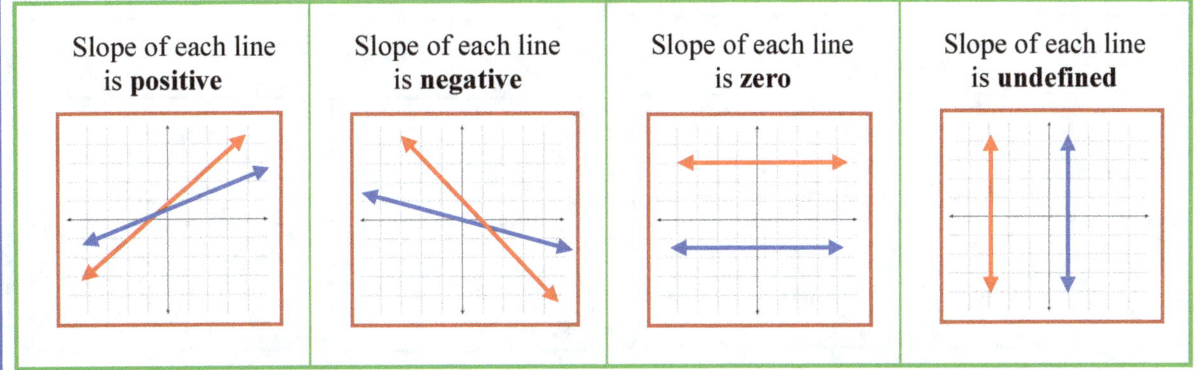

Find the slope of the line through each pair of points.

1. (1,2), (3,4)	2. (0,2), (–2,3)	3. (0,0), (–1, –2)
4. (2,4), (3,4)	5. (3,2), (3,– 2)	6. (4,0), (–5, –3)
7. (1,2), (1,4)	8. (3,1), (0,4)	9. (0,0), (–1, –2)
10. (1,1), (–1, –4)	11. (2,1), (0,1)	12. (7,5), (–1, –3)

Find the slope of each line through the given pair of points on the graph.

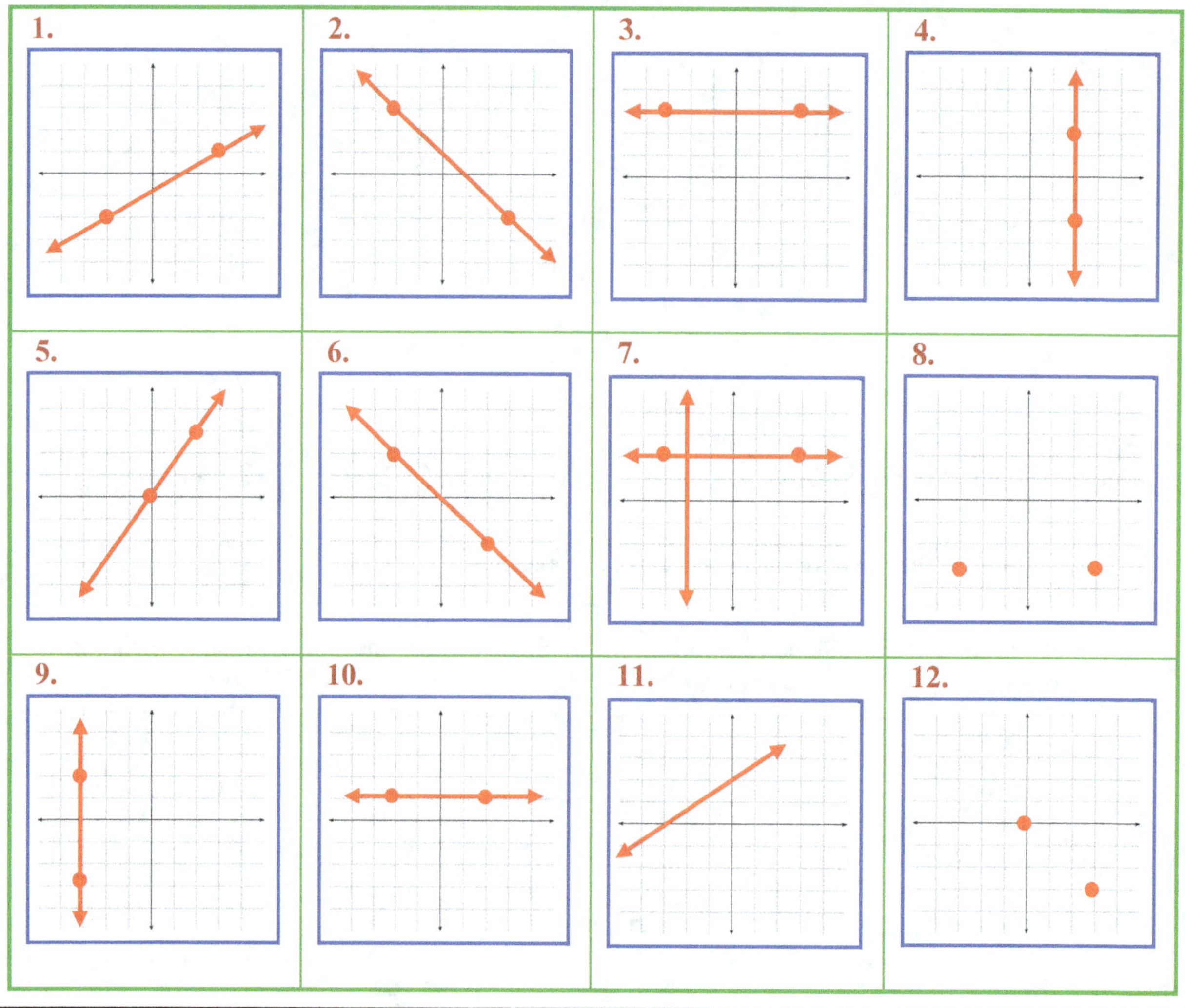

Find the slope of each line from the equation of the line.

1. $x + y = 3$	2. $-2x + y = 4$	3. $y - 5x = 4$
4. $2x + 2y = 6$	5. $3x - 2y = 8$	6. $3y + 6 = x$
7. $x + 3y = 6$	8. $5x + 2y = 8$	9. $2x - y = 4$
10. $x - 4y = 12$	11. $6x - 2y = -4$	12. $\frac{1}{4}x + y = 1$
13. $x + 3y = 0$	14. $x - \frac{1}{2}y = \frac{1}{4}$	15. $\frac{1}{2}x = 3 - \frac{3}{2}y$
16. $y = 2x + 3$	17. $y = -x + 1$	18. $y = 5 - 4x$
19. $y = 6x$	20. $-x = 4 + 2y$	21. $3y + 5 = 2x$
22. $y = 3 + 5x$	23. $5y = 10x$	24. $x = -y + 1$
25. $0.2x = -0.4y - 0.8$	26. $7 - 2x = -y$	27. $\frac{1}{4}y = x + 1$
28. $2y = 8 + 3x$	29. $4x = \frac{1}{2} - \frac{1}{2}y$	30. $y = \frac{1}{2} - \frac{1}{2}x$

■ WRITING EQUATIONS OF LINES

We can write the equation of the line if we have the slope and the y-intercept of that line, or if we have the slope and the coordinates of one point of that line, or if we have the coordinates of any two points of that given line.

Example 1 Write the equation of the line in slope-y-intercept and in standard form if the slope of the line is **m = 2** and the y-intercept is **b = 3.**

SOLUTION ▶ The equation of the line in slope-y-intercept form is

$$y = m x + b \qquad \text{were} \quad m = 2 \quad \text{and} \quad b = 3$$

$$y = 2x + 3 \qquad \text{the slope-y-intercept form of the line}$$

$$y = 2x + 3$$

$$-2x + y = 3 \qquad \text{the standard form of the line}$$

Example 2 Write the equation of the line in slope-point, in slope-y-intercept and in standard forms if the slope of the line is **m = 2** and the line passes through the point **(1,2).**

SOLUTION ▶ The equation of the line in slope-point form is

$$y - 2 = 2(x - 1) \qquad \text{the slope-point form of the line}$$

The equation of the line in the slope-y-intercept form is

$$y - 2 = 2(x - 1)$$

$$y = 2x - 2 + 2$$

$$y = 2x \qquad \text{the slope-y-intercept form of the line}$$

$$y = 2x$$

$$-2x + y = 0 \qquad \text{the standard form of the line}$$

Example 3 Write the equation of the line passes through the points **(2,3)** and **(1,4).**

SOLUTION ▶ We will find the slope and use the equation of the line in slope-point form.

$$m = \frac{y_2 - y_1}{x_2 - x_1} = \frac{4 - 3}{1 - 2} = \frac{1}{-1} = -1$$

We will use this slope and choose the point **(1,4)** to write the slope-point form.

$$y - 4 = -1(x - 1) \qquad \text{the slope-point form of the line}$$

$$y = -x + 5 \qquad \text{the slope-y-intercept form of the line}$$

$$x + y = 5 \qquad \text{the standard form of the line}$$

Example 4 Write the equation of the line in slope-y-intercept and in standard form if the slope of the line is $m = 0$ and the y-intercept is $b = 4$.

SOLUTION ▶ The equation of the line in slope-y-intercept form is

$y = m\,x + b$ were $m = 0$ and $b = 4$

$y = 0 \cdot x + 4$ the slope-y-intercept form of the line

$y = 4$ the slope-y-intercept form and same time the standard form

Example 5 Write the equation of the line in slope-point, in slope-y-intercept and in standard forms if the slope of the line is $m = 0$ and the line passes through the point $(1,3)$.

SOLUTION ▶ The equation of the line in slope-point form is

$y - 3 = 0(x - 1)$ the slope-point form of the line

The equation of the line in the slope-y-intercept form is

$y - 3 = 0$

$y = 0 + 3$

$y = 3$ the slope-y-intercept form and same time the standard form

Example 6 Write the equation of the line if the slope is $m = $ **undefined** and the line passes through the point $(1,3)$.

SOLUTION ▶ If $m = $ **undefined** then line is a vertical to x-axis line and the equation is

$x = 1$

Example 7 Write the equations of the horizontal and the vertical lines passes through the point $(-2,3)$.

SOLUTION ▶ If the line is a **horizontal line** then the slope $m = 0$.
The equation of the line will be

$y - 3 = 0(x + 2)$ the slope-point form of the line

$y = 3$ the slope-y-intercept form and same time the standard form

If the line is a **vertical line** then the slope $m = $ **undefined**.
The equation of the line will be

$x = -2$

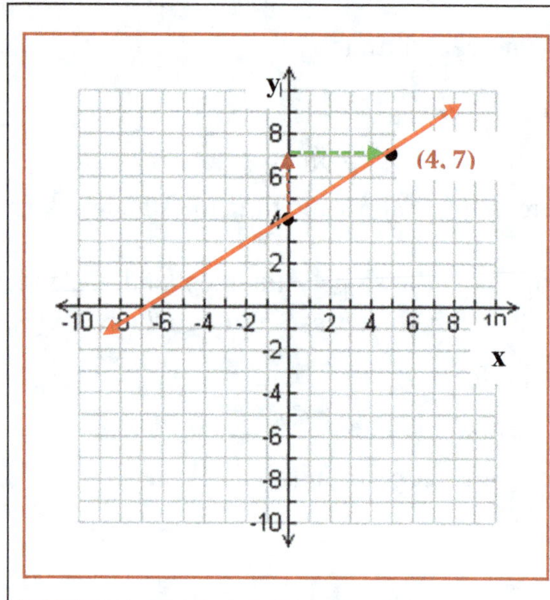

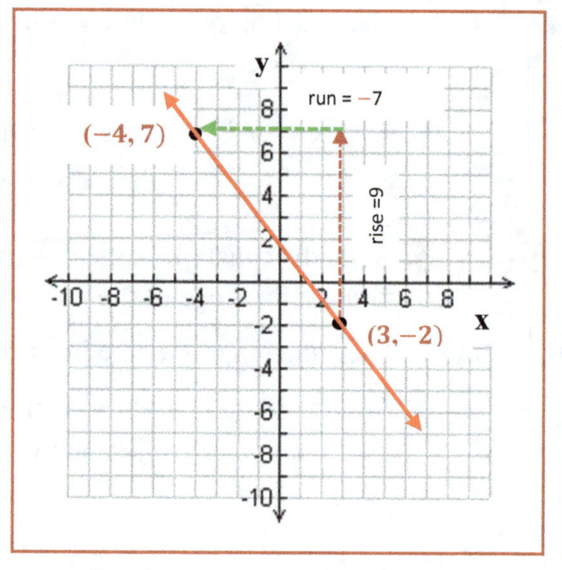

SOLUTION ▶ The equations of the lines in slope-y-intercept form see

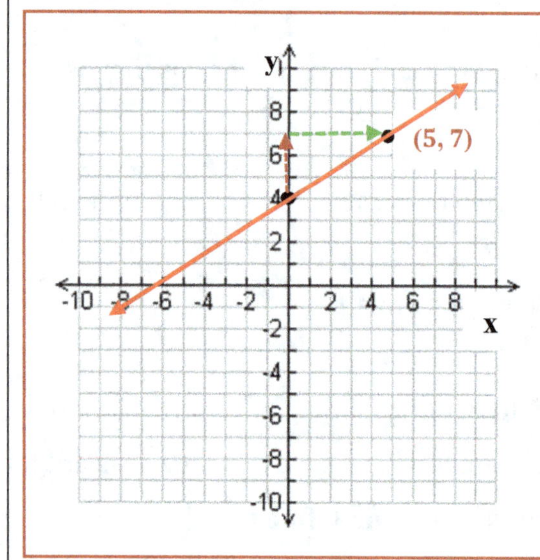

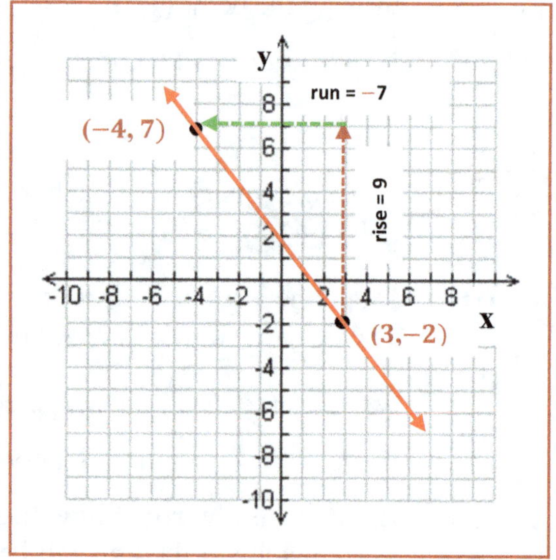

$$y = \frac{\text{rise}}{\text{run}}\, x + b$$

$$y = \frac{3}{5}x + 4 \qquad \text{the slope-y-intercept form}$$

$$-3x + 5y = 20 \qquad \text{the standard form}$$

$$y - 7 = \frac{9}{-7}(x + 4) \quad \text{the slope-point form}$$

$$y = -\frac{9}{7}x + \frac{13}{7} \qquad \text{the slope-y-intercept form}$$

$$9x + 7y = 13 \qquad \text{the standard form}$$

Write the equation of each line with the given slope **m** and y-intercept **b** in the slope-y-intercept form and in the standard form

1. $m = 2, b = 3$	2. $m = -2, b = 3$	3. $m = \frac{1}{2}, b = 3$
4. $m = 3, b = -2$	5. $m = -4, b = -2$	6. $m = \frac{2}{3}, b = -2$
7. $m = 1, b = 0$	8. $m = -2, b = 0$	9. $m = 0, b = 0$
10. $m = 0, b = 3$	11. $m = 0, b = -2$	12. $m = 0, b = \frac{2}{3}$

Write the equation of each line with the given slope **m** and the given point **(x,y)** in the **slope-point**, the **slope-y-intercept**, and the **standard forms**

1. $m = 1, (3,4)$	2. $m = 3, (0,2)$	3. $m = -2, (0,0)$
4. $m = 3, (-1,-2)$	5. $m = 0, (3,-2)$	6. $m = $ undefined, $(-5,-3)$
7. $m = \frac{1}{2}, (3,4)$	8. $m = -\frac{2}{3}, (0,-3)$	9. $m = 0, (0,0)$
10. $m = -\frac{1}{3}, (-1,-9)$	11. $m = 0.25, (-4,-4)$	12. $m = $ undefined, $(0,-2)$

Write the equation of each line with the given two points in the **slope-point**, the **slope-y-intercept**, and the **standard forms**

1. $(1,2), (3,4)$	2. $(0,2), (-2,3)$	3. $(0,0), (-1,-2)$
4. $(2,4), (3,4)$	5. $(3,2), (3,-2)$	6. $(4,0), (-5,-3)$
7. $(1,2), (1,4)$	8. $(3,1), (0,4)$	9. $(0,0), (-1,-2)$
10. $(1,1), (-1,-4)$	11. $(2,1), (0,1)$	12. $(7,5), (-1,-3)$

PRACTICE ■ WRITING EQUATIONS OF THE LINES

Write the equations of the lines with the graphs shown below

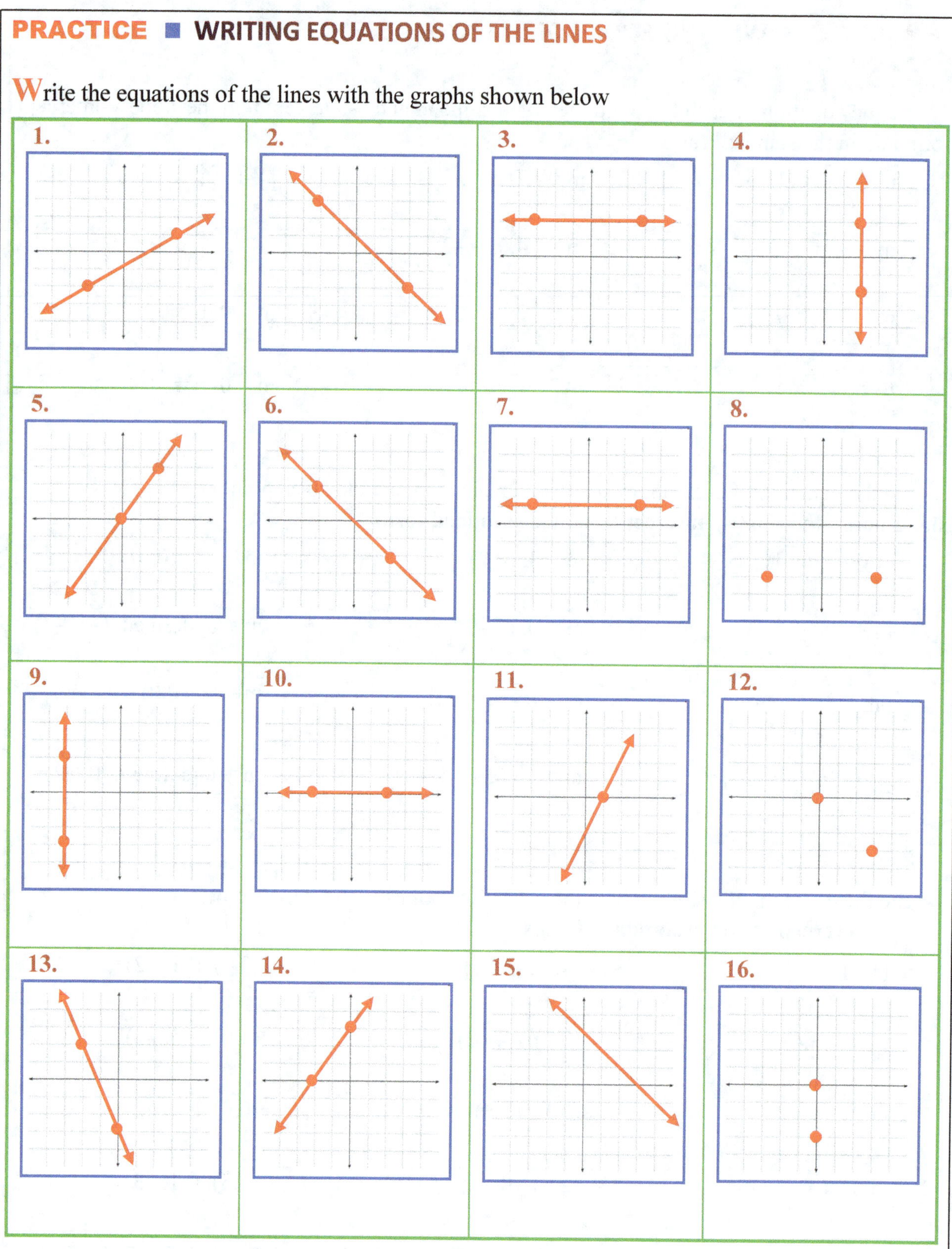

5.3. Graphs of Lines

- **POINTS METHOD**
- **X-Y-INTERCEPTS METHOD**
- **SLOPE-Y-INTERCEPT METHOD**
- **GRAPHS OF HORIZONTAL AND VERTICAL LINES**

To draw the graph of any line we will use the basic postulate of Plane (or Euclidian) Geometry.

POSTULATE

> ▶ **There is exactly one line passing through two distinct points**

So, if we have a linear equation then we can graph corresponding line on the coordinate plane by choosing any two points (two ordered pair of numbers) of that line and drawing the line passes through those two points.

■ POINTS METHOD

Points' method includes **2 steps**: making the **x-y-chart** for any two values of independent variable **x** and corresponding two values of dependent variable **y**, and then drawing the **graph** through the points corresponding to those ordered pairs **(x,y)**.

This method works perfectly for **slanted lines** (lines with nonzero **x-** and **y**-terms of equation) and dos note works for **horizontal lines** ($y = a$) and for **vertical lines** ($x = b$).

Example 1 Graph the line $y = 2x + 1$

SOLUTION▶ **Step 1.** Making the **x-y-chart** for any two values of independent variable **x** and corresponding two values of dependent variable **y**

Step 2. Drawing the graph through the points corresponding to those ordered pairs **(0,1); (1,3)**

<table>
<tr><td>Step 1</td><td>Step 2</td></tr>
<tr><td>

x	y	y = 2x+1
0	1	$1 = 2\cdot0+1$
1	3	$3 = 2\cdot1+1$

The points are:
    (0,1); (1,3)

</td><td>

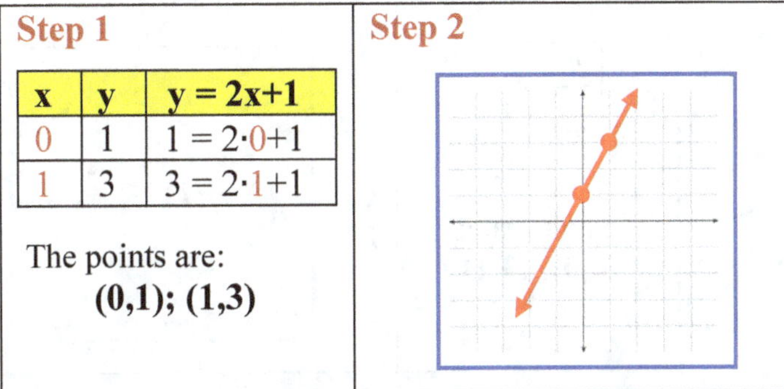

</td></tr>
</table>

Example 2 Graph the line $y = -\dfrac{2}{3}x + 3$

SOLUTION ▶ **Step 1.** Making the **x-y-chart**
Step 2. Drawing the graph

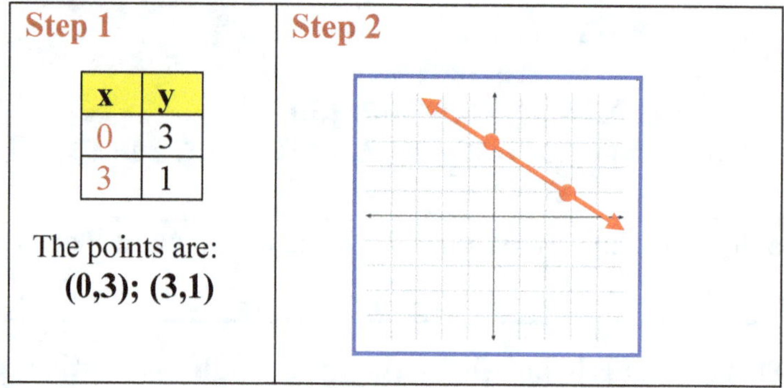

Step 1	Step 2
x **y** 0 | 3 3 | 1 The points are: **(0,3); (3,1)**	

Example 2 Graph the line **2x + 4y = 12**

SOLUTION ▶ **Step 1.** Making the **x-y-chart**
Step 2. Drawing the graph

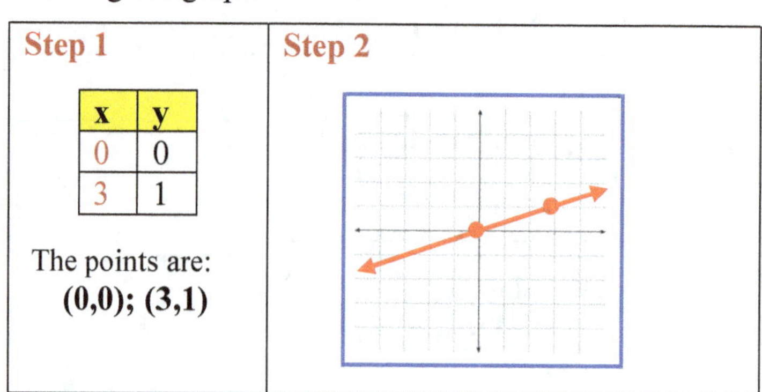

Step 1	Step 2
x **y** 0 | 3 2 | 2 The points are: **(0,3); (2,2)**	

Example 3 Graph the line $y = \dfrac{1}{3}x$

SOLUTION ▶ **Step 1.** Making the **x-y-chart**
Step 2. Drawing the graph

Step 1	Step 2
x **y** 0 | 0 3 | 1 The points are: **(0,0); (3,1)**	

Graph each line by **points method**.

1. $y = x + 2$

x	y

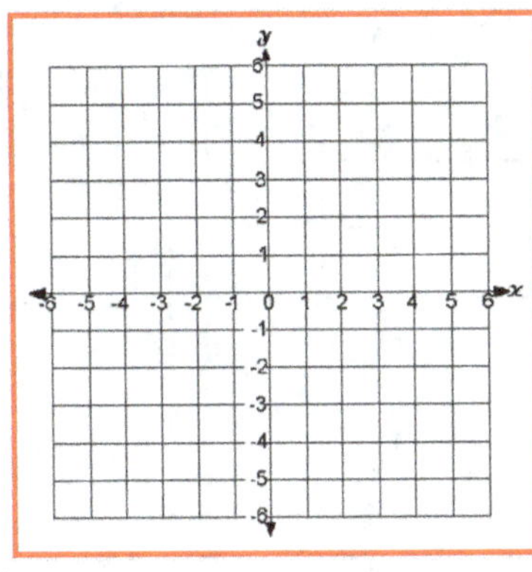

2. $y = -x - 1$

x	y

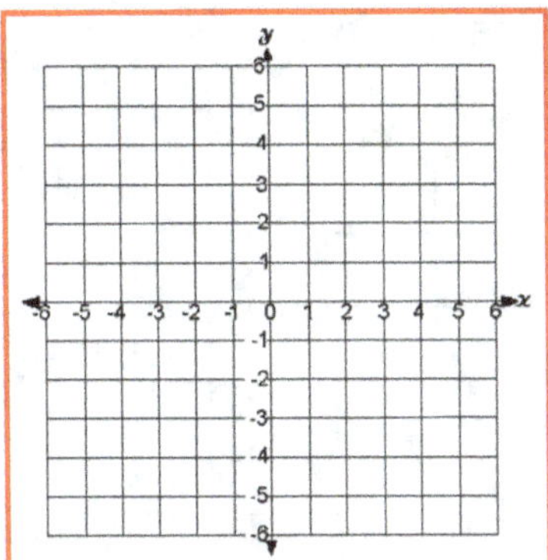

3. $y + 2x = 3$

x	y

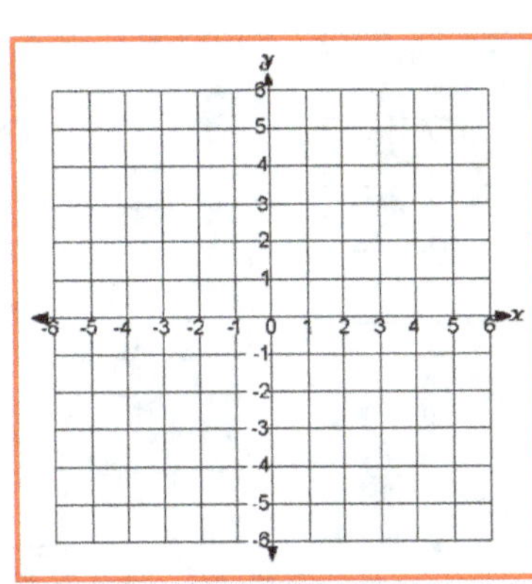

4. $x - y = -4$

x	y

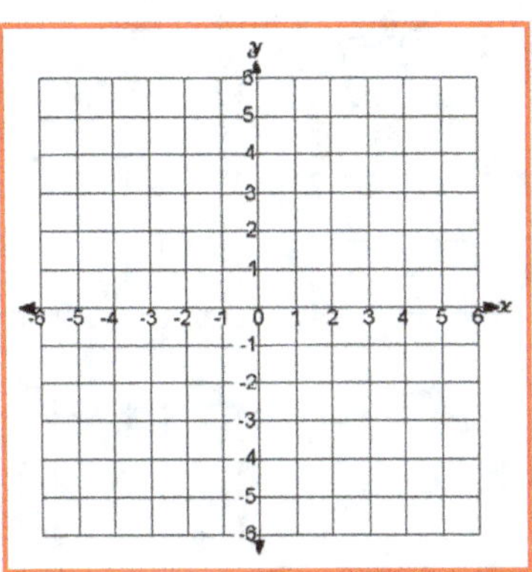

5. $y = \dfrac{2}{3}x - 4$

x	y

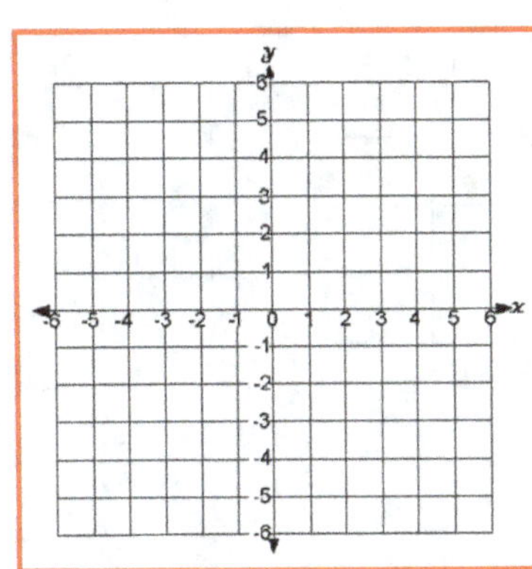

6. $\dfrac{3}{2}x + \dfrac{1}{4}y = 1$

x	y

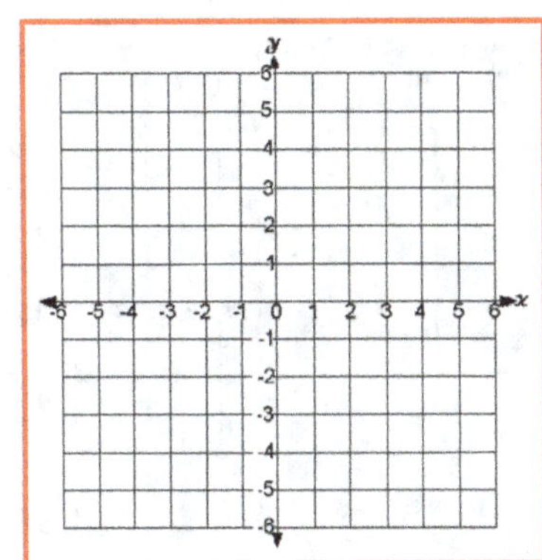

Graph each line by **points method**.

7. $y = 3 - 2x$

x	y

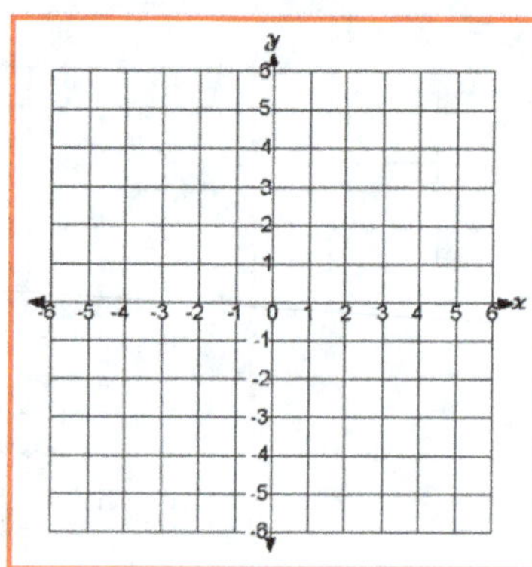

8. $y = -4x$

x	y

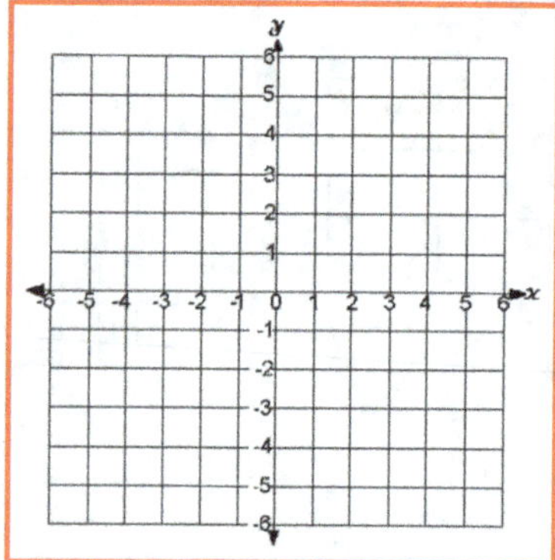

9. $x + 2y = 8$

x	y

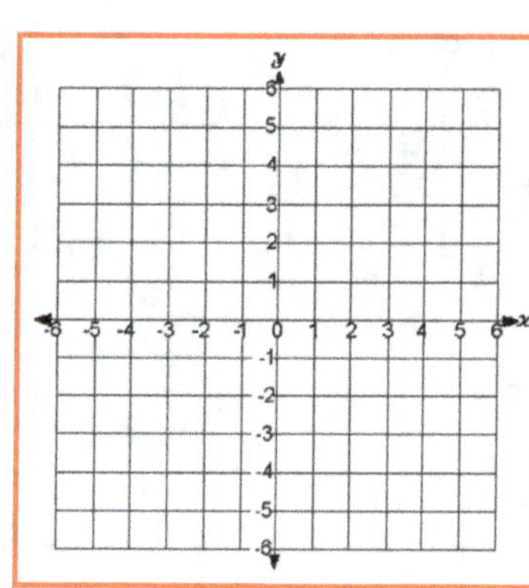

10. $x - 3y = 6$

x	y

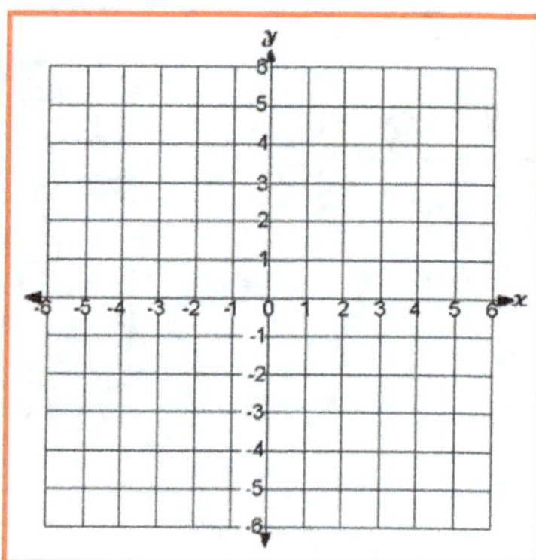

11. $y = \frac{1}{4}x - 3$

x	y

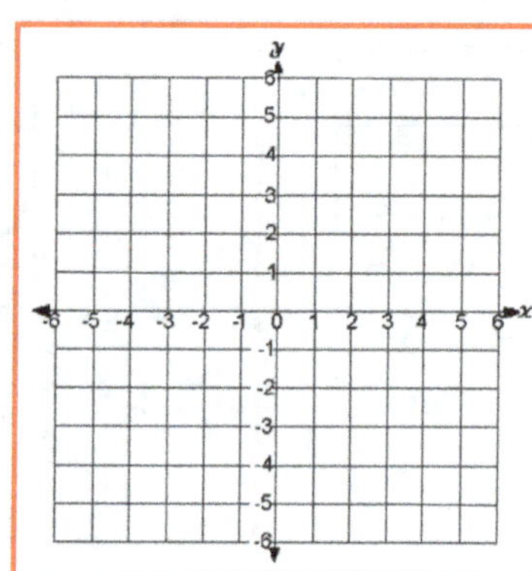

12. $x + \frac{1}{2}y = 4$

x	y

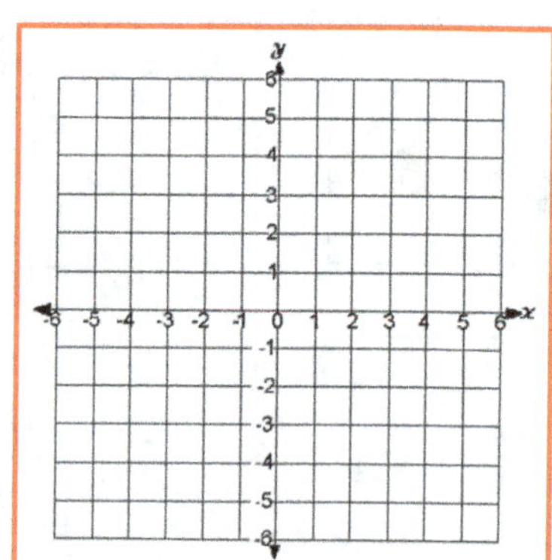

■ X-Y-INTERCEPT METHOD

This method includes 2 steps: making the "0-0 diagonal" **x-y-chart** for values **x = 0** (y-intercept) and **y = 0** (x-intercept), and then drawing the graph through the points corresponding to those ordered pairs $(x_{intercept}, 0)$ and $(0, y_{intercept})$.

This method also works perfectly for slanted lines with nonzero y-intercept, especially for equations in standard form, and dos note works for horizontal and vertical lines.

Example 1 Graph the line **y = 2x + 4.**

SOLUTION ▶ **Step 1.** Making the "0-0 diagonal" **x-y-chart** for values **x = 0** (y-intercept) and **y = 0** (x-intercept)

Step 2. Drawing the graph through the points corresponding to those ordered pairs $(x_{intercept}, 0)$ and $(0, y_{intercept})$

<table>
<tr><td colspan="2">

Step 1

x	y	y = 2x+4
0	4	$4 = 2 \cdot 0 + 4$
-2	0	$0 = 2 \cdot (-2) + 4$

The points are:
    (0,4); (-2,0)

</td><td>

Step 2

</td></tr>
</table>

Example 2 Graph the line **2x + 3y = 6.**

SOLUTION ▶ **Step 1.** Making the "0-0 diagonal" **x-y-chart** for values **x = 0** (y-intercept) and **y = 0** (x-intercept)

Step 2. Drawing the graph through the points corresponding to those ordered pairs $(x_{intercept}, 0)$ and $(0, y_{intercept})$

<table>
<tr><td colspan="2">

Step 1

x	y	2x + 3y = 6
0	2	$2 \cdot 0 + 3y = 6$
3	0	$2x + 3 \cdot 0 = 6$

The points are:
    (0,2); (3,0)

</td><td>

Step 2

</td></tr>
</table>

SOLUTION ▶ **Step 1.** Making the "0-0 diagonal" **x-y-chart** for values **x = 0** (y-intercept) and **y = 0** (x-intercept)

Step 2. Drawing the graph through the points corresponding to those ordered pairs **($x_{intercept}$, 0)** and **(0, $y_{intercept}$)**

Step 1			Step 2
x	**y**	$y = \frac{1}{3}x + 1$	
0	1	$y = \frac{1}{3} \cdot 0 + 1$	
-3	0	$0 = \frac{1}{3}(-3) + 1$	

The points are:
(0,1); (-3,0)

Example 4 Graph the line **x − y = 3.**

SOLUTION ▶ **Step 1.** Making the "0-0 diagonal" **x-y-chart** for values **x = 0** (y-intercept) and **y = 0** (x-intercept)

Step 2. Drawing the graph through the points corresponding to those ordered pairs **($x_{intercept}$, 0)** and **(0, $y_{intercept}$)**

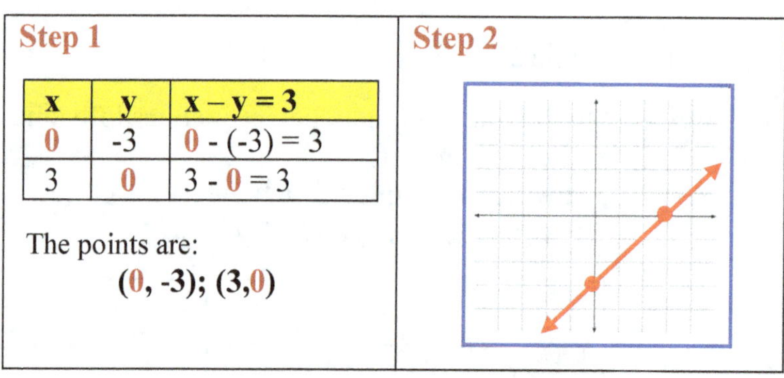

Step 1			Step 2
x	**y**	**x − y = 3**	
0	-3	0 - (-3) = 3	
3	0	3 - 0 = 3	

The points are:
(0, -3); (3,0)

Drawing the graph of a linear equation by **x-y-intercept method** or by **points' method** give us the same result or the same line on coordinate plane.

Usually **x-y-intercept method** is more useable if equation of the line is in standard form. **Points' method** is more easy to use if equation of the line is in slope-y-intercept form.

We also can draw the graph of linear equation by the **slope-y-intercept method**.

Graph each line by **x-y-intercepts method**.

1. $x + 4y = 4$

x	y
0	
	0

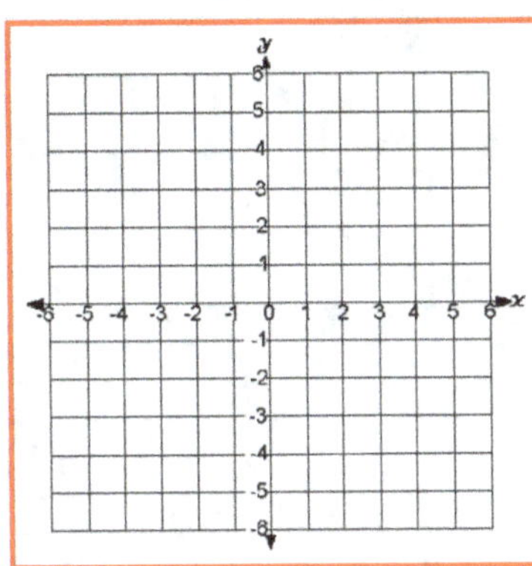

2. $-x + y = 3$

x	y

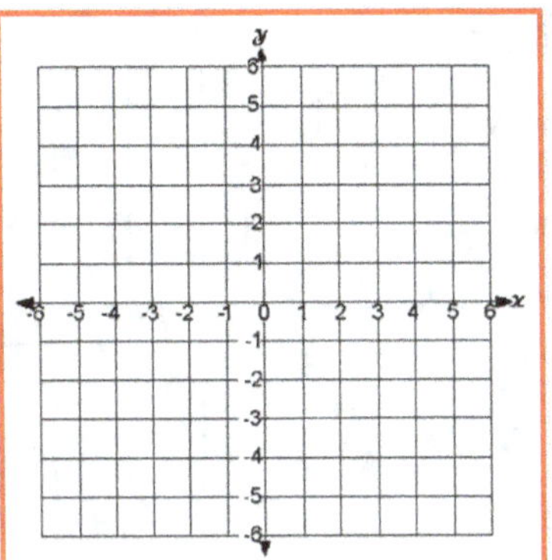

3. $y + 2x = 2$

x	y

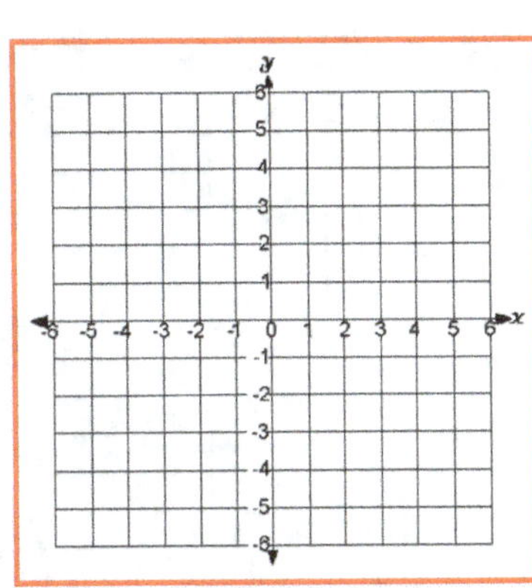

4. $x - y = -4$

x	y

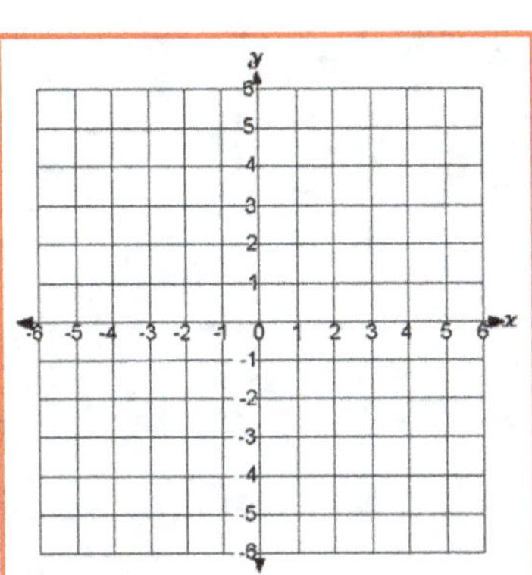

5. $y = \frac{1}{3}x - 1$

x	y

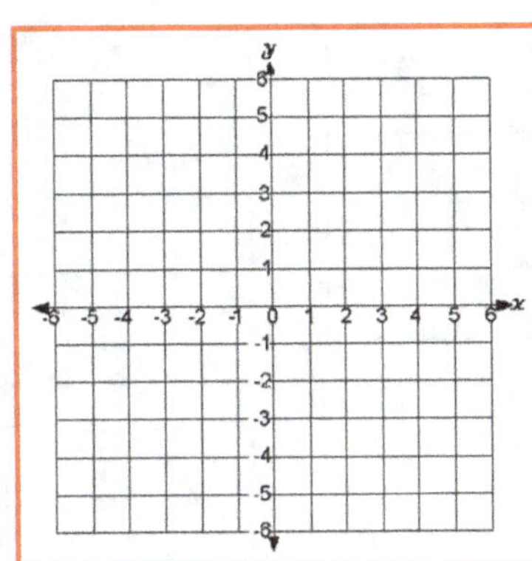

6. $y = \frac{1}{2}x + 2$

x	y

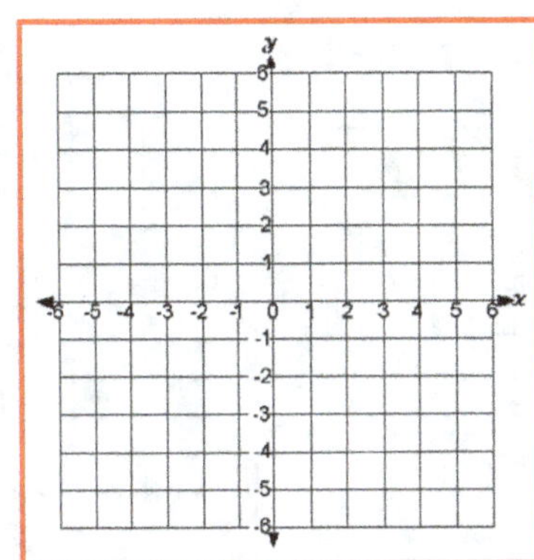

Graph each line by **x-y-intercepts method**.

7. $y = 4 - 2x$

x	y

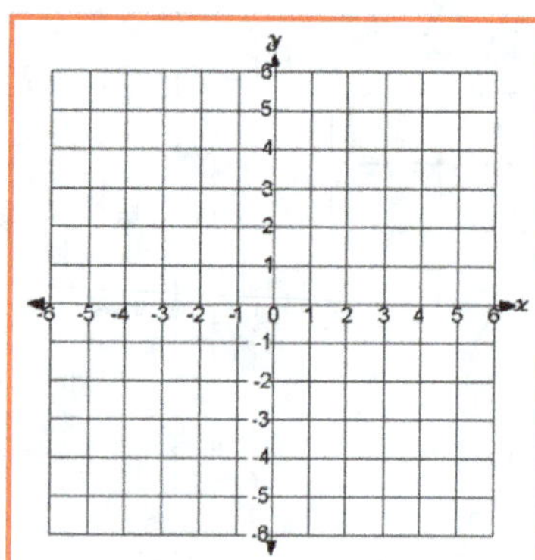

8. $y = -x + 3$

x	y

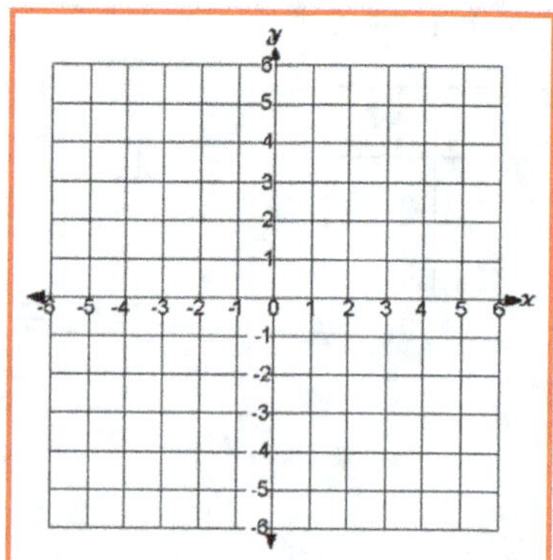

9. $x + 2y = 4$

x	y

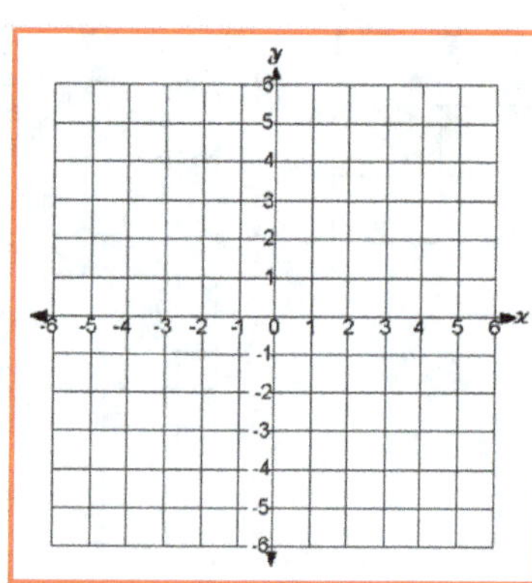

10. $3x - 2y = 6$

x	y

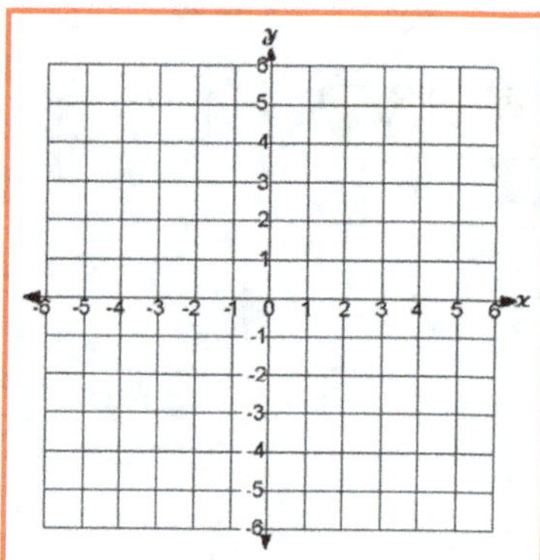

11. $y = \dfrac{1}{4}x - 1$

x	y

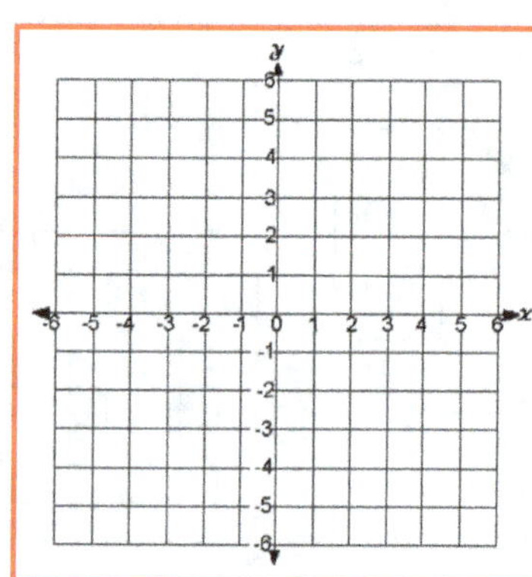

12. $x + \dfrac{1}{2}y = 1$

x	y

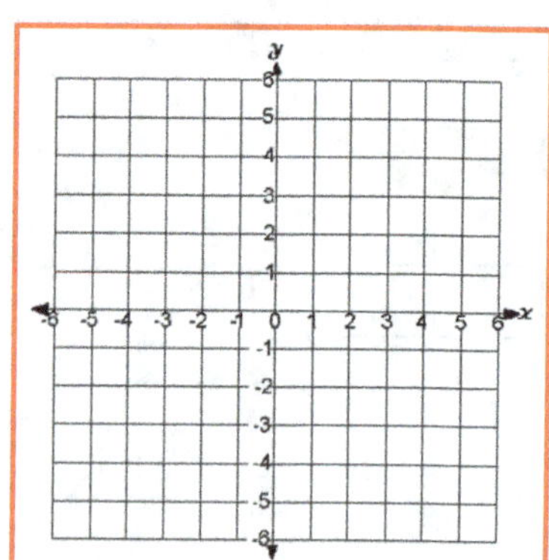

■ SLOPE-Y-INTERCEPT METHOD

This method includes **2 steps** if equation is in the standard form: converting equation of the line from standard form to slope-y-intercept form; drawing the line using **y-intercept point** and the second point which we can find by slope or by moving to up or dawn corresponding to the **rise** and moving to right or left corresponding to **run**. And this method includes only **1step** if equation is in the slope-y-intercept form: drawing the line using **y-intercept point** and the second point which we can find by slope or by moving to up or dawn corresponding to the **rise** and moving to right or left corresponding to **run**.

This method works perfectly for **slanted lines** and dos note works for **horizontal lines** and **vertical lines**.

Example 1   Graph the line **x + y = 2.**

SOLUTION ▶   **Step 1.**   Converting equation from standard form to slope-y-intercept form
Step 2.   Drawing the line using **y-intercept point (0,2)** and the second point which we can find by slope **m = -1** or by moving to dawn **1** unit (**rise**) and moving to right **1** unite (**run**).

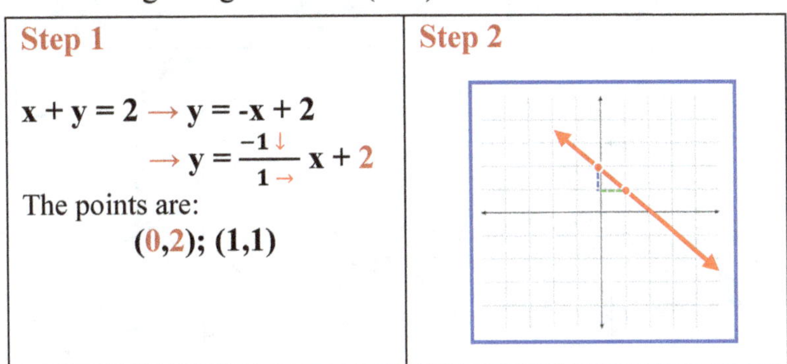

Example 2   Graph the line **y = $\frac{2}{3}$ x - 1.**

SOLUTION ▶   **Step 1.**   Drawing the line by using **y-intercept point (0,-1)** and the second point which we can find by slope **m = 2/3** or by moving to up **2** unit (**rise**) and moving to right **3** unite (**run**).

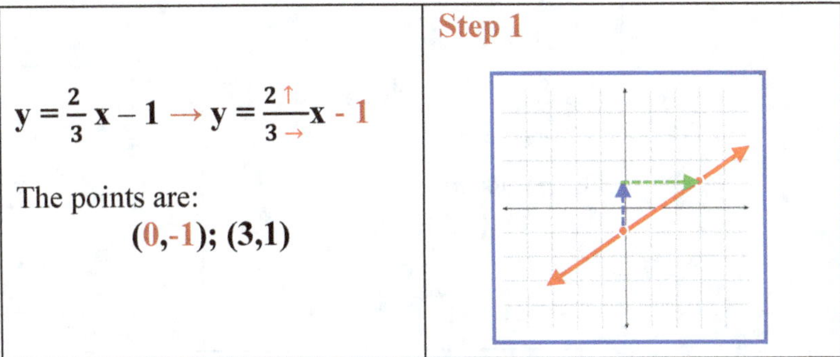

Example 3 Graph the line **3x + 4y = 8.**

SOLUTION ▶ **Step 1.** Converting the equation
  Step 2. Drawing the line

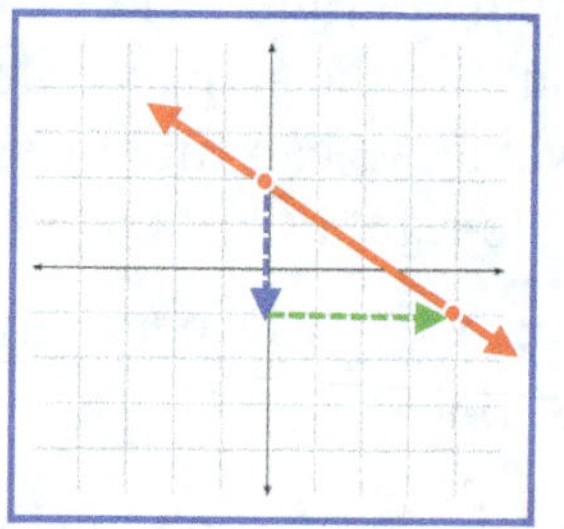

Step 1	Step 2
$3x + 4y = 8 \rightarrow 4y = -3x + 8$ $\rightarrow y = \dfrac{-3\downarrow}{4\rightarrow}x + 2$ The points are: **(0,2); (4,-1)**	

Example 4 Graph the line **y = 3x − 2.**

SOLUTION ▶ **Step 1.** Drawing the line

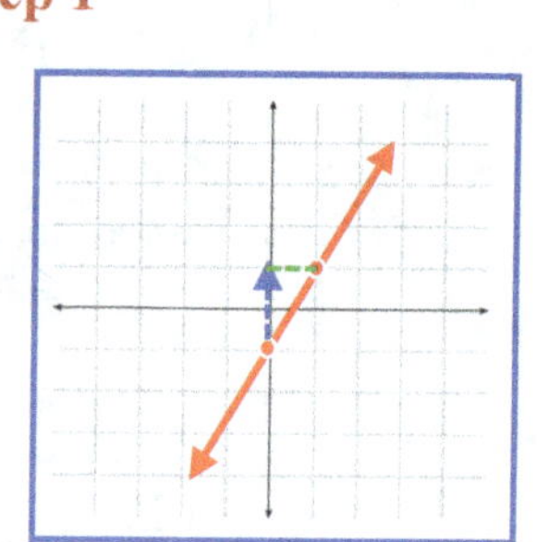

Step 1
$y = 3x - 2 \rightarrow y = \dfrac{3\uparrow}{1\rightarrow}x - 2$ The points are: **(0,-2); (1,1)**

Example 5 Graph the line $y = \dfrac{3}{4}x$.

SOLUTION ▶ **Step 1.** Drawing the line

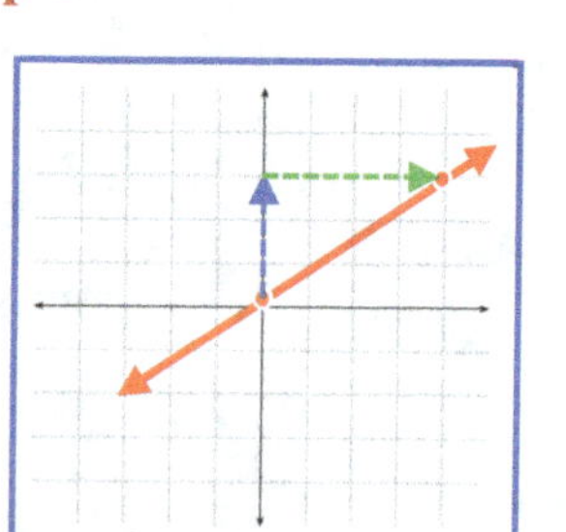

Step 1
$y = \dfrac{3}{4}x \rightarrow y = \dfrac{3\uparrow}{4\rightarrow}x + 0$ The points are: **(0,0); (4,3)**

Graph each line by **slope-y-intercept method**.

1. $y = \frac{1}{2}x - 4$

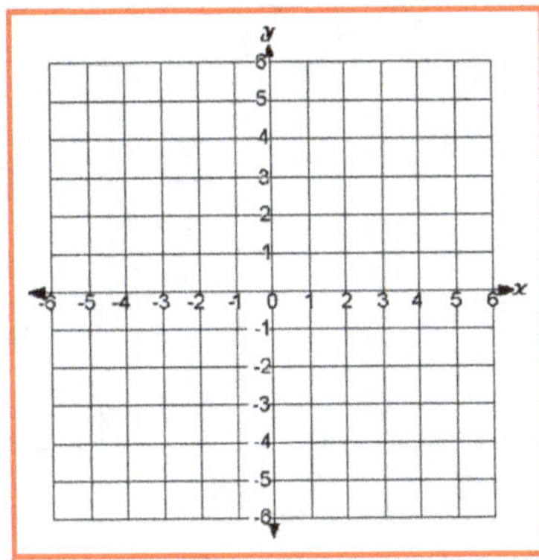

2. $y = \frac{2}{3}x + 2$

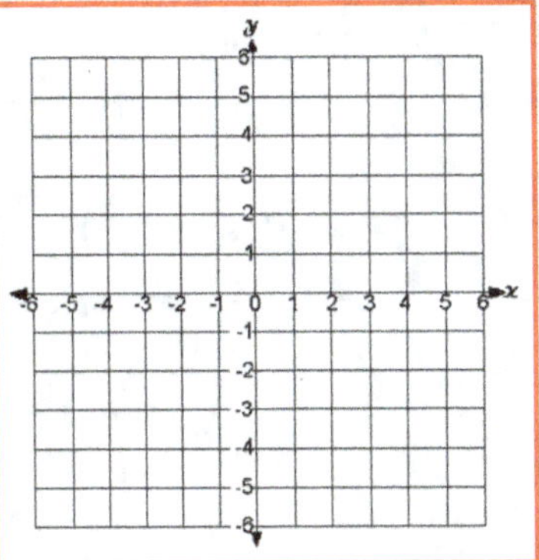

3. $y = -\frac{3}{4}x$

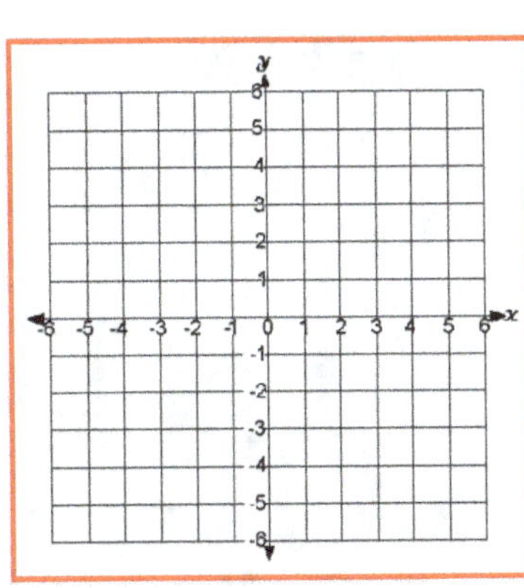

4. $y = -\frac{2}{3}x + 3$

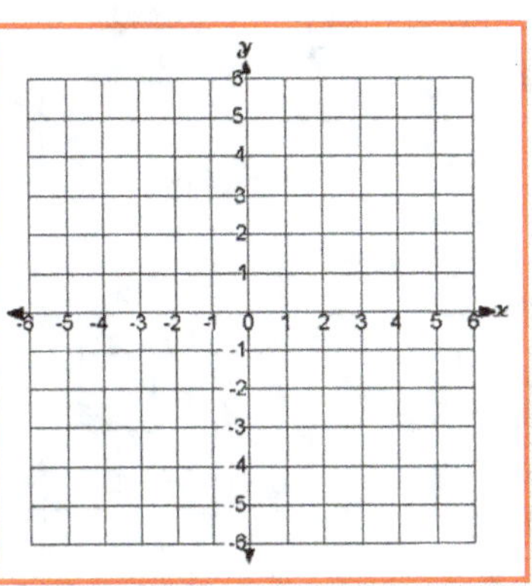

5. $y = \frac{1}{3}x - 5$

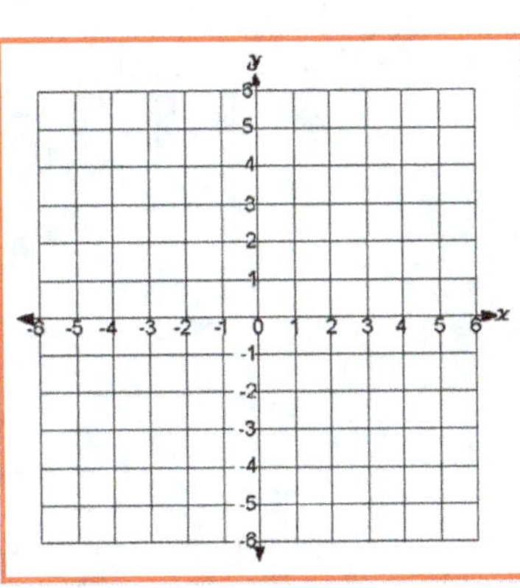

6. $y = \frac{1}{4}x + 1$

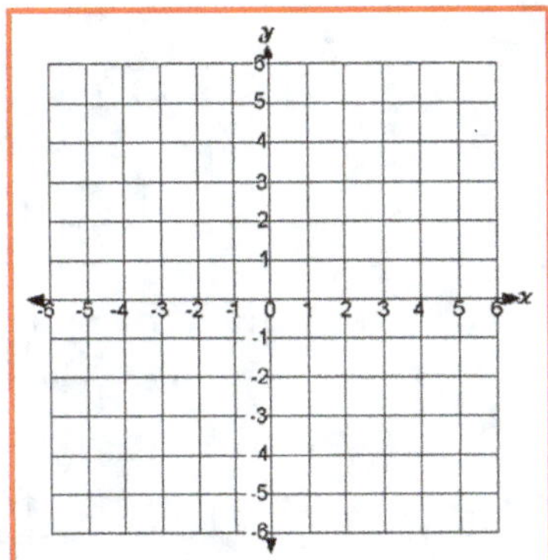

Graph each line by **slope-y-intercept method**.

7. $y = 2 - 2x$

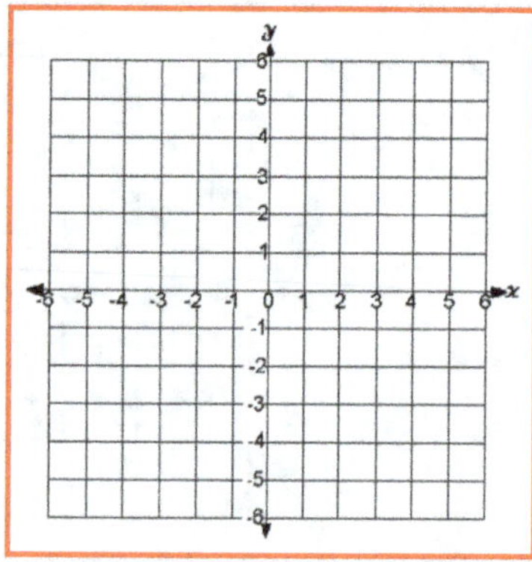

8. $y = -x + 4$

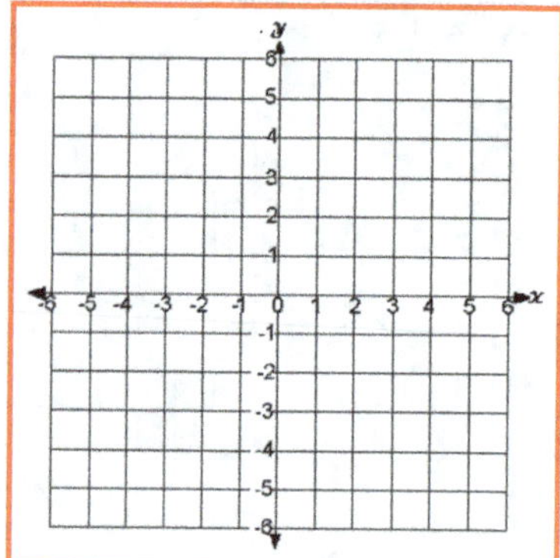

9. $x + 4y = 4$

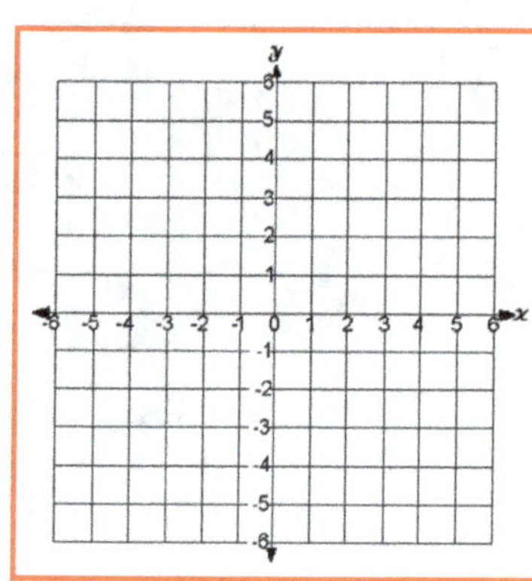

10. $3x - y = 2$

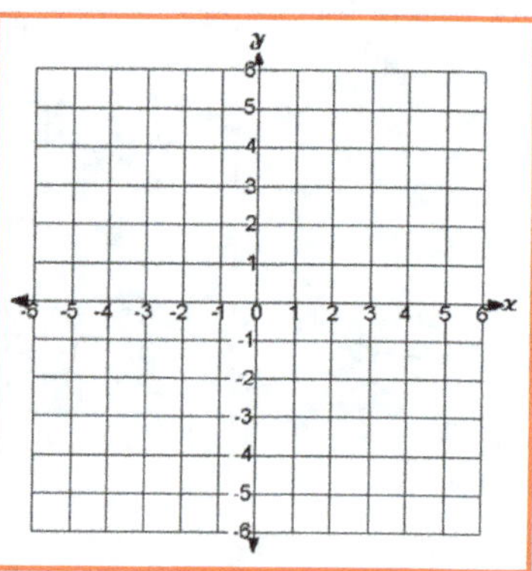

11. $2x + y = 2$

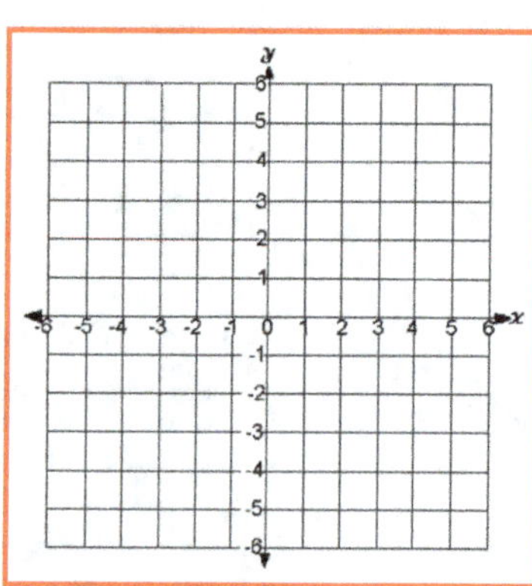

12. $x - 2y = 4$

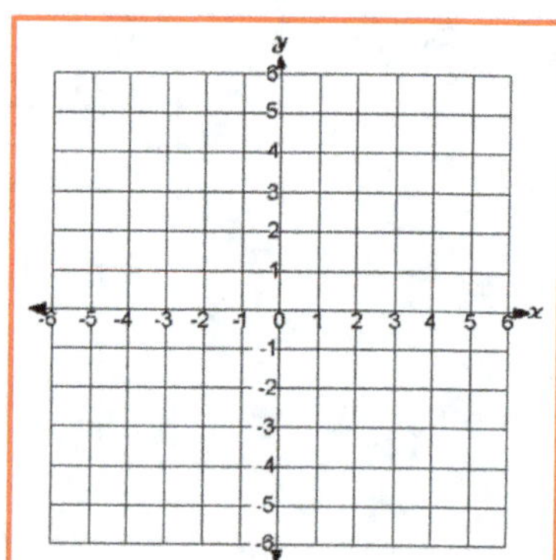

■ GRAPHS OF HORIZONTAL AND VERTICAL LINES

The line with equation **y = a** is a **horizontal line** (parallel to x-axis) which passes through the point **(0,a)**. The slope of any horizontal line is **m = 0**.

The line with equation **x = b** is a **vertical line** (perpendicular to x-axis) which passes through the point **(b,0)**. The slope of any vertical line is **m = undefined**.

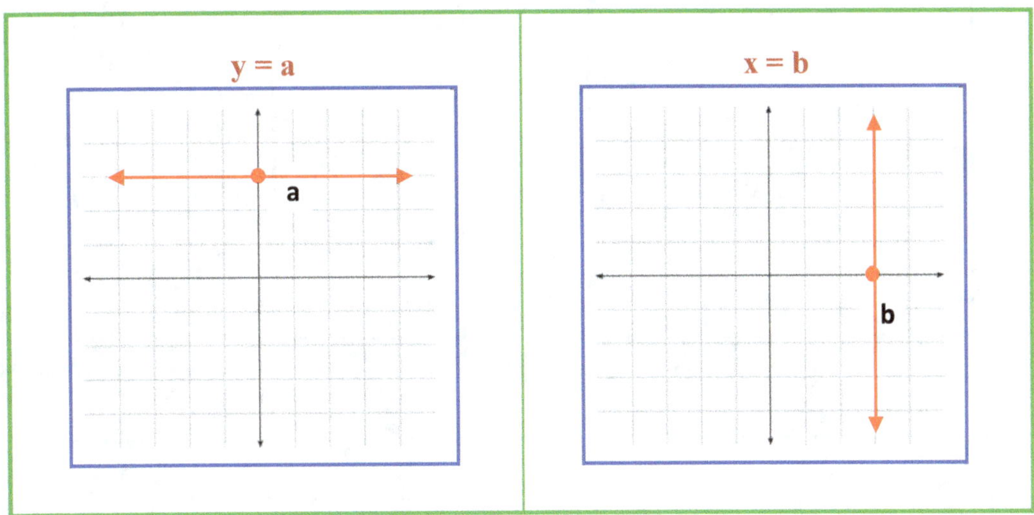

Example 1 Graph the lines **y = -3** and **x = -2**.

SOLUTION ▶ Line **y = -3** is a **horizontal line** and line **x = -2** is a **vertical line**.

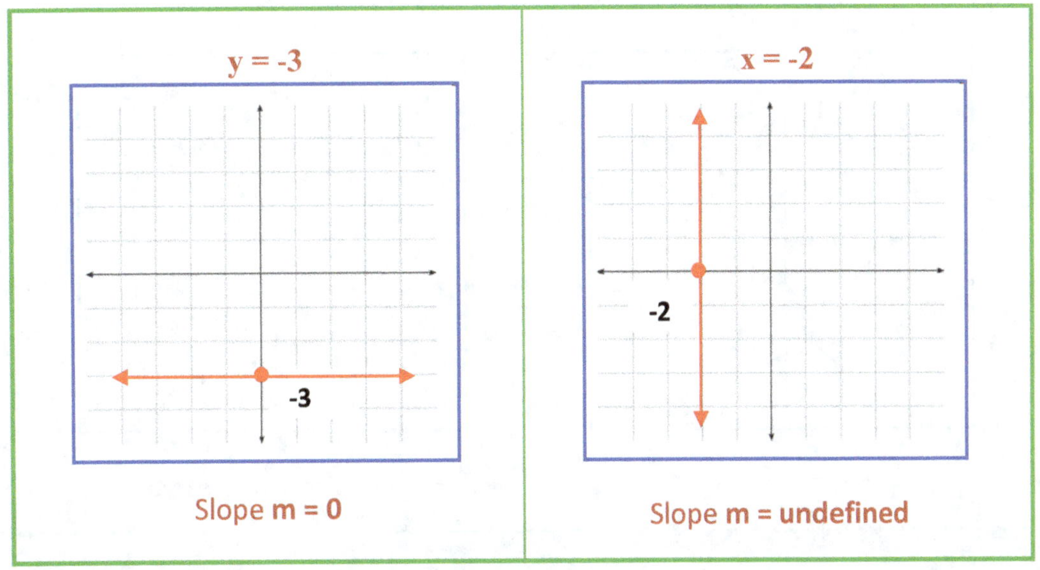

 Graph the lines **y = 0** and **x = 0**.

SOLUTION ▶ Line **y = 0** is a **horizontal line** and line **x = 0** is a **vertical line**.

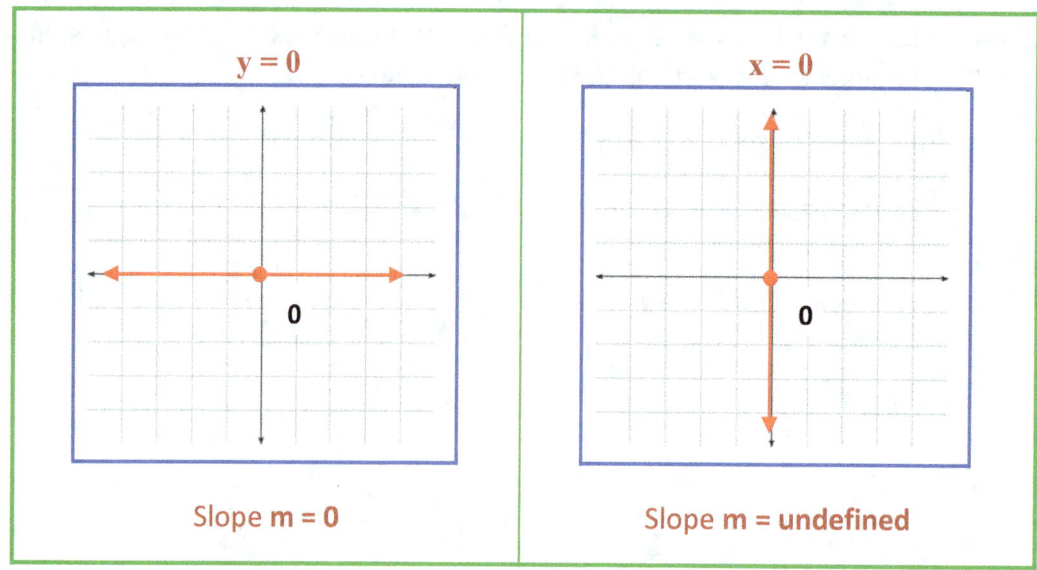

 Graph the lines **y = 2** and **x = 3**.

SOLUTION ▶ Line **y = 2** is a **horizontal line** and line **x = 3** is a **vertical**

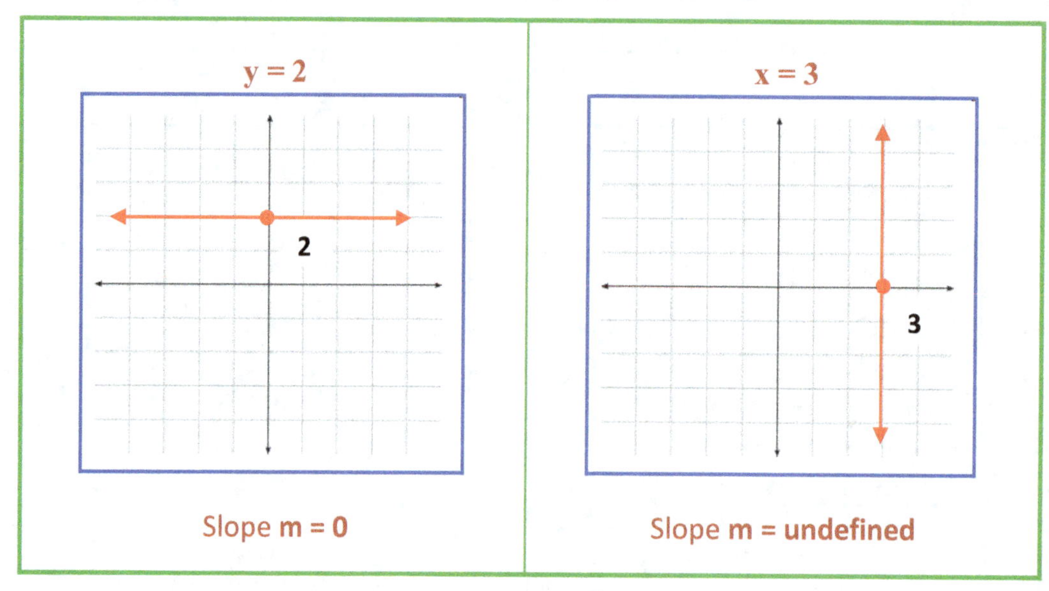

Graph each **horizontal** or **vertical** line.

1. $y = -4$

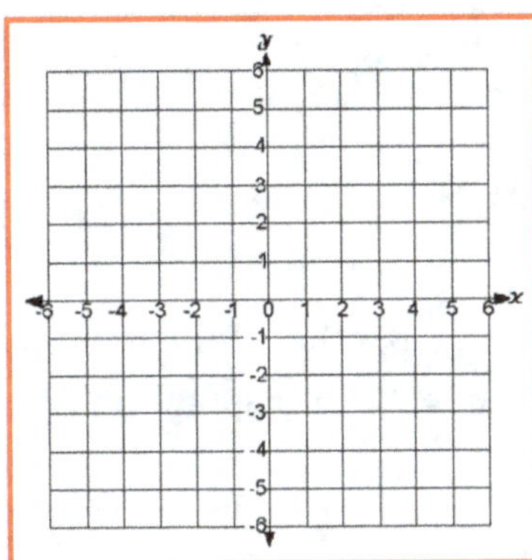

2. $y = 5$

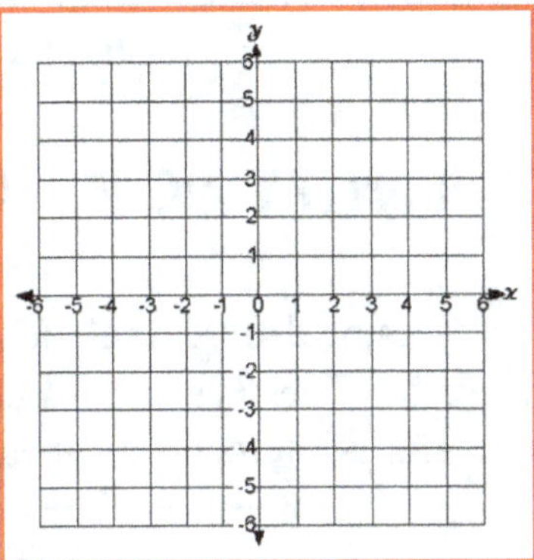

3. $y = -1$

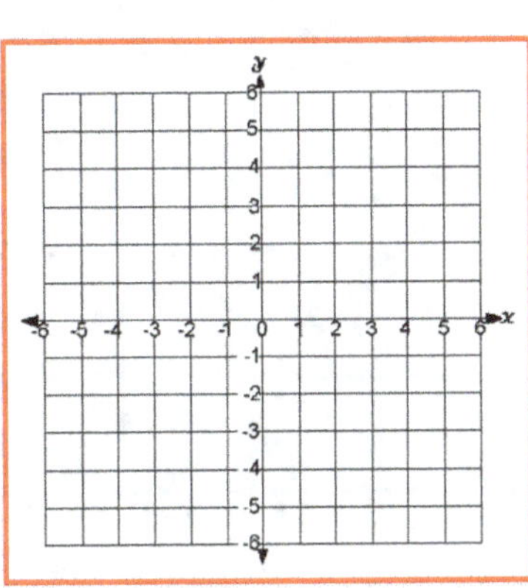

4. $x = 3$

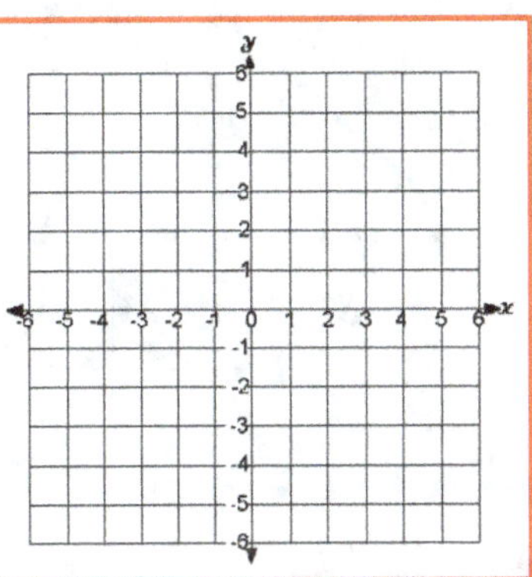

5. $x = 3$

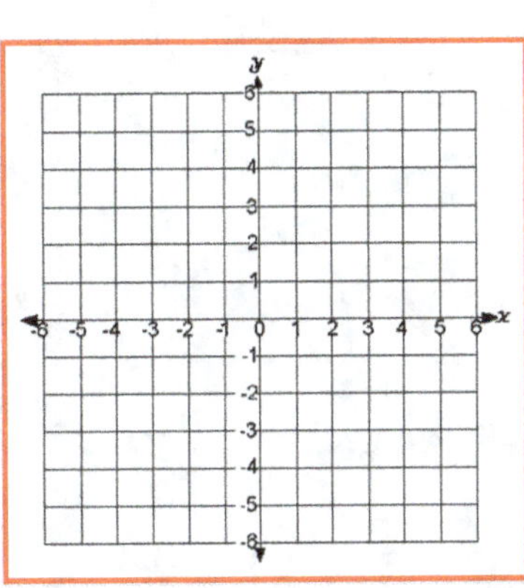

6. $x = -4$

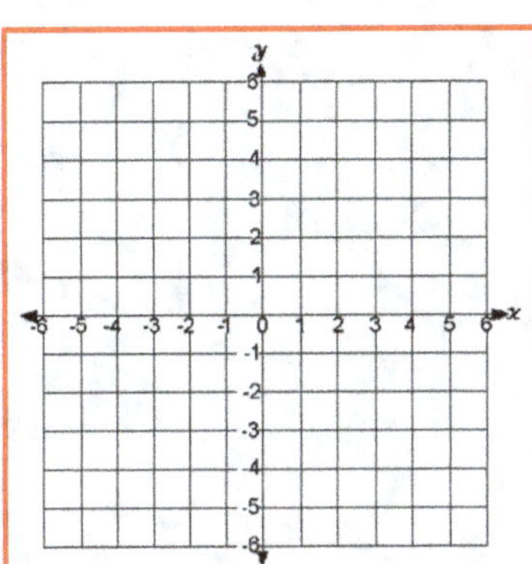

5.4. Parallel, Coincident, Perpendicular, and Intersecting Lines

- **IDENTIFYING RELATIONS OF LINES**
- **PARALLEL AND PERPENDICULAR LINES**

■ IDENTIFYING RELATIONS OF LINES

There can be different relations between graphs of lines: parallel, coincident, perpendicular, and intersecting lines (see graphs below). For each case we have corresponding relation between slopes and y-intercepts of lines.

Parallel lines	Coincident lines	Perpendicular lines	Intersecting Lines
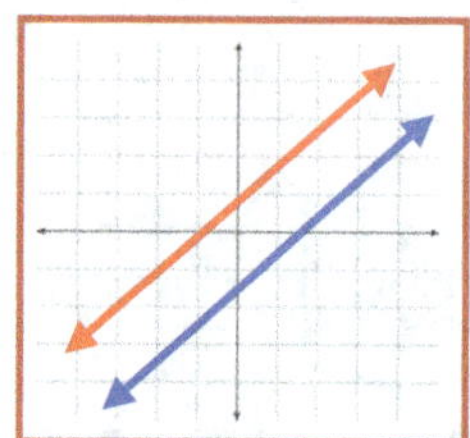	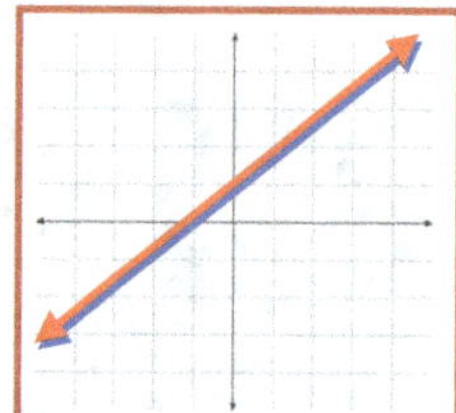	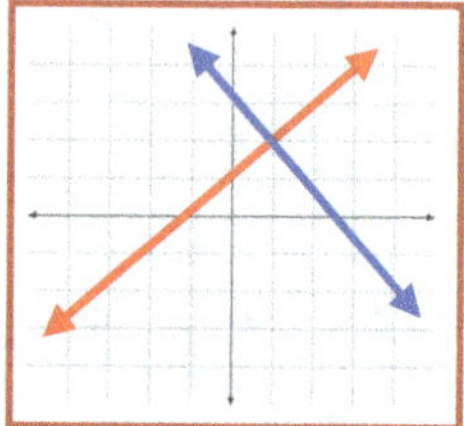	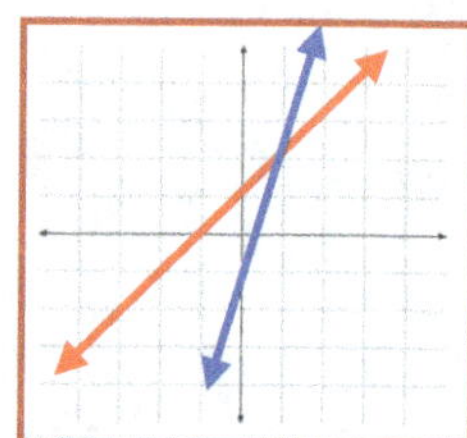
$m_1 = m_2$ & $b_1 \neq b_2$	$m_1 = m_2$ & $b_1 = b_2$	$m_1 \cdot m_2 = -1$ or $m_2 = -1/m_1$	$m_1 \neq m_2$ & $m_1 \cdot m_2 \neq -1$

Examples

Parallel lines	Coincident lines	Perpendicular lines	Intersecting Lines
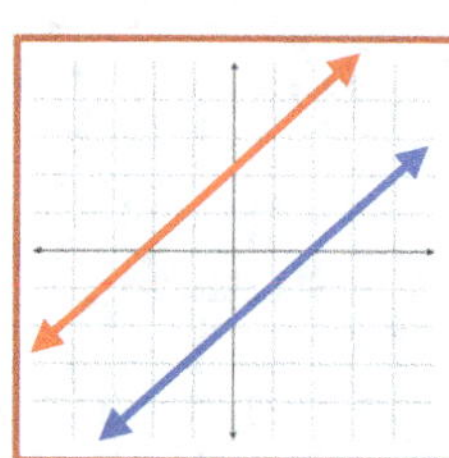	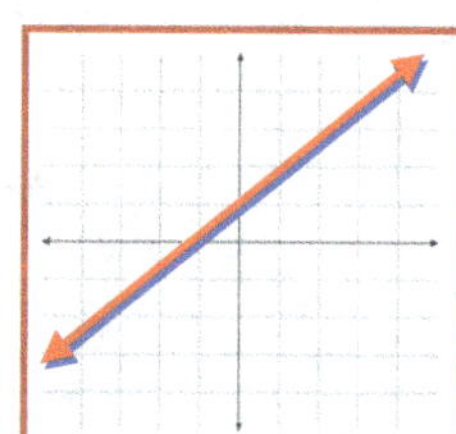	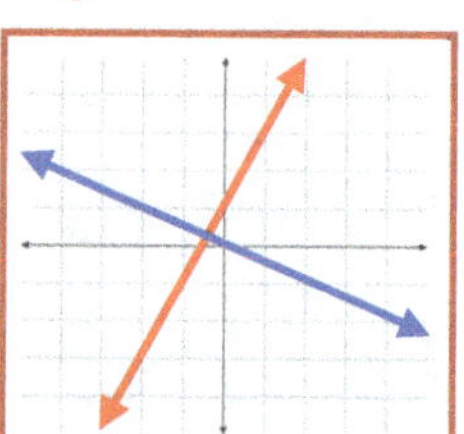	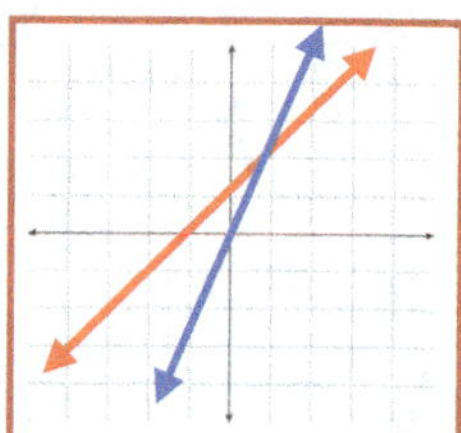
$y = x + 2$ & $y = x - 2$ $m_1 = m_2 = 1$ & $b_1 = 2 \neq -2 = b_2$	$y = x + 1$ & $y - x = 1$ $m_1 = m_2 = 1$ & $b_1 = b_2 = 1$	$y = 2x + 1$ & $y = -\frac{1}{2}x$ $m_1 m_2 = 2(-\frac{1}{2}) = -1$	$y = x + 1$ & $y = 2x$ $m_1 = 1 \neq 2 = m_2$ & $m_1 m_2 = 2 \neq -1$

■ PARALLEL AND PERPENDICULAR LINES

If we have a line and a point out of that line on a coordinate plane then we can draw the parallel and the perpendicular lines to that line passes through that point. We also can write the corresponding equations of those parallel and perpendicular lines.

Example Let given line and given point are **y = 2x + 2** and **(2,1)**. Write the equations of parallel and perpendicular lines to the given line **y = 2x + 2** and passes the given point **(2,1)**. Draw the graphs of those parallel and perpendicular lines.

SOLUTION▶ To write the equations of those parallel and perpendicular and then draw those lines we will use the relations of slopes and the slope-point forms of lines.

<table>
<tr><th>Equation of the parallel line</th><th>Equation of the perpendicular line</th></tr>
<tr><td>

Parallel line has the slope equal to the slope of the given line $m_{||} = m = 2$ and passes through the given point **(2,1)**.
So, the slope-point equation of the parallel line will be

$$y_{||} - 1 = 2(x - 2)$$

The slope-y-intercept equation of the parallel line will be

$$y_{||} - 1 = 2(x - 2)$$
$$y_{||} - 1 = 2x - 4$$
$$\phantom{y_{||} - 1 = }{+1} {+1}$$
$$y_{||} = 2x - 3$$

</td><td>

Perpendicular line has the slope equal to $m_\perp = -1/m = -1/2$ and passes through the given point **(2,1)**.
So, the slope-point equation of the parallel line will be

$$y_\perp - 1 = -\frac{1}{2}(x - 2)$$

The slope-y-intercept equation of the parallel line will be

$$y_\perp - 1 = -\frac{1}{2}x + 1$$
$${+1} \phantom{-\frac{1}{2}x}{+1}$$
$$y_\perp = -\frac{1}{2}x + 2$$

</td></tr>
<tr><td>

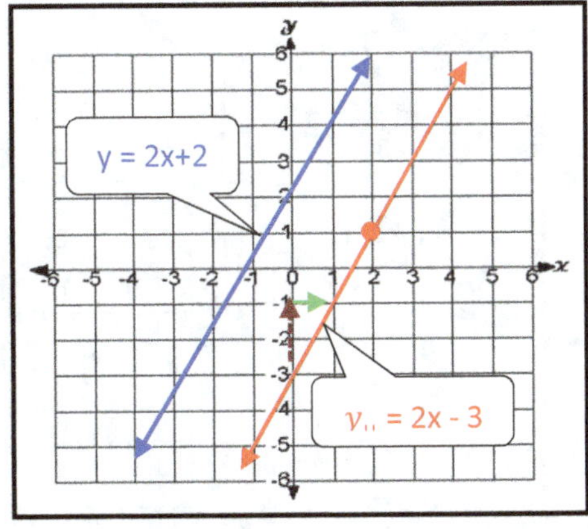

$$y_{||} = \frac{2}{1}x - 3$$

</td><td>

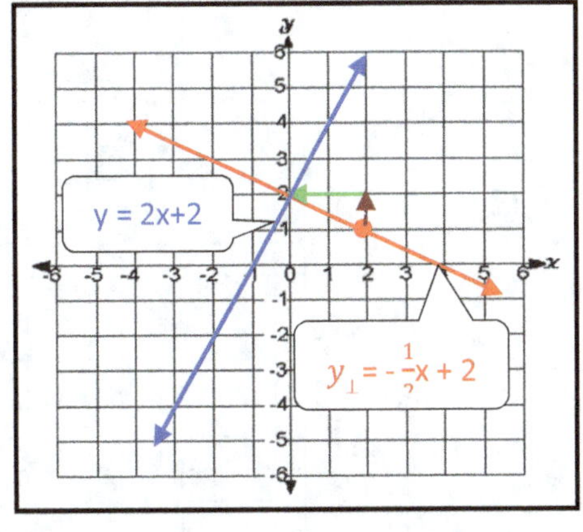

$$y_\perp = -\frac{1}{2}x + 2$$

</td></tr>
</table>

Identify each pair of lines as **parallel, coincident, perpendicular,** and **intersecting** lines.

1. $y = 2x + 1$ $y = 2x - 3$

2. $y = -3x + 2$ $y = \frac{1}{3}x + 8$

3. $y = x + 1$ $x - y = -1$

4. $y = -x + 2$ $y = 4x + 2$

5. $y = 4x + 1$ $-8x + 2y = 2$

6. $y = -2x + 3$ $y = -\frac{1}{2}x + 1$

7. $x + 2y = 6$ $y - 2x = -1$

8. $y = \frac{1}{3}x + 1$ $-x + 3y = 3$

9. $y = -3x + 1$ $x - 3y = 2$

10. $y - x = 1$ $y = -x + 3$

Write the equations of **parallel** and **perpendicular** lines to the given line and passes through the given point.

Graph the **given line** and the **given point** as well as the **parallel** and the **perpendicular** lines for each case on the same coordinate plane.

1. $y = x + 2$

   $(0,0)$

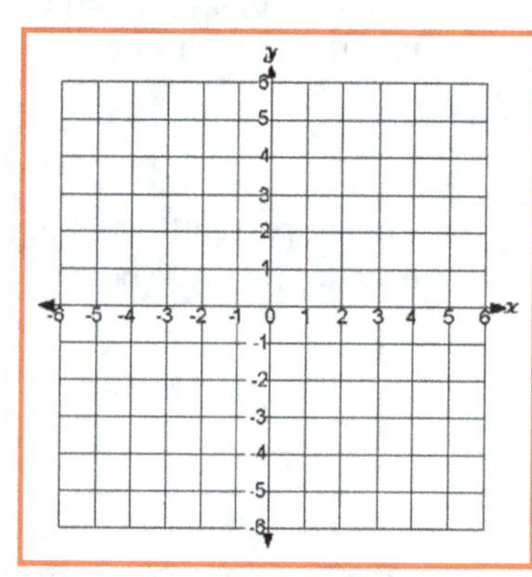

2. $y = -x + 3$

   $(0,0)$

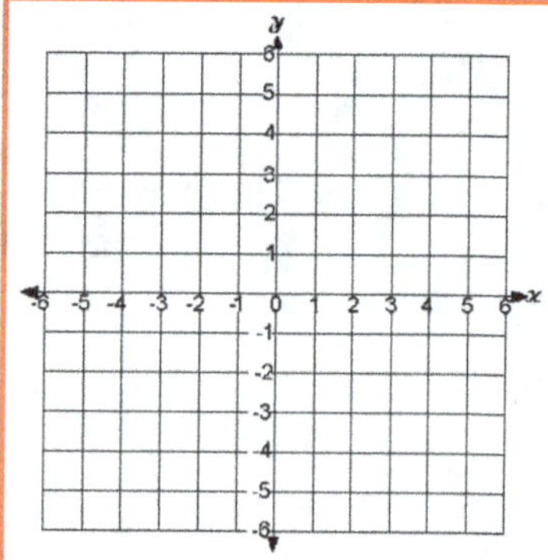

3. $y = -\frac{3}{4}x$

   $(2,3)$

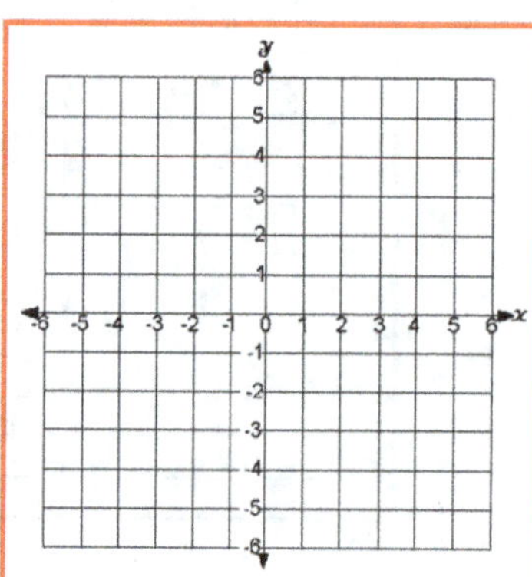

4. $y = \frac{2}{3}x + 2$

   $(-2,0)$

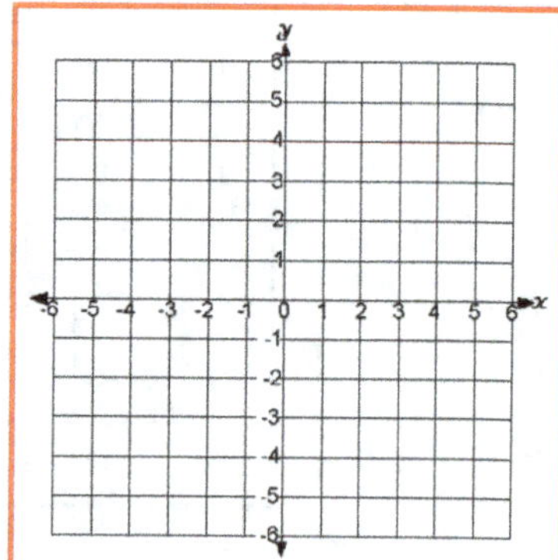

$\mathbf{W}$rite the equations of **parallel** and **perpendicular** lines to the given line and passes through the given point.

Graph the **given line** and the **given point** as well as the **parallel** and the **perpendicular** lines for each case on the same coordinate plane.

5. $y = 2 - 2x$

   (3,3)

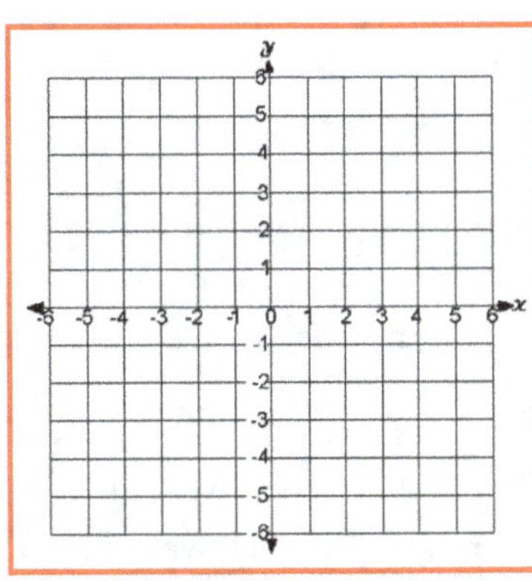

6. $y = -x + 4$

   (0,1)

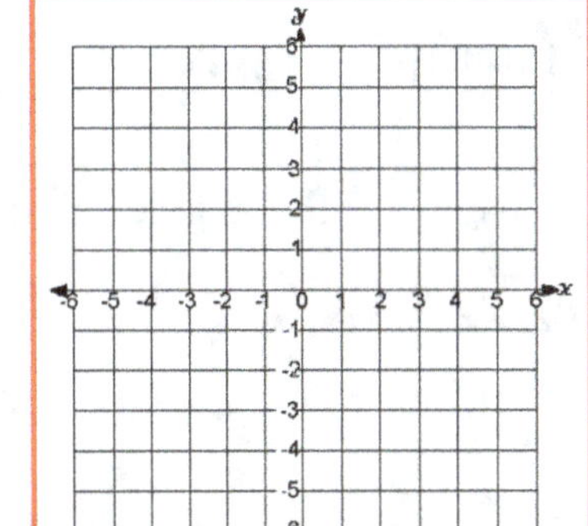

7. $x + y = 4$

   (2,2)

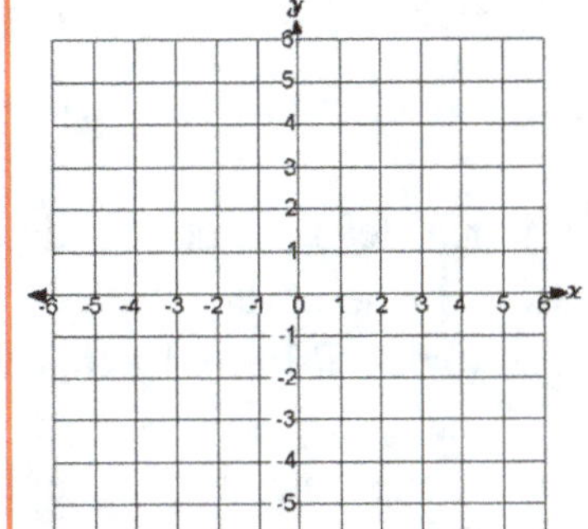

8. $2x - y = 3$

   (4,0)

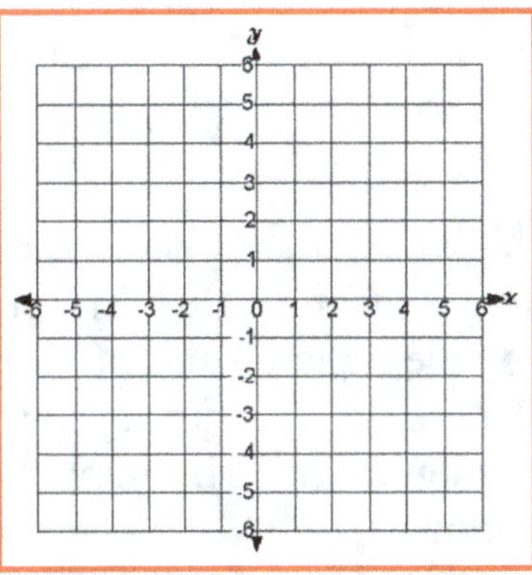

9. $2x + y = 4$

   (−2,2)

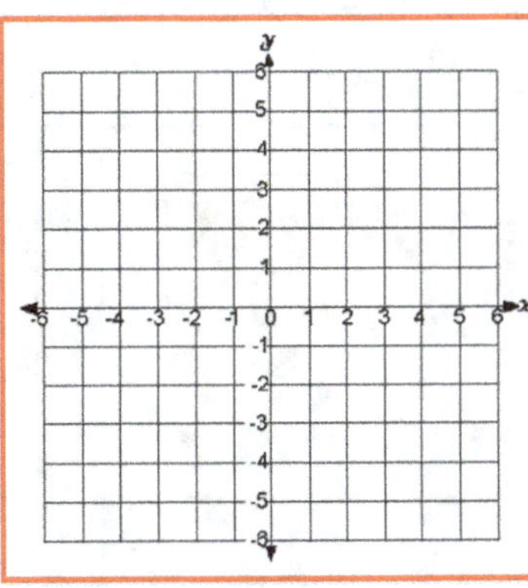

10. $x - 2y = 4$

   (2,1)

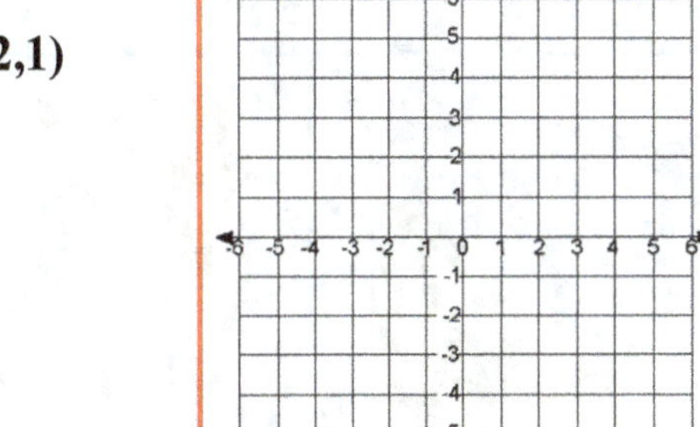

5.5. **Linear Inequalities on the Coordinate Plane**

- **DEFINITIONS**
- **GRAPHING LINEAR INEQUALITIES ON PLANE**

■ DEFINITIONS

We use the following definition of a linear inequality with two variables.

IMPORTANT DEFINITION

Definition ▶ A **linear inequality with two variables** x and y is an inequality that can be written as:

$$y > mx + b; \ y < mx + b; \ y \geq mx + b; \ \text{or} \ y \leq mx + b, \ \text{where} \ m, b \in R$$

$$ax + by > c; \ ax + by < c; \ ax + by \geq c; \ \text{or} \ ax + by \leq c, \ \text{were} \ a, b, c \in R$$

Examples The inequalities $y > 2x + 3$, $y < -x + 2$, $y \geq \frac{1}{2}x + 3$, $y \leq 0.3x - 2$, $x + y \leq 4$, $3x - y > -5$, $x > 5$, and $y \leq 1$ are **linear inequalities with two variables**.

The set of solutions of **a linear inequality with two variables** is a set of ordered pairs **(x, y)** that satisfies the inequality. The graph of each set of solutions is the part of the **xy−plane** over/under or on the right/left side of the **corresponding line** (or **associated line**). Each set of solutions has one **dashed** (if inequality sign is > or <) or one **solid** (if inequality sign is ≥ or ≤) **associated line** as its **boundary**. The graph of a linear inequality with two variables is the graph of the set of solutions.

Sets of solutions of the inequalities $y < m_1x + b_1$, $y > m_2x + b_2$, $y \geq m_3x + b_3$, and $y \leq m_4x + b_4$ are parts of **xy−plane** under or over the *slanted* associated lines $y = m_1x + b_1$, $y = m_2x + b$, $y = m_3x + b_3$, and $y = m_4x + b_4$ (see figure below).

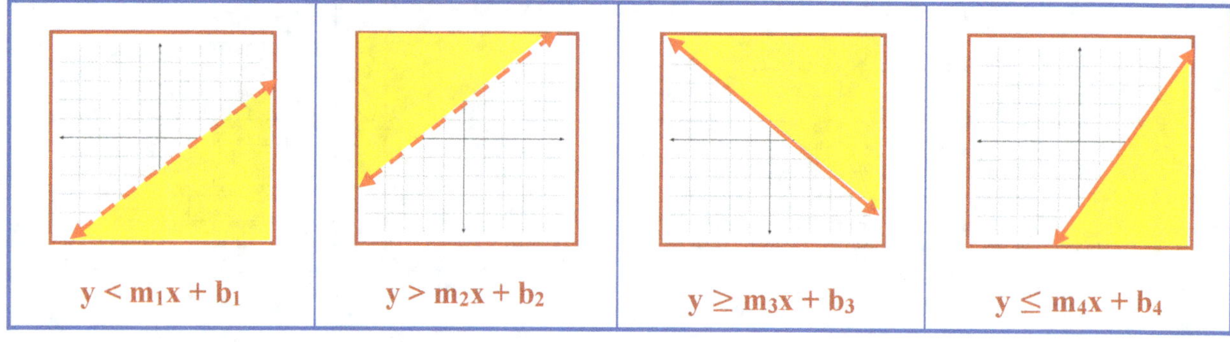

■ GRAPHING LINEAR INEQUALITIES ON A PLANE

Sets of solutions of the inequalities $y < b_1$, $y > b_2$, $y \geq b_3$, and $y \leq b_4$ are parts of **xy–plane** under or over the *horizontal* lines $y = b_1$, $y = b_2$, $y = b_3$, and $y = b_4$ (see figure below).

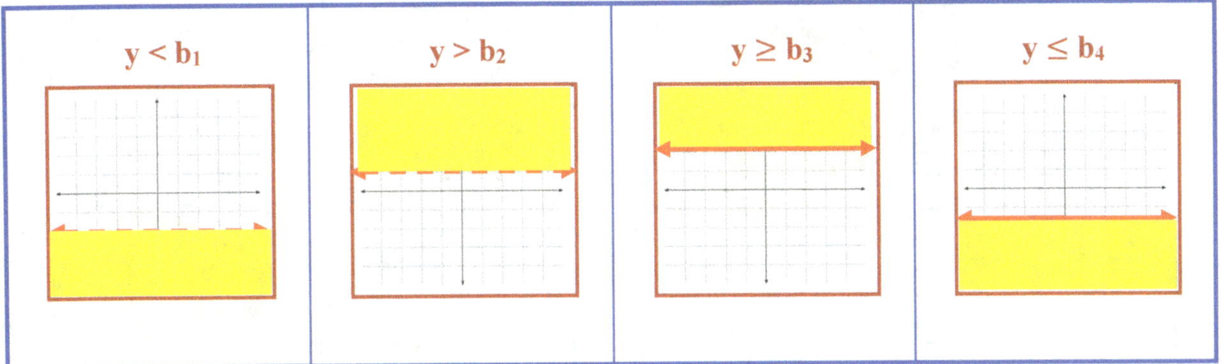

Sets of solutions of the inequalities $x < b_1$, $x > b_2$, $x \geq b_3$, and $x \leq b_4$ are parts of **xy–plane** on the left or on the right sides of the *vertical* lines $x = b_1$, $x = b_2$, $x = b_3$, and $x = b_4$ (see figure).

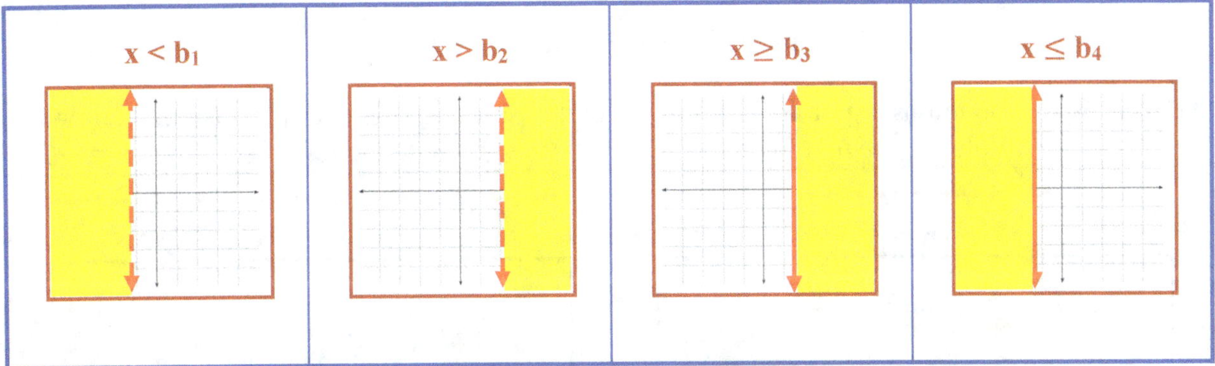

Example 1 Graph the linear inequalities $x + y > 3$ and $y \leq \frac{1}{2}x - 5$ on the xy-plane.

SOLUTION ▶ We will rewrite the standard inequality in slope-y-intercept form: $x + y > 3 \rightarrow$ $y > -x + 3$. The inequality has a > sign, so the **associated line is dashed,** and solution set is part of the xy-plane **over** that line. This dashed associated line is a boundary line for **solution's region** on the **xy-plane**.

The second inequality $y \leq \frac{1}{2}x - 5$ is in slope-y-intercept form. The inequality has a ≤ sign , so the **associated line is solid,** and solution set is part of the xy-plane **under** that **solid** line. This solid associated line is a boundary line for the **solution's region**.

$$x + y > 3 \quad \rightarrow \quad y > -x + 3$$

$$y \leq \frac{1}{2}x - 5$$

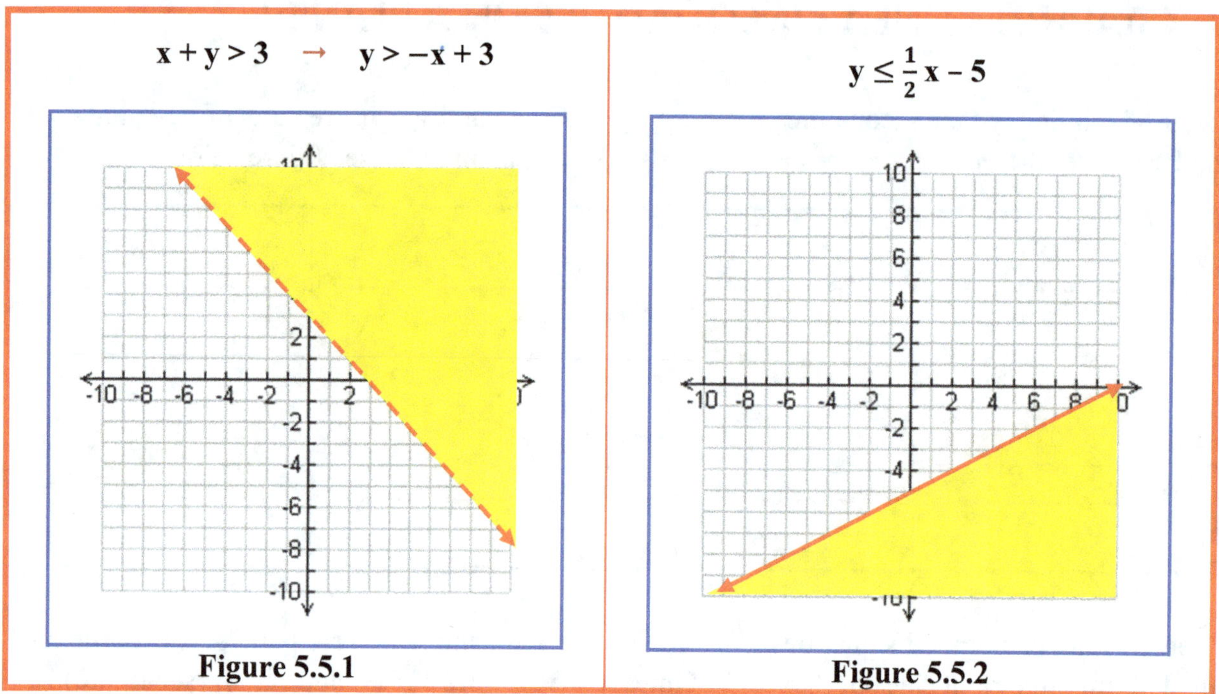

Figure 5.5.1

Figure 5.5.2

Example 2   Graph the linear inequalities $y > 4$ and $x \leq -6$.

SOLUTION ▶   Solutions region for $y > 4$ is the part of the xy-plane over the dashed line $y = 4$. Solutions region for $x \leq -6$ is part of xy-plane to the left of solid line $x = -6$.

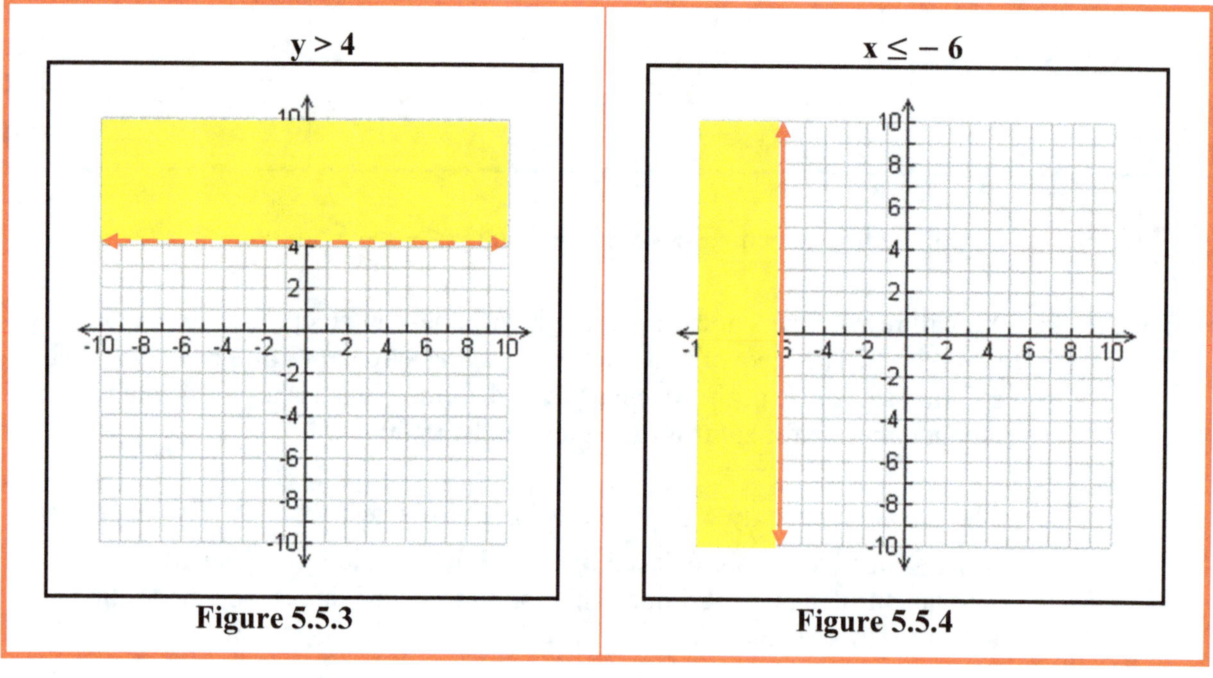

Figure 5.5.3

Figure 5.5.4

Below we will sketch the graphs of solution sets of some other linear inequalities.

Examples G1 Sketch the graphs of linear inequalities $y > 2x + 3$; $y < -x - 2$; $x \geq 3$; $y \leq -2$.

SOLUTION ▶ Each set of solutions is the part of xy-plane with **associated line** as the **boundary**

$$y > 2x + 3 \qquad y < -x - 2 \qquad x \geq 3 \qquad y \leq -2$$

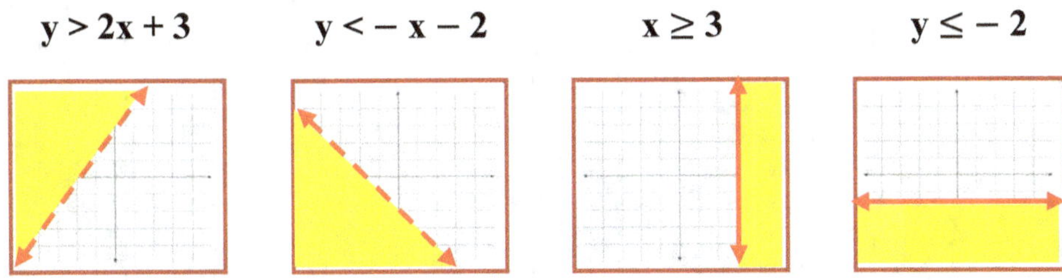

Figure 5.5.5

Examples G2 Sketch the graphs of the inequalities $x + y > 1$; $4x + 2y < -6$; $-x \leq -2$; $y \geq 3$.

SOLUTION ▶ Each set of solutions is the part of xy-plane with **associated line** as the **boundary**

$$x + y > 1 \rightarrow \qquad 4x + 2y < -6 \rightarrow \qquad -x \leq -2 \rightarrow$$
$$y > -x + 1 \qquad\qquad y < -2x - 3 \qquad\qquad x \geq 2 \qquad\qquad y \geq 3$$

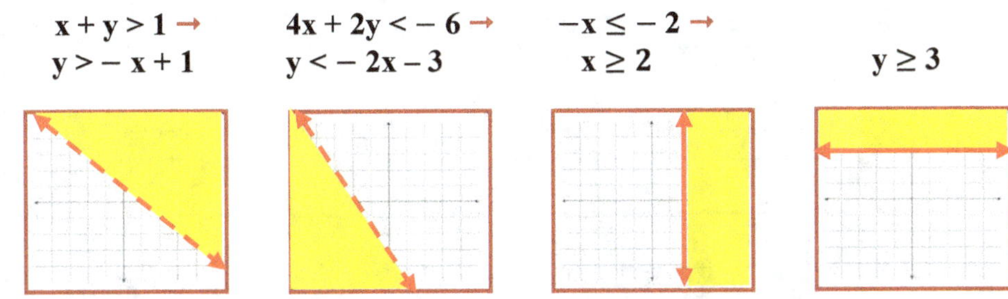

Figure 5.5.6

Examples G3 Sketch the graphs of linear inequalities $x - y > 0$; $y < -\dfrac{3}{2}$; $x \geq 0$; $y \leq -\dfrac{1}{3}x$.

SOLUTION ▶ Each set of solutions is the part of xy-plane with **associated line** as the **boundary**

$$x - y > 0 \rightarrow \qquad y < -\dfrac{3}{2} \qquad\qquad x \geq 0 \qquad\qquad y \geq -\dfrac{1}{3}x$$
$$y < x$$

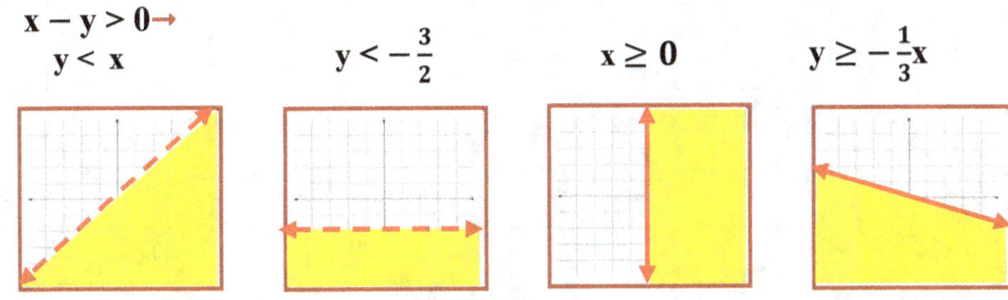

Figure 5.5.7

Graph the **solution set** on the coordinate plane.

1. $y > -2x$

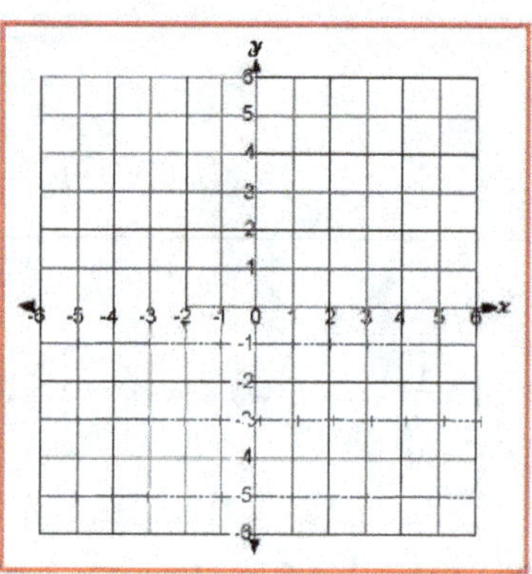

2. $y < -x + 3$

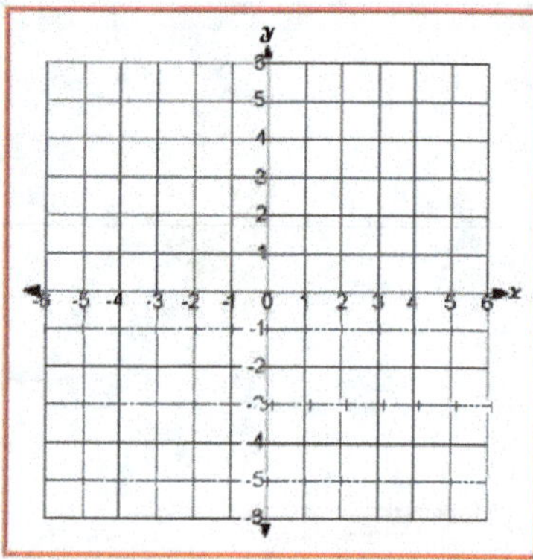

1. $3x + y \geq 4$

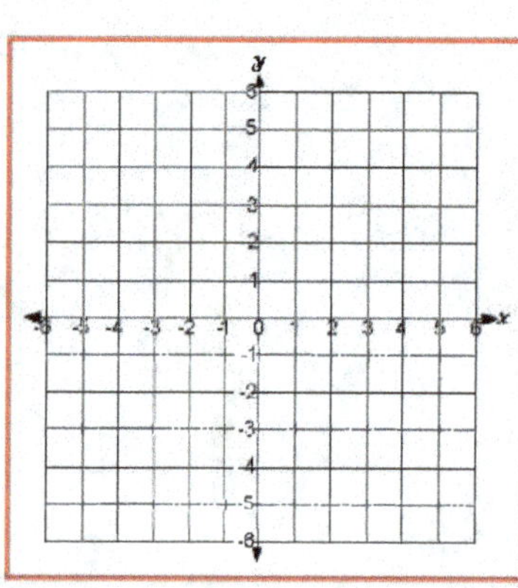

2. $2x - y \leq 3$

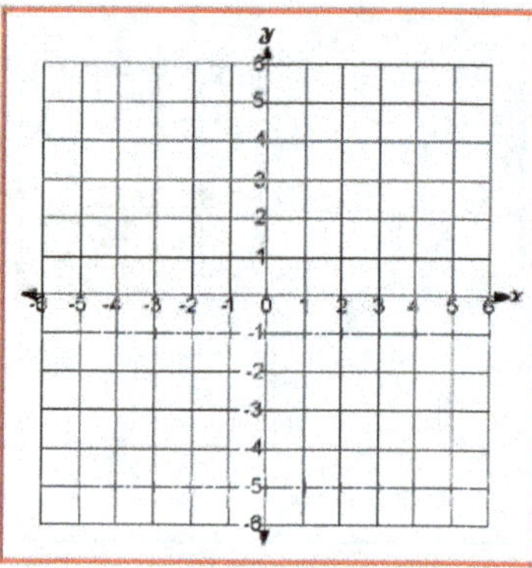

3. $x + 2y > 4$

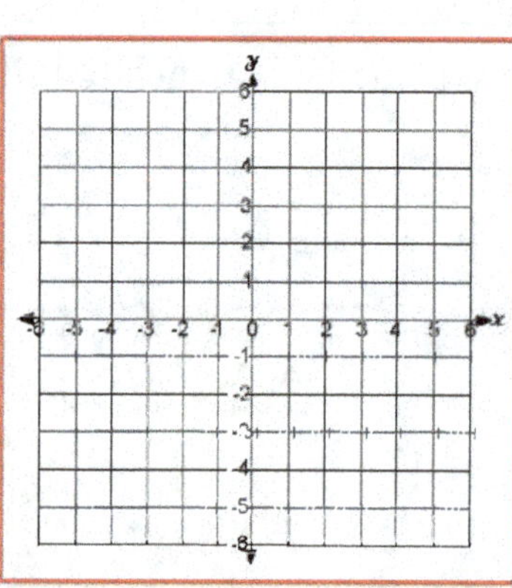

4. $x - 2y < 0$

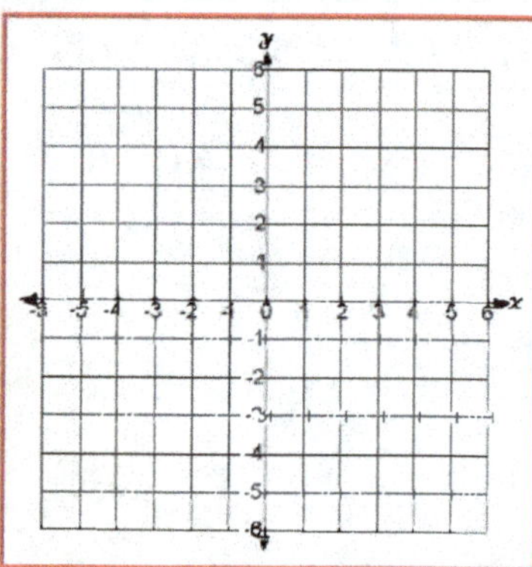

1. Find the ordered pairs of numbers (or coordinates) associated with the following points

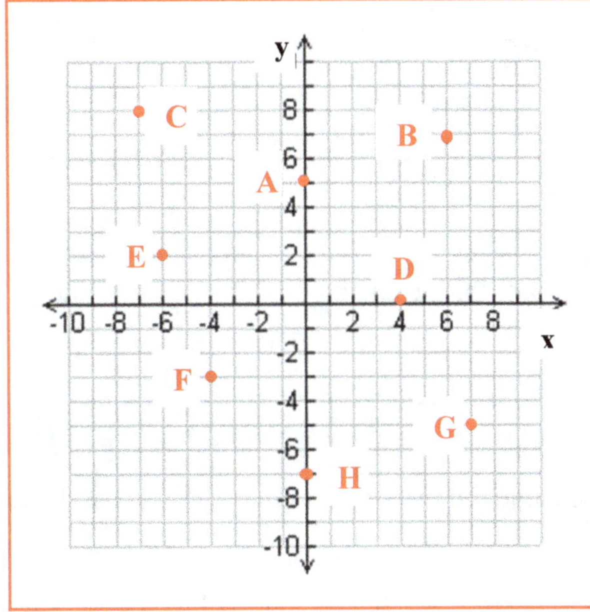

2. Rewrite each equation in slope-y-intercept form.

$x + 2y = 4$	$-x + 3y = 6$	$x - 4y = 8$
$3x + y = 1$	$4x - 2y = 8$	$2y + 12 = -4x$

3. Rewrite each equation in standard form.

$y = x + 2$	$y = -2x + 3$	$y = 6 - 3x$
$y = x$	$-x = 4 + 2y$	$y + 5 = -4x$

4. Rewrite the slope-point form for the equation of the line with the given slope and point.

$m = 2,\ (1,4)$	$m = -1,\ (0,3)$	$m = -2,\ (5,0)$
$m = -\frac{1}{2},\ (2,-3)$	$m = \frac{2}{3},\ (-6,-3)$	$m = \frac{1}{4},\ (-8, 2)$

5. Rewrite each slope-point equation in slope-y-intercept form and then in standard form.

$y - 1 = 4(x + 1)$	$y - 4 = 3(x - 3)$	$y + 2 = 4(x + 1)$
$y - 3 = 2(x - 1)$	$y - 0.5 = 0.5(x - 1)$	$y + 6 = \frac{1}{2}(x - 2)$

1. Rewrite each equation in slope-y-intercept form.

$2x + y = 4$	$-2x + 3y = 12$	$x - 3y = 9$
$3x + 2y = 4$	$3x - 4y = 4$	$4y + 8 = -4x$

2. Rewrite each equation in standard form.

$y = 3x + 2$	$2y = -2x + 3$	$y = 1 - 4x$
$y = 5x$	$-2x = 1 + y$	$2y + 3 = -x$

3. Rewrite the slope-point form for the equation of the line with the given slope and point.

$m = 3,\ (1,1)$	$m = -2,\ (0,4)$	$m = -2,\ (0,0)$
$m = -\dfrac{3}{5},\ (5,-3)$	$m = \dfrac{2}{3},\ (-3,-1)$	$m = \dfrac{3}{4},\ (-4,2)$

4. Rewrite each slope-point equation in slope-y-intercept form and then in standard form.

$y - 2 = 2(x + 1)$	$y - 1 = 4(x - 1)$	$y + 3 = 5(x + 1)$
$y - 3 = 2(x - 1)$	$y + 0.3 = 0.7(x - 1)$	$y + 1 = 3(x - 2)$

5. Graph each line by **points**, **x-y-intercept**, and **slope-y-intercept methods**.

$2x + 4y = -4$

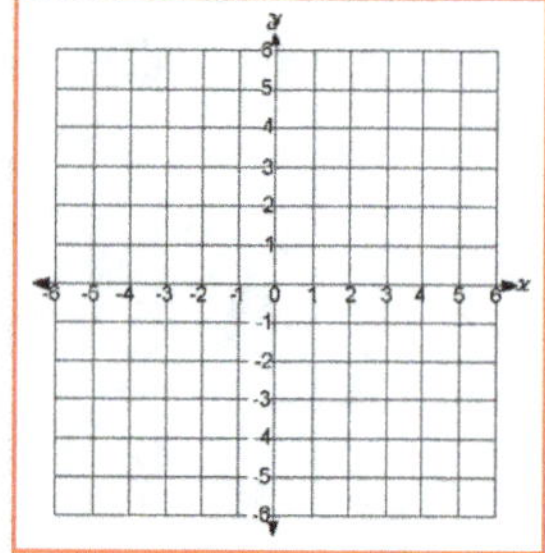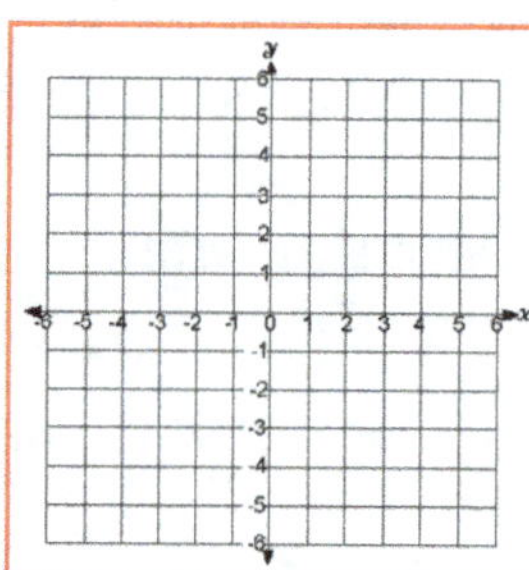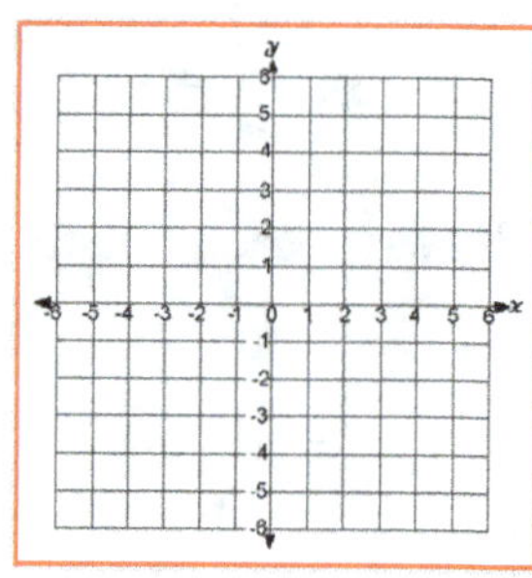

$y = \dfrac{2}{3}x + 1$

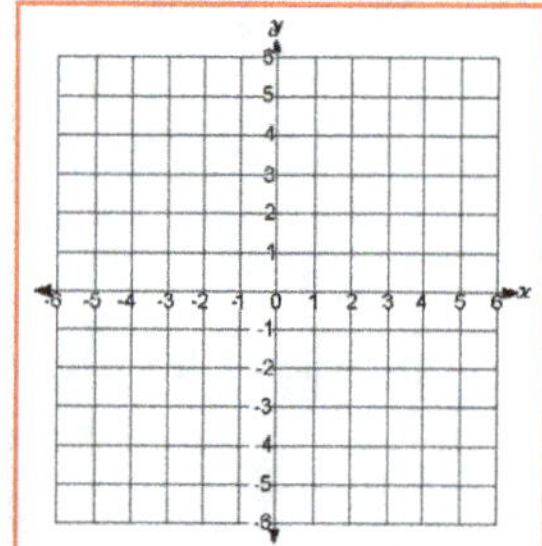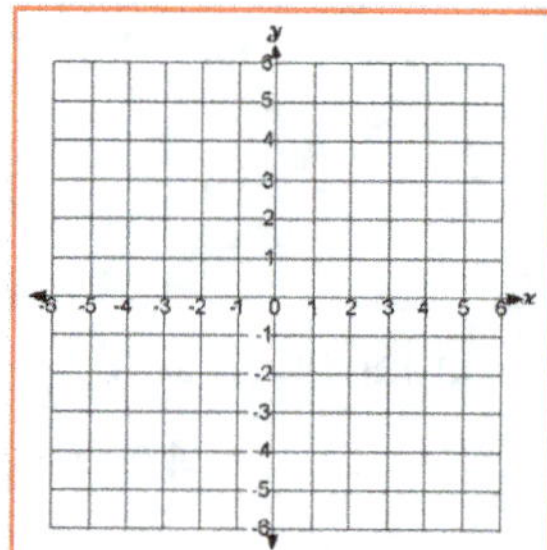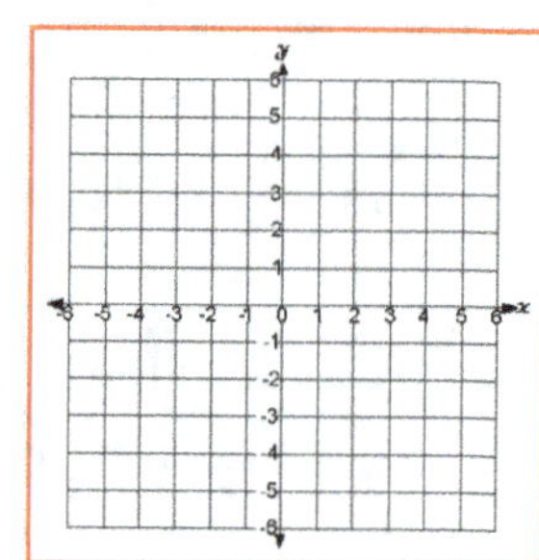

6. Graph the solution sets of following inequalities.

a) $x + y > 2$

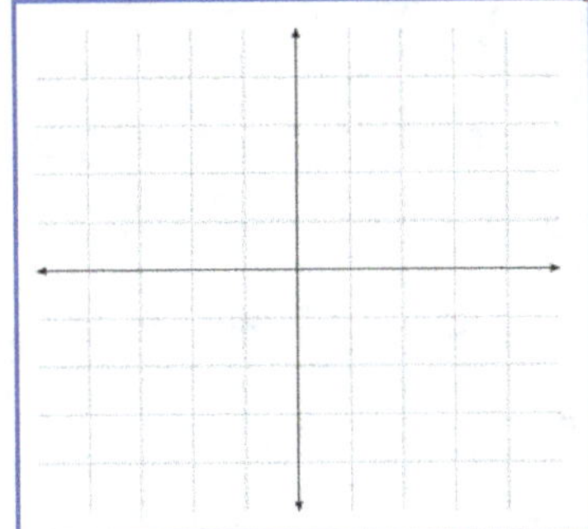

b) $2x + 3y \leq 3$

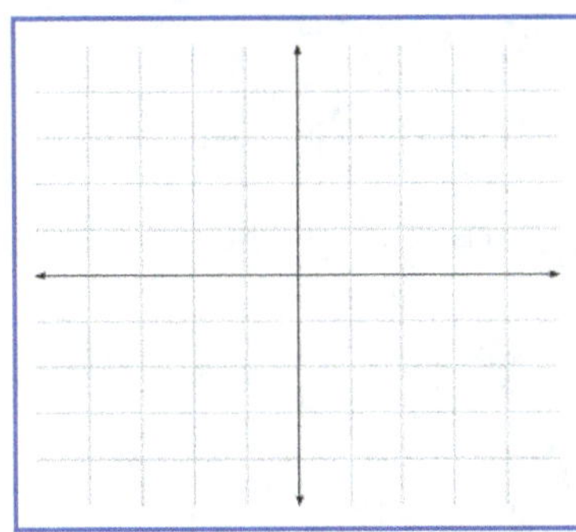

7.   Graph the solution sets of following inequalities.

a) $y < -x + 2$

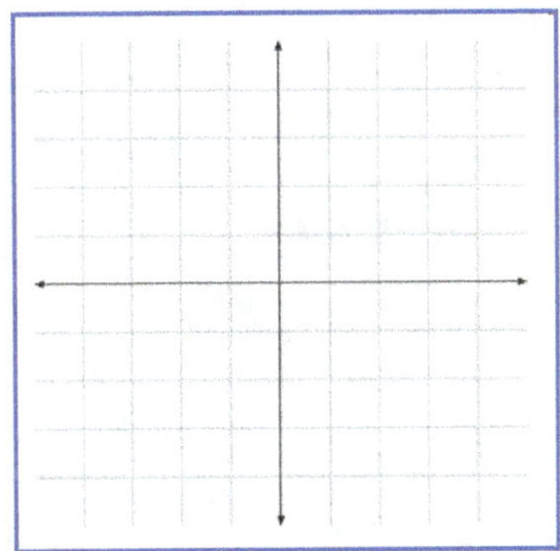

b) $y \geq 2x$

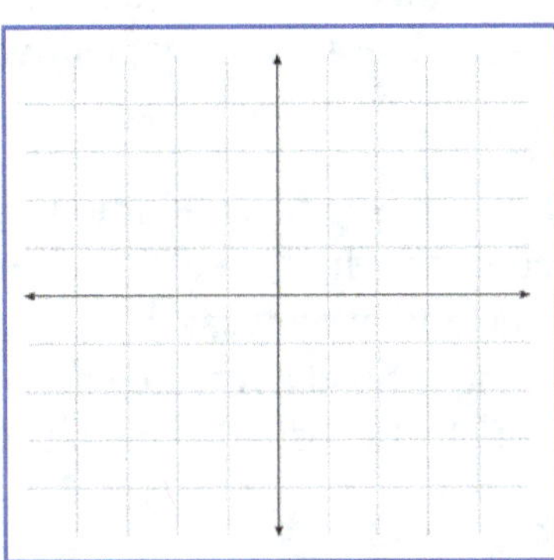

8.   Graph the solution sets of following inequalities.

a) $y \geq 2$

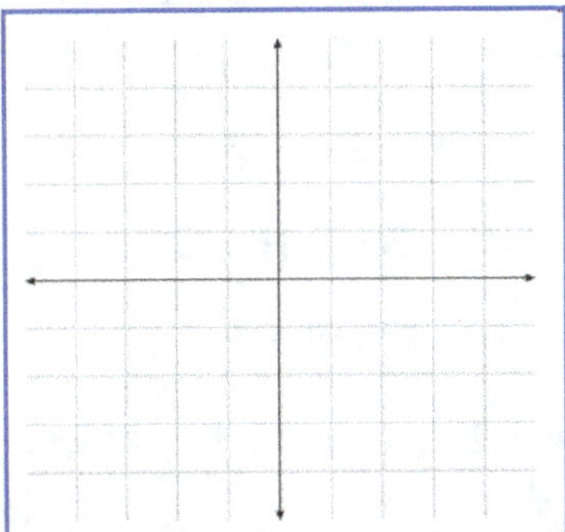

b) $x < -3$

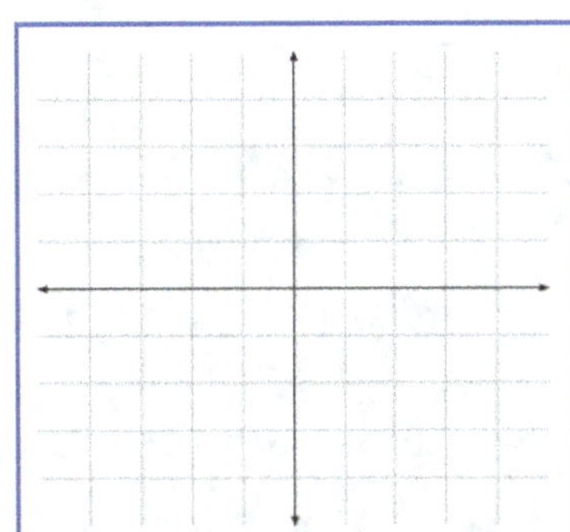

EXTRA PRACTICE
Equations, Inequalities, and Systems

Hayk Yegoryan & Rubik Yegoryan

INTERMEDIATE ALGEBRA

Solve each linear or absolute value equation and inequality.

1. $6 - 3(y - 1) = -9$

2. $2 + 4(x - 3) - 2x = 2(1 + x)$

3. $2 = -1 + \dfrac{3}{7}y$

4. $\dfrac{x}{3} = \dfrac{x + 2}{5}$

5. $\dfrac{2x}{3} - \dfrac{x - 3}{6} = \dfrac{x + 2}{4} + 1$

6. $-x + 0.72x = 5.6$

7. Find **b** for which the equation $2x - 3 = b + 4$ has only positive solution.

8. Solve $F = \dfrac{9}{5}C + 32$ for **C**.

9. Mary deposits **$10,000** into two accounts with **6%** and **7%** interests. At the end of the year the total income is **$670**. How much was originally deposited in each account?

10. $2x - 4 > 12$

11. $-5(x - 2) \le -15$

12. $1 + 4(x - 2) - x > 2(1 + x)$

13. $\dfrac{x}{4} \le \dfrac{x + 1}{3}$

14. $\dfrac{11}{4} < \dfrac{3}{4}x + 2 \le 5$

15. $4x + 1 < 13$ and $\dfrac{1}{2}x + 2 \ge 1$

16. $x + 1 \le 0.3$ or $3(x + 1) > 9$

17. $x - 2 \le 3x \;\cap\; 2x - 8 < -2$

18. $4|x - 5| = 12$

19. $3 + |1 - 2x| = 8$

20. $|2x + 5| = -3$

21. $|1.2x - 3.6| + 0.7 = 1.9$

22. $|x| = x$

23. $2|x| = x$

24. $|x + 2| = x - 2$

25. $\dfrac{1}{2}|x| - 1 = -\dfrac{1}{2}x$

26. $\left|\dfrac{x - 2}{2}\right| = x + 2$

27. $|x - |x + 2|| = 4$

28. $-|x - 1| + \dfrac{1}{2} < \dfrac{1}{4}$

29. $|x| \le x$

30. $\left|\dfrac{x - 1}{2}\right| \ge x + 2$

31. $x - |x + 2|| < 4$

32. At what values of **b** does the inequality $||x - 2| + b| \le 5$ have a unique solution?

33. $2 \le |2 + x| \le 9$

34. $2x - 1| \le 3$ or $-|x + 2| < -4$

35. $|x + 1| > 1$ and $|x| \le 6$

36. $-|x + 2| \ge 0$ and $2 + x < |8 + 3x|$

37. $|2x - 1| \le 3 \;\cup\; -|x + 2| < -4$

Solve each quadratic or complex equation and inequality.

38. $x^2 + 2x = 0$

39. $x^2 - 4 = 0$

40. $(x - 4)^2 + 25 = 0$

41. $4x^2 + 9 = 0$

42. $x^2 - 2x - 8 = 0$

43. $x^2 + 5x - 14 = 0$

44. $x^2 + 6x + 8 = 0$

45. $\dfrac{1}{5}x^2 + \dfrac{1}{5}x - 6 = 0$

46. $x^2 + 2.6x + 1.2 = 0$

47. $2x^2 + 5x + 3 = 0$

48. $3x^2 - 7x + 4 = 0$

49. $\dfrac{3}{2}x^2 - \dfrac{1}{2}x + 1 = 0$

50. $-x^4 - 2x^2 = 2$

51. $x^4 - 625 = 0$

52. $2x^{2/3} + x^{1/3} - 3 = 0$

53. The product of two consecutive positive odd numbers is 117. Find the numbers.

54. Find the value of **k** so that the equation $3x^2 - 6x + k = 0$ will have
  a) two real solutions,
  b) two complex solutions,
  c) one real (or two same) solution(s).

55. One root of the equation $2x^2 + bx - 10 = 0$ is **2**. Find **b**.

56. $25x^2 + 36k^2 = 0$, **k** is **constant** and k > 0

57. $(2 + i)(3 + yi) = 11 - 13i$

58. $x(2 + i) + 2(3 + yi) = 12 + 5i$

59. $(x - 1)^2 < 16$

60. $(x + 2)^2 \geq -1$

61. $x^2 - 4x > 12$

62. $x^2 + 6x < 40$

63. $-x^2 - 2x \leq 2$

64. $x^2 - 625 > 0$

65. $2x^{2/3} + x^{1/3} - 3 < 0$

66. The product of two consecutive positive odd integers is ≤ 24. Find the numbers.

67. Find the value of **k** so that the inequality $x^2 - 6x + k \leq 0$ will have
  a) one real solution
  b) infinitely many solutions
  c) no solutions.

68. Roots of quadratic inequality, with leading coefficient **a = 1**, are in interval **(3, –5)**. Write the quadratic inequality.

69. Unique root of inequality $-2x^2 + bx - 8 \geq 0$ is **2**. Find **b**.

70. Find interval of all possible values of **x**.

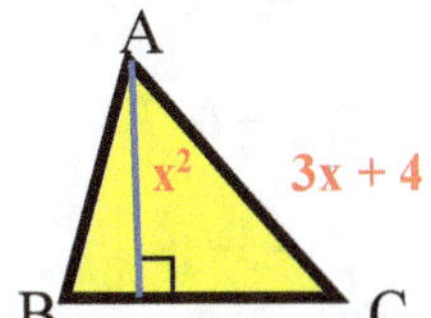

EXTRA PRACTICE
Equations, Inequalities, and Systems

continued **2**

Hayk Yegoryan & Rubik Yegoryan

INTERMEDIATE ALGEBRA

Solve each polynomial equation and inequality.

71.	$x^4 - 16 = 0$	92.	$x^4 - 81 < 0$
72.	$x^4 - 256 = 0$	93.	$x^4 - 4 \geq 0$
73.	$x^4 - 9x^2 = 0$	94.	$x^4 - 16x^2 > 0$
74.	$x^5 - 64x^3 = 0$	95.	$x^6 - 9x^4 \leq 0$
75.	$x^4 - 5x^2 - 36 = 0$	96.	$x^4 + 2x^2 - 24 > 0$
76.	$x^4 + 6x^2 + 9 = 0$	97.	$x^4 + 6x^2 + 9 < 0$
77.	$x^4 - 6x^2 + 9 = 0$	98.	$x^4 - 4x^2 + 4 \leq 0$
78.	$x^4 - 3x^2 - 4 = 0$	99.	$x^4 - 5x^2 - 36 > 0$
79.	$x^3 - 27 = 0$	100.	$x^5 - 32 < 0$
80.	$x^3 + 8 = 0$	101.	$x^4 + 1 < 0$
81.	$x^5 - 32 = 0$	102.	$x^5 - 32 \geq 0$
82.	$4x^6 - 64x^2 = 0$	103.	$4x^6 - 64x^2 < 0$
83.	$x^8 - 81x^4 = 0$	104.	$x^8 + 3x^4 \leq 0$
84.	$x^5 - 16x^3 = 0$	105.	$x^5 - x^3 > 0$
85.	$x^7 + 4x^5 = 0$	106.	$x^9 + 5x^5 < 0$
86.	$x^4 - 4x^2 + 4x^2 - 16 = 0$	107.	$x^4 - 9x^2 + 3x^2 - 27 > 0$
87.	$x^3 - x^2 - 3x^2 + 3 = 0$	108.	$x^3 - 5x^2 + 6x < 0$
88.	$x^3 + 5x^2 + 6x = 0$	109.	$x^4 + 6x^2 + 5 > 0$
89.	$x^5 + x^4 - 9x^3 - 9x^2 = 0$	110.	$x^5 + 2x^4 - 9x^3 - 18x^2 < 0$
90.	$5x^3 + 2x^2 - 3x = 0$	111.	$3x^3 + 2x^2 - 5x > 0$
91.	$2x^4 - 3x^3 - 5x^2 = 0$	112.	$5x^4 - 3x^3 - 2x^2 < 0$

Solve each fractional equation and inequality.

113. $\dfrac{3}{x+4} = \dfrac{1}{4}$

114. $\dfrac{x-3}{x-4} = \dfrac{3}{4}$

115. $\dfrac{4}{x} = \dfrac{x}{9}$

116. $\dfrac{x+2}{3} = \dfrac{10}{x+1}$

117. $\dfrac{x-1}{2x} = \dfrac{x}{x+1}$

118. $\dfrac{x-2}{3x} = \dfrac{x}{x+2}$

119. $\dfrac{x+1}{2} = \dfrac{5}{x-2}$

120. $\dfrac{x^2-6x+8}{x-2} = -\dfrac{4}{x}$

121. $\dfrac{1}{x+1} + \dfrac{1}{x+2} = \dfrac{5}{x^2+3x+2}$

122. $\dfrac{\frac{1}{x+4}+\frac{1}{x-4}}{\frac{1}{x-4}-\frac{1}{x+4}} = \dfrac{9}{x}$

123. $\dfrac{\frac{1}{x+5}+\frac{1}{x-5}}{\frac{1}{x-5}-\frac{1}{x+5}} = \dfrac{5}{x}$

124. $\dfrac{1-\frac{x}{x-\frac{1}{x}}}{1+\frac{x}{x-\frac{1}{x}}} = \dfrac{1}{x}$

125. Together Nick and Rick can paint a car in 16 hours. Alone Rick needs 24 hours longer than Nick needs. How long would it take for each Nick and Rick to paint the car?

126. $\dfrac{1}{3} - \dfrac{1}{x} < \dfrac{1}{2}$

127. $\dfrac{1}{x+1} - \dfrac{1}{2} < 0$

128. $\dfrac{2}{x-5} + \dfrac{1}{3} \le \dfrac{1}{2}$

129. $\dfrac{2}{x+1} - \dfrac{1}{x} < 1$

130. $\dfrac{x}{2x+3} - \dfrac{1}{x} < 0$

131. $\dfrac{1}{1-x} > x+1$

132. $\dfrac{1}{x-1} > \dfrac{1}{x+1}$

133. $\dfrac{3x^2-2x-5}{x+1} < 0$

134. $\dfrac{5x^2-5x+2}{x^2-x-2} \ge 5$

135. $\dfrac{1}{x^2} > \dfrac{1}{(x+1)^2}$

136. $\dfrac{1}{x+1} + \dfrac{1}{x+2} \le \dfrac{5}{x^2+3x+2}$

137. $\dfrac{1}{x+1} + \dfrac{2}{x-2} > \dfrac{3}{x^2-x-2}$

138. $\dfrac{\frac{x+1}{x-1}+\frac{x-1}{x+1}}{\frac{x+1}{x-1}-\frac{x-1}{x+1}} < -\dfrac{1}{x} + \dfrac{5-x}{2x}$

139. $\dfrac{\frac{1}{x+4}+\frac{1}{x-4}}{\frac{1}{x-4}-\frac{1}{x+4}} \ge \dfrac{9}{x}$

Solve each radical equation and inequality.

140. $\sqrt{x} = 10$

141. $\sqrt{-x} = 1$

142. $\sqrt{5 + x} = 2$

143. $\sqrt{x - 2} = -4$

144. $2\sqrt{3x - 2} = 4$

145. $3\sqrt{x - 3} - 3 = 5$

146. $\sqrt{6x - 1} + 7 = 2$

147. $\sqrt{x - 2} - x = -4$

148. $\sqrt{6 - x} = 4 - x$

149. $\sqrt{(x + 3)(x - 2)} = x$

150. $\sqrt{2x^2 - 9} = x$

151. $\sqrt{10x^2 - 16} = 3x$

152. $\sqrt{3x - 8} + 2 = x$

153. $\sqrt{2x - 7} = 2\sqrt{x} - \sqrt{x + 5}$

154. $\sqrt{3x + 12} = \sqrt{4x + 13} + \sqrt{x + 1}$

155. $\sqrt{5x - 1} = \sqrt{x - 5} + 2\sqrt{x + 1}$

156. $\sqrt{x^2 + 9x + 14} = x + 4$

157. $\sqrt{x} < 5$

158. $\sqrt{-x} > 2$

159. $\sqrt{4 + x} \leq 2$

160. $\sqrt{x - 1} \geq -2$

161. $\sqrt{x - 1} \leq 3$

162. $3\sqrt{2x - 1} < 9$

163. $\sqrt{x - 3} - 1 > 5$

164. $\sqrt{4x - 3} + 5 < 3$

165. $\sqrt{x - 2} - x \geq -4$

166. $\sqrt{(x + 3)(x - 2)} \leq x$

167. $\sqrt{2x^2 - 9} > x$

168. $\sqrt{5x^2 - 16} < 2x$

169. $\sqrt{3x - 8} + 2 \geq x$

170. $\sqrt{2x - 7} < \sqrt{x + 5}$

171. $\sqrt{2x - x^2} > 1 - x$

172. $\sqrt{x^3 + 2x + 1} > x + 1$

173. $\sqrt{2x^2 - 6x + 1} \geq x - 2$

<table>
<tr><td>PART
2</td><td>PRE-TEST
Equations, Inequalities, and Graphs</td></tr>
</table>

Hayk Yegoryan & Rubik Yegoryan

INTERMEDIATE ALGEBRA

1. Solve

    a) $5(x-2)+5-4x=-4$

    b) $\dfrac{2x}{3}-1=\dfrac{x}{6}$

    c) $\dfrac{x}{4x-1}=\dfrac{1}{3}$

2. Solve

    a) $4x-1>-(x-3)+1$

    b) $2\le 2+x\le 9$

    c) $x<2x-1\le 11$

    d) $2x-1\le 3\quad$ or $-x+2<-4$

    e) $-x+2>0$ and $2+x<6+3x$

3. Solve

    a) $|2-x|+5=6$

    b) $|x+1|+6=2$

4. Solve

    a) $|1+x|\ge 2$

    b) $|2x-1|\le 5$

    c) $|x+2|>2$ and $|x|\le 5$

    d) $|x-1|\le 3$ or $|x+1|>6$

5. Solve

    a) $x^2-7x-8=0$

    b) $x^2+7x+12\le 0$

6. Simplify each expression and write all the answers in **a + b*i*** form

   a) $(1 + \sqrt{-16})(1 + \sqrt{-4}) + i^5$

   b) $\dfrac{1 + \sqrt{-2}}{1 + \sqrt{-1}}$

7. Solve

   a) $x^3 - x^2 + 9x - 9 = 0$

   b) $4x^3 + 3x^2 - 7x > 0$

   c) $2x^6 - 18x^4 \leq 0$

8. Solve

   a) $\dfrac{1}{x + \sqrt{3}} + \dfrac{1}{x - \sqrt{3}} = 1$

   b) Find **x** (triangles are similar)

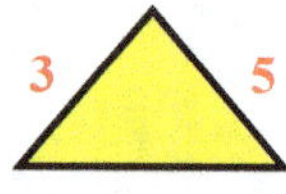

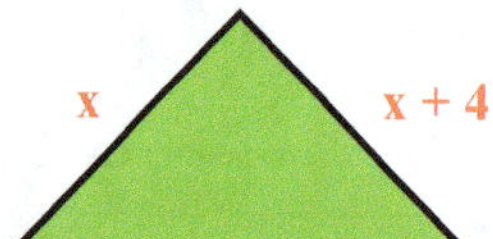

9. Solve each inequality and represent the solution set in the graph and in the interval notations

   a) $\dfrac{1}{x} - \dfrac{1}{x + 1} < 0$

   b) $\dfrac{3}{x} + 2 \geq \dfrac{5}{x^2}$

10. Solve

   a) $\sqrt{3x + 1} - \sqrt{2x} = 1$

   b) $\sqrt{2x + 1} - \sqrt{x} < 1$

11. Graph these lines using **the point's method.**

a) $x + 3y = 6$

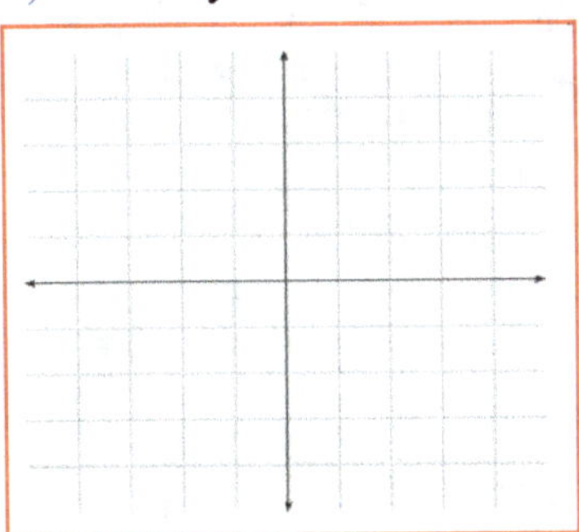

b) $y = 1 - 2x$

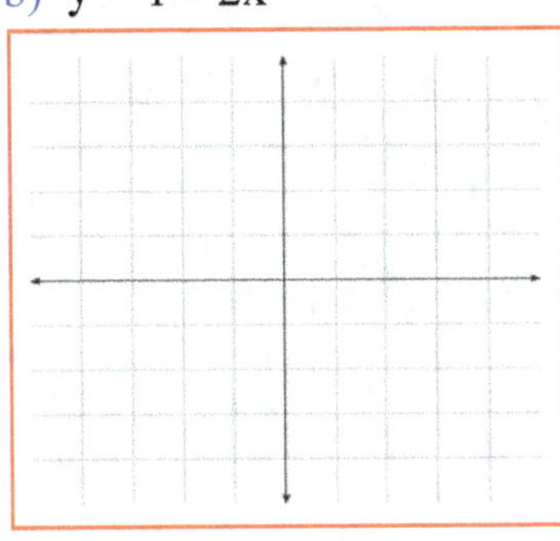

12. Graph these lines using **x- and y-intercept's method.**

a) $4x + 3y = 12$

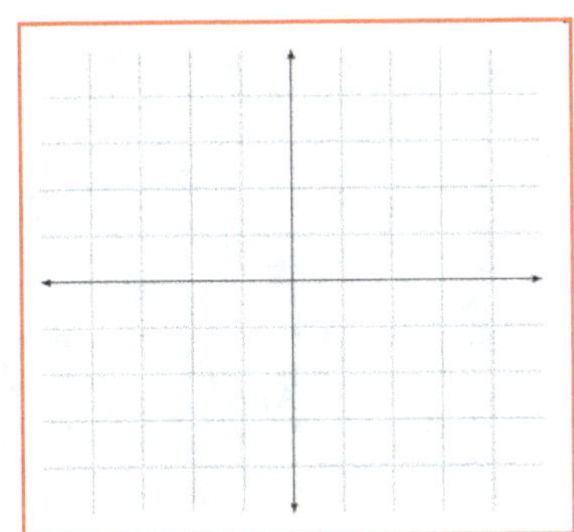

b) $y = 3x - 3$

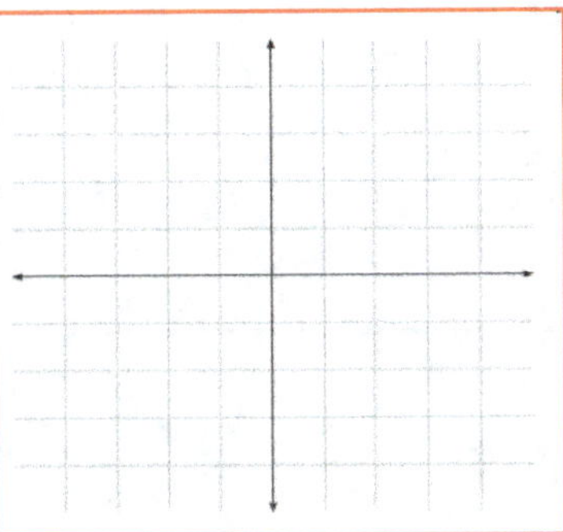

13. Graph these lines using **the slope – y– intercept's method**.

a) $x + y = 3$

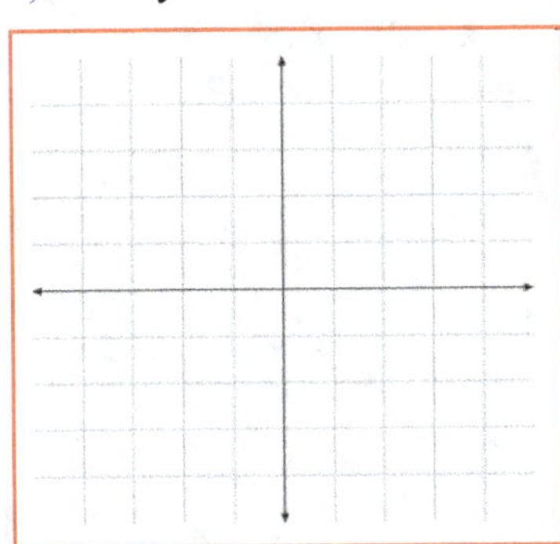

b) $y = \dfrac{1}{2}x - 2$

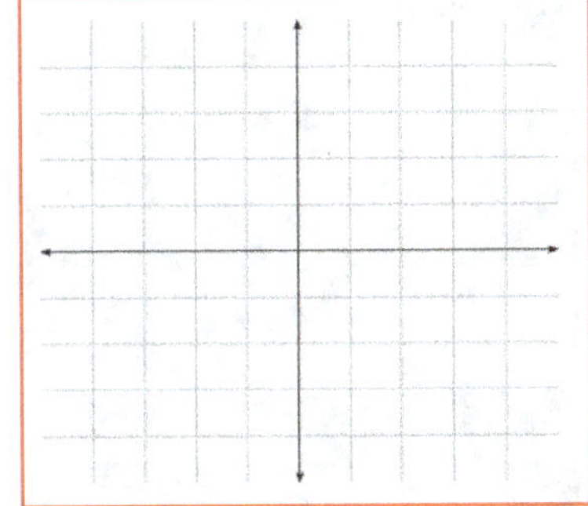

14. Graph the solution sets of following inequalities.

a) $x + 3y > 6$

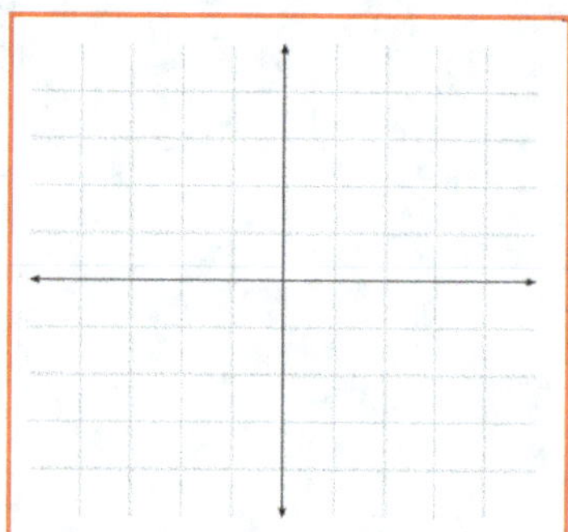

b) $2x + y \leq 1$

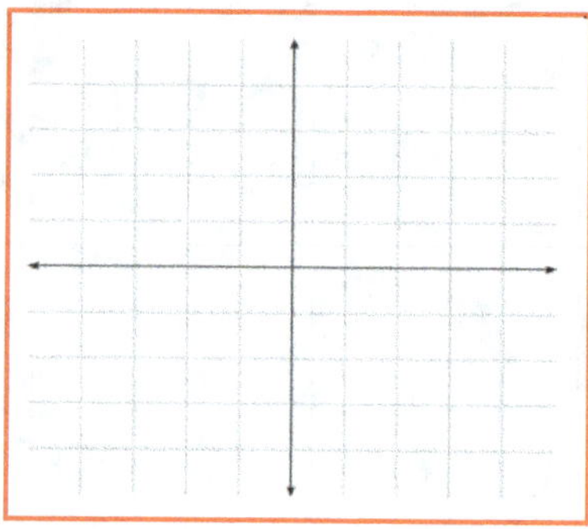

15. Graph the solution sets of following inequalities.

a) $y < -x + 4$

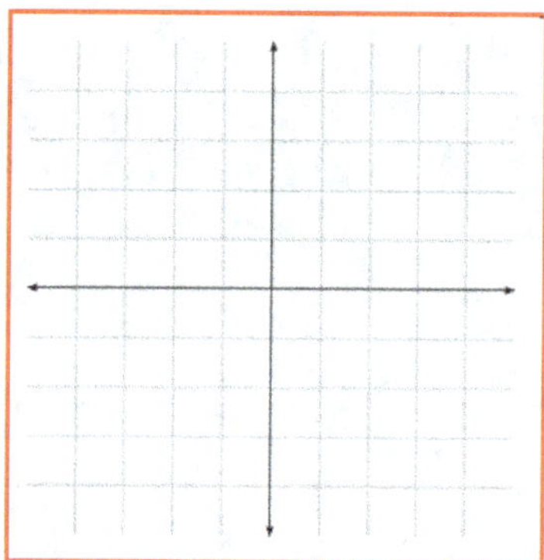

b) $y \geq 3x$

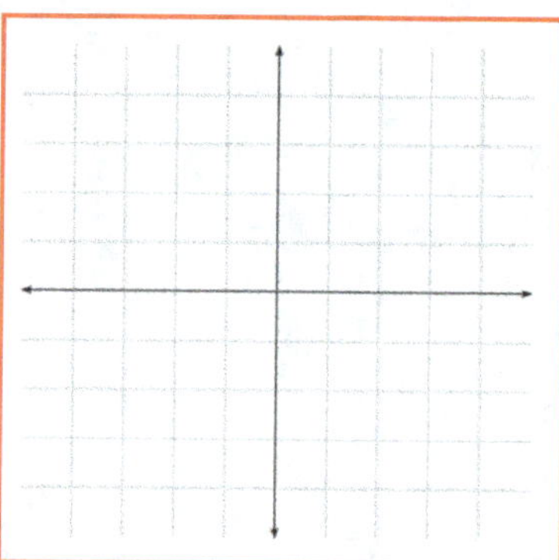

16. Graph the solution sets of following inequalities.

a) $y \geq 3$

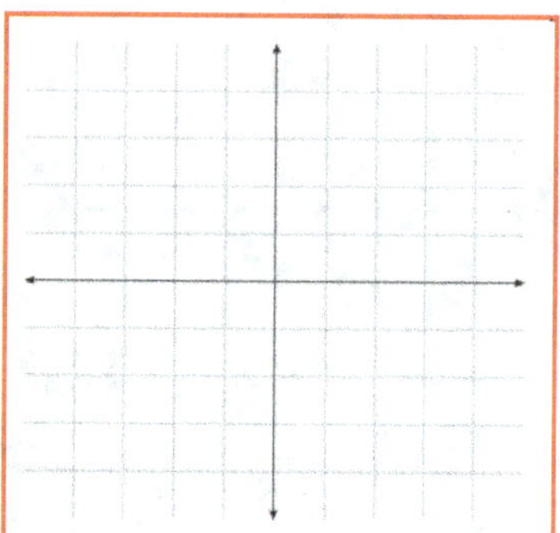

b) $x < -2$

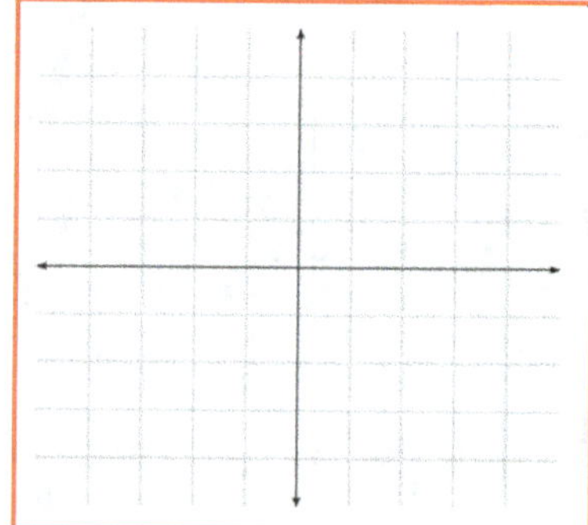

PART 3

LINEAR SYSTEMS

Chapter 6. Systems of Equations and Inequalities

6.1.   Systems of Linear Equations
6.2.   Cramer's Rules
6.3.   Systems of Linear Inequalities

Chapter 6   Classwork
Chapter 6   Homework

Part 3. Pre-Test

Chapter 6. Systems of Equations and Inequalities

6.1. **Systems of Linear Equations**
6.2. **Cramer's Rules**
6.3. **Systems of Linear Inequalities**

Chapter 6   Classwork
Chapter 6   Homework

Part 3. Pre-Test

$$\begin{cases} a_{11}x + a_{12}y = b_1 \\ a_{21}x + a_{22}y = b_2 \end{cases}$$

$$\begin{cases} a_{11}x + a_{12}y + a_{13}z = b_1 \\ a_{21}x + a_{22}y + a_{23}z = b_1 \\ a_{31}x + a_{32}y + a_{33}z = b_1 \end{cases}$$

$$\det A = \begin{vmatrix} a_1 & b_1 & c_1 \\ a_2 & b_2 & c_2 \\ a_3 & b_3 & c_3 \end{vmatrix} = \begin{vmatrix} a_1 & b_1 & c_1 \\ a_2 & b_2 & c_2 \\ a_3 & b_3 & c_3 \end{vmatrix} \begin{matrix} a_1 & b_1 \\ a_2 & b_2 \\ a_3 & b_3 \end{matrix} = (a_1b_2c_3 + b_1c_2a_3 + c_1a_2b_3) - (a_3b_2c_1 + b_3c_2a_1 + c_3a_2b_1)$$

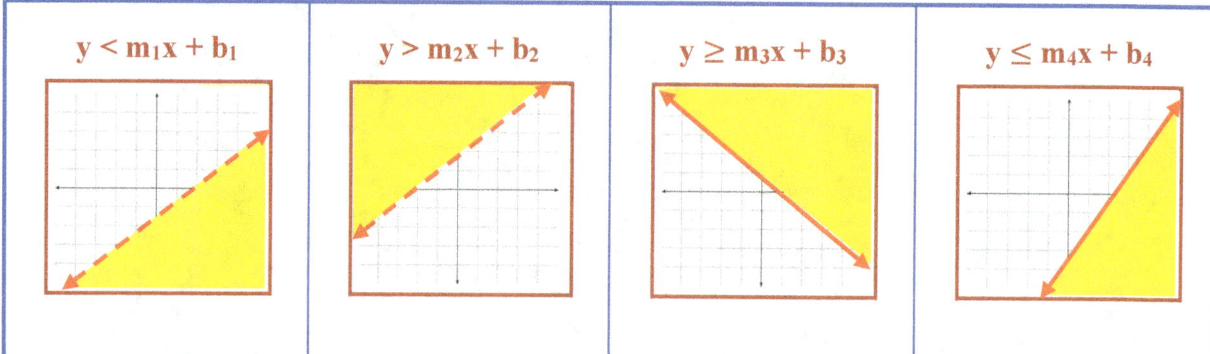

6.1. Systems of Linear Equations

- **DEFINITIONS**
- **ALGEBRAIC METHODS TO SOLVE THE SYSTEMS**

■ DEFINITIONS

The theory of systems of linear equations is a very frequently used subject in mathematics, physics, computer science, business, biology, engineering, architecture, and the social sciences.

We will use the following definitions.

IMPORTANT DEFINITIONS

Definition ▶ A *system of m linear equations with n variables* (unknowns) $x_1, x_2, x_3, \ldots, x_n$ (or "**m x n system**") is a set of **m** linear equations each with **n** unknowns:

$$\begin{cases} A_{11}x_1 + A_{12}x_2 + A_{13}x_3 + \ldots + A_{1n}x_n = B_1 \\ A_{21}x_1 + A_{22}x_2 + A_{23}x_3 + \ldots + A_{2n}x_n = B_2 \\ A_{31}x_1 + A_{32}x_2 + A_{33}x_3 + \ldots + A_{3n}x_n = B_3 \\ \\ \\ A_{m1}x_1 + A_{m2}x_2 + A_{m3}x_3 + \ldots + A_{mn}x_n = B_m \end{cases} \tag{1}$$

where A_{ij} ($i = 1,2,3\ldots,m$; $j = 1,2,3,\ldots, n$) and B_i are known constants, $x_1, x_2, x_3, \ldots, x_i, \ldots, x_n$ are unknowns which we want to find, and $A_{i1}x_1 + A_{i2}x_2 + A_{i3}x_3 + \ldots + A_{in}x_n = B_i$ is the ith equation of the system.

- A **solution** of linear system is a set of numbers $s_1, s_2, s_3, \ldots, s_n$, which satisfy each equation of the system when $x_1 = s_1, x_2 = s_2, x_3 = s_3, \ldots, x_n = s_n$ are substituted.
- If the system has no solutions, then that system is called **inconsistent**.
- If the system has one or more solutions, then that system is called **consistent**.
- If $B_1 = B_2 = B_3 = \ldots = B_m = 0$, then the system is called **homogeneous**.
- If $x_1 = x_2 = x_3 = \ldots = x_n = 0$ is a solution to a homogeneous system, then that solution is called the **trivial solution**.
- A solution of a homogeneous system in which not all $x_1, x_2, x_3, \ldots, x_n$ are zero is called a **nontrivial solution**.

$\mathbf{W}$e will also use the following definitions for a system of **2** linear equations with **2** variables and for a system of **3** linear equations with **3** variables.

Definition ▶ A *system of 2 linear equations with 2 variables* **x** and **y** (or **"2 x 2 system"**) is a set of **2** linear equations each with **2** unknowns:

$$\begin{cases} a_{11}x + a_{12}y = b_1 \\ a_{21}x + a_{22}y = b_2 \end{cases}$$

where a_{11}, a_{12}, a_{21}, a_{22}, b_1, and b_2 are known constants, **x** and **y** are unknowns which we want to find, and $a_{11}x + a_{12}y = b_1$ is the **first** equation, and $a_{21}x + a_{22}y = b_2$ is the **second** equation of the **system.**

Definition ▶ A *system of 3 linear equations with 3 variables* **x, y,** and **z** (or **"3 x 3 system"**) is a set of **3** linear equations each with **3** unknowns:

$$\begin{cases} a_{11}x + a_{12}y + a_{13}z = b_1 \\ a_{21}x + a_{22}y + a_{23}z = b_1 \\ a_{31}x + a_{32}y + a_{33}z = b_1 \end{cases}$$

where a_{11}, a_{12}, a_{13}, a_{21}, a_{22}, a_{23}, a_{31}, a_{32}, a_{33}, b_1, b_2, and b_3 are known constants, **x, y,** and **z** are unknowns which we want to find, and $a_{11}x + a_{12}y + a_{13}z = b_1$ is the **first** equation, $a_{21}x + a_{22}y + a_{23}z = b_2$ is the **second** equation, and $a_{31}x + a_{32}y + a_{33}z = b_3$ is the **third** equation of the system.

Examples The system of equations shown below is a **2 x 2 system**.

$$\begin{cases} 4x + 3y = 10 \\ 2x - 5y = -8 \end{cases}$$

The system of equations shown below is a **3 x 3 system**.

$$\begin{cases} x + 2y + 3z = 14 \\ 2x - 5y + 4z = 4 \\ 7x + y - 2z = 3 \end{cases}$$

$\mathbf{B}$ellow we will study **3** algebraic methods to solve the systems of linear equations: **substitution, algebraic elimination (or addition),** and **graph methods,** as well as several methods related to matrix algebra: **Cramer's Rule, Gaussian elimination, Gauss-Jordan elimination,** and **inverse matrix methods.**

■ ALGEBRAIC METHODS TO SOLVE THE SYSTEMS

In this section we study **substitution**, **elimination**, and **graph methods** to solve systems of **2** linear equations with **2** variables (2x2 systems) and systems of **3** linear equations with **3** variables (3x3 systems).

■ ■ SUBSTITUTION METHOD

To solve a system of **2** linear equations with **2** variables by substitution we have to use the following steps:

> **Step 1** ▶ **Solve one equation for one of the variables** (for **x** or for **y**).
>
> You have to choose the equation which will make substitution easier.
> If one of the equations has a variable with a coefficient of 1 or -1 you should choose that equation.
>
> **Step 2** ▶ **Substitute that variable in the second equation and solve**.
>
> If, after substituting and combining like terms, the second equation becomes an equation with one variable, then we have to solve that equation to find the value of the variable (**x** or **y**) and substitute the result into the first equation to find the other variable (**y** or **x**). In this case the equation has a **unique** (or **only one**) **solution**.
>
> If, after substituting and combining like terms, the second equation becomes just a true number identity, then we have to write that we have **infinitely many solutions**.
>
> If, after substituting and combining like terms, the second equation becomes just a false number statement, then we have to write that we have **no solutions**.
>
> **Step 3** ▶ **Check the solution in both original equations** if substitution method gives a unique solution.

Example 1 Solve the system by the substitution method.

$$\begin{cases} 2x + 3y = 13 \\ y = x + 1 \end{cases}$$

SOLUTION ▶

The second equation, **y = x + 1,** is already solved for **y**, so we substitute **x + 1** for **y** in the first equation, **2x + 3y = 13.**

$$\begin{cases} 2x + 3(x+1) = 13 \\ y = x + 1 \end{cases}$$

Step 2. Substitute **y** from the second equation into the first.

$$\begin{cases} 2x + 3x + 3 = 13 \\ y = x + 1 \end{cases}$$

$$\begin{cases} 5x + 3 = 13 \\ y = x + 1 \end{cases}$$

Combine like terms in the first equation and solve.

$$\begin{cases} 5x = 13 - 3 \\ y = x + 1 \end{cases}$$

$$\begin{cases} 5x / 5 = 10 / 5 \\ y = x + 1 \end{cases}$$

$$\begin{cases} x = 2 \\ y = x + 1 \end{cases}$$

$$\begin{cases} x = 2 \\ y = 2 + 1 \end{cases}$$

$$\begin{cases} x = 2 \\ y = 3 \end{cases}$$

$x = 2, y = 3$ is a possible solution of the original system.

$$\begin{cases} 2 \cdot 2 + 3 \cdot 3 = 13 \quad \text{True} \\ 3 = 2 + 1 \quad \text{True} \end{cases}$$

Step 3. Check

$$\begin{cases} x = 2 \\ y = 3 \end{cases}$$

$(2, 3)$ is the unique solution of the original system.

Example 2 Solve the system by the substitution method.

$$\begin{cases} 4x + y = 9 \\ x - y = 1 \end{cases}$$

SOLUTION ▶

We will choose the second equation, $x - y = 1$, and solve for **x** in terms of **y**.

$$\begin{cases} 4x + y = 9 \\ x - y = 1 \end{cases}$$

Step 1. Solve the second equation for **x** in terms of **y**.

$$\begin{cases} 4x + y = 9 \\ x = y + 1 \end{cases}$$

$$\begin{cases} 4(y + 1) + y = 9 \\ x = y + 1 \end{cases}$$

Step 2. Substitute **x** from the second equation into the first.

$$\begin{cases} 4y + 4 + y = 9 \\ x = y + 1 \end{cases}$$ Combine like terms in the first equation and solve.

$$\begin{cases} 5y + 4 = 9 \\ x = y + 1 \end{cases}$$

$$\begin{cases} 5y\,/\,5 = 5\,/\,5 \\ x = y + 1 \end{cases}$$

$$\begin{cases} y = 1 \\ x = y + 1 \end{cases}$$ Substitute $y = 1$ in the second equation and find x.

$$\begin{cases} y = 1 \\ x = 2 \end{cases}$$ **(2, 1)** is a possible solution of the original system.

$$\begin{cases} 4 \cdot 2 + 1 = 9 \quad \text{True} \\ 2 - 1 = 1 \quad \text{True} \end{cases}$$ **Step 3.** Check

$$\downarrow$$

$$\begin{cases} x = 2 \\ y = 1 \end{cases}$$ **(2, 1)** is the unique solution of the original system.

Example 3 Solve the system by the substitution method.

$$\begin{cases} 4x - 4y = 8 \\ x = y + 1 \end{cases}$$

SOLUTION ▶

The second equation, $x = y + 1$, is already solved for **x**, so we substitute $y + 1$ for **x** in the first equation, $4x - 4y = 8$.

$$\begin{cases} 4(y + 1) - 4y = 8 \\ x = y + 1 \end{cases}$$ **Step 2.** Substitute **x** from the second equation into the first.

$$\begin{cases} 4y + 4 - 4y = 8 \\ x = y + 1 \end{cases}$$ Combine like terms in the first equation.

$$\begin{cases} 4 = 8 \quad \text{False} \\ x = y + 1 \end{cases}$$

$$\downarrow$$

No solution

 Solve the system by the substitution method.

$$\begin{cases} 3x - 6\,y = 3 \\ x = 2y + 1 \end{cases}$$

SOLUTION▶

The second equation, $x = 2y + 1$, is already solved for **x**, so we substitute $2y + 1$ for **x** in the first equation, $3x - 6\,y = 3$.

$$\begin{cases} 3(2y + 1) - 6y = 3 \\ x = 2y + 1 \end{cases}$$ **Step 2.** Substitute **x** from the second equation into the first.

$$\begin{cases} 6y + 3 - 6y = 3 \\ x = 3y + 1 \end{cases}$$ Combine like terms in the first equation.

$$\begin{cases} 3 = 3 \quad \text{True} \\ x = y + 1 \end{cases}$$

Infinitely many solutions or $x \in R, y \in R$

To solve a system of **3** linear equations with **3** variables by substitution we have to use the same steps:

 Solve the system by the substitution method.

$$\begin{cases} 2x + 3y + 4z = 20 \\ 5x - y - z = 0 \\ x + y = 3 \end{cases}$$

SOLUTION▶

We will choose the third equation, $x + y = 3$, and solve for **x** in terms of **y.**

$$\begin{cases} 2x + 3y + 4z = 20 \\ 5x - y - z = 0 \\ x + y = 3 \end{cases}$$ **Step 1.** Solve the third equation for **x** in the terms of **y.**

$$\begin{cases} 2x + 3y + 4z = 20 \\ 5x - y - z = 0 \\ x = 3 - y \end{cases}$$ **Step 2.** Substitute **x** from the third equation into the second.

$$\begin{cases} 2x + 3y + 4z = 20 \\ 5(3 - y) - y - z = 0 \\ x = 3 - y \end{cases}$$ Open parentheses and combine like terms.

$$\begin{cases} 2x + 3y + 4z = 20 \\ 15 - 6y - z = 0 \\ x = 3 - y \end{cases}$$ Solve the second equation for **z** in the terms of **y**.

$$\begin{cases} 2x + 3y + 4z = 20 \\ z = 15 - 6y \\ x = 3 - y \end{cases}$$ Substitute **z** from the second equation into the first.
Substitute **x** from the third equation into the first.

$$\begin{cases} 2(3 - y) + 3y + 4(15 - 6y) = 20 \\ z = 15 - 6y \\ x = 3 - y \end{cases}$$ Open parentheses.

$$\begin{cases} 6 - 2y + 3y + 60 - 24y = 20 \\ z = 15 - 6y \\ x = 3 - y \end{cases}$$ Combine like terms.

$$\begin{cases} 66 - 23y = 20 \\ z = 15 - 6y \\ x = 3 - y \end{cases}$$ Solve for **y**.

$$\begin{cases} -23y/(-23) = -46/(-23) \\ z = 15 - 6y \\ x = 3 - y \end{cases}$$

$$\begin{cases} y = 2 \\ z = 15 - 6y \\ x = 3 - y \end{cases}$$ **Step 2.** Substitute **y = 2** into the 2nd and 3rd equations.

$$\begin{cases} y = 2 \\ z = 3 \\ x = 1 \end{cases}$$ **(1, 2, 3)** is the candidate solution.

Step 3. Check

$$\begin{cases} 2 \cdot 1 + 3 \cdot 2 + 4 \cdot 3 = 20 \\ 5 \cdot 1 - 2 - 3 = 0 \\ 1 + 2 = 3 \end{cases}$$

$$\begin{cases} 20 = 20 \\ 0 = 0 \\ 3 = 3 \end{cases}$$ True
True
True

$$\Downarrow$$

$$\begin{cases} x = 1 \\ y = 2 \\ z = 3 \end{cases}$$ **(1, 2, 3)** is the unique solution of the original system.

■ ■ ELIMINATION METHOD

To solve a system of **2** linear equations with **2** variables by elimination we have to use the following steps:

> **Step 1** ▶ **Rewrite both equations of the system in standard form:**
> $$\begin{cases} Ax + By = C \\ Dx + Ey = F \end{cases}$$
>
> **Step 2** ▶ **Eliminate one of the variables (x or y).**
> You have to multiply one or both equations of the given linear system by appropriate numbers so that the sum of two equations will eliminate **x** or **y**.
>
> If, after eliminating and combining like terms, the resulting equation becomes an equation with one variable then we have to solve that equation to find the variable (**x** or **y**) and substitute the result into the first or second equation to find the other variable (**y** or **x**). In this case, the system has a **unique** (or **only one**) **solution**.
>
> If, after eliminating and combining like terms, the resulting equation becomes just a true number identity, then we have to write that we have **infinitely many solutions**.
>
> If, after eliminating and combining like terms, the resulting equation becomes just a false number statement, then we have to write that we have **no solutions**.
>
> **Step 3** ▶ **Check the solution in both original equations** if the elimination method gives a unique solution.

Example 1 Solve the system by the algebraic elimination method.

$$\begin{cases} 2x + 3y = 8 \\ \quad 4y = 2x + 6 \end{cases}$$

SOLUTION ▶

The first equation, **2x + 3y = 8,** is standard form. However, we have to rewrite the second equation, **4y = 2x + 6,** in standard form, too.

$$\begin{cases} 2x + 3y = 8 \\ \quad 4y = 2x + 6 \end{cases}$$ **Step 1.** Rewrite both equations in standard form.

$$\begin{cases} 2x + 3y = 8 \\ -2x + 4y = 6 \end{cases}$$ **Step 2.** Eliminate **x** by adding both equations.

$$\begin{cases} \qquad 7y = 14 \\ -2x + 4y = \ 6 \end{cases}$$ Solve the resulting equation **7y = 14**.

$$\begin{cases} \qquad 7y/7 = 14/7 \\ -2x + 4y = 6 \end{cases}$$

338

$$\begin{cases} y = 2 \\ -2x + 4y = 6 \end{cases}$$

Substitute solution **y = 2** into the second equation.

$$\begin{cases} y = 2 \\ -2x + 4 \cdot 2 = 6 \end{cases}$$

Solve the second equation for **x.**

$$\begin{cases} y = 2 \\ -2x = 6 - 8 \end{cases}$$

$$\begin{cases} y = 2 \\ -2x\,/(-2) = -2\,/(-2) \end{cases}$$

$$\begin{cases} y = 2 \\ x = 1 \end{cases}$$

x = 1, y = 2 is the candidate solution for the system

Step 3. Check

$$\begin{cases} 2 \cdot 1 + 3 \cdot 2 = 8 \\ 4 \cdot 2 = 2 \cdot 1 + 6 \end{cases}$$

$$\begin{cases} 8 = 8 \\ 8 = 8 \end{cases}$$

True

True

$$\begin{cases} x = 1 \\ y = 2 \end{cases}$$

(1, 2) is the unique solution of the original system.

Example 2 Solve the system by the algebraic elimination method.

$$\begin{cases} 5x + 2y = 12 \\ 2x + 3y = 7 \end{cases}$$

SOLUTION ▶

Both equations of the system are in standard form. We have to multiply both equations of the given linear system by appropriate numbers so that the sum of the two equations will eliminate **y**.

$$\begin{cases} 5x + 2y = 12 & | \cdot (3) \\ 2x + 3y = 7 & | \cdot (-2) \end{cases}$$

Step 2. Multiply the first equation by **3.**
Multiply the second equation by **– 2.**

$$\begin{cases} 15x + 6y = 36 \\ -4x - 6y = -14 \end{cases}$$

Add the 1st and the 2nd equations and write as the 2nd.

$$\begin{cases} 15x + 6y = 36 \\ 11x = 22 \end{cases}$$

Solve the 2nd equation for **x.**

$$\begin{cases} 15x + 6y = 36 \\ 11x\,/11 = 22/11 \end{cases}$$

$$\begin{cases} 15x + 6y = 36 \\ x = 2 \end{cases}$$ Substitute $x = 2$ into the 1st equation and solve.

$$\begin{cases} 15{\cdot}2 + 6y = 36 \\ x = 2 \end{cases}$$

$$\begin{cases} 6y = 36 - 30 \\ x = 2 \end{cases}$$

$$\begin{cases} 6y/\,6 = 6/\,6 \\ x = 2 \end{cases}$$

$$\begin{cases} y = 1 \\ x = 2 \end{cases}$$ $x = 2,\ y = 1$ is candidate solution of the system.

Step 3. Check

$$\begin{cases} 5{\cdot}2 + 2{\cdot}1 = 12 \\ 2{\cdot}2 + 3{\cdot}1 = \ \ 7 \end{cases}$$ True

True

$$\begin{cases} x = 2 \\ y = 1 \end{cases}$$ **(2, 1)** is the unique solution of the original system.

Example 3 Solve the system by the algebraic elimination method.

$$\begin{cases} 4x - 4y = 8 \\ x - \ \ y = 1 \end{cases}$$

SOLUTION▶

Both equations of the system are in standard form. We have to multiply both equations of the given linear system by appropriate numbers so that the sum of the two equations will eliminate **y**.

$$\begin{cases} 4x - 4y = 8 \mid \cdot (1) \\ x - \ \ y = 1 \mid \cdot (-4) \end{cases}$$ **Step 2.** Multiply the first equation by **1**.

Multiply the second equation by $-\,$**4**.

$$\begin{cases} 4x - 4y = 8 \\ -4x + 4y = -4 \end{cases}$$ Add the 1st and the 2nd equations and write as the 2nd.

$$\begin{cases} 4x - 4y = 8 \\ 0 = 4 \end{cases}$$ False statement.

No solution

Example 4 Solve the system by the algebraic elimination method.

$$\begin{cases} 3x - 6y = 3 \\ x - 2y = 1 \end{cases}$$

SOLUTION▶

Both equations of the system are in standard form. We have to multiply both equations of the given linear system by appropriate numbers so that the sum of the two equations will eliminate **y**.

$$\begin{cases} 3x - 6y = 3 \quad | \cdot (1) \\ x - 2y = 1 \quad | \cdot (-3) \end{cases}$$ **Step 2.** Multiply the first equation by **1**.
Multiply the second equation by **− 3**.

$$\begin{cases} 3x - 6y = 3 \\ -3x + 6y = -3 \end{cases}$$ Add the 1st and the 2nd equations and write as the 2nd.

$$\begin{cases} 3x - 6y = 3 \\ 0 = 0 \end{cases}$$ True statement

⇩

infinitely many solutions or $x = k, \ y = \dfrac{k-1}{2}, \ k \in R$ or $\left(k, \dfrac{k-1}{2}\right), k \in R$

To solve systems with **3** linear equations and **3** variables we have to use the same steps:

Example 5 Solve the system by the algebraic elimination method.

$$\begin{cases} 2x + 3y + 4z = 20 \\ 5x - 2y - z = -2 \\ x + y + z = 6 \end{cases}$$

SOLUTION▶

We will add the second and the third equations to eliminate **z**.

$$\begin{cases} 2x + 3y + 4z = 20 \\ 5x - 2y - z = -2 \quad | \cdot 4 \\ x + y + z = 6 \end{cases}$$ **Step 2.** Add the 1st and 4 times the 2nd and write as the 1st.
Add the 2nd and 3rd equations and write as the 2nd.

$$\begin{cases} 22x - 5y = 12 \\ 6x - y = 4 \quad | \cdot (-5) \\ x + y + z = 6 \end{cases}$$ Add the 1st and -5 times the 2nd and write as the 1st.

$$\begin{cases} -8x = -8 \\ 6x - y = 4 \\ x + y + z = 6 \end{cases}$$ Solve the 1st for x.

$$\begin{cases} -8x \,/ -8 & = -8 \,/ -8 \\ 6x - y & = 4 \\ x + y + z = 6 \end{cases}$$

$$\begin{cases} x = 1 \\ 6x - y & = 4 \\ x + y + z = 6 \end{cases}$$

Substitute $x = 1$ into the **2nd** and solve for **y.**

$$\begin{cases} x = 1 \\ 6 \cdot 1 - y & = 4 \\ x + y + z = 6 \end{cases}$$

Substitute $x = 1$ into the **2nd** and solve for **y.**

$$\begin{cases} x = 1 \\ y = 2 \\ 1 + 2 + z = 6 \end{cases}$$

Substitute $x = 1$ and $y = 2$ into the **3rd** and solve for **z.**

$$\begin{cases} x = 1 \\ y = 2 \\ 3 + z = 6 \end{cases}$$

$$\begin{cases} x = 1 \\ y = 2 \\ z = 6 - 3 \end{cases}$$

$$\begin{cases} x = 1 \\ y = 2 \\ z = 3 \end{cases}$$

$(1, 2, 3)$ is the candidate solution of the given system.

Step 3. Check

$$\begin{cases} 2 \cdot 1 + 3 \cdot 2 + 4 \cdot 3 = 20 \\ 5 \cdot 1 - 2 \cdot 2 - 3 = -2 \\ 1 + 2 + 3 = 6 \end{cases}$$

$$\begin{cases} 20 = 20 \\ -2 = -2 \\ 6 = 6 \end{cases}$$

True
True
True

$$\begin{cases} x = 1 \\ y = 2 \\ z = 3 \end{cases}$$

$(1, 2, 3)$ is the unique solution of the original system.

■ ■ GRAPH METHOD

Solving a system of **2** linear equations with **2** variables by graphing is done as follows:

> **Step 1** ▶ **Graph each linear equation of the system on the same coordinate plane.**
>
> **Step 2** ▶ **Find the common points for those lines.**
> a). If, after graphing, the lines corresponding to the equations of system intersect each other at one point, then the given system has **one solution** and the pair of coordinates of the intersection point is the unique solution of the system. In this case the **system is consistent** and the **equations are independent**. See **Figure 6.1.1**(a). Sometimes we cannot determine the exact coordinates of the point that represents the solution, especially if those coordinates are not integers. In this case we will just approximate the solution.
> b). If, after graphing, both lines pass the same points of the plane (the graphs are of the same line), then the system has **infinitely many solutions**. In this case the **system is consistent** and the **equations are dependent**. See **Figure 6.1.1**(b).
> c). If, after graphing, both lines are parallel to each other, then the given system has **no solution**. In this case, the **system is inconsistent** and the equations **are independent**. See Figure **6.1.1**(c).
>
> 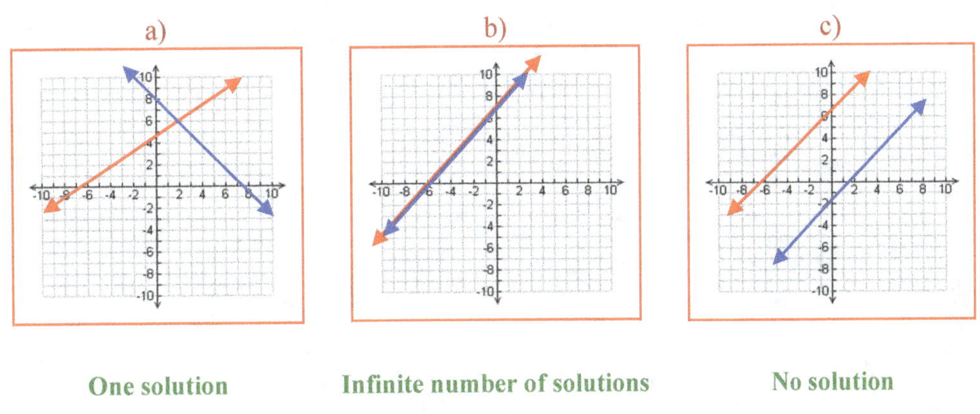
>
> One solution Infinite number of solutions No solution
>
> **Figure 6.1.1**
>
> **Step 3** ▶ **Check the solution in each of the initial equations** if graphing method gives a unique solution.

Example 1 Solve the system by graphing.

$$\begin{cases} 3x + y = 6 \\ y = 2x + 1 \end{cases}$$

SOLUTION ▶

Step 1. We will graph each line by slope-y-intercept method (but we can also graph these lines by points or by x-y-intercept methods).
To graph by the slope-y-intercept method, we have to solve the first equation for **y** and then graph both lines.

$$\begin{cases} 3x + y = 6 \\ y = 2x + 1 \end{cases}$$

$$\begin{cases} y = -3x + 6 \\ y = 2x + 1 \end{cases} \qquad \textbf{y-intercept is 6, slope is } -3.$$

Or we can rewrite the slopes of the lines in terms of rise over run.

$$\begin{cases} y = \dfrac{-3}{1}x + 6 \\ y = \dfrac{2}{1}x + 1 \end{cases}$$

Graphs of the lines are shown below.

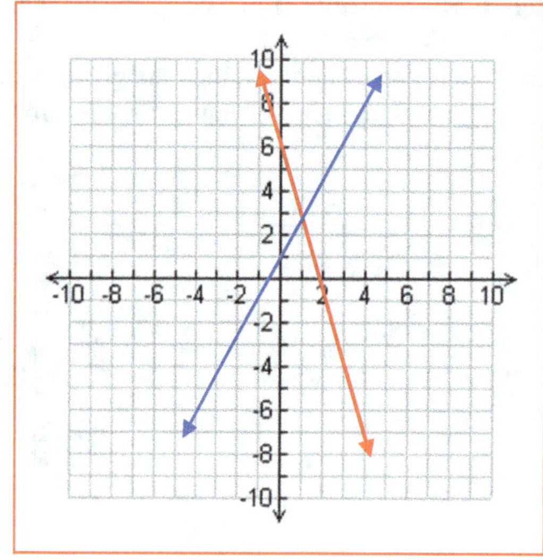

Figure 6.1.2

Step 2. Lines intersect at the point **(1, 3).** So the system has a possible unique solution **x = 1** and **y = 3** or **(1, 3).**

Step 3. Check

$$\begin{cases} 3 \cdot 1 + 3 = 6 \\ 3 = 2 \cdot 1 + 1 \end{cases}$$

$$\begin{cases} 6 = 6 \qquad\qquad \text{True} \\ 3 = 3 \qquad\qquad \text{True} \end{cases}$$

$$\Downarrow$$

$$\begin{cases} x = 1 \\ y = 3 \end{cases} \qquad \textbf{(1, 3)} \text{ is the unique solution of the original system.}$$

Example 2 Solve the system by graphing.

$$\begin{cases} 3x + 3y = 6 \\ x + y = 2 \end{cases}$$

SOLUTION ▶

Step 1. We will graph each line by x-y-intercept method.
To graph by x-y-intercept method we have to make x-y-intercepts chart for each equation of the system.

$$\begin{cases} 3x + 3y = 6 \\ \\ x + y = 2 \end{cases}$$

x	y
0	6/3 = 2
6/3 = 2	0

x	y
0	2
2	0

Both lines are the same line (graphs of the lines are shown below).

Infinite number of solutions or $x = k, \; y = 2 - k, \; k \in R$

or $(k, 2 - k), k \in R$

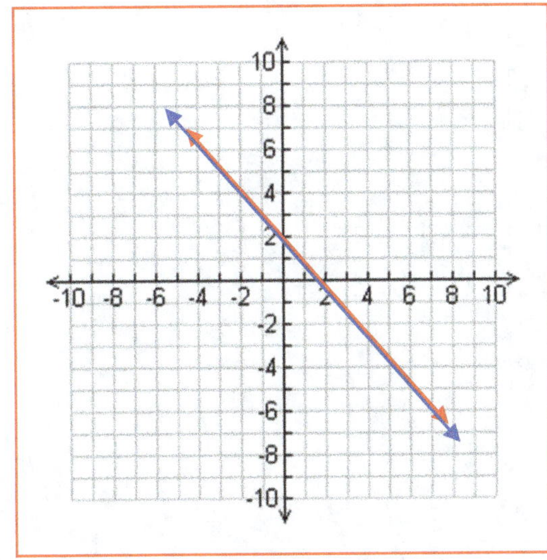

Figure 6.1.3

Example 3 Solve the system by graphing.

$$\begin{cases} x + 2y = 6 \\ \frac{1}{2}x + y = 1 \end{cases}$$

SOLUTION ▶

Step 1. We will graph each line by x-y-intercept method.
To graph by x-y-intercept method we have to make x-y-intercepts chart for each equation of the system.

$\begin{cases} x + 2y = 6 \\[2em] \frac{1}{2}x + y = 1 \end{cases}$

x	Y
0	6/2 = 3
6	0

x	y
0	1
2	0

Both lines are parallel lines (graphs of the lines are shown below).

No Solution or $x \in \emptyset, \ y \in \emptyset$

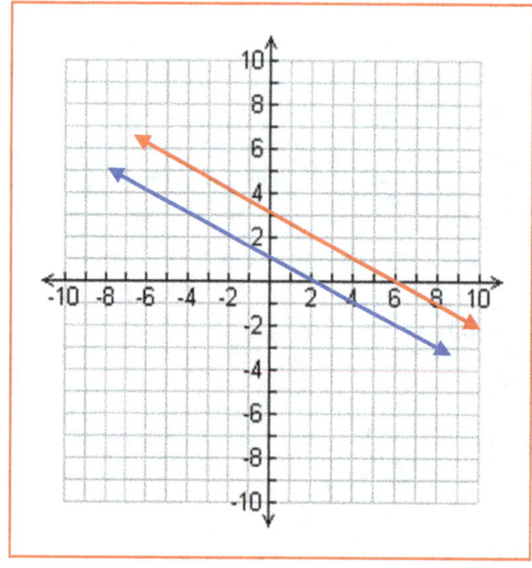

Figure 6.1.4

Example 4 Find the number of solutions for each system without graphing.

$$\begin{cases} 2x + 3y = 12 \\ x - y = 1 \end{cases} \qquad \begin{cases} x + \dfrac{y}{2} = 1 \\ 2x + y = 4 \end{cases} \qquad \begin{cases} 6x + y = 9 \\ 2x + \dfrac{y}{3} = 3 \end{cases}$$

SOLUTION ▶

We have to solve each equation of each system for variable **y** and then compare the slopes of the lines in each system.

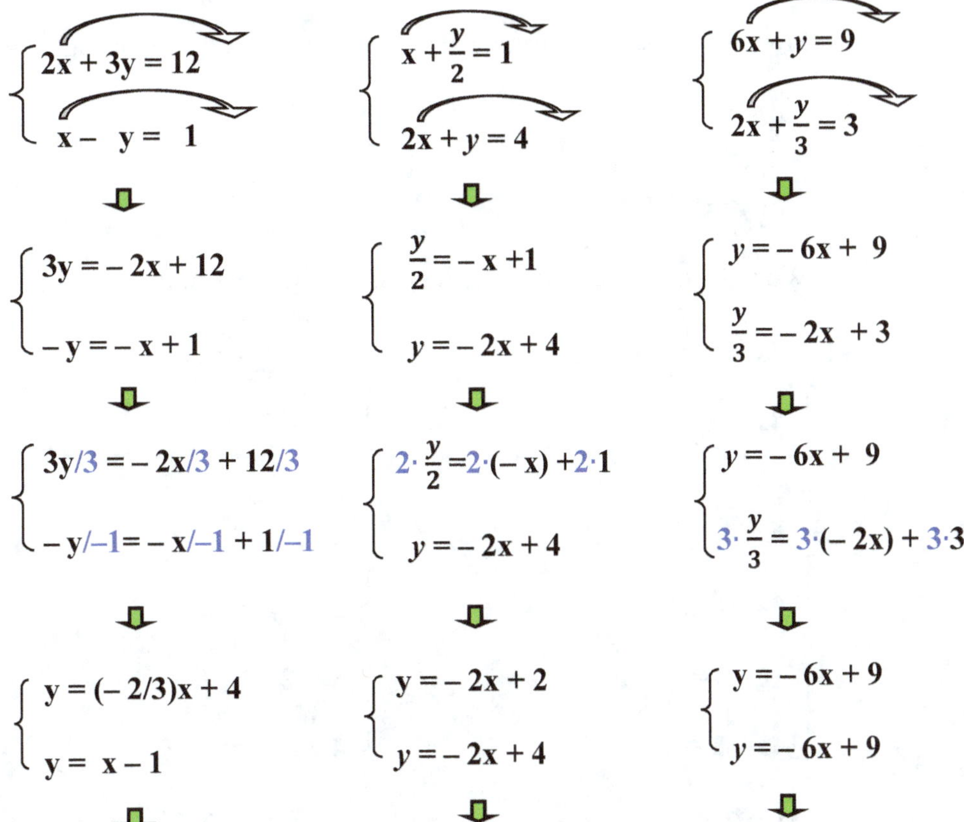

The system has **one solution** because slopes are different and lines are intersecting in one point.

The system has **no solution** because slopes are the same but y-intercepts are different numbers (lines are parallel lines)

The system has **an infinite number of solutions** because the slopes are the same and the y-intercepts are the same (lines are on top of each other)

Example 1 Find x and y.

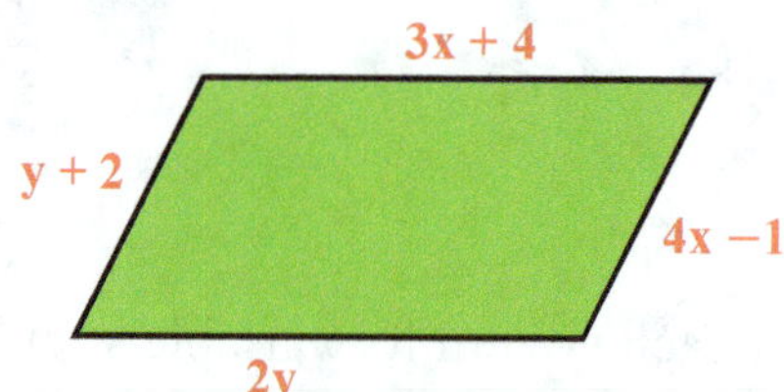

SOLUTION ▶

From the properties of parallelograms we can write the following system of equations:

$$\begin{cases} 3x + 4 = 2y \\ 4x - 1 = y + 2 \end{cases} \Rightarrow \begin{cases} 3x + 4 = 2y \\ 4x - 3 = y \end{cases} \Rightarrow$$

We will solve the system by the substitution method.

$$\begin{cases} 3x + 4 = 2(4x - 3) \\ 4x - 3 = y \end{cases} \Rightarrow \begin{cases} 3x + 4 = 8x - 6 \\ 4x - 3 = y \end{cases} \Rightarrow$$

$$\begin{cases} 6 + 4 = 8x - 3x \\ 4x - 3 = y \end{cases} \Rightarrow \begin{cases} 10 = 5x \\ 4x - 3 = y \end{cases} \Rightarrow$$

$$\begin{cases} 10 / 5 = 5x / 5 \\ 4x - 3 = y \end{cases} \Rightarrow \begin{cases} x = 2 \\ 4 \cdot 2 - 3 = y \end{cases} \Rightarrow$$

$$\begin{cases} x = 2 \\ y = 5 \end{cases} \Rightarrow \begin{cases} 3 \cdot 2 + 4 = 2 \cdot 5 \quad \text{true} \\ 4 \cdot 2 - 1 = 5 + 2 \quad \text{true} \end{cases}$$

$$\Downarrow$$

$$\begin{cases} x = 2 \\ y = 5 \end{cases}$$

$(2, 5)$ is the unique solution of the system

 John invests **$ 20,000** in two bank accounts. The first bank account earns **10%** annually, and the second bank account earns **5%** annually. John's total interest earned from both bank accounts in a year is **$1,800**. How much did John invest in each account?

SOLUTION▶

Let **x** be the amount invested at **10%** and **y** be the amount invested at **5%**. We can now write the following system of equations:

$$\begin{cases} x + y = 20{,}000 \\ 0.10x + 0.05y = 1{,}800 \end{cases}$$

We will multiply the second equation by **100** and then solve the new equivalent system of equations by the substitution method.

$$\begin{cases} x + y = 20{,}000 \\ 0.10x + 0.05y = 1{,}800 \mid \cdot 100 \end{cases} \Rightarrow \begin{cases} x + y = 20{,}000 \\ 10x + 5y = 180{,}000 \end{cases} \Rightarrow$$

$$\begin{cases} x = 20{,}000 - y \\ 10(20{,}000 - y) + 5y = 180{,}000 \end{cases} \Rightarrow \begin{cases} x = 20{,}000 - y \\ 200{,}000 - 10y + 5y = 180{,}000 \end{cases} \Rightarrow$$

$$\begin{cases} x = 20{,}000 - y \\ 200{,}000 - 5y = 180{,}000 \end{cases} \Rightarrow \begin{cases} x = 20{,}000 - y \\ 200{,}000 - 5y = 180{,}000 \end{cases} \Rightarrow$$

$$\begin{cases} x = 20{,}000 - y \\ 5y = 200{,}000 - 180{,}000 \end{cases} \Rightarrow \begin{cases} x = 20{,}000 - y \\ 5y = 20{,}000 \end{cases} \Rightarrow$$

$$\begin{cases} x = 20{,}000 - y \\ 5y/5 = 20{,}000/5 \end{cases} \Rightarrow \begin{cases} x = 20{,}000 - 4{,}000 \\ y = 4{,}000 \end{cases} \Rightarrow$$

$$\begin{cases} x = 16{,}000 \\ y = 4{,}000 \end{cases}$$

So, John invested **$16,000** in the first bank account (with **10%** income) and **$4,000** in the second bank account (with **5%** income).

Solve the systems of linear equations by any method.

1. $\begin{cases} x + y = 2 \\ x - y = 0 \end{cases}$

2. $\begin{cases} 2x - 2y = 4 \\ y = 3x - 4 \end{cases}$

3. $\begin{cases} x + y = 3 \\ x - y = 1 \end{cases}$

4. $\begin{cases} x - y = -1 \\ y = 5x - 3 \end{cases}$

5. $\begin{cases} x + y = 5 \\ x - y = 3 \end{cases}$

6. $\begin{cases} 4x - y = 7 \\ y = 4x - 3 \end{cases}$

7. $\begin{cases} \dfrac{x}{2} + \dfrac{y}{3} = 6 \\ \dfrac{2x}{3} - \dfrac{3y}{9} = 1 \end{cases}$

8. $\begin{cases} \dfrac{x}{6} + \dfrac{y}{4} = \dfrac{2}{3} \\ \dfrac{2x}{3} - \dfrac{y}{9} = \dfrac{4}{9} \end{cases}$

9. $\begin{cases} 2x - y = 5 \\ 4x = 2y + 10 \end{cases}$

10. $\begin{cases} x + y + z = 6 \\ 2x + 3y - z = 5 \\ 5x + y - 2z = 1 \end{cases}$

11. Solve the system by the graph method.

$\begin{cases} y = 3x + 1 \\ y = -2x + 6 \end{cases}$

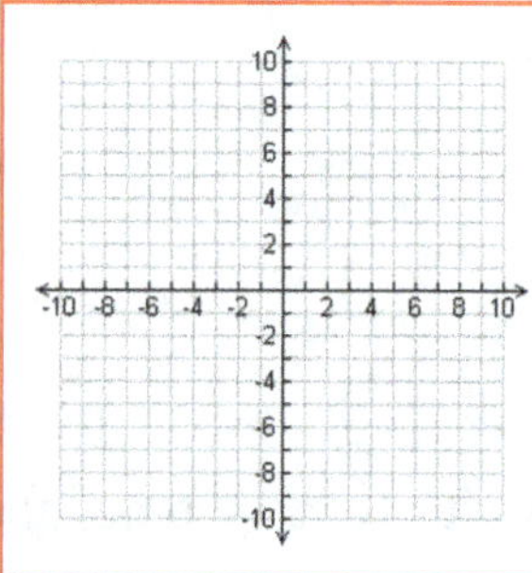

12. Write the corresponding system.

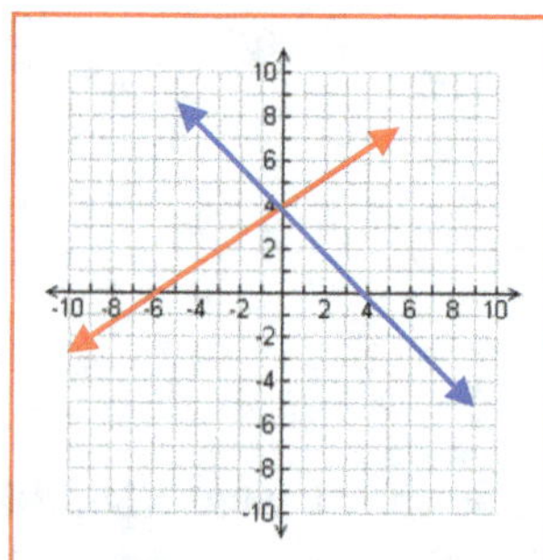

- **MATRICES OF SYSTEMS**
- **CRAMER'S RULE FOR 2 x 2 SYSTEMS**
- **CRAMER'S RULE FOR 3 x 3 SYSTEMS**

■ MATRICES OF SYSTEMS

For any system of **n** linear equations with **n** variables we can write a corresponding **augmented matrix** and **coefficient matrix**.

System of Linear Equations ▶

$$\begin{cases} A_{11}x_1 + A_{12}x_2 + A_{13}x_3 + \ldots + A_{1n}x_n = B_1 \\ A_{21}x_1 + A_{22}x_2 + A_{23}x_3 + \ldots + A_{2n}x_n = B_2 \\ A_{31}x_1 + A_{32}x_2 + A_{33}x_3 + \ldots + A_{3n}x_n = B_3 \\ \quad \vdots \\ A_{n1}x_1 + A_{n2}x_2 + A_{n3}x_3 + \ldots + A_{nn}x_n = B_n \end{cases}$$

Augmented Matrix of System ▶

$$\left[\begin{array}{ccccc|c} A_{11} & A_{12} & A_{13} & \ldots & A_{1n} & B_1 \\ A_{21} & A_{22} & A_{23} & \ldots & A_{2n} & B_2 \\ A_{31} & A_{32} & A_{33} & \ldots & A_{3n} & B_3 \\ \vdots & & \vdots & & \vdots & \vdots \\ A_{n1} & A_{n2} & A_{n3} & \ldots & A_{nn} & B_n \end{array}\right]$$

Coefficient Matrix of System ▶

$$\left[\begin{array}{ccccc} A_{11} & A_{12} & A_{13} & \ldots & A_{1n} \\ A_{21} & A_{22} & A_{23} & \ldots & A_{2n} \\ A_{31} & A_{32} & A_{33} & \ldots & A_{3n} \\ \vdots & & \vdots & & \vdots \\ A_{n1} & A_{n2} & A_{n3} & \ldots & A_{nn} \end{array}\right]$$

Example Write the **augmented matrix** and the **coefficient matrix** of the system.

$$\begin{cases} 5x + 2y = 12 \\ 2x + 3y = 7 \end{cases}$$

SOLUTION ▶ The augmented matrix is $\left[\begin{array}{cc|c} 5 & 2 & 12 \\ 2 & 3 & 7 \end{array}\right]$. The coefficient matrix is $\left[\begin{array}{cc} 5 & 2 \\ 2 & 3 \end{array}\right]$.

■ CRAMER'S RULE FOR 2x2 SYSTEMS

We can use determinants of systems to solve systems of **2** equations with **2** variables (or **2x2** systems) and systems of **3** equations with **3** variables (or **3x3** systems). The corresponding methods are called **Cramer's rule for 2 x 2 systems** and **Cramer's rule for 3 x 3 systems**.

Cramer's Rule 1 ▶ The solution of a system with **2** equations and **2** variables (**2x2 system**)

$$\begin{cases} a_1x + b_1y = c_1 \\ a_2x + b_2y = c_2 \end{cases} \qquad\qquad \textbf{(1)}$$

is

$$x = \frac{D_x}{D}, \quad y = \frac{D_y}{D} \quad \text{or} \quad \text{is an ordered pair of numbers} \ \left(\frac{D_x}{D}, \frac{D_y}{D}\right) \ (D \neq 0)$$

where

$$A = \begin{bmatrix} a_1 & b_1 & \bigm| c_1 \\ a_2 & b_2 & \bigm| c_2 \end{bmatrix} \qquad \text{is the } \textbf{augmented matrix} \text{ of system}$$

and

$$D = \begin{vmatrix} a_1 & b_1 \\ a_2 & b_2 \end{vmatrix} = a_1b_2 - a_2b_1 \quad \text{is the determinant of } \textbf{coefficient} \text{ matrix,}$$

$$D_x = \begin{vmatrix} c_1 & b_1 \\ c_2 & b_2 \end{vmatrix} = c_1b_2 - c_2b_1 \quad \text{is the determinant of } \textbf{x-replacing} \text{ matrix,}$$

$$D_y = \begin{vmatrix} a_1 & c_1 \\ a_2 & c_2 \end{vmatrix} = a_1c_2 - a_2c_1 \quad \text{is the determinant of } \textbf{y-replacing} \text{ matrix.}$$

- If $D \neq 0$, then the linear system has **one solution** (or **unique solution**).

- If $D = 0$ and at least one of the other determinants, D_x or D_y, is not **0**, then the linear system has **no solutions** (**inconsistent** system).

- If $D = 0$ and the determinants D_x and D_y are also 0 ($D = 0$, $D_x = 0$, and $D_y = 0$), then the linear system has **infinitely many solutions** (**dependent** or **coinciding** system).

Example 1 Use Cramer's rule to solve the system

$$\begin{cases} 5x + 2y = 12 \\ 2x + 3y = 7 \end{cases}$$

SOLUTION ▸ System augmented matrix is $\mathbf{A} = \begin{bmatrix} 5 & 2 & | & 12 \\ 2 & 3 & | & 7 \end{bmatrix}$.
The determinants $\mathbf{D}$, $\mathbf{D_x}$, and $\mathbf{D_y}$ are

$$\mathbf{D} = \begin{vmatrix} 5 & 2 \\ 2 & 3 \end{vmatrix} = 5 \cdot 3 - 2 \cdot 2 = 11$$

$$\mathbf{D_x} = \begin{vmatrix} 12 & 2 \\ 7 & 3 \end{vmatrix} = 12 \cdot 3 - 7 \cdot 2 = 22$$

$$\mathbf{D_y} = \begin{vmatrix} 5 & 12 \\ 2 & 7 \end{vmatrix} = 5 \cdot 7 - 2 \cdot 12 = 11$$

So, system has **exactly one solution** ($\mathbf{D} \neq \mathbf{0}$):

$$x = \frac{\mathbf{D_x}}{\mathbf{D}} = \frac{22}{11} = 2, \quad y = \frac{\mathbf{D_y}}{\mathbf{D}} = \frac{11}{11} = 1 \qquad \text{or} \qquad (x, y) = (2, 1).$$

Example 2 Use Cramer's rule to solve the system

$$\begin{cases} 3x - 2y = 0 \\ 2x + y = 7 \end{cases}$$

SOLUTION ▸ System matrix is $\mathbf{A} = \begin{bmatrix} 3 & -2 & | & 0 \\ 2 & 1 & | & 7 \end{bmatrix}$. The determinants $\mathbf{D}$, $\mathbf{D_x}$, and $\mathbf{D_y}$ are

$$\mathbf{D} = \begin{vmatrix} 3 & -2 \\ 2 & 1 \end{vmatrix} = 3 \cdot 1 - 2 \cdot (-2) = 7$$

$$\mathbf{D_x} = \begin{vmatrix} 0 & -2 \\ 7 & 1 \end{vmatrix} = 0 \cdot 1 - 7 \cdot (-2) = 14$$

$$\mathbf{D_y} = \begin{vmatrix} 3 & 0 \\ 2 & 7 \end{vmatrix} = 3 \cdot 7 - 2 \cdot 0 = 21$$

So, the system has **exactly one solution** ($\mathbf{D} \neq \mathbf{0}$):

$$x = \frac{\mathbf{D_x}}{\mathbf{D}} = \frac{14}{7} = 2, \quad y = \frac{\mathbf{D_y}}{\mathbf{D}} = \frac{21}{7} = 3 \qquad \text{or} \qquad (x, y) = (2, 3).$$

 Use Cramer's rule to solve the system

$$\begin{cases} x + 3y = 4 \\ 2x + 6y = 7 \end{cases}$$

SOLUTION ▶ The system matrix is $A = \begin{bmatrix} 1 & 3 & | & 4 \\ 2 & 6 & | & 7 \end{bmatrix}$. The determinants D, D_x, and D_y are

$$D = \begin{vmatrix} 1 & 3 \\ 2 & 6 \end{vmatrix} = 1 \cdot 6 - 2 \cdot 3 = 0$$

$$D_x = \begin{vmatrix} 4 & 3 \\ 7 & 6 \end{vmatrix} = 4 \cdot 6 - 7 \cdot 3 = 3 \neq 0$$

$$D_y = \begin{vmatrix} 1 & 4 \\ 2 & 7 \end{vmatrix} = 1 \cdot 7 - 2 \cdot 4 = -1 \neq 0$$

So, system has **no solutions** ($D = 0$, $D_x \neq 0$, $D_y \neq 0$).

Or $x \in \emptyset, y \in \emptyset,$

 Use Cramer's rule to solve the system

$$\begin{cases} x - 3y = 1 \\ 3x - 9y = 3 \end{cases}$$

SOLUTION ▶ The system matrix is $A = \begin{bmatrix} 1 & -3 & | & 1 \\ 3 & -9 & | & 3 \end{bmatrix}$. The determinants D, D_x, and D_y are

$$D = \begin{vmatrix} 1 & -3 \\ 3 & -9 \end{vmatrix} = 1 \cdot (-9) - 3 \cdot (-3) = 0$$

$$D_x = \begin{vmatrix} 1 & -3 \\ 3 & -9 \end{vmatrix} = 1 \cdot (-9) - 3 \cdot (-3) = 0$$

$$D_y = \begin{vmatrix} 1 & 1 \\ 3 & 3 \end{vmatrix} = 1 \cdot 3 - 3 \cdot 1 = 0$$

So, the system has **infinitely many solution** ($D = 0$, $D_x = 0$, $D_y = 0$):

$$x = k, \quad y = (k-1)/3 \quad \text{or} \quad (x, y) = \left(k, \frac{k-1}{3} \right), \quad k \in R$$

■ CRAMER'S RULE FOR 3x3 SYSTEMS

In this subsection we will use **Cramer's rule for 3 x 3 systems** to solve the systems of **3** equations with **3** variables (or **3 x 3** systems).

Cramer's Rule 2 ▶ The solution of a system with **3** equations and **3** variables (**3x3 system**)

$$\begin{cases} a_1x + b_1y + c_1z = d_1 \\ a_2x + b_2y + c_2z = d_2 \\ a_3x + b_3y + c_3z = d_3 \end{cases} \tag{1}$$

is

$$x = \frac{D_x}{D}, \quad y = \frac{D_y}{D}, \quad z = \frac{D_z}{D} \quad \text{or is an ordered set of numbers} \quad \left(\frac{D_x}{D}, \frac{D_y}{D}, \frac{D_z}{D}\right) \quad (D \neq 0)$$

where

$$A = \begin{bmatrix} a_1 & b_1 & c_1 & d_1 \\ a_2 & b_2 & c_2 & d_2 \\ a_3 & b_3 & c_3 & d_3 \end{bmatrix} \quad \text{is the } \textbf{augmented matrix} \text{ of system}$$

and

$$D = \begin{vmatrix} a_1 & b_1 & c_1 \\ a_2 & b_2 & c_2 \\ a_3 & b_3 & c_3 \end{vmatrix} \quad \text{is the determinant of the } \textbf{coefficient} \text{ matrix,}$$

$$D_x = \begin{vmatrix} d_1 & b_1 & c_1 \\ d_2 & b_2 & c_2 \\ d_3 & b_3 & c_3 \end{vmatrix} \quad \text{is the determinant of } \textbf{x column replaced} \text{ matrix,}$$

$$D_y = \begin{vmatrix} a_1 & d_1 & c_1 \\ a_2 & d_2 & c_2 \\ a_3 & d_3 & c_3 \end{vmatrix} \quad \text{is the determinant of } \textbf{y column replaced} \text{ matrix,}$$

$$D_z = \begin{vmatrix} a_1 & b_1 & d_1 \\ a_2 & b_2 & d_2 \\ a_3 & b_3 & d_3 \end{vmatrix} \quad \text{is the determinant of } \textbf{z column replaced} \text{ matrix,}$$

- If $D \neq 0$, then the linear system has **one solution** (or **unique solution**).
- If $D = 0$ and at least one of the other determinants, D_x, D_y, or D_z, is not 0, then the linear system has **no solution** (**inconsistent** system).
- If $D = 0$ and the other determinants D_x, D_y, and D_z are all 0 ($D = 0$, $D_x = 0$, $D_y = 0$ and $D_z = 0$), then the linear system has **infinitely many solutions** (**dependent** or **coinciding** system).

Example 1 Use Cramer's rule to solve the system

$$\begin{cases} x + 2y + z = 8 \\ 3x + y + 2z = 11 \\ -x + y - z = -2 \end{cases}$$

SOLUTION ▶ The system augmented matrix is $A = \begin{bmatrix} 1 & 2 & 1 & \vline & 8 \\ 3 & 1 & 2 & \vline & 11 \\ -1 & 1 & -1 & \vline & -2 \end{bmatrix}$.

The determinants D, D_x, D_y, and D_z are

$$D = \begin{vmatrix} 1 & 2 & 1 \\ 3 & 1 & 2 \\ -1 & 1 & -1 \end{vmatrix} \begin{matrix} 1 & 2 \\ 3 & 1 \\ -1 & 1 \end{matrix} = [1 \cdot 1 \cdot (-1) + 2 \cdot 2 \cdot (-1) + 1 \cdot 3 \cdot 1] -$$

$$- [(-1) \cdot 1 \cdot 1 + 1 \cdot 2 \cdot 1 + (-1) \cdot 3 \cdot 2] = 3$$

$$D_x = \begin{vmatrix} 8 & 2 & 1 \\ 11 & 1 & 2 \\ -2 & 1 & -1 \end{vmatrix} \begin{matrix} 8 & 2 \\ 11 & 1 \\ -2 & 1 \end{matrix} = [8 \cdot 1 \cdot (-1) + 2 \cdot 2 \cdot (-2) + 1 \cdot 11 \cdot 1] -$$

$$- [(-2) \cdot 1 \cdot 1 + 1 \cdot 2 \cdot 8 + (-1) \cdot 11 \cdot 2] = 3$$

$$D_y = \begin{vmatrix} 1 & 8 & 1 \\ 3 & 11 & 2 \\ -1 & -2 & -1 \end{vmatrix} \begin{matrix} 1 & 8 \\ 3 & 11 \\ -1 & -2 \end{matrix} = [1 \cdot 11 \cdot (-1) + 8 \cdot 2 \cdot (-1) + 1 \cdot 3 \cdot (-2)]$$

$$- [(-1) \cdot 11 \cdot 1 + (-2) \cdot 2 \cdot 1 + (-1) \cdot 3 \cdot 8] = 6$$

$$D_z = \begin{vmatrix} 1 & 2 & 8 \\ 3 & 1 & 11 \\ -1 & 1 & -2 \end{vmatrix} \begin{matrix} 1 & 2 \\ 3 & 1 \\ -1 & 1 \end{matrix} = [1 \cdot 1 \cdot (-2) + 2 \cdot 11 \cdot (-1) + 8 \cdot 3 \cdot 1] -$$

$$- [(-1) \cdot 1 \cdot 8 + 1 \cdot 11 \cdot 1 + (-2) \cdot 3 \cdot 2] = 9$$

So, the system has **exactly one solution** $(D \neq 0)$:

$$x = \frac{D_x}{D} = \frac{3}{3} = 1, \ y = \frac{D_y}{D} = \frac{6}{3} = 2, \ z = \frac{D_z}{D} = \frac{9}{3} = 3 \quad \text{or} \quad (x, y, z) = (1, 2, 3).$$

 Use Cramer's rule to solve the system

$$\begin{cases} x + y + z = 3 \\ x + y - z = 1 \\ 2x + 2y + 2z = 5 \end{cases}$$

SOLUTION ▶ The system matrix is $A = \begin{bmatrix} 1 & 1 & 1 & \bigm| & 3 \\ 1 & 1 & -1 & \bigm| & 1 \\ 2 & 2 & 2 & \bigm| & 5 \end{bmatrix}$.

The determinants D, D_x, D_y, and D_z are

$$D = \begin{vmatrix} 1 & 1 & 1 \\ 1 & 1 & -1 \\ 2 & 2 & 2 \end{vmatrix} \begin{matrix} 1 & 1 \\ 1 & 1 \\ 2 & 2 \end{matrix} = [1 \cdot 1 \cdot 2 + 1 \cdot (-1) \cdot 2 + 1 \cdot 1 \cdot 2] -$$

$$- [2 \cdot 1 \cdot 1 + 2 \cdot (-1) \cdot 1 + 2 \cdot 1 \cdot 1] = 0$$

$$D_x = \begin{vmatrix} 3 & 1 & 1 \\ 1 & 1 & -1 \\ 5 & 2 & 2 \end{vmatrix} \begin{matrix} 3 & 1 \\ 1 & 1 \\ 5 & 2 \end{matrix} = [3 \cdot 1 \cdot 2 + 1 \cdot (-1) \cdot 5 + 1 \cdot 1 \cdot 2] -$$

$$- [5 \cdot 1 \cdot 1 + 2 \cdot (-1) \cdot 3 + 2 \cdot 1 \cdot 1] = 2$$

$$D_y = \begin{vmatrix} 1 & 3 & 1 \\ 1 & 1 & -1 \\ 2 & 5 & 2 \end{vmatrix} \begin{matrix} 1 & 3 \\ 1 & 1 \\ 2 & 5 \end{matrix} = [1 \cdot 1 \cdot 2 + 3 \cdot (-1) \cdot 2 + 1 \cdot 1 \cdot 5]$$

$$- [2 \cdot 1 \cdot 1 + 5 \cdot (-1) \cdot 1 + 2 \cdot 1 \cdot 3] = -2$$

$$D_z = \begin{vmatrix} 1 & 1 & 3 \\ 1 & 1 & 1 \\ 2 & 2 & 5 \end{vmatrix} \begin{matrix} 1 & 1 \\ 1 & 1 \\ 2 & 2 \end{matrix} = [1 \cdot 1 \cdot 5 + 1 \cdot 1 \cdot 2 + 3 \cdot 1 \cdot 2] -$$

$$- [2 \cdot 1 \cdot 3 + 2 \cdot 1 \cdot 1 + 5 \cdot 1 \cdot 1] = 0$$

The system of equations has **no solutions** because $D = 0$ and at least one of componental determinants is not zero ($D_x \neq 0$ and $D_y \neq 0$).

 Use Cramer's rule to solve the system

$$\begin{cases} x + y \quad\;\; = 3 \\ x \quad\;\; - z = 1 \\ \quad y + z = 2 \end{cases}$$

SOLUTION ▶ The system matrix is $A = \begin{bmatrix} 1 & 1 & 0 & \;3 \\ 1 & 0 & -1 & \;1 \\ 0 & 1 & 1 & \;2 \end{bmatrix}$.

The determinants $D, D_x, D_y,$ and D_z are

$$D = \begin{vmatrix} 1 & 1 & 0 \\ 1 & 0 & -1 \\ 0 & 1 & 1 \end{vmatrix} \begin{matrix} 1 & 1 \\ 1 & 0 \\ 0 & 1 \end{matrix} = [1 \cdot 0 \cdot 1 + 1 \cdot (-1) \cdot 0 + 0 \cdot 1 \cdot 1] -$$

$$- [0 \cdot 0 \cdot 0 + 1 \cdot (-1) \cdot 1 + 1 \cdot 1 \cdot 1] = 0$$

$$D_x = \begin{vmatrix} 3 & 1 & 0 \\ 1 & 0 & -1 \\ 2 & 1 & 1 \end{vmatrix} \begin{matrix} 3 & 1 \\ 1 & 0 \\ 2 & 1 \end{matrix} = [3 \cdot 0 \cdot 1 + 1 \cdot (-1) \cdot 2 + 0 \cdot 1 \cdot 1] -$$

$$- [2 \cdot 0 \cdot 0 + 1 \cdot (-1) \cdot 3 + 1 \cdot 1 \cdot 1] = 0$$

$$D_y = \begin{vmatrix} 1 & 3 & 0 \\ 1 & 1 & -1 \\ 0 & 2 & 1 \end{vmatrix} \begin{matrix} 1 & 3 \\ 1 & 1 \\ 0 & 2 \end{matrix} = [1 \cdot 1 \cdot 1 + 3 \cdot (-1) \cdot 0 + 0 \cdot 1 \cdot 2]$$

$$- [0 \cdot 1 \cdot 0 + 2 \cdot (-1) \cdot 1 + 1 \cdot 1 \cdot 3] = 0$$

$$D_z = \begin{vmatrix} 1 & 1 & 3 \\ 1 & 0 & 1 \\ 0 & 1 & 2 \end{vmatrix} \begin{matrix} 1 & 1 \\ 1 & 0 \\ 0 & 1 \end{matrix} = [1 \cdot 0 \cdot 2 + 1 \cdot 1 \cdot 0 + 3 \cdot 1 \cdot 1] -$$

$$- [0 \cdot 0 \cdot 3 + 1 \cdot 1 \cdot 1 + 2 \cdot 1 \cdot 1] = 0$$

The system has **infinitely many solutions** $(D = 0, D_x = 0, D_y = 0, D_z = 0)$:

$x = k, \; y = 3 - k, \; z = k - 1$ or $(x, y, z) = (k, \; 3 - k, k - 1), \; k \in R$

Solve each of the following systems using **Cramer's rules**

1. $\begin{cases} x + y = 2 \\ 2x + y = 3 \end{cases}$

2. $\begin{cases} x - y = 1 \\ 2x - 4y = 0 \end{cases}$

3. $\begin{cases} x + 2y = 4 \\ 2x + y = 4 \end{cases}$

4. $\begin{cases} x - y = 2 \\ 2x + y = 1 \end{cases}$

5. $\begin{cases} 2x + y = 2 \\ x + 3y = 1 \end{cases}$

6. $\begin{cases} x - 2y = -2 \\ y - x = 0 \end{cases}$

7. $\begin{cases} x + y = 4 \\ x + 2y = 7 \end{cases}$

8. $\begin{cases} 2x + y = 5 \\ x - y = 1 \end{cases}$

9. $\begin{cases} x + 2y = 3 \\ 3x + y = 4 \end{cases}$

10. $\begin{cases} 2x - 3y = 3 \\ 6y - 4x = 6 \end{cases}$

11. $\begin{cases} \frac{1}{2}x + \frac{1}{3}y = \frac{3}{2} \\ \frac{1}{3}x + \frac{1}{2}y = \frac{11}{6} \end{cases}$

12. $\begin{cases} y = \frac{3}{4}x - 1 \\ -3x + 4y = -4 \end{cases}$

13. $\begin{cases} x - y = 1 \\ y + z = 0 \\ x + z = -1 \end{cases}$

14. $\begin{cases} x + 2y = 3 \\ 4y + 2x = 8 \\ x + z = 1 \end{cases}$

15. $\begin{cases} x + 2y + 3z = 6 \\ 2x + 4y - z = 5 \\ 3x - 2y + z = 2 \end{cases}$

16. $\begin{cases} \frac{1}{6}x + \frac{1}{4}y - \frac{1}{8}z = 1 \\ \frac{1}{3}x - \frac{1}{2}y - \frac{2}{3}z = 3 \\ \frac{1}{3}x + \frac{1}{2}y - \frac{1}{4}z = 2 \end{cases}$

6.3. Systems of Linear Inequalities

- ■ **DEFINITIONS**
- ■ **GRAPHING LINEAR INEQUALITIES ON PLANE**
- ■ **SYSTEMS OF LINEAR INEQUALITIES**

■ DEFINITIONS

We use the following definition of a linear inequality with two variables.

IMPORTANT DEFINITION

Definition ▶ A **linear inequality with two variables x** and **y** is an inequality that can be written as:

$$y > mx + b;\ y < mx + b;\ y \geq mx + b;\ \text{or } y \leq mx + b,\ \text{where } m, b \in \mathbb{R} \text{ or as}$$

$$ax + by > c;\ ax + by < c;\ ax + by \geq c;\ \text{or } ax + by \leq c,\ \text{where } a, b, c \in \mathbb{R}$$

Examples The inequalities $y > 2x + 3$, $y < -x + 2$, $y \geq \frac{1}{2}x + 3$, $y \leq 0.3x - 2$, $x + y \leq 4$, $3x - y > -5$, $x > 5$, and $y \leq 1$ are **linear inequalities with two variables**.

The set of solutions of **a linear inequality with two variables** is a set of ordered pairs **(x, y)** that satisfies the inequality. The graph of each set of solutions is the part of the **xy−plane** over/under or on the right/left side of the **corresponding line** (or **associated line**). Each set of solutions has one **dashed** (if inequality sign is > or <) or one **solid** (if inequality sign is ≥ or ≤) **associated line** as its **boundary**. The graph of a linear inequality with two variables is the graph of the set of solutions.

Sets of solutions of the inequalities $y < m_1x + b_1$, $y > m_2x + b_2$, $y \geq m_3x + b_3$, and $y \leq m_4x + b_4$ are the parts of the **xy−plane** under or over the *slanted* associated lines $y = m_1x + b_1$, $y = m_2x + b$, $y = m_3x + b_3$, and $y = m_4x + b_4$ (see figure below).

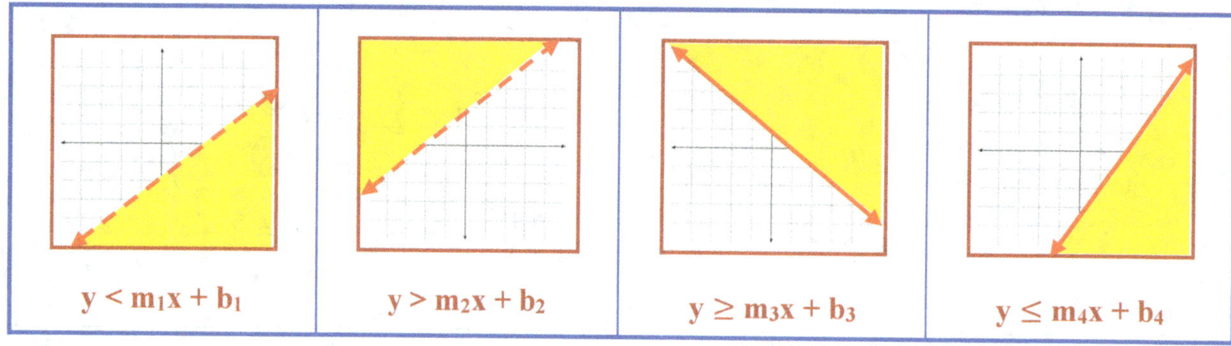

■ GRAPHING LINEAR INEQUALITIES ON A PLANE

Sets of solutions of the inequalities $y < b_1$, $y > b_2$, $y \geq b_3$, and $y \leq b_4$ are the parts of the **xy−plane** under or over the *horizontal* lines $y = b_1$, $y = b_2$, $y = b_3$, and $y = b_4$ (see figure below).

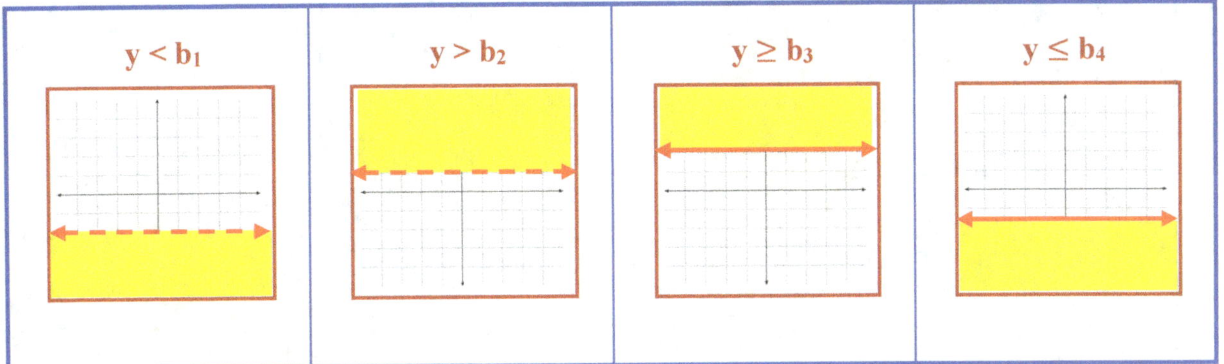

Sets of solutions of the inequalities $x < b_1$, $x > b_2$, $x \geq b_3$, and $x \leq b_4$ are the parts of **xy−plane** on the left or on the right sides of the *vertical* lines $x = b_1$, $x = b_2$, $x = b_3$, and $x = b_4$ (see figure).

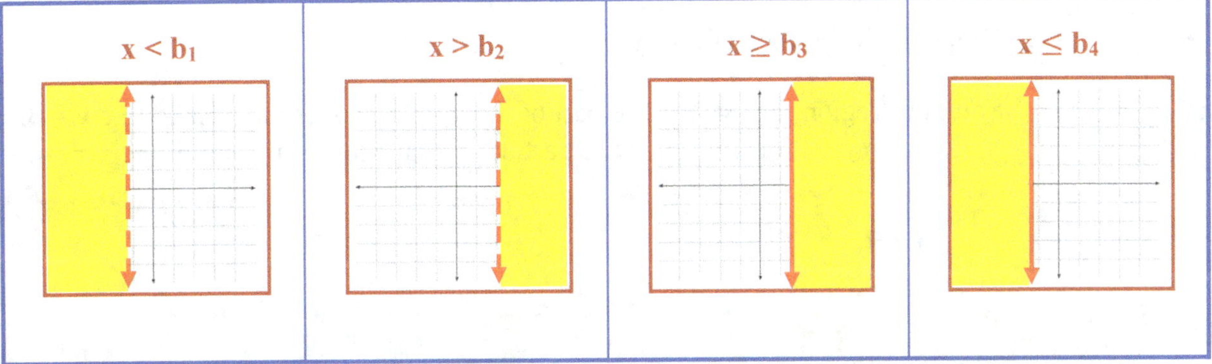

Example 1 Graph the linear inequalities $x + y > 3$ and $y \leq \frac{1}{2}x - 5$ on the xy-plane.

SOLUTION▶ We will rewrite the standard inequality in slope-y-intercept form: $x + y > 3 \rightarrow$
$y > -x + 3$. The inequality has a > sign, so the **associated line is dashed,** and
the solution set is part of the xy-plane **over** that line. This dashed associated line
is a boundary line for the **solution's region** on the **xy-plane**.

The second inequality $y \leq \frac{1}{2}x - 5$ is in slope-y-intercept form. The inequality
has a $\leq$ sign , so the **associated line is solid,** and the solution set is the part of
the xy-plane **under** that **solid** line. This solid associated line is a boundary line
for the **solution's region**.

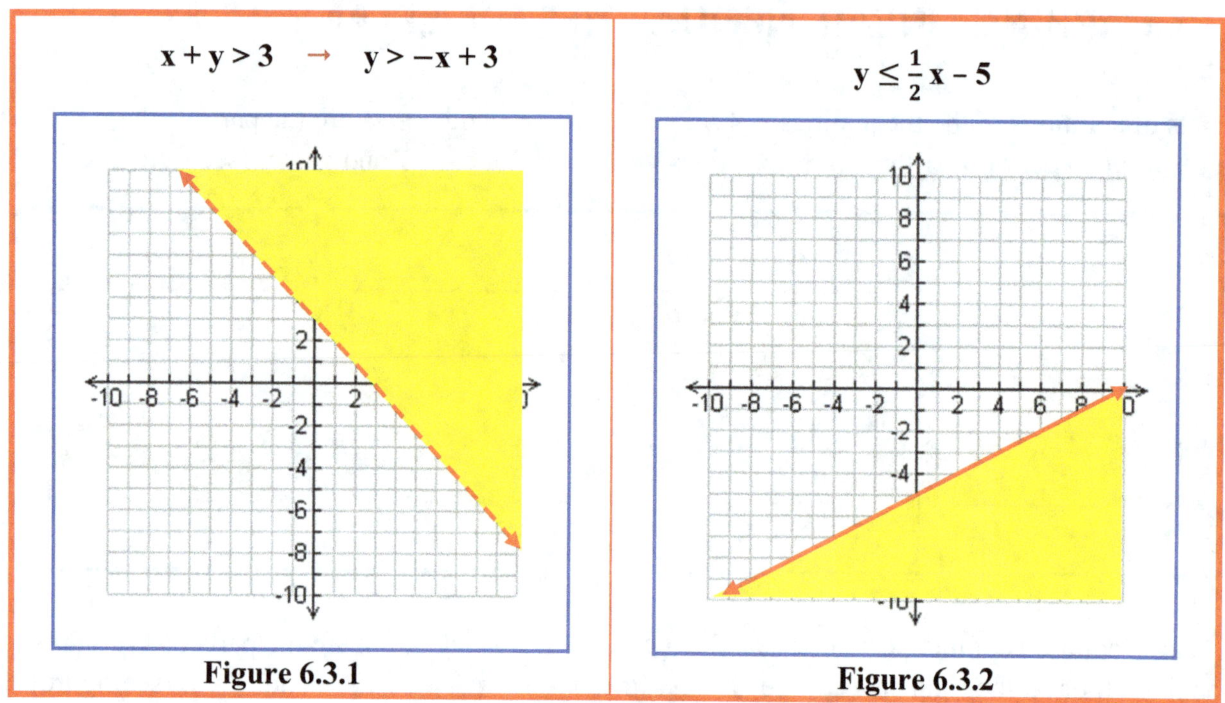

Figure 6.3.1

Figure 6.3.2

 Example 2   Graph the linear inequalities **y > 4** and **x ≤ − 6**.

SOLUTION ▶ Solutions region for **y > 4** is the part of the xy-plane over the dashed line **y = 4**. Solutions region for **x ≤ −6** is the part of xy-plane left of the solid line **x = −6**.

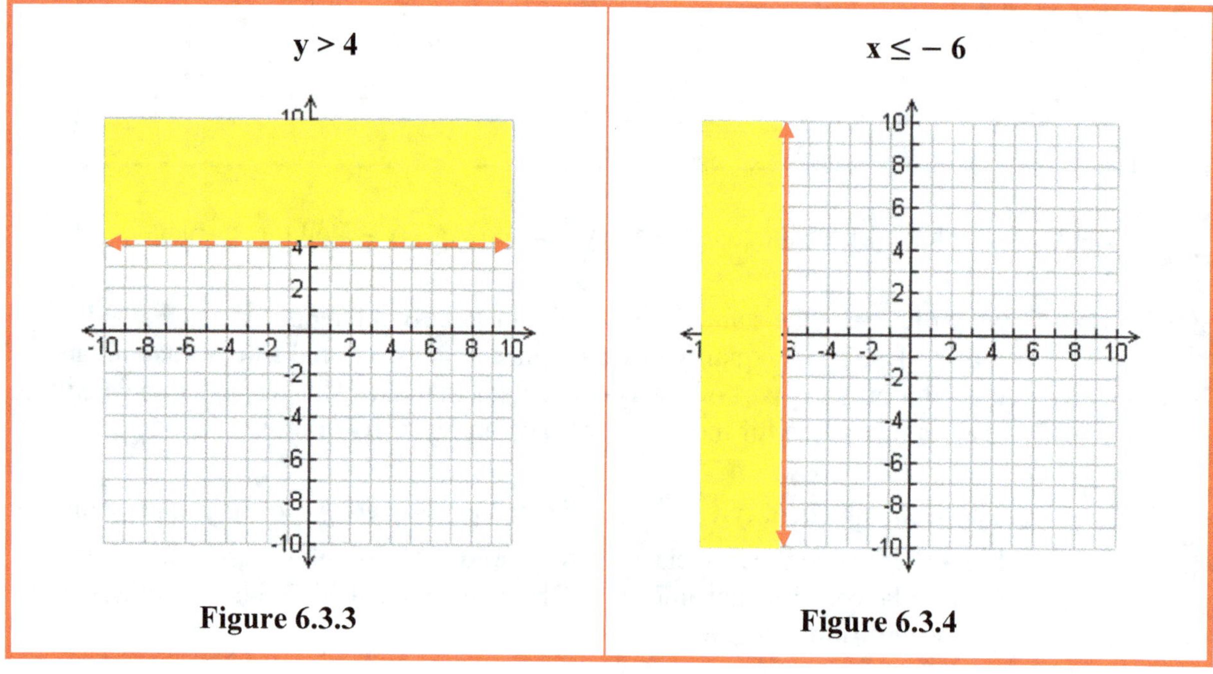

Figure 6.3.3

Figure 6.3.4

$\mathbf{B}$elow we will sketch the graphs of solution sets of some other linear inequalities.

Examples G1 Sketch the graphs of linear inequalities $y > 2x + 3$; $y < -x - 2$; $x \geq 3$; $y \leq -2$.

SOLUTION ▶ Each set of solutions is the part of xy-plane with **associated line** as the **boundary**

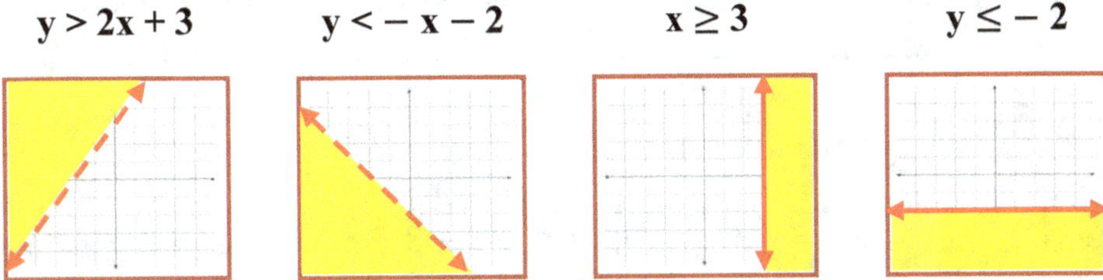

Figure 6.3.5

Examples G2 Sketch the graphs of the inequalities $x + y > 1$; $4x + 2y < -6$; $-x \leq -2$; $y \geq 3$.

SOLUTION ▶ Each set of solutions is the part of xy-plane with **associated line** as the **boundary**

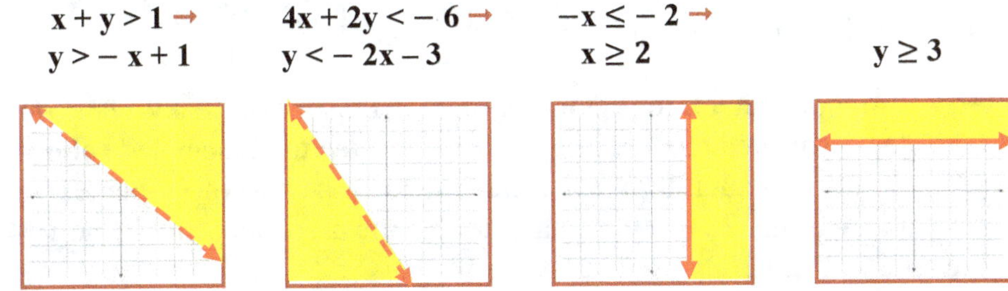

Figure 6.3.6

Examples G3 Sketch the graphs of linear inequalities $x - y > 0$; $y < -\dfrac{3}{2}$; $x \geq 0$; $y \leq -\dfrac{1}{3}x$.

SOLUTION ▶ Each set of solutions is the part of xy-plane with **associated line** as the **boundary**

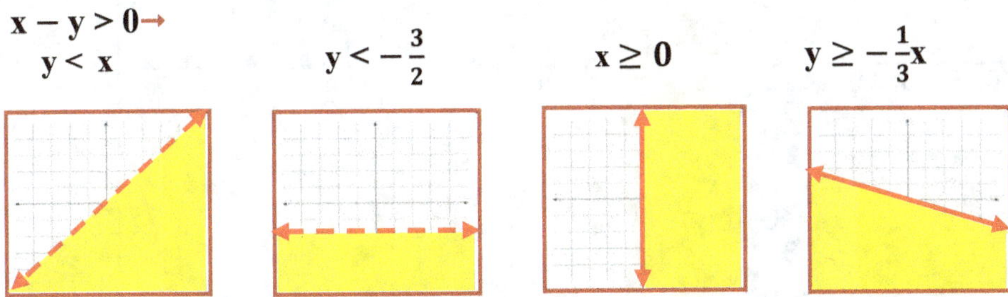

Figure 6.3.7

■ SYSTEMS OF LINEAR INEQUALITIES

We use the following definition of system of linear inequalities.

Definition ▶ A **system of linear inequalities** is a system with two or more linear inequalities.

Examples

The systems $\begin{cases} x > 2 \\ y \leq 3 \end{cases}$; $\begin{cases} x + 2y > 4 \\ y \leq 3x - 2 \end{cases}$; $\begin{cases} y > -3 \\ y < -x + 3 \\ x - y \geq -3 \end{cases}$ are **systems of linear inequalities.**

The set of solutions of **a system of linear inequalities** is the set of ordered pairs **(x, y)** that satisfies all inequalities of the system. The graph of each set of solutions is the part of the **xy−plane** over/under or on the right/left side, or between the **associated lines**.

Example 1

Graph the sets of solutions of the systems $\begin{cases} x > 2 \\ y \leq 3 \end{cases}$ and $\begin{cases} 2x + 2y > 4 \\ y \leq 3x - 2 \end{cases}$.

SOLUTION ▶ Solutions region for the first system is the part of the xy-plane on the **right** of the associated **dashed** line **x = 2** and **under** the associated **solid** line **y = 3.**
Solutions region for the second system is the part of xy-plane **over** the associated **dashed** line **2x + 2y = 4** and **under** the associated **solid** line $y \leq 3x - 2$

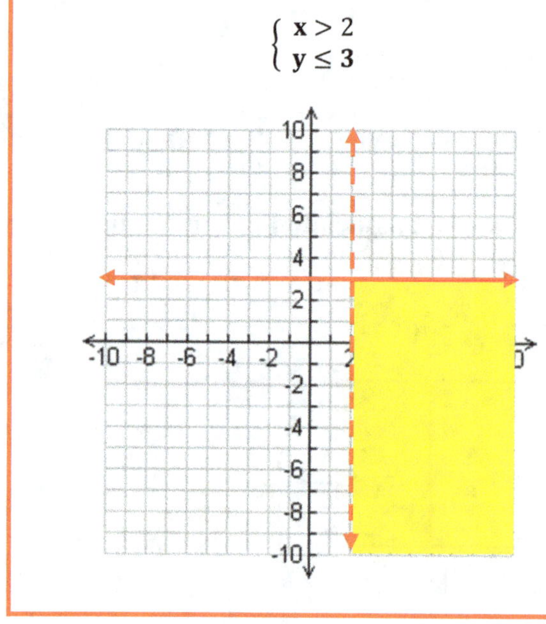

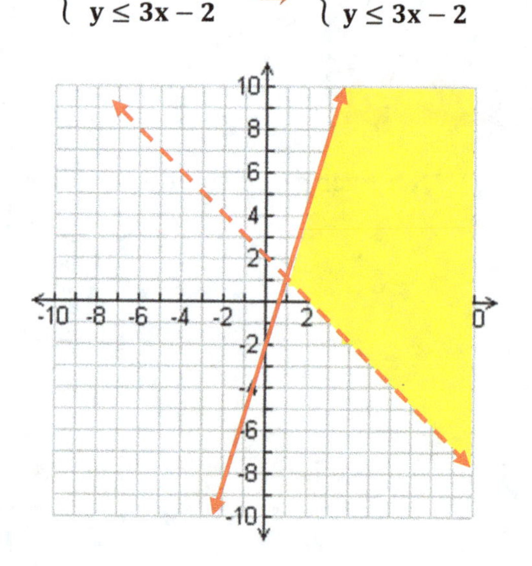

Graph the sets of solutions of the systems $\begin{cases} x > 1 \\ y \le 8 \\ y > 2 \\ x \le 6 \end{cases}$ and $\begin{cases} y > -3 \\ y < -x + 3 \\ x - y \ge -3 \end{cases}$.

SOLUTION ▶ Solutions region for the first system is the part of the xy-plane on the **right** side of the associated **dashed** line **x = 1,** on the **left** side of **solid** line **x = 6,** **under** the **solid** line **y = 8,** and **over** the associated **dashed** line **y = 2.**
Solutions region for the second system is the part of the xy-plane **over** the associated **dashed** line **y = −3,** **under** the associated **dashed** line **y = x + 3,** and **under** the **solid** line **x − y ≥ −3 (y ≤ x + 3).**

$\begin{cases} x > 1 \\ y \le 8 \\ y > 2 \\ x \le 6 \end{cases}$

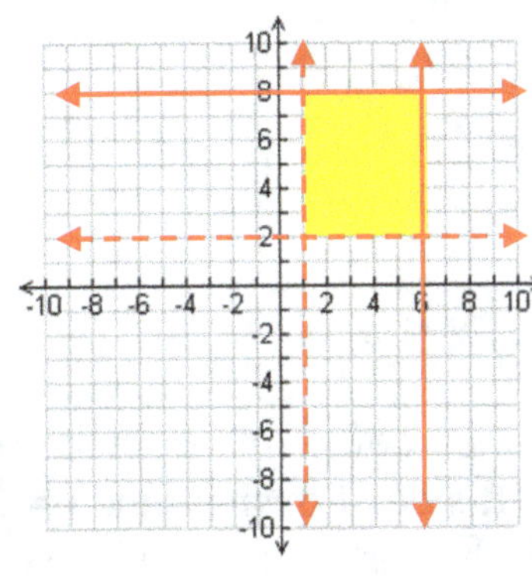

$\begin{cases} y > -3 \\ y < -x + 3 \\ x - y \ge -3 \end{cases} \rightarrow \begin{cases} y > -3 \\ y < -x + 3 \\ y \le x + 3 \end{cases}$

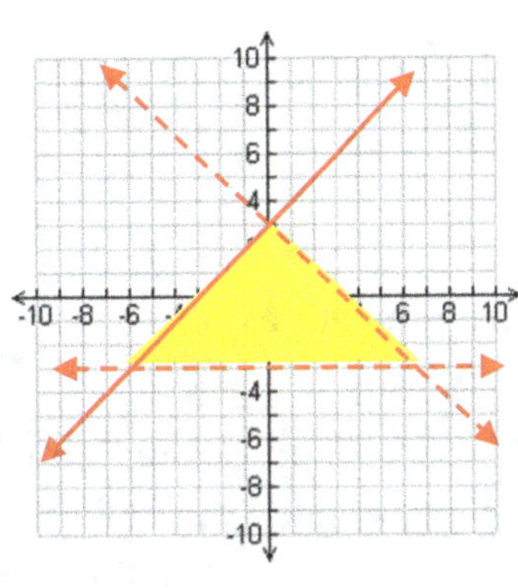

Example 3   Write the systems of inequalities corresponding to the graphs below.

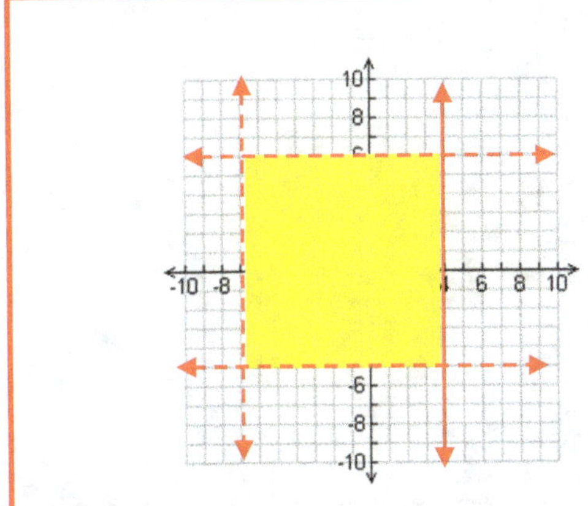

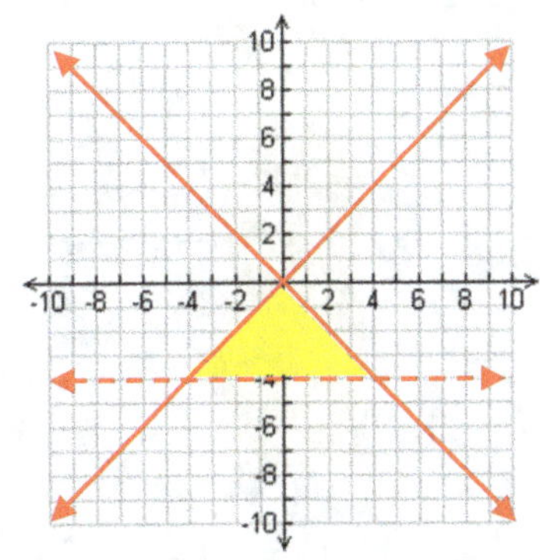

SOLUTION ▶ The systems of inequalities corresponding to the first and second systems of graphs are the following:

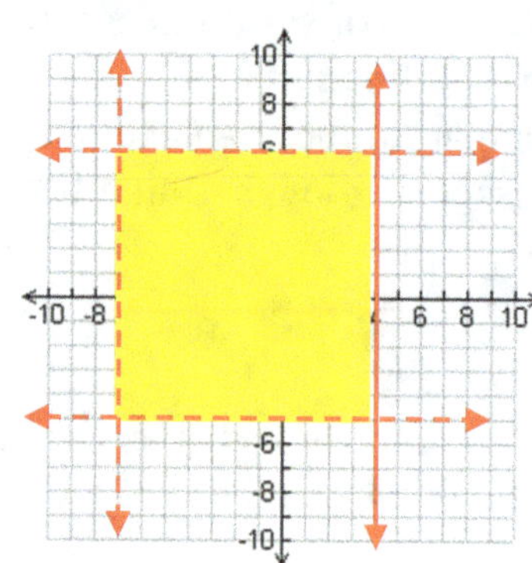

$$\begin{cases} x > -7 \\ x \le 4 \\ y > -5 \\ y < 6 \end{cases}$$

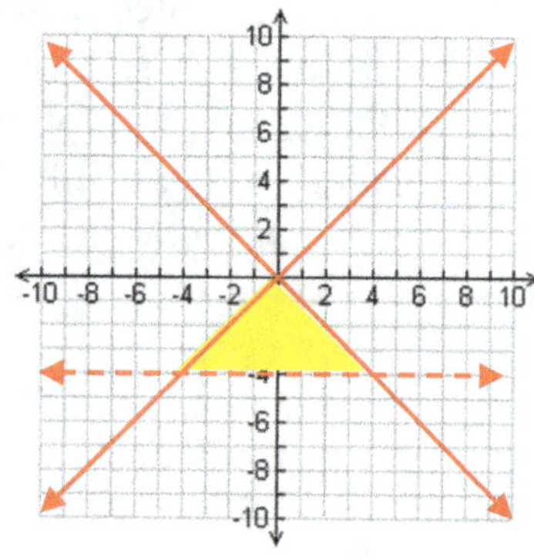

$$\begin{cases} y > -4 \\ y \le -x \\ y \le x \end{cases}$$

Example 4

Graph the sets of solutions of the systems $\begin{cases} x > -3 \\ x < 3 \end{cases}$ and $\begin{cases} y > -3 \\ y < 3 \end{cases}$.

SOLUTION ▶ The solutions region for the first system is the part of xy-plane on the **right** of the associated **dashed** line $x = -3$, and on the **left** side of **dashed** line $x = 3$.
The solutions region for the second system is the part of the xy-plane **over** the associated **dashed** line $y = -3$, **under** the associated **dashed** line $y = 3$.

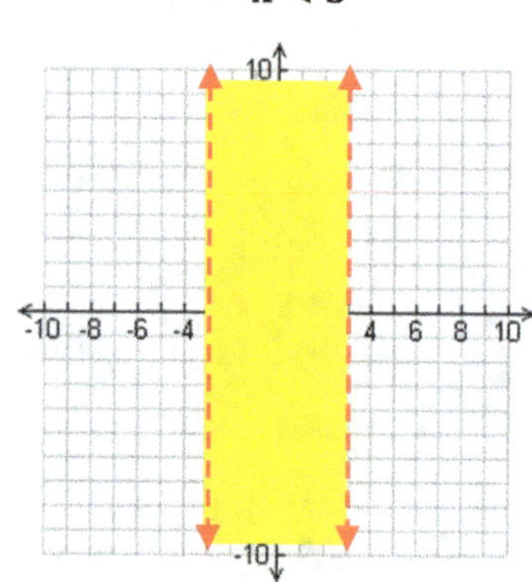

$$\begin{cases} x > -3 \\ x < 3 \end{cases}$$

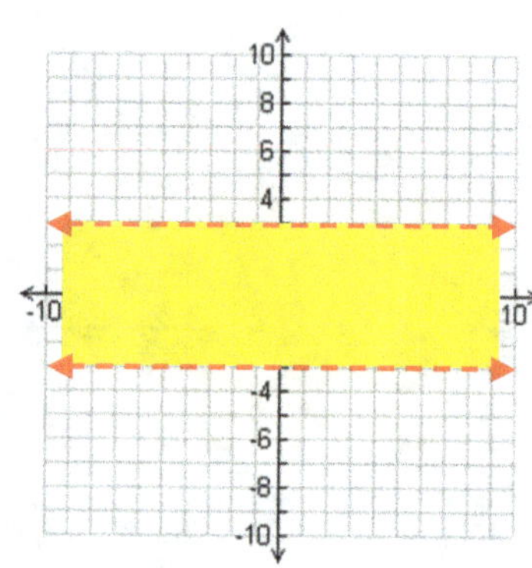

$$\begin{cases} y > -3 \\ y < 3 \end{cases}$$

Sketch the graph of the set of solutions for each inequality or for each system of inequalities.

1. $y < 2x - 4$

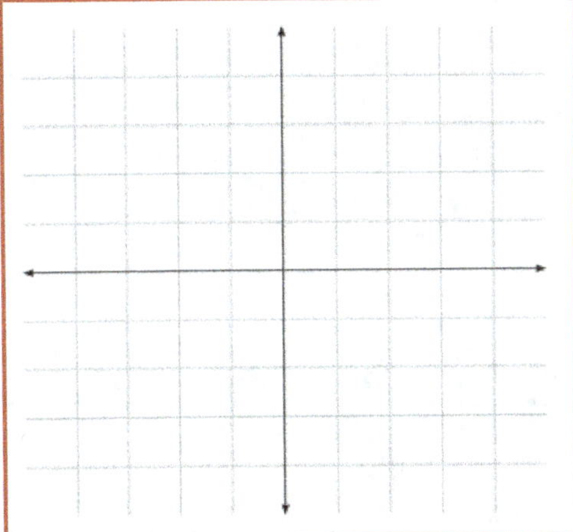

2. $2x + y \geq -3$

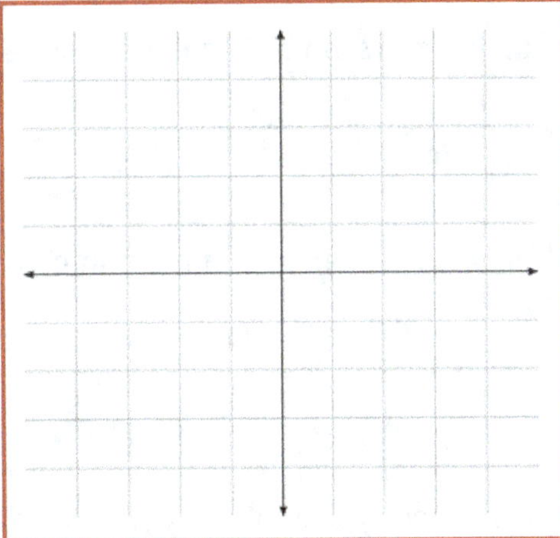

3. $\begin{cases} 2x - y > 1 \\ y \geq -\frac{1}{2}x + 2 \end{cases}$

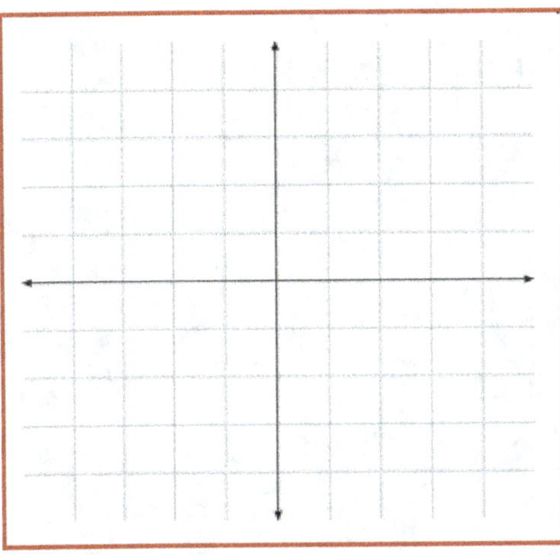

4. $\begin{cases} x > -4 \\ y \leq x + 3 \\ -x + y > -3 \\ x \leq 4 \end{cases}$

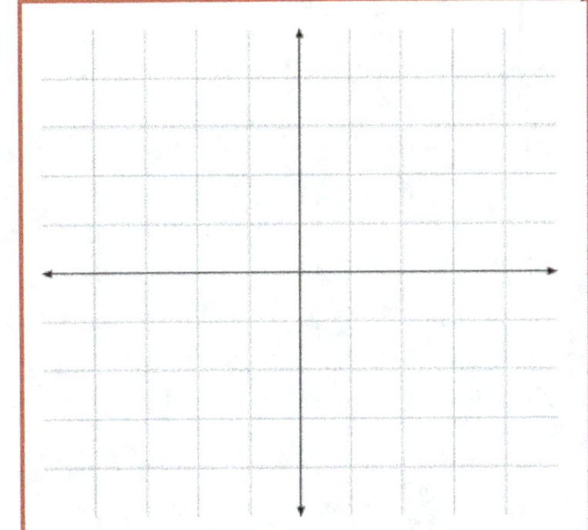

1. Solve the system by substitution method.

$$\begin{cases} x + y = 5 \\ 2x + y = 6 \end{cases}$$

2. Solve the system by graph method.

$$\begin{cases} 2x + y = 3 \\ x + y = 2 \end{cases}$$

3. Solve the system by elimination method.

$$\begin{cases} x - y + z = 1 \\ x + y + z = 3 \\ 2x + 3y - z = 4 \end{cases}$$

4. Solve each system by the corresponding Cramer's Rule.

$$\begin{cases} x + y = 5 \\ 2x + y = 6 \end{cases} \qquad \begin{cases} x - y + z = 1 \\ x + y + z = 3 \\ 2x + 3y - z = 4 \end{cases}$$

5. Sketch the graph of the set of solutions for each system of inequalities.

$$\begin{cases} 2x - y > 1 \\ y \geq -\frac{1}{2}x + 2 \end{cases} \qquad \begin{cases} 2x - y > 1 \\ y \geq -\frac{1}{2}x + 2 \end{cases}$$

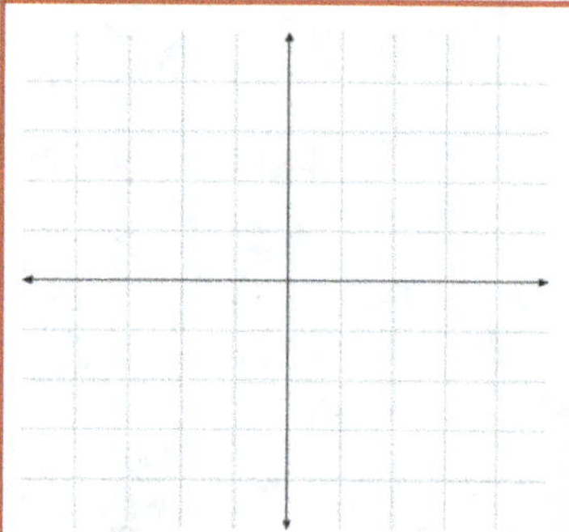
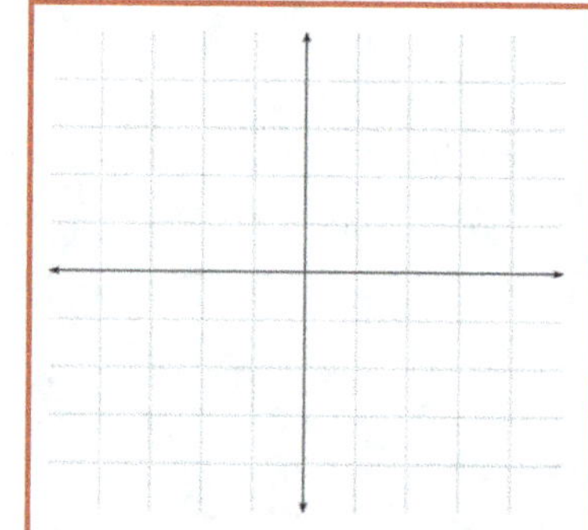

1. Solve the system by substitution method.

$$\begin{cases} x + \ y = 2 \\ x + 2y = 3 \end{cases}$$

2. Solve the system by graph method.

$$\begin{cases} x + y = 3 \\ x - y = 1 \end{cases}$$

3. Solve the system by elimination method.

$$\begin{cases} x + \ y - z = 1 \\ x + 2y + z = 4 \\ 2x + \ y - z = 2 \end{cases}$$

4. Solve each system by the corresponding Cramer's Rule.

$$\begin{cases} x + \ y = 3 \\ x + 2y = 4 \end{cases} \qquad \begin{cases} x - y - z = 1 \\ x + y + z = 5 \\ 2x - y - z = 4 \end{cases}$$

5. Sketch the graph of the set of solutions for each system of inequalities.

$$\begin{cases} x + y < 1 \\ y \geq x + 1 \end{cases} \qquad \begin{cases} x - y > 1 \\ 2y + x \geq 2 \end{cases}$$

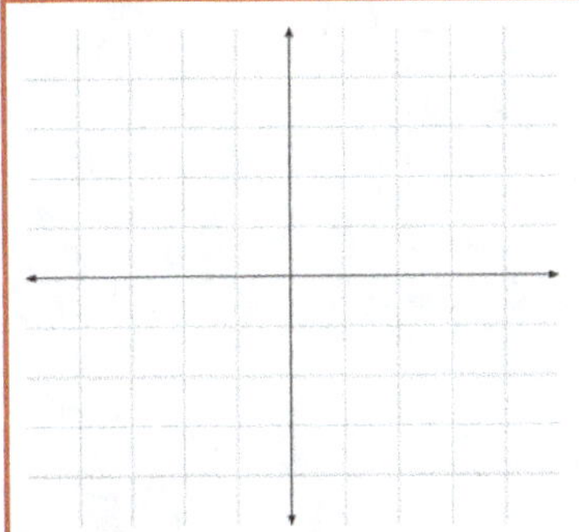

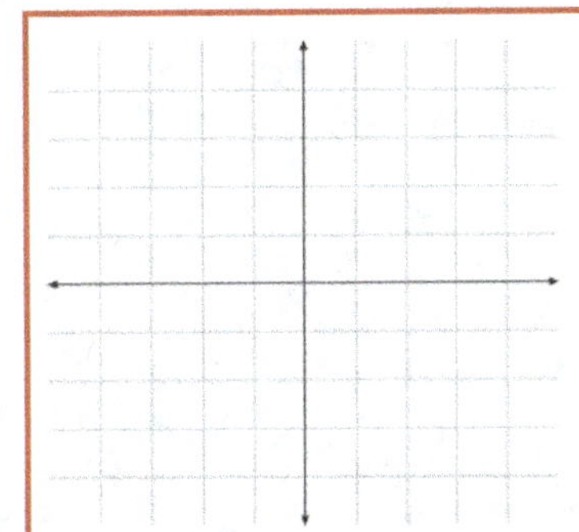

Hayk Yegoryan & Rubik Yegoryan

ELEMENTARY ALGEBRA

1. Solve the system by substitution method.

$$\begin{cases} x + y = 2 \\ 2x + 5y = 7 \end{cases}$$

2. Solve the system by graph method.

$$\begin{cases} 2x - y = -3 \\ x + y = 3 \end{cases}$$

3. Solve the system by elimination method.

$$\begin{cases} x - y + z = 2 \\ x + y + 2z = 3 \\ 2x + 3y - z = 1 \end{cases}$$

4. Solve each system by the corresponding Cramer's Rule.

$$\begin{cases} x + y = 3 \\ x + 2y = 5 \end{cases} \qquad \begin{cases} x - y + z = 0 \\ x + y + z = 2 \\ 2x + 3y - z = 5 \end{cases}$$

5. Sketch the graph of the set of solutions for each system of inequalities.

a) $$\begin{cases} 2x - y > -3 \\ y \geq -x + 2 \end{cases}$$
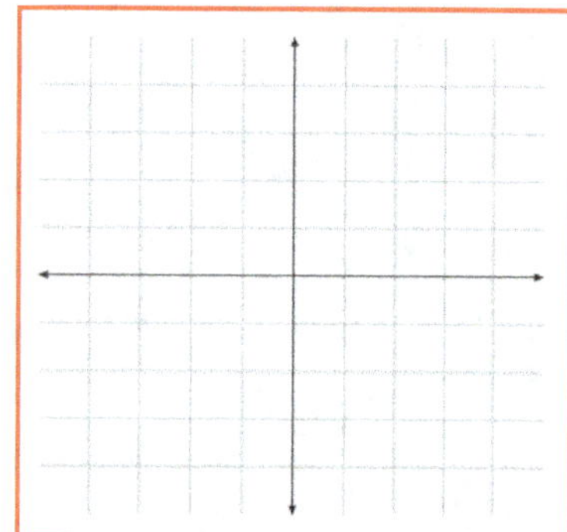

b) $$\begin{cases} x - y > -1 \\ y \leq -x + 1 \end{cases}$$
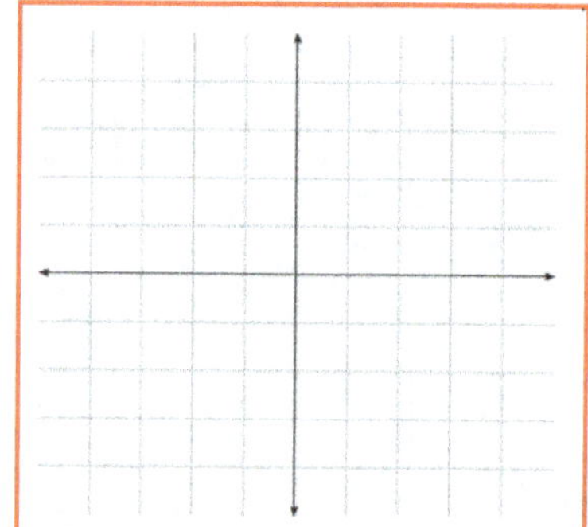

PART 4

FUNCTIONS AND GRAPHS

Chapter 7. Algebraic Functions

7.1. Functions
7.2. Transformations of Functions
7.3. Polynomial and Fractional Functions

Chapter 7 Classwork
Chapter 7 Homework

Part 4. Pre-Test

Chapter 7. Algebraic Functions

7.1. Functions
7.2. Transformations of Functions
7.3. Polynomial and Fractional Functions

Chapter 7   Classwork
Chapter 7   Homework

Part 4. Pre-Test

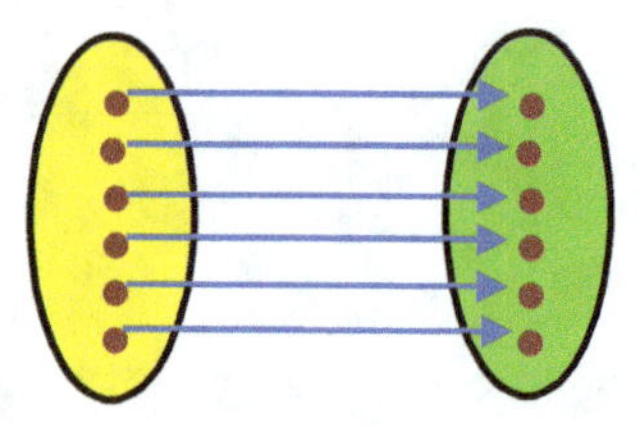
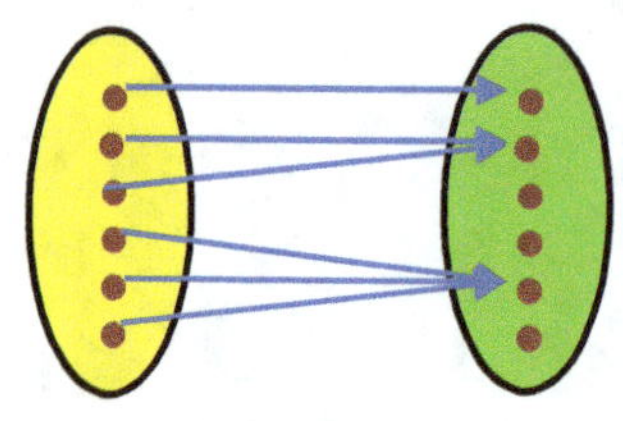
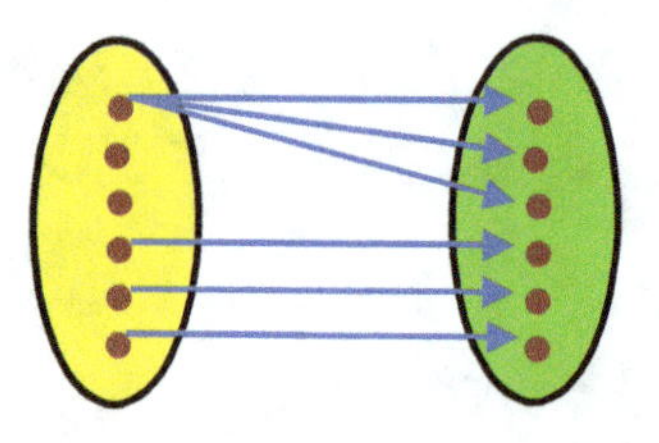

y = x

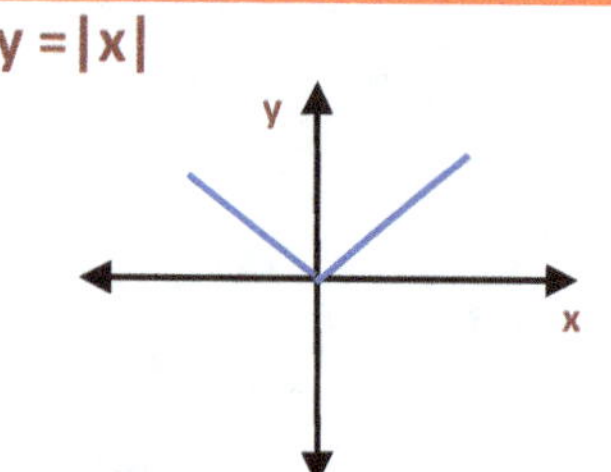

y = |x|

y = x^2

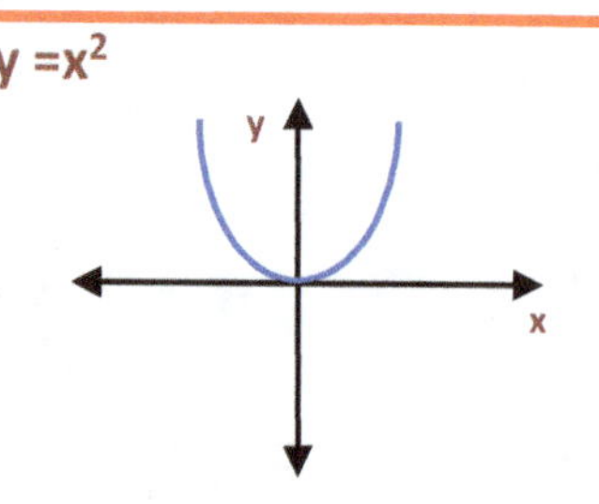

y = x^3

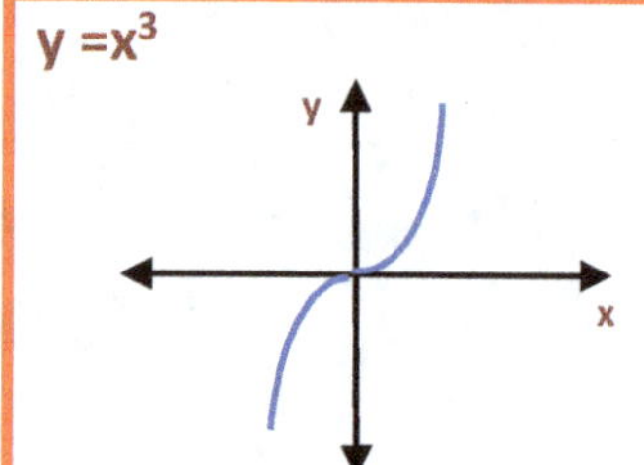

y = $\sqrt{x}$

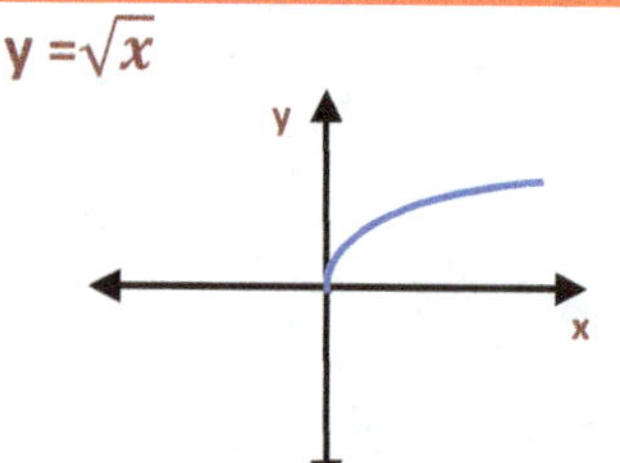

y = 1/x

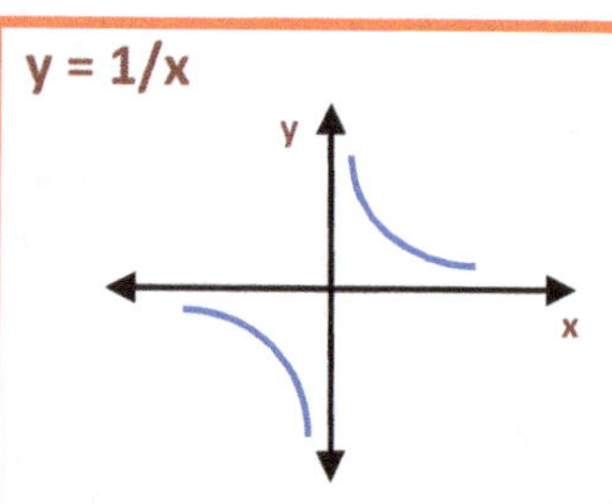

7.1. Functions

- **RELATIONS AND FUNCTIONS**
- **FINDING THE DOMAIN AND THE RANGE**
- **EVEN AND ODD FUNCTIONS**
- **ALGEBRAIC OPERATIONS WITH FUNCTIONS**
- **COMPOSITE FUNCTIONS**
- **BASIC ALGEBRAIC FUNCTIONS**
- **INVERSE FUNCTIONS**
- **FINDING THE DOMAIN AND THE RANGE**

■ RELATIONS AND FUNCTIONS

A **relation** is just relationship between two sets of numbers **X** and **Y**, or a **relation** is any set of ordered pairs of numbers. There are **3** possible types of relation between number sets **X** and **Y** shown on the diagrams below: **1−1 relation**, **n−1 relation**, and **1−n relation**.

The **function** is a special type of relation. A relation is a **function** if and only if there is one and exactly one **y** that corresponds to the given **x**.

Only **2** relations, **1−1 relation** and **n−1 relation**, are functions (**1−n relation** is not a function).

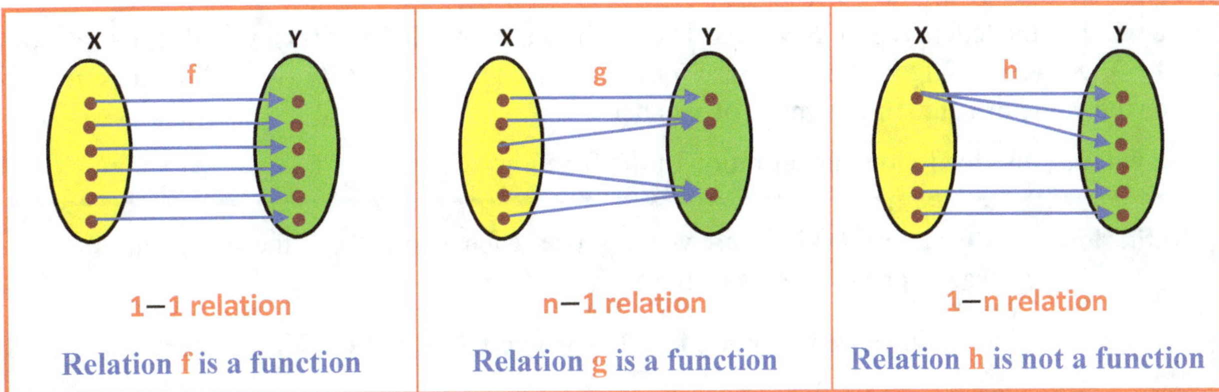

There are different ways of looking at the **function** and its definition. We will use the following definition of the **function** in terms of the relation.

> **Definition** ▶ The **relation** between number sets **X** and **Y** is **a function** if and only if there is one and only one corresponding **y** in set **Y** for each element **x** from set **X**.

Determine which of the following relations on the diagrams represents **functions**.

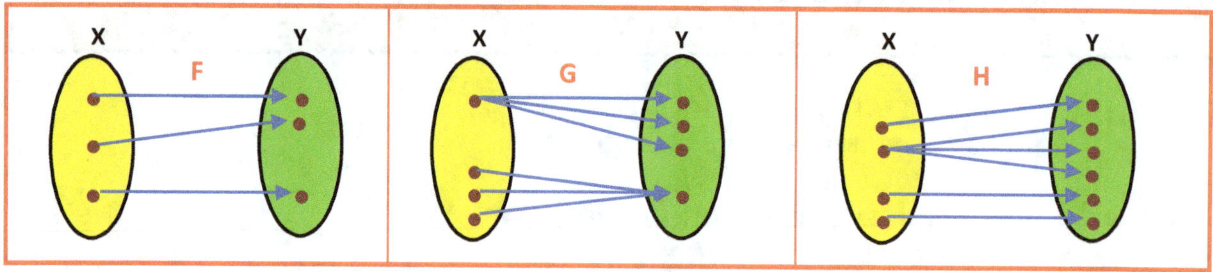

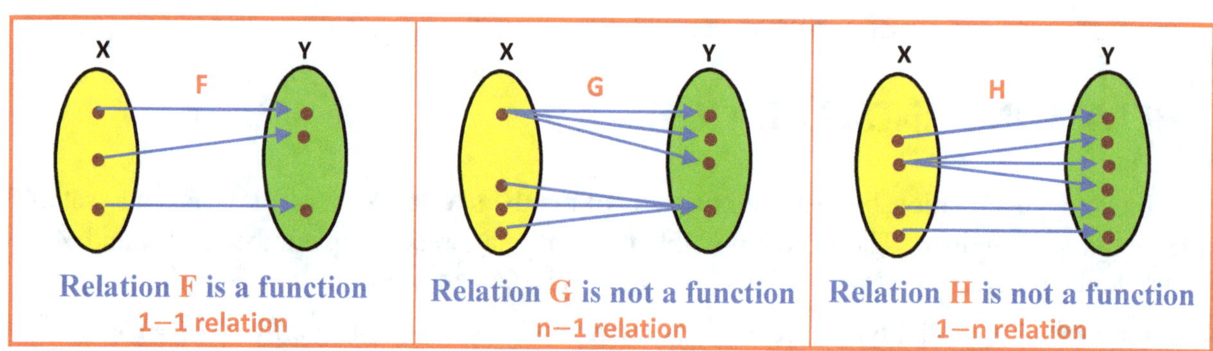

So, a **function** F is a relation (or relation rule) between elements **x** of the set **X** and elements **y** of the set **Y** with **only one** corresponding **y** for each **x**.

We will use the letters **f, g, h, F, G,** and **H** as well as the formulas **y = f(x), y = g(x), y = h(x), y = F(x), y = G(x),** and **H(x)** to represent functions. The function **f(x)** we will read as "**f** of **x**", or the value of **f** at **x,** or the image of **x** under **f.**

We will also use the following definition of the function.

Definition ▶ A **function f(x)** is a **rule** which relates each element **x** of the set **X** with exactly one element **y** of the set **Y.**

Sometimes we will write **f: x → y** where **x ∈ X** and **y ∈ Y.**

The variable **x** is called an **independent variable** or the **argument** of the function **f.** The variable **y** is called a **dependent variable** or the **value of the function.**

The set **X,** set of all the possible values of the argument **x,** is called the **Domain** of the function (**D_f = X**).

The set **Y,** set of all the possible values of the function **f(x),** is called the **Range** of the function (**R_f = Y**).

Visually relations we can imagine **as machines** are shown below.

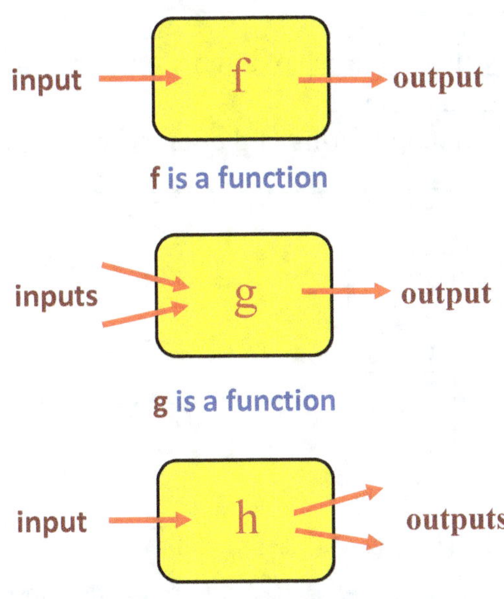

1. Coffee maker machine works as a function, because the output is only coffee.
2. Orange juice maker machine works as a relation, but not as a function, because for each input orange machine makes two types of outputs: the orange juice and the rest parts of the orange.

A relation is a **function** if for every **input** there is exactly one corresponding **output**.

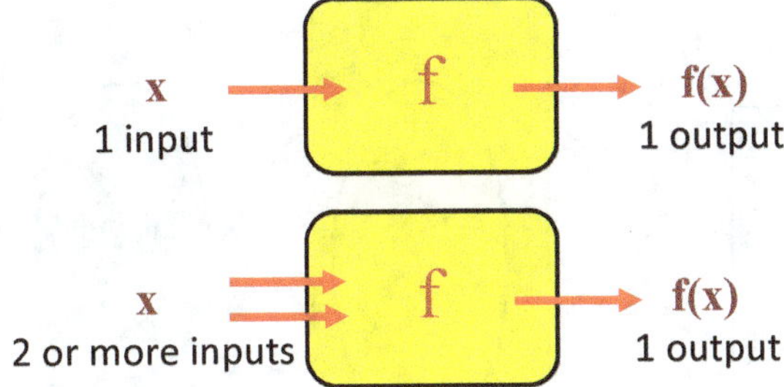

Below we give another definition of the function in terms of input and output.

> **Definition** ▶ A **function** **f(x)** is a **relation rule** between two quantities, called the **input x** and the **output f(x),** where for each input **x**, there is exactly one corresponding output **f(x).**

REPRESENTATION FORMS OF A FUNCTION

We will use the following representation forms of a function: **diagram form**, **table or set of ordered pairs**, **formula notation**, and **graph**.

Below you can see different representation forms of the same function **y = f(x) = x + 3**.

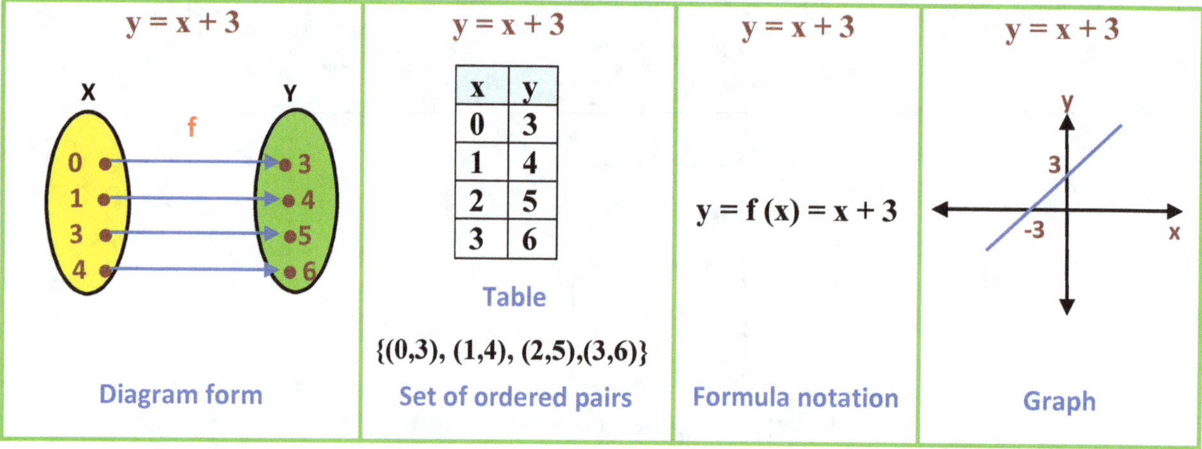

If we have one of these forms of a function then we can represent that function in other forms.

Example 1 Represent the function **y = x²** in table, ordered pair, diagram, and graph forms.

SOLUTION▶

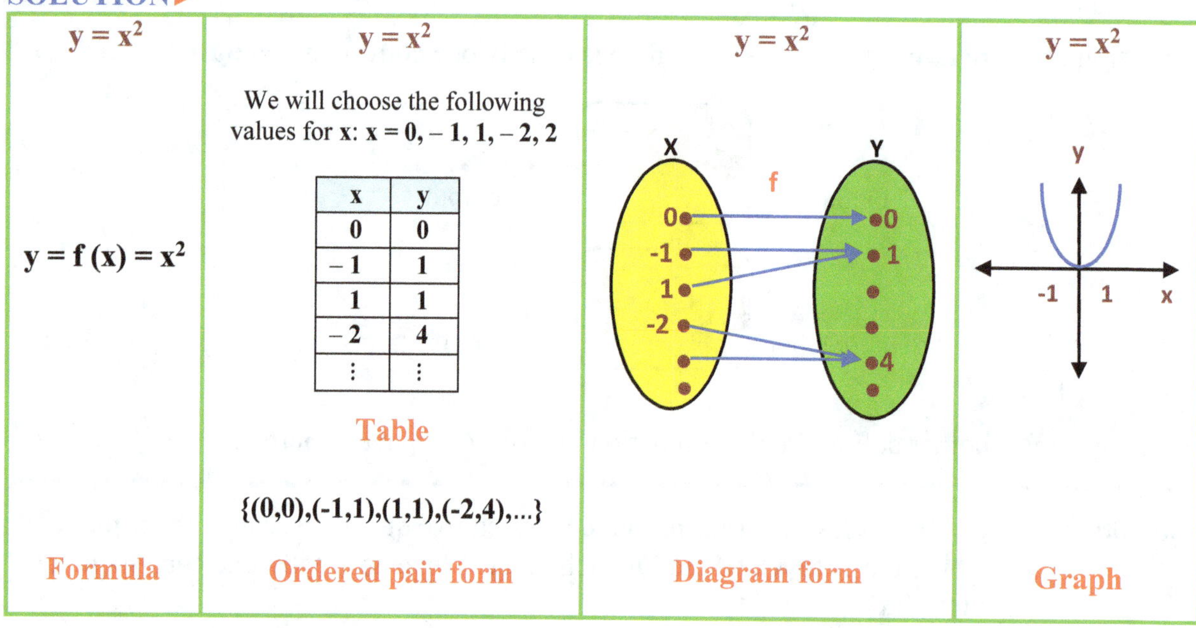

Example 2   Represent the table | x | 0 | 4 | 9 |
| y | 0 | 2 | 3 | in ordered pair, formula, diagram, and graph forms.

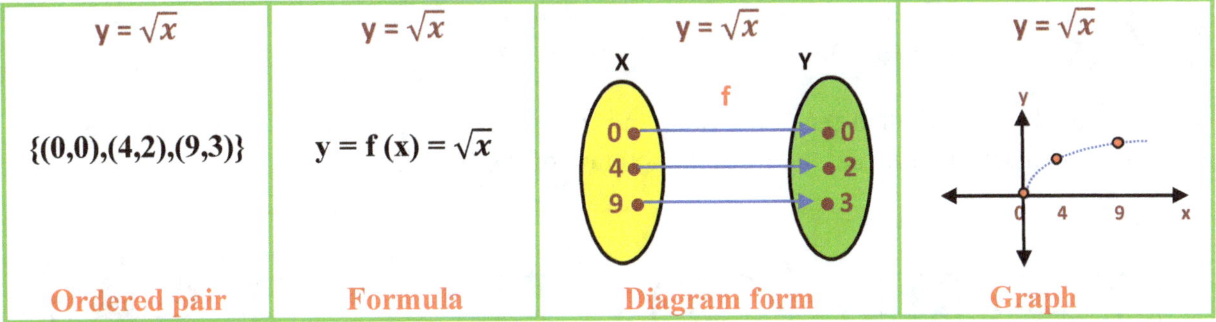

Ordered pair	Formula	Diagram form	Graph
$y = \sqrt{x}$ $\{(0,0),(4,2),(9,3)\}$	$y = \sqrt{x}$ $y = f(x) = \sqrt{x}$	$y = \sqrt{x}$	$y = \sqrt{x}$

Example 3   Represent the set of ordered pairs $\{(0,0), (2,8), (3,27)\}$ in table, formula, diagram, and graph forms.

SOLUTION▶

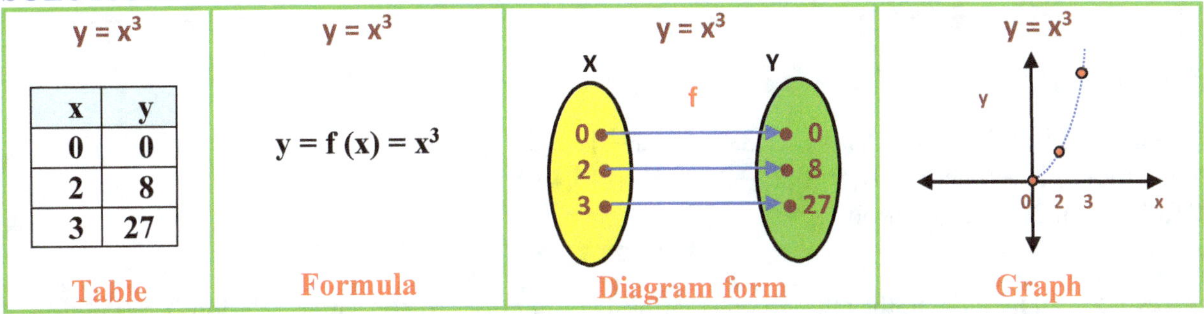

Table	Formula	Diagram form	Graph
$y = x^3$	$y = x^3$	$y = x^3$	$y = x^3$

Example 4   Represent the diagram below in table, ordered pair, formula, and graph forms.

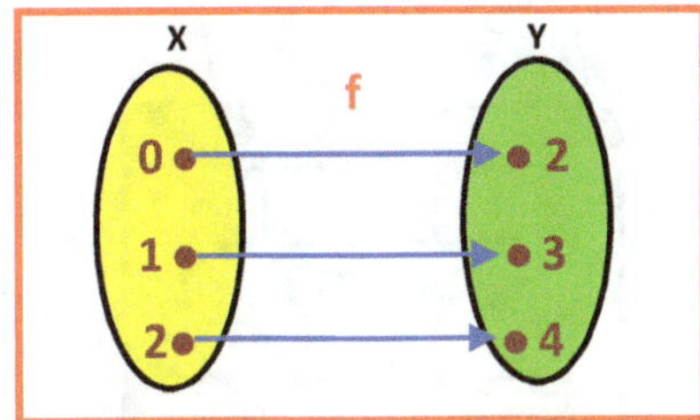

SOLUTION▶

Table	Ordered pair form	Formula notation	Graph
	$\{(0,0), (1,3), (2,4)\}$	$y = f(x) = x + 2$	

We can determine whether the given relation is a function. Bellow we will study some examples for the cases if we have diagram of relation, table of relation, set of **x** and **y** pairs of relation, or graphs of relation.

 Identify Functions by Diagrams

EXAMPLE 1: Decide whether the given relation is a function. If the given relation is a function, give the domain, the range, and the formula notation of the function.

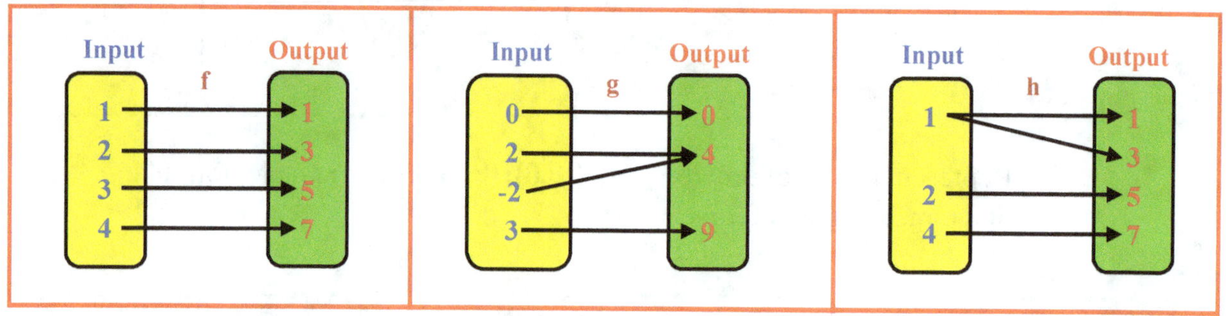

SOLUTION ▶

- Relation **f** is a function. $D_f = \{1,2,3,4\}$; $R_f = \{1,3,5,7\}$; $f(x) = 2x-1$.
- Relation **g** is a function. $D_g = \{0,2,-2,3\}$; $R_f = \{0,4,9\}$; $g(x) = x^2$.
- Relation **h** is not a function.

EXAMPLE 2: Decide whether the given relation is a function. If the given relation is a function, give the domain, the range, and the formula notation of the function.

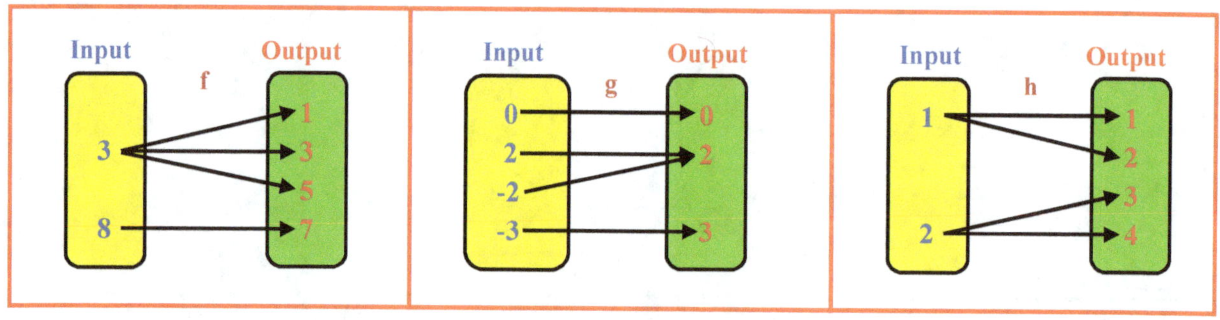

SOLUTION ▶

- Relation **f** is not a function.
- Relation **g** is a function. $D_g = \{0, 2, -2, -3\}$; $R_f = \{0,2,3\}$; $g(x) = |x|$.
- Relation **h** is not a function.

Determine whether the given relation is a function. If the given relation is a function, give the domain, the range, and the formula notation of the function

<table>
<tr><td colspan="2" align="center">f</td><td colspan="2" align="center">g</td><td colspan="2" align="center">h</td><td colspan="2" align="center">j</td></tr>
<tr><td>Input
x</td><td>Output
y</td><td>Input
x</td><td>Output
y</td><td>Input
x</td><td>Output
y</td><td>Input
x</td><td>Output
y</td></tr>
<tr><td>1</td><td>3</td><td>0</td><td>0</td><td>0</td><td>1</td><td>1</td><td>2</td></tr>
<tr><td>2</td><td>4</td><td>1</td><td>1</td><td>1</td><td>2</td><td>1</td><td>3</td></tr>
<tr><td>3</td><td>5</td><td>2</td><td>4</td><td>−1</td><td>2</td><td>1</td><td>4</td></tr>
<tr><td>4</td><td>6</td><td>1</td><td>−1</td><td>2</td><td>5</td><td>4</td><td>5</td></tr>
<tr><td>5</td><td>7</td><td>3</td><td>−4</td><td>−2</td><td>5</td><td>5</td><td>8</td></tr>
</table>

SOLUTION▶

- Relation **f** is a function. $D_f = \{1, 2, 3, 4, 5\}$; $R_f = \{3, 4, 5, 6, 7\}$; $f(x) = x + 2$.
- Relation **g** is not a function. For $x = 1$ we have two corresponding y: $y = 1$, and $y = -1$.
- Relation **h** is a function. $D_h = \{0, 1, -1, 2, -2\}$; $R_h = \{0, 2, 5\}$; $h(x) = x^2 + 1$.
- Relation **j** is not a function. For $x = 1$ we have three corresponding y: $y = 2$, $y = 3$, and $y = 4$.

Example Identify Functions by Sets of Ordered Pairs (x,y)

Determine whether the given relation is a function. If the given relation is a function, give the domain, the range, and the formula notation of the function
a). **f:** $\{(1,2), (3,4), (5,6), (7,8)\}$
b). **g:** $\{(1,2), (5,4), (1,6), (-1, -3)\}$

SOLUTION▶

a). Relation **f** is a function. $D_f = \{1, 3, 5, 7\}$; $R_f = \{2,4,6,8\}$; $f(x) = x + 1$.
b). Relation **g** is not a function. For $x = 1$ we have two corresponding y: $y = 2$, $y = 6$.

Example Identify Functions by Formulas (or by Equations)

Determine whether the given equation (relations) is a function: **a)** $3x+2y = 6$; **b)** $x = 4$; **c)** $y = 2$.

SOLUTION▶

We will make the **xy** table (or diagram, or set of ordered pairs) with different values of **x** for each equation. **a)** The equation $3x + 2y = 6$ always has one and only one **y** for each value of **x**. So, the relation $3x + 2y = 6$ is a function. **b)** The equation **x = 4** has infinitely many corresponding values of **y** for **x = 4**. So, the relation **x = 4** is not a function. **c)** The equation **y = 2** always has the same one and only one **y = 2** for each value of **x**. So, the relation **y = 2** is a function.

VERTICAL LINE TEST FOR FUNCTIONS

We will use vertical line test to determine whether the given relation is a function in case if we have a graph of the relation.

VERTICAL LINE TEST ▶ A graph represents a function if no vertical line intersects the graph at more than one point.

If there is as minimum one vertical line with two or more intersects with the given graph, then the given graph represents a relation, but not a function.

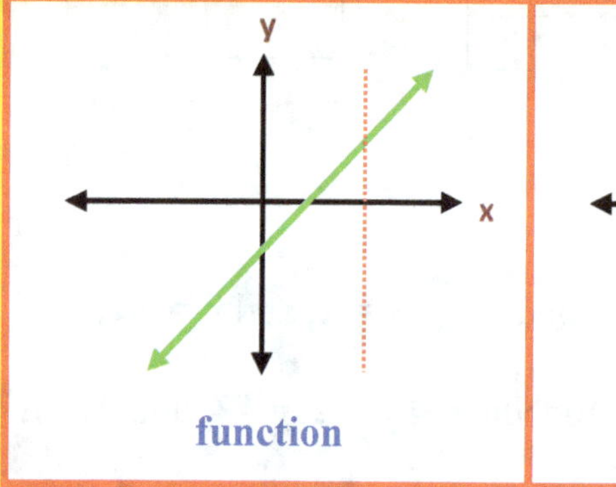

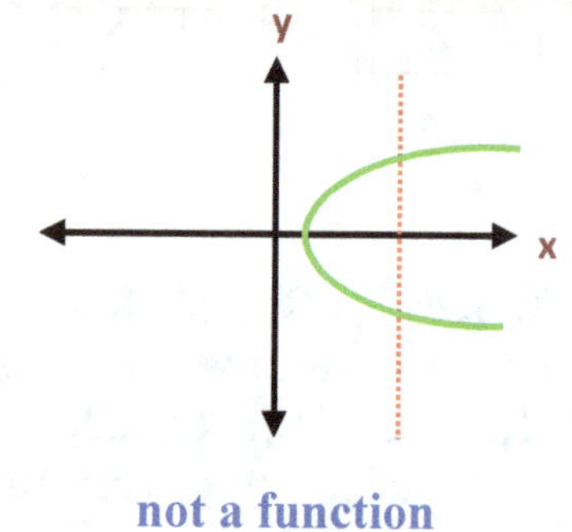

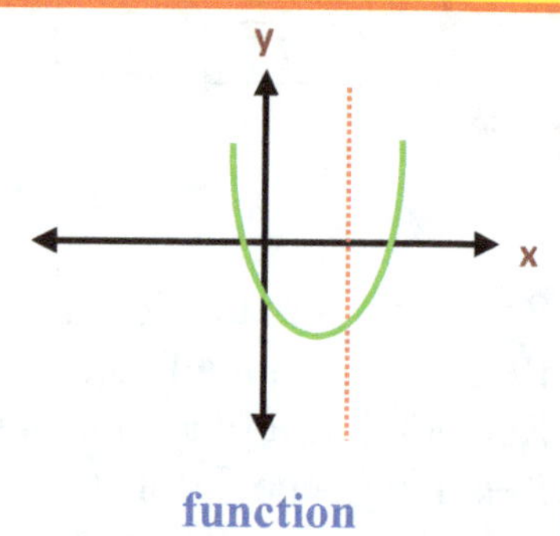

Examples **Identify Functions by using the Vertical Line Test**

Determine whether the graph represents a function

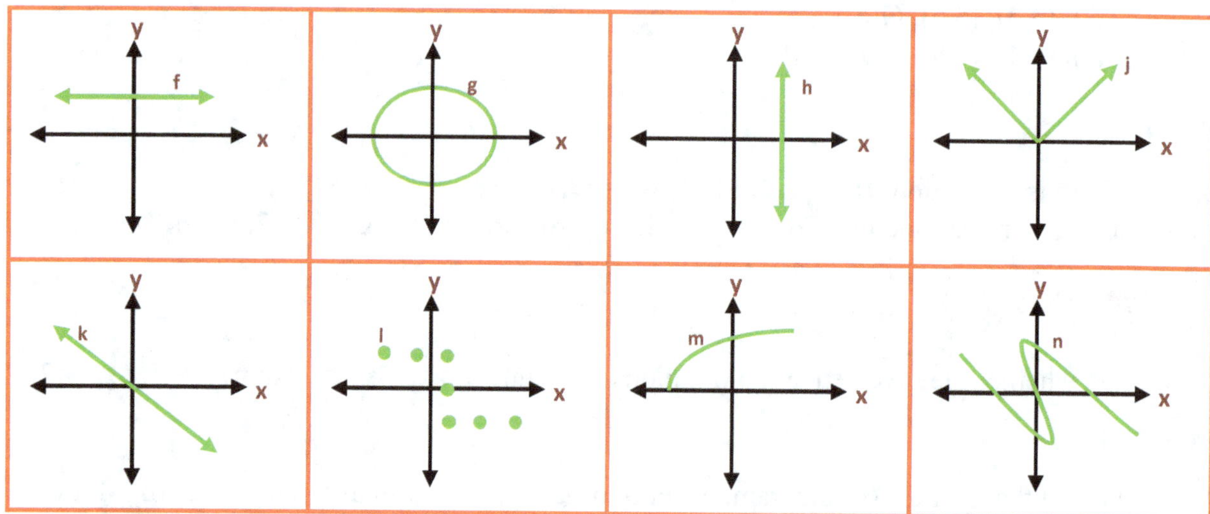

SOLUTION ▶

f is a function	g is not a function	h is not a function	j is a function
k is a function	l is not a function	m is a function	n is not a function

FUNCTION VALUES

Let the function **f** defines by the formula $f(x) = x^3 + 2$. This formula tells us that **f** first cubes the input **x** and then adds **2** to the cube of input.

$$\text{output} = f(\text{input}) = (\text{input})^3 + 2$$

To find the function value **f(a)** of the given function **f(x)** for the given value **x = a** of argument (input) we have to plug-in the given **a** into the function formula.

For example, to evaluate the outputs **f(0), f(1), f(– 1),** and **f(2)** of the function $f(x) = x^3 + 2$ we have to plug-in the inputs **x = 0, 1, – 1, 2,** into the function's formula $f(x) = x^3 + 2$.

$$f(0) = 0^3 + 2 = 0 + 2$$
$$f(1) = 1^3 + 2 = 1 + 2 = 3$$
$$f(-1) = (-1)^3 + 2 = -1 + 2 = 1$$
$$f(2) = 2^3 + 2 = 8 + 2 = 10$$

The **domain D_f** of the function **f** is the set of all possible values of input **x**.
The **range R_f** of the function **f** is the set of all possible values of output **f(x)**.

Assume the function **g** defines by the formula **g(x) = 2x +1.**

This formula tells us that **output** = **g(input)** = **2(input) + 1.** To find the values **g(0), g(1), g(–1), g(2), g(–2),** and **g(3)** of outputs for the inputs **0, 1, – 1, 2, – 3,** and **3,** we have to use the function formula **g(x) = 2x + 1.**

$$g(0) = 2 \cdot 0 + 1 = 1$$
$$g(1) = 2 \cdot 1 + 1 = 3$$
$$g(-1) = 2 \cdot (-1) + 1 = -1$$
$$g(2) = 2 \cdot 2 + 1 = 5$$
$$g(-2) = 2 \cdot (-2) + 1 = -3$$
$$g(3) = 2 \cdot 3 + 1 = 7$$

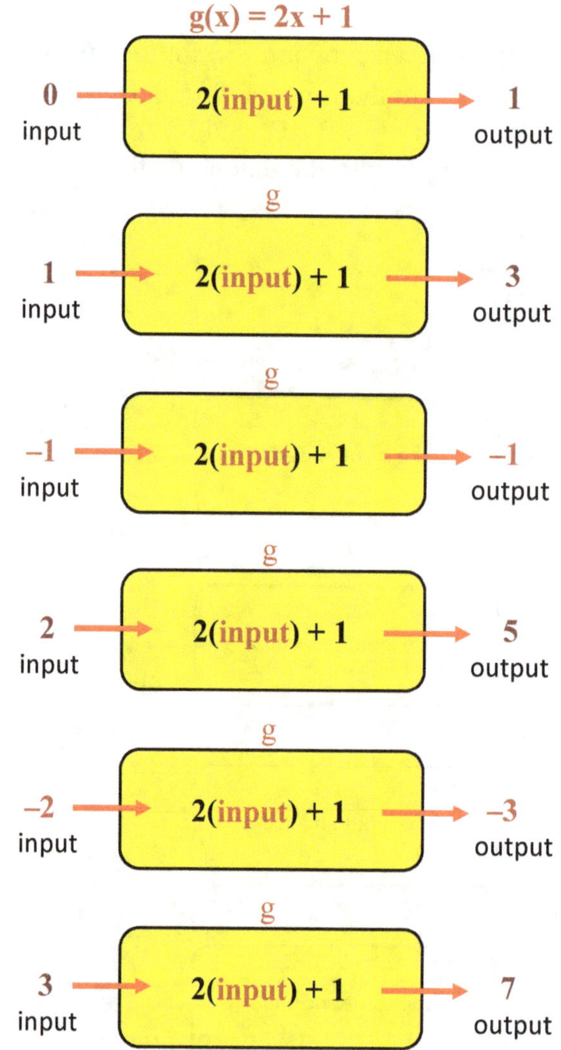

The **domain D_g** of the function **g** is the set of all possible values of input **x.**

The **range R_g** of the function **g** is the set of all possible values of output **g(x).**

■ FINDING THE DOMAIN AND RANGE

We will use the following definitions for **domain** and **range** of a function.

> **Definition** ▶ The set **X** of all possible values of the argument **x** (input) is the **domain** of the function **f (D_f = X)**.
>
> **Definition** ▶ The set **Y** of all possible values of the function **y = f(x)** (output) is the **range** of the function **f (R_f = Y)**.

Usually we have restrictions for argument **x** if we have a function with **even power root** or with **fraction** terms (or both together) containing variable **x**. In each of these cases domain of a function **f(x)** is all real numbers minus set of restrictions for x: D_f = **R−** {**restrictions for x**}. If argument **x** of a function **f(x)** has not any restriction, then domain of a function will be the set of all real numbers: D_f = **R** or D_f = **(−∞, ∞)**.

Example 1 Find the domain of functions:

a) $f(x) = 2x + 1$
b) $g(x) = x^2 - 2$
c) $h(x) = \sqrt{x - 1}$
d) $k(x) = \dfrac{1}{x + 1}$

SOLUTION▶

a) $D_f = (-\infty, \infty)$, because $f(x) = 2x + 1$ does not have any restriction for **x**.
b) $D_g = (-\infty, \infty)$, because $g(x) = x^2 - 2$ does not have any restriction for **x**.
c) $D_h = [1, \infty)$, because $h(x) = \sqrt{x - 1}$ exists if and only if $x \geq 1$.
d) $D_k = (-\infty, -1) \cup (-1, \infty)$, because $k(x) = \dfrac{1}{x + 1}$ exists if $x \neq -1$.

Example 2 Find the domain of the function $f(x) = x + \sqrt{x - 3} + 1/(x - 4)$

SOLUTION▶

$D_f = [3, 4) \cup (4, \infty)$, because $f(x) = x + \sqrt{x - 3} + 1/(x - 4)$ exists if $x - 3 \geq 0$ (or $x \geq 3$) and $x - 4 \neq 0$ (or $x \neq 4$).

 Find the domain of the function $g(x) = 1/[x(x-4)]$

 ▶

$$D_g = (-\infty, 0) \cup (0, 4) \cup (4, \infty), \text{ because } g(x) = \frac{1}{x(x-4)} \text{ exists if } x \neq 0, 4.$$

The graph of a function can help us to find the **domain** and the **range** of that function. On graph level domain of a function is a projection of graph on the **x-axis** and the range is the projection of the function on the **y-axis** (see below).

$D = [2, 7]; \ R = [3,8]$

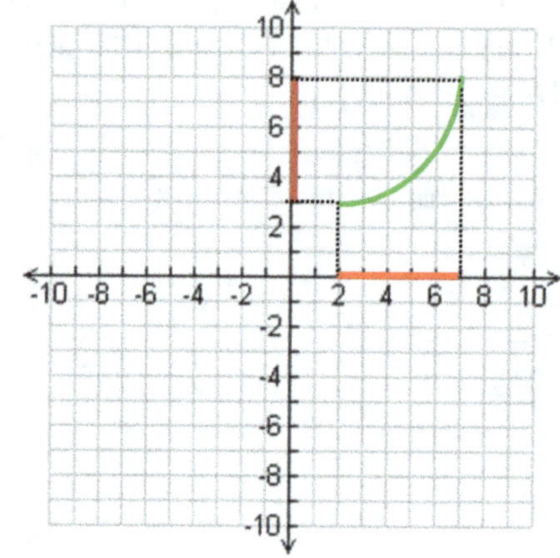

 Find the domain and the range of the function $f(x) = \sqrt{x+2}$

 ▶ The graph of $f(x) = \sqrt{x+2}$ shows that $D_f = [-2, \infty); \ R_f = [0, \infty)$.

$D = [-2, \infty]; \ R = [0, \infty]$

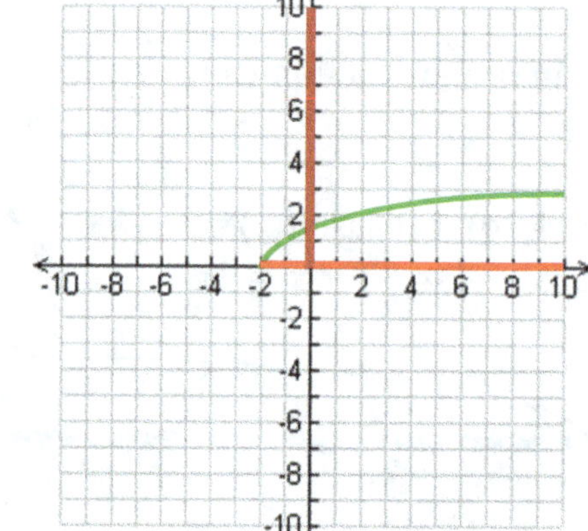

x	y
-2	0
-1	1
2	2

 Find the domain and the range of the function $g(x) = |x - 2| - 3$

The graph of $g(x) = |x - 2| - 3$ shows that $D_g = (-\infty, \infty)$; $R_g = [-3, \infty)$.

$D_g = (-\infty, \infty)$; $R_g = [-3, \infty)$

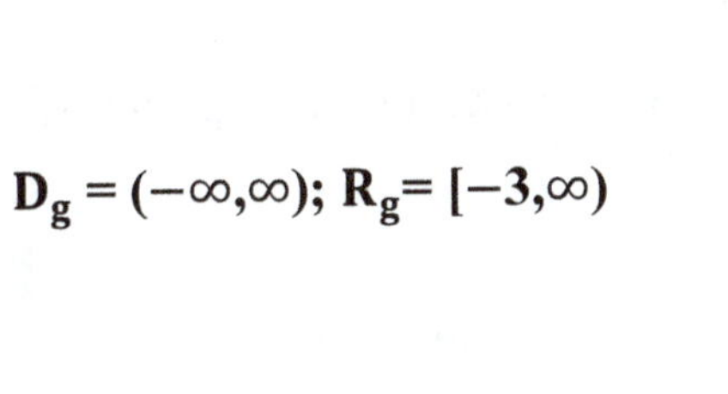

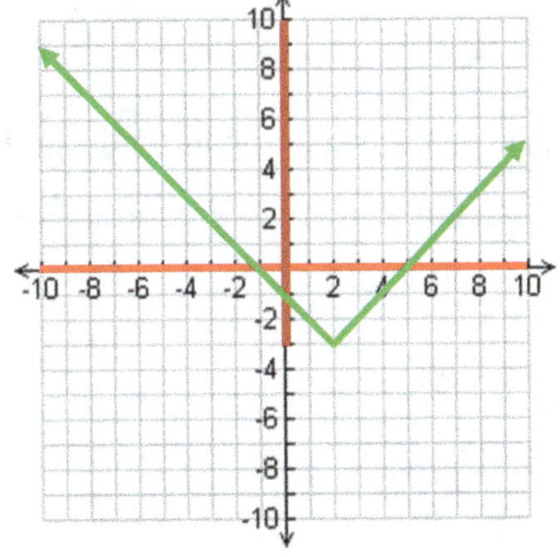

x	y
-1	0
0	-1
2	-3
3	0

 Find the domain and the range of the function $h(x) = (x - 3)^2 + 2$

The graph of $h(x) = (x - 3)^2 + 2$ shows that $D_h = (-\infty, \infty)$; $R_h = [2, \infty)$.

$D_h = (-\infty, \infty)$; $R_h = [2, \infty)$

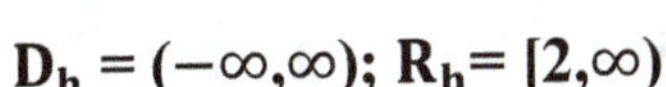

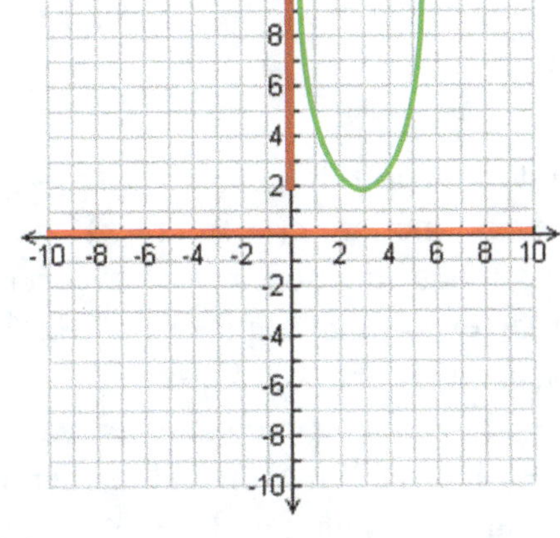

x	y
1	6
2	3
3	2
5	6

■ EVEN AND ODD FUNCTIONS

$\mathbf{W}$e will use the following definitions of the **even** and **odd function**.

Definition ▶ A function $\mathbf{f(x)}$ is **an even function** if $\mathbf{f(-x) = f(x)}$.

Definition ▶ A function $\mathbf{f(x)}$ is **an odd function** if $\mathbf{f(-x) = -f(x)}$.

$\mathbf{A}$ graph of an even function is a symmetric graph around **y-axis**.

$\mathbf{A}$ graph of an odd function is a symmetric graph around the **origin** of **xy-system** of coordinate.

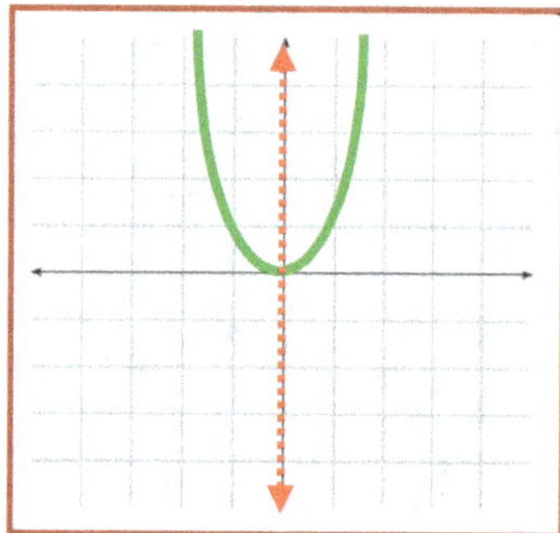

$$\mathbf{f(x) = x^2} \text{ is an even function}$$

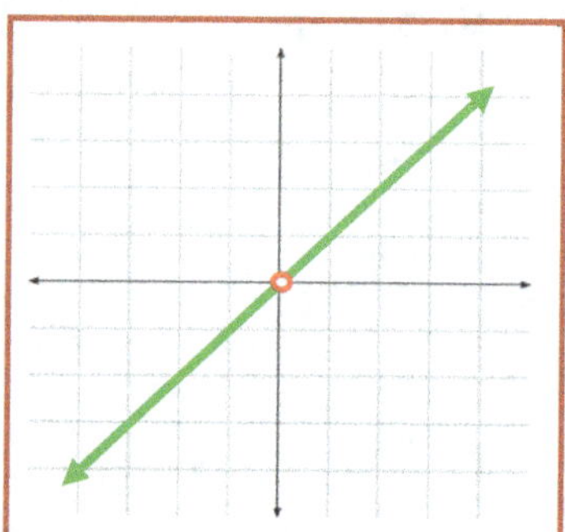

$$\mathbf{g(x) = x} \text{ is an odd function}$$

Examples Classify functions as **even**, **odd**, or **neither**.

a) $f(x) = x$

b) $g(x) = x^2$

c) $h(x) = x^3$

d) $j(x) = x^4$

e) $k(x) = x^5$

f) $F(x) = (x - 1)^3$

g) $G(x) = (x + 1)^2$

h) $H(x) = |x|$

i) $K(x) = x^2 + |x| + 1$

j) $J(x) = x^3 + x$

SOLUTION▶

a) $f(x)$ is an odd function: $f(-x) = (-x) = -(x) = -f(x)$

b) $g(x)$ is an even function: $g(-x) = (-x)^2 = x^2 = g(x)$

c) $h(x) = x^3$ is an odd function: $h(-x) = (-x)^3 = -x^3 = -f(x)$

d) $j(x) = x^4$ is an even function: $j(-x) = (-x)^4 = x^4 = j(x)$

e) $k(x) = x^5$ is an odd function: $k(-x) = (-x)^5 = -x^5 = -k(x)$

f) $F(x) = (x - 1)^3$ is not an even and neither an odd function

g) $G(x) = (x + 1)^2$ is not an even and neither an odd function

h) $H(x) = |x|$ is an even function: $H(-x) = |-x| = x = H(x)$

i) $K(x) = x^2 + |x| + 1$ is an even function: $K(-x) = (-x)^2 + |-x| + 1 = K(x)$

j) $J(x) = x^3 + x$ is an odd function: $J(-x) = (-x)^3 + (-x) = -(x^3 + x) = -f(x)$

■ ALGEBRAIC OPERATIONS WITH FUNCTIONS

We can add, subtract, multiply, and divide any two or more functions.

Example 1   Let $f(x) = x + 1$ and $g(x) = x - 1$. Find $f(x) + g(x)$; $f(x) - g(x)$; $f(x) \cdot g(x)$;

$$\frac{f(x)}{g(x)}; \text{ and } \frac{g(x)}{f(x)}.$$

SOLUTION ▶

$$f(x) + g(x) = (x + 1) + (x - 1) = 2x$$

$$f(x) - g(x) = (x + 1) - (x - 1) = 2$$

$$f(x) \cdot g(x) = (x + 1) \cdot (x - 1) = x^2 - 1$$

$$\frac{f(x)}{g(x)} = \frac{x + 1}{x - 1}$$

$$\frac{g(x)}{f(x)} = \frac{x - 1}{x + 1}$$

Example 2   Let $f(x) = x + 2$, $g(x) = x$, $h(x) = x - 2$. Find $f(x) + g(x) + h(x)$; $f(x) - g(x)h(x)$;

$$f(x) \cdot h(x) + g(x); \frac{f(x)}{g(x)}; \frac{f(x)}{h(x)}; \text{ and } \frac{f(x) + g(x)}{h(x)}.$$

SOLUTION ▶

$$f(x) + g(x) + h(x) = (x + 2) + x + (x - 2) = 3x$$

$$f(x) - g(x)h(x) = (x + 2) - x(x - 2) = -x^2 + 3x + 2$$

$$f(x) \cdot h(x) + g(x) = (x + 2) \cdot (x - 2) + x = x^2 + x - 4$$

$$\frac{f(x)}{g(x)} = \frac{x + 2}{x}$$

$$\frac{f(x)}{h(x)} = \frac{x + 2}{x - 2}$$

$$\frac{f(x) + g(x)}{h(x)} = \frac{2x + 2}{x - 2}$$

■ COMPOSITE FUNCTIONS

We will use the following definitions for the compositions of two functions **f(x)** and **g(x)**.

Definition ▶ $f(x) \circ g(x) = f(g(x))$

Definition ▶ $g(x) \circ f(x) = g(f(x))$

Example 1   Let **f(x) = x + 1** and **g(x) = x − 1**.
Find **f(x)∘f(x); f(x)∘g(x); g(x)∘f(x); g(x)∘g(x);**

SOLUTION ▶

If $f(x) = x + 1$, then

$$\mathbf{f(x) \circ f(x)} = f(f(x)) = f(x) + 1 = x + 1 + 1 = \mathbf{x + 2}$$

If $f(x) = x + 1$ and $g(x) = x − 1$, then

$$\mathbf{f(x) \circ g(x)} = f(g(x)) = g(x) + 1 = x − 1 + 1 = \mathbf{x}$$

If $f(x) = x + 1$ and $g(x) = x − 1$, then

$$\mathbf{g(x) \circ f(x)} = g(f(x)) = f(x) − 1 = x + 1 − 1 = \mathbf{x}$$

If $g(x) = x − 1$, then

$$\mathbf{g(x) \circ g(x)} = g(g(x)) = g(x) − 1 = x − 1 − 1 = \mathbf{x − 2}$$

Example 2   Let **f(x) = x²** and **g(x) = 2x**.
Find **f(x)∘f(x); f(x)∘g(x); g(x)∘f(x); g(x)∘g(x);**

SOLUTION ▶

If $f(x) = x^2$, then

$$\mathbf{f(x) \circ f(x)} = f(f(x)) = [f(x)]^2 = [x^2]^2 = \mathbf{x^4}$$

If $f(x) = x^2$ and $g(x) = 2x$, then

$$\mathbf{f(x) \circ g(x)} = f(g(x)) = [g(x)]^2 = [2x]^2 = \mathbf{4x^2}$$

If $f(x) = x^2$ and $g(x) = 2x$, then

$$\mathbf{g(x) \circ f(x)} = g(f(x)) = 2f(x) = 2(x^2) = \mathbf{2x^2}$$

If $g(x) = 2x$, then

$$\mathbf{g(x) \circ g(x)} = g(g(x)) = 2g(x) = 2(2x) = \mathbf{4x}$$

■ BASIC ALGEBRAIC FUNCTIONS

For the whole theory of functions it is very important to know the **basic algebraic functions** and their graphs (the list is shown below).

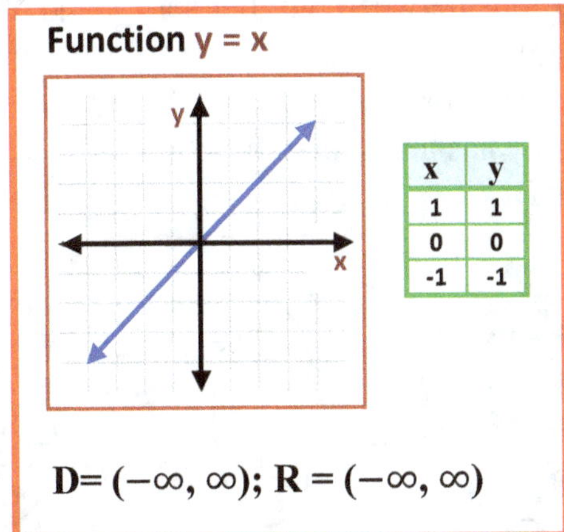

Function y = x

x	y
1	1
0	0
-1	-1

$D = (-\infty, \infty); R = (-\infty, \infty)$

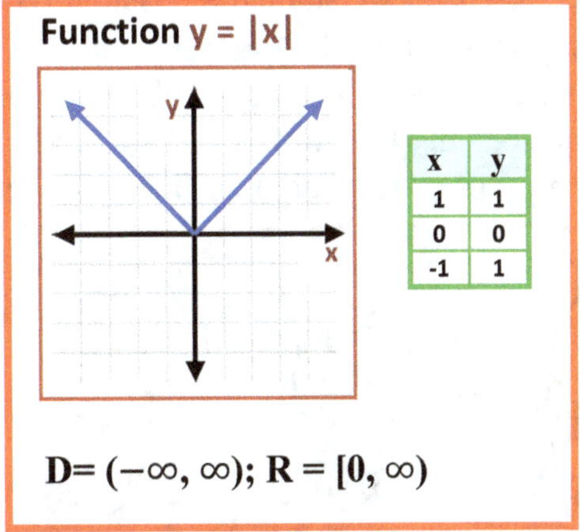

Function y = |x|

x	y
1	1
0	0
-1	1

$D = (-\infty, \infty); R = [0, \infty)$

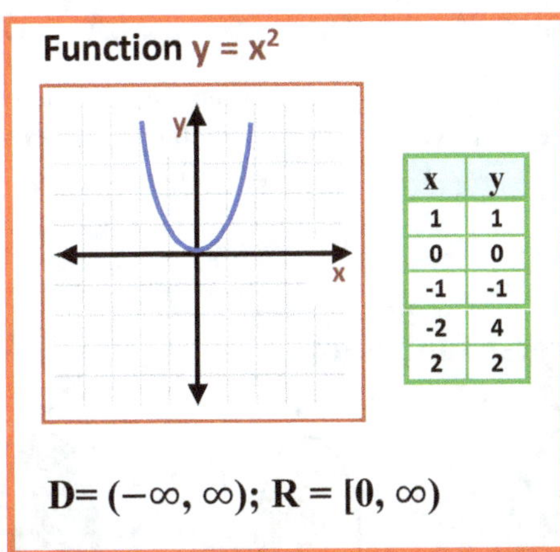

Function y = x²

x	y
1	1
0	0
-1	-1
-2	4
2	2

$D = (-\infty, \infty); R = [0, \infty)$

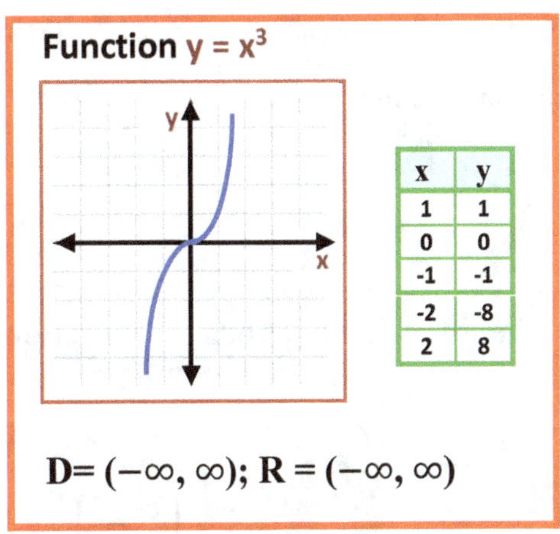

Function y = x³

x	y
1	1
0	0
-1	-1
-2	-8
2	8

$D = (-\infty, \infty); R = (-\infty, \infty)$

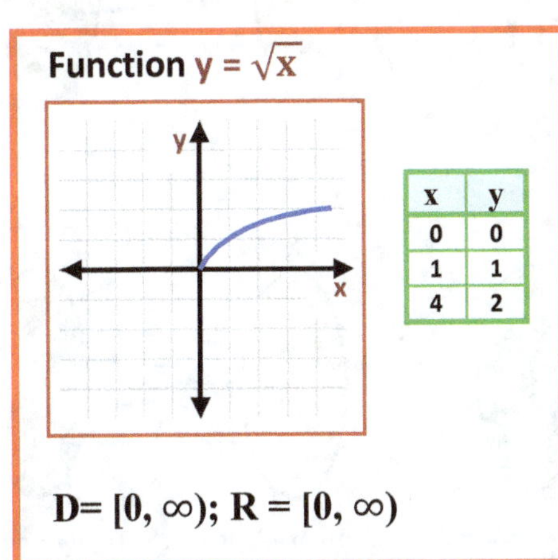

Function y = √x

x	y
0	0
1	1
4	2

$D = [0, \infty); R = [0, \infty)$

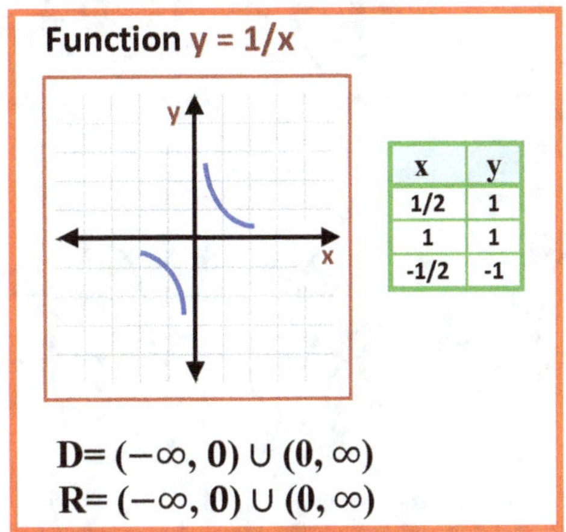

Function y = 1/x

x	y
1/2	1
1	1
-1/2	-1

$D = (-\infty, 0) \cup (0, \infty)$
$R = (-\infty, 0) \cup (0, \infty)$

■ INVERSE FUNCTIONS

ONE-TO-ONE RELATIONS

In the section **3.1.1** we studied **3** possible types of relations between number sets **X** and **Y** shown on the diagrams below: **1−1 relation, n−1 relation,** and **1−n relation.**
On the same diagrams we can see also possible **inverse** relations between sets **Y** and **X.**
As we can see only function **f** is a 1−1 relation and only **f** has an inverse function **f⁻¹**.
The very important facts are that $D_{f^{-1}} = R_f$ and $R_{f^{-1}} = D_f$.

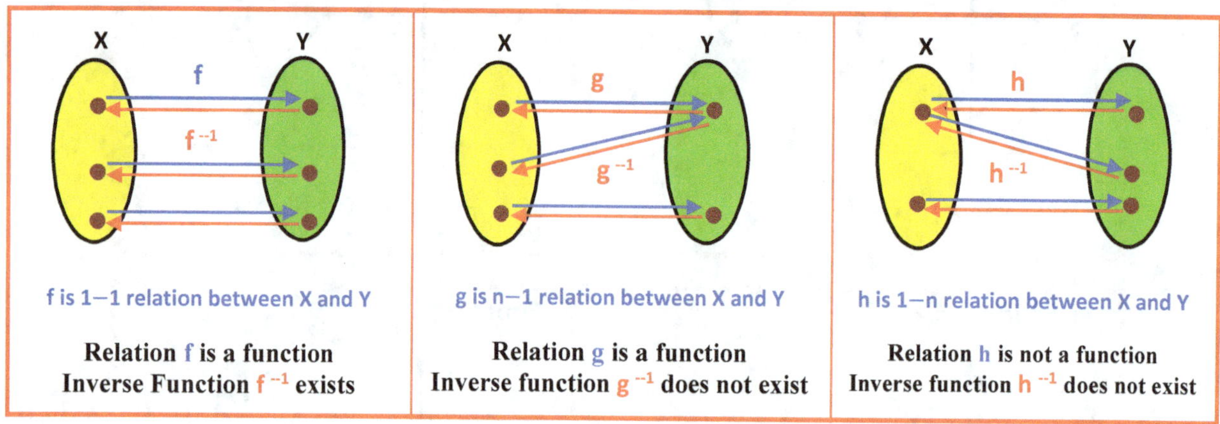

f is 1−1 relation between X and Y

Relation f is a function
Inverse Function f⁻¹ exists

g is n−1 relation between X and Y

Relation g is a function
Inverse function g⁻¹ does not exist

h is 1−n relation between X and Y

Relation h is not a function
Inverse function h⁻¹ does not exist

Example

Which of the following relations is a **function** with **inverse function**?

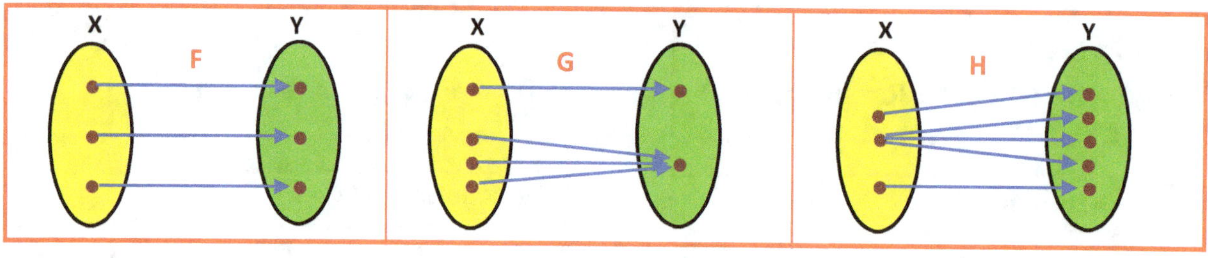

SOLUTION ▶

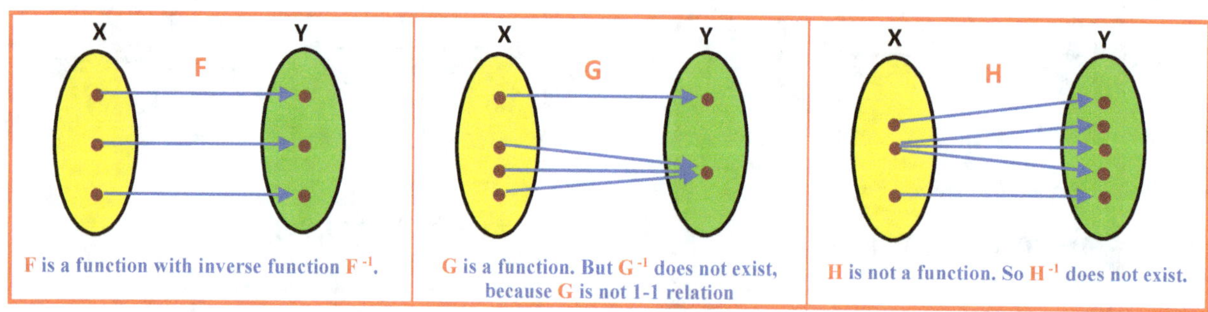

F is a function with inverse function F⁻¹.

G is a function. But G⁻¹ does not exist, because G is not 1-1 relation

H is not a function. So H⁻¹ does not exist.

HORIZONTAL LINE TEST FOR FUNCTIONS

We will use the **horizontal line test** to determine whether the given relation is 1-1 relation in case if we have a graph of the relation.

HORIZONTAL LINE TEST ▶ **A** graph represents a 1-1 relation if no horizontal line intersects the graph of function at more than one point.
That means the function corresponding to the given graph has an inverse function.

If there is as minimum one horizontal line with two or more intersects with the given graph of function, then the given graph is not a graph of 1-1 relation.
That means the function has not an inverse function.

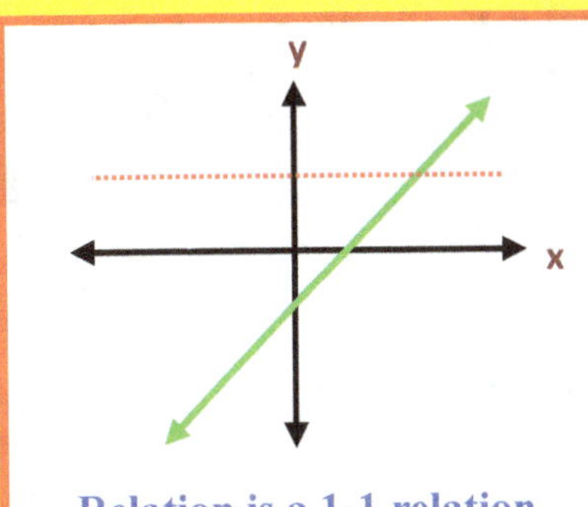

Relation is a 1-1 relation

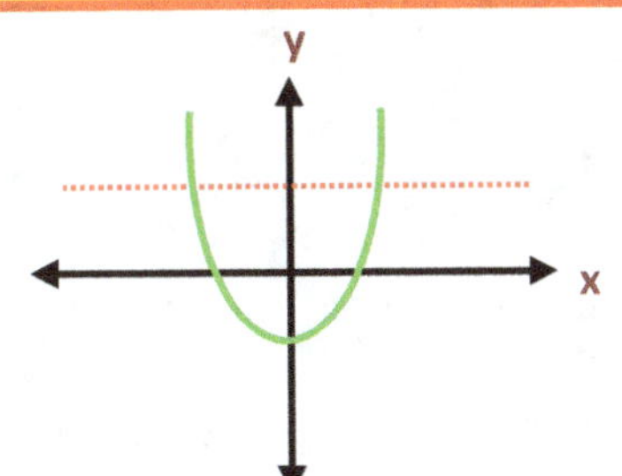

Relation is not a 1-1 relation

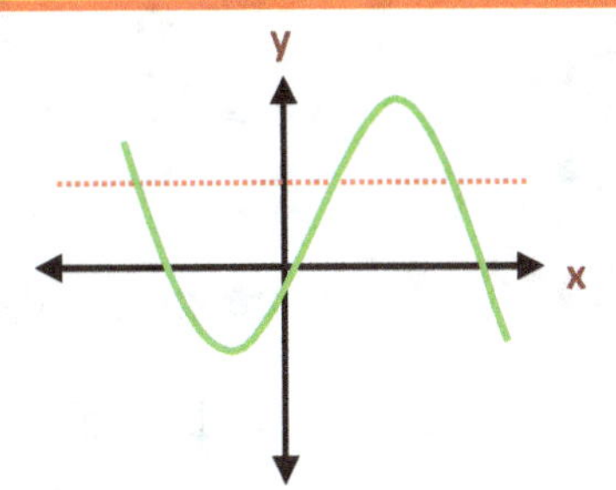

Relation is not a 1-1 relation

Examples Identify 1-1 relation by using the Horizontal Line Test

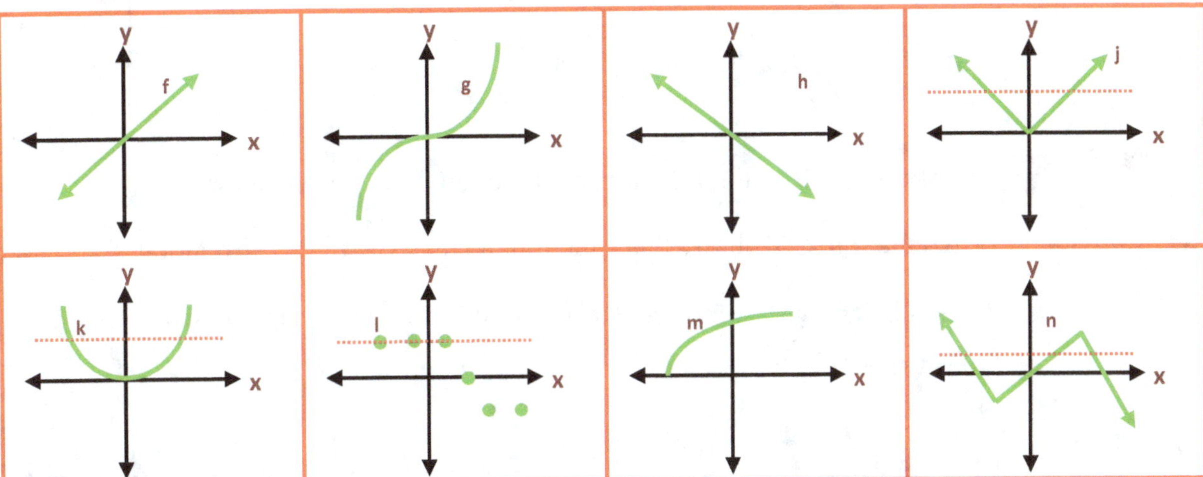

Solution

f is a 1-1 relation	g is a 1-1 relation	h is a 1-1 relation	j is not a 1-1 relation
k is not a 1-1 relation	l is not a 1-1 relation	m is a 1-1 relation	n is not a 1-1 relation

FINDING INVERSE FUNCTIONS

To find the inverse of the given function we have, first of all, to check is that function **1-1** relation, then rename **x** to **y**, **y** to **x**, and solve that equation for **y**. The resulted equation will represent the inverse function of the given function.

Example 1 Let $f(x) = x + 1$. Find the inverse function $f^{-1}(x)$ if that exists.

SOLUTION ▶ Horizontal line test shows that the function $f(x) = x + 1$ is a 1-1 relation.

To find $f^{-1}(x)$ we have to rename **x** to **y**, **y** to **x** and solve that equation for **y**.

$$y = f(x) = x + 1$$
$$x = y + 1$$
$$y = x - 1 \Rightarrow f^{-1}(x) = x - 1$$

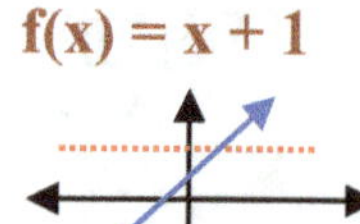

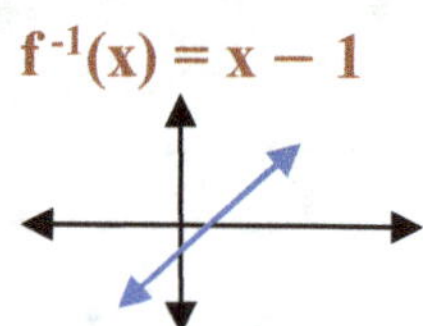

Example 2 Let $g(x) = \sqrt{x}$. Find the inverse function $g^{-1}(x)$ if that exists.

SOLUTION ▶ Horizontal line test shows that function $g(x) = \sqrt{x}$ is a 1-1 relation. And $x \geq 0$ because $g(x) = \sqrt{x}$.

To find $g^{-1}(x)$ we have to rename **x** to **y**, **y** to **x** and solve that equation for **y**.

$$y = g(x) = \sqrt{x}, \; x \geq 0$$
$$x = \sqrt{y}$$
$$y = x^2, \; x \geq 0 \Rightarrow g^{-1}(x) = x^2, \; x \geq 0$$

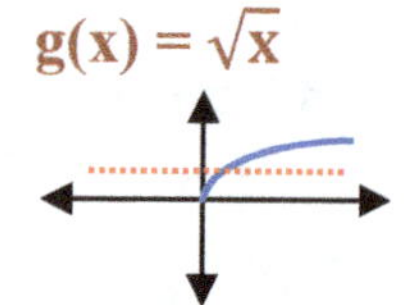

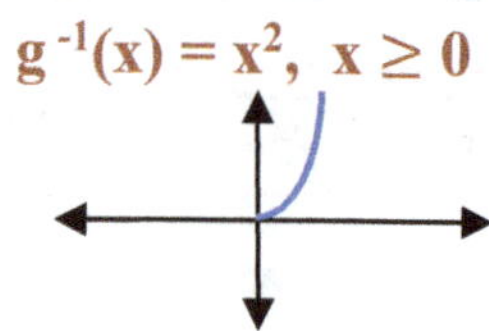

Example 3 Let $h(x) = 2x - 4$. Find the inverse function $h^{-1}(x)$ if that exists.

SOLUTION ▶ Horizontal line test shows that function $h(x) = 2x - 4$ is a 1-1 relation.

To find $h^{-1}(x)$ we have to rename **x** to **y**, **y** to **x** and solve that equation for **y**.

$$y = h(x) = 2x - 4$$
$$x = 2y - 4$$
$$2y = x + 4$$
$$y = \frac{x}{2} + 2 \Rightarrow h^{-1}(x) = \frac{x}{2} + 2$$

$$h(x) = 2x - 4 \qquad h^{-1}(x) = \frac{x}{2} + 2$$

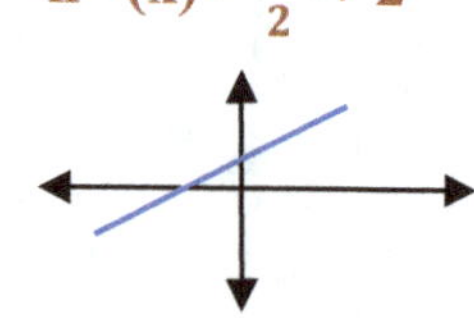

GRAPHS OF INVERSE FUNCTIONS

$\mathbf{W}$e know that $\mathbf{D}_{f^{-1}} = \mathbf{R}_f$ and $\mathbf{R}_{f^{-1}} = \mathbf{D}_f$. It means that the graphs of any function $\mathbf{f}$ and its inverse function $\mathbf{f}^{-1}$ are the symmetric graphs to each other related with the axis $\mathbf{y = x}$.

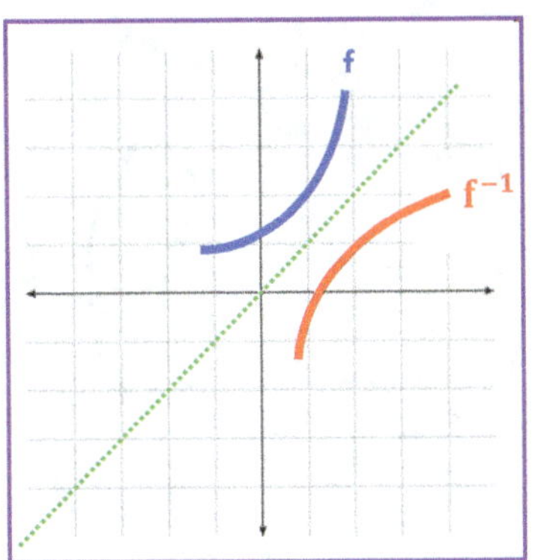

Example 1   Graph $\mathbf{f^{-1}(x)}$, $\mathbf{g^{-1}(x)}$, and $\mathbf{h^{-1}(x)}$, if $\mathbf{f(x) = \sqrt{x}}$, $\mathbf{g(x) = 2x + 1}$, $\mathbf{h(x) = x^3}$

SOLUTION▶

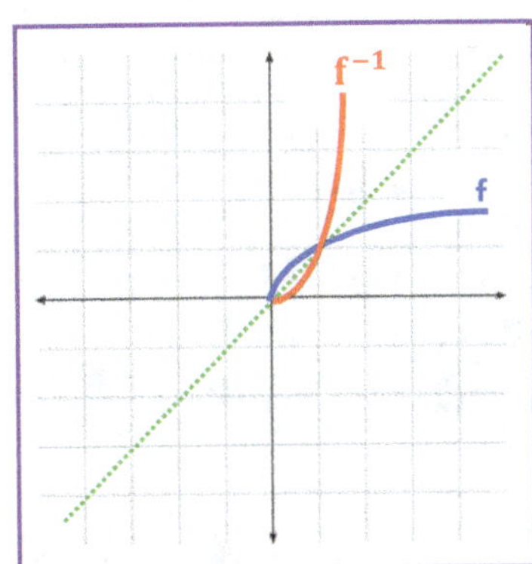
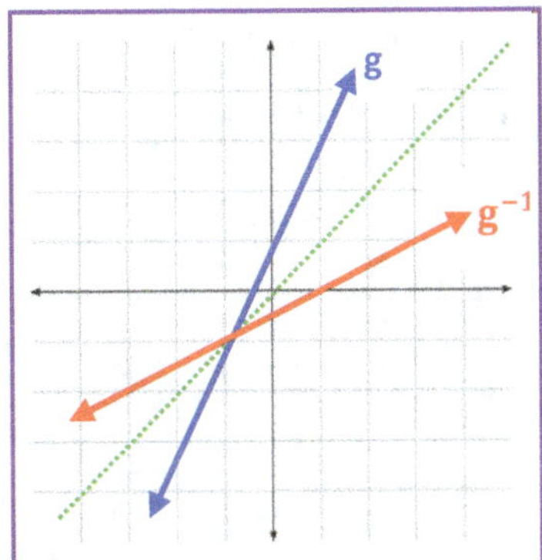
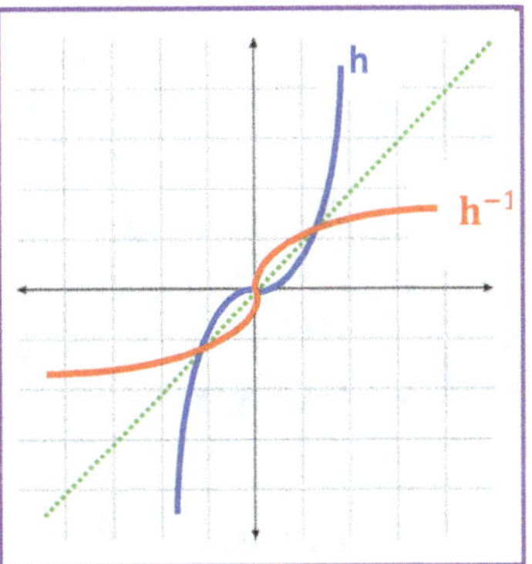

Example 2   Graph $\mathbf{f^{-1}(x)}$, $\mathbf{g^{-1}(x)}$, and $\mathbf{h^{-1}(x)}$, if $\mathbf{f(x) = \sqrt{x} + 1}$, $\mathbf{g(x) = 3x - 3}$, $\mathbf{h(x) = -x^3}$

SOLUTION▶

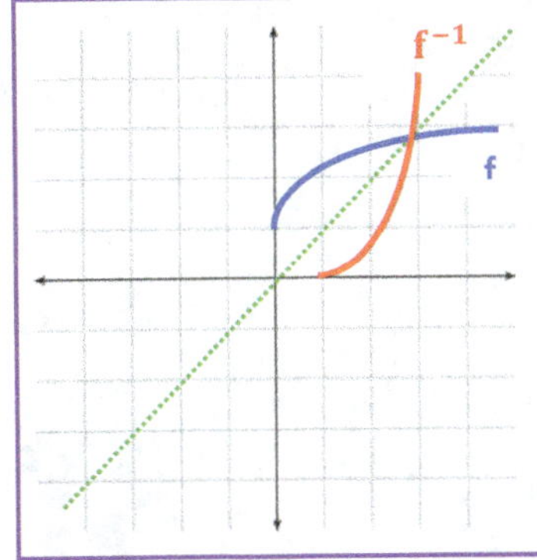
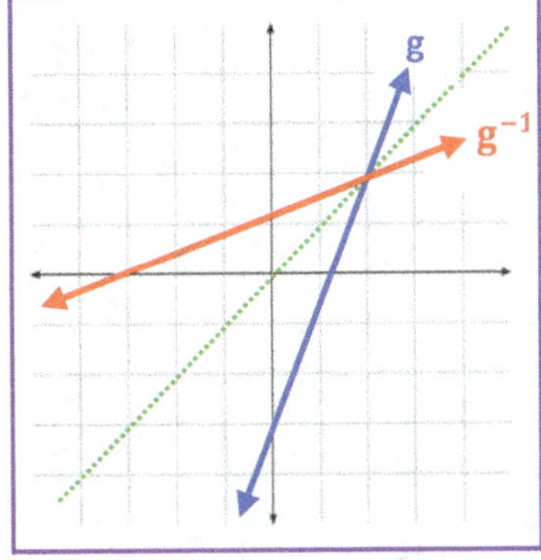
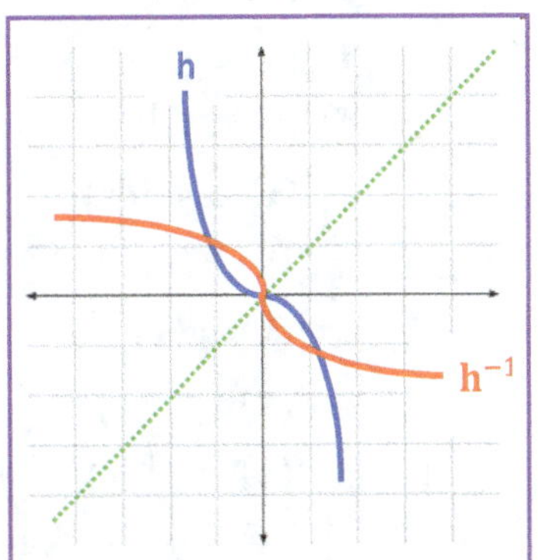

1. Determine which of the following relations represents **functions**.

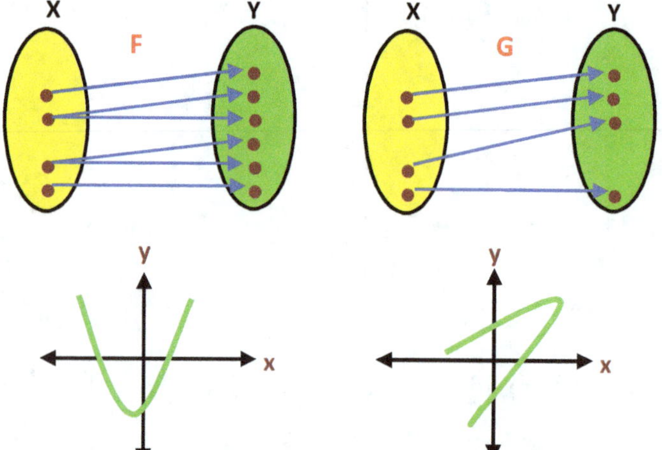

2. Find the domain and range of the function $f(x) = \sqrt{x + 1} - 2$

3. Let $f(x) = 2x + 1$, $g(x) = x$. Find the following:

a) $f(x) + 2g(x)$
b) $f(x) - g(x)$
c) $f(x)\,g(x)$
d) $f(x) \circ g(x)$
e) $g(x) \circ f(x)$
f) $f(x) \circ f(x)$
g) $g(x) \circ g(x)$
h) $f^{-1}(x)$
i) $g^{-1}(x)$
j) $f^{-1}(x) \circ f(x)$
k) $f^{-1}(x) \circ g(x)$
l) $f^{-1}(x) \circ f^{-1}(x)$
m) $f^{-1}(x) \circ g^{-1}(x)$
n) $g^{-1}(x) \circ f(x)$
o) $g^{-1}(x) \circ g(x)$
p) $g^{-1}(x) \circ f^{-1}(x)$
q) $g^{-1}(x) \circ g^{-1}(x)$

4. Classify functions as **even**, **odd**, or **neither**

a) $f(x) = |x| + 5$
b) $g(x) = x^5 + x^3 + x$
c) $h(x) = (x + 2)^2 - 1$

- **LIST OF TRANSFORMATIONS**
- **TRANSFORMATIONS OF BASIC FUNCTIONS**
- **COMBINATIONS OF TRANSFORMATIONS**

■ LIST OF TRANSFORMATIONS

In this section we will study **transformations of functions**: vertical and horizontal translations, vertical and horizontal stretching/compressing, and reflection about x- and y-axes.

TRANSLATION

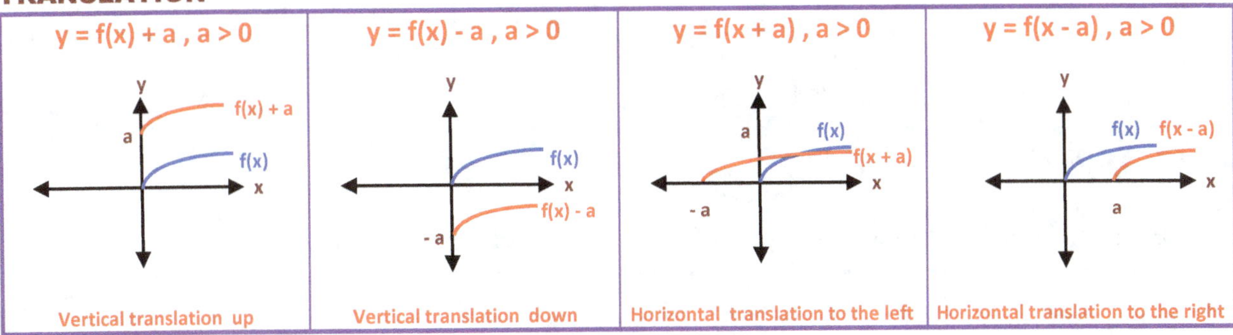

STRETCHING

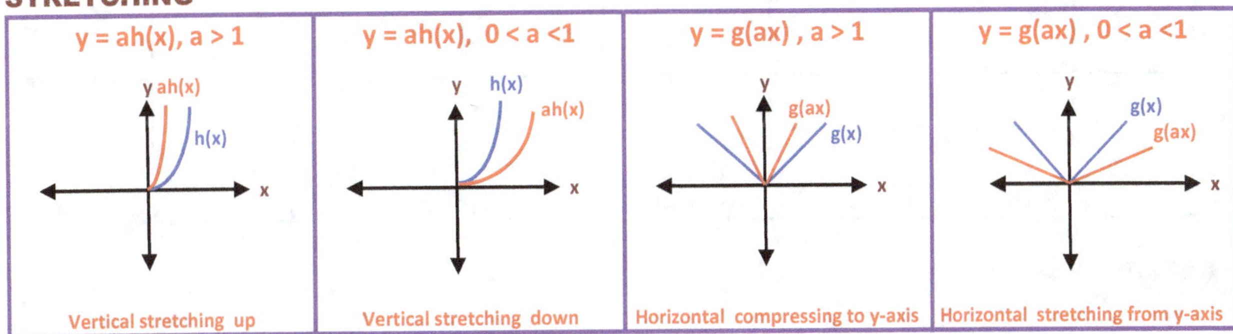

REFLECTION

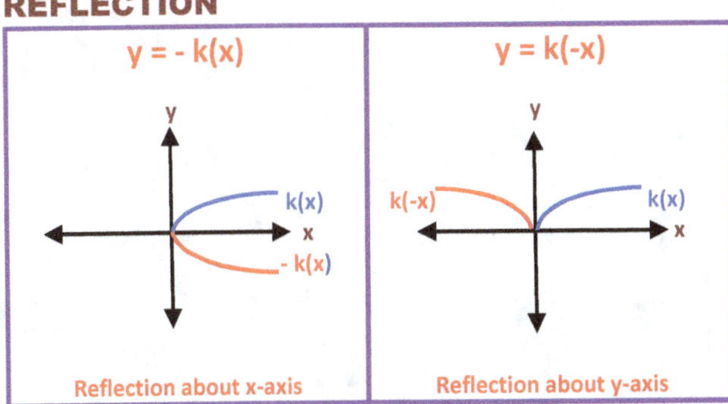

■ TRANSFORMATIONS OF BASIC FUNCTIONS

As examples we will study transformations of several basic algebraic functions.

TRANSLATION

Translations of the basic quadratic function $y = x^2$ are shown below.

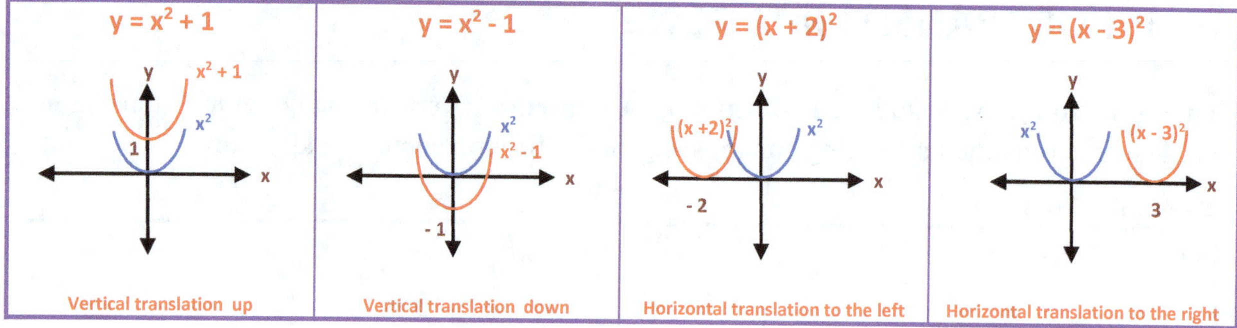

STRETCHING

Stretching of the basic function $y = |x|$ with different coefficients is shown below.

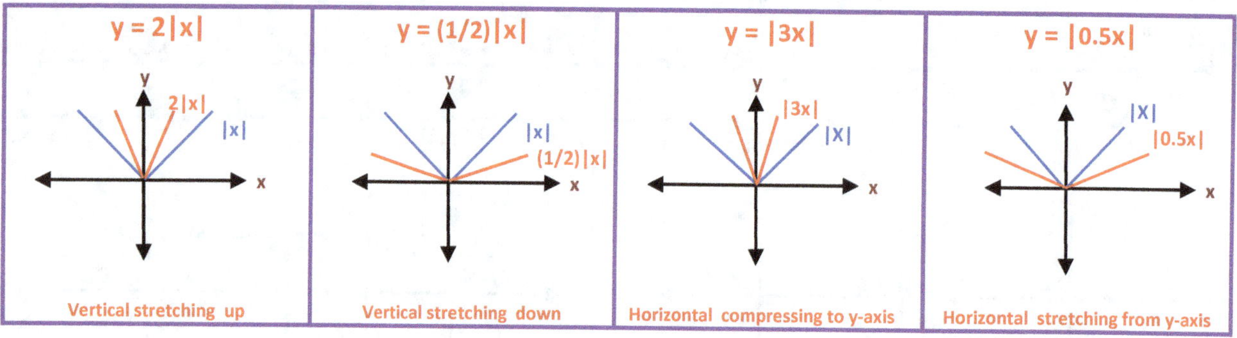

REFLECTION

Reflections of the basic functions $y = x$ and $y = \sqrt{x}$ are shown below.

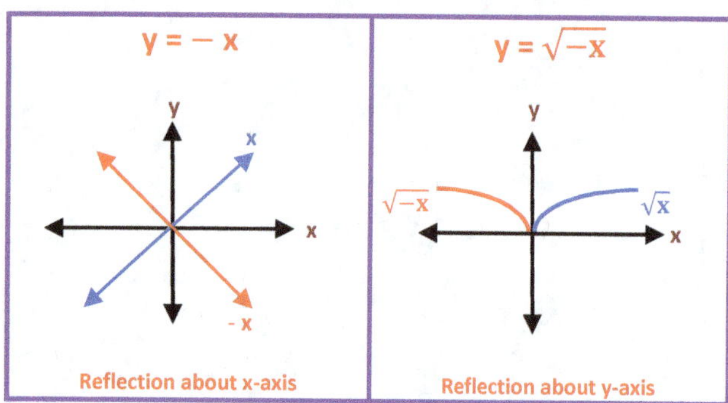

■ COMBINATIONS OF TRANSFORMATIONS

In this section we will study several examples with combinations of different transformations.

Example 1 Sketch the graphs $f(x) = (x - 2)^3 + 3$; $g(x) = 2(x + 1) - 2$; $h(x) = -|x + 2| - 1$

SOLUTION ▶ To sketch the graph $f(x) = (x - 2)^3 + 3$ we use the basic algebraic function $y = x^3$ with **2** transformations: translation **2** units to the right and translation **3** units up.
To sketch the graph $g(x) = 2(x + 1) - 2$ we use the basic algebraic function $y = x$ with **3** transformations: translation **1** unit to the left, translation **2** units down, and stretching up with coefficient **2**.
To sketch the graph $h(x) = -|x + 2| - 1$ we use the basic algebraic function $y = |x|$ with **3** transformations: translation **2** units to the left, translation **1** unit down, and reflection about **x**.

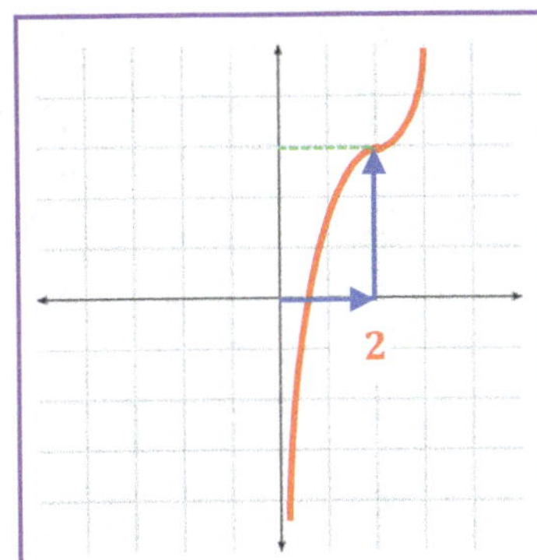
$f(x) = (x - 2)^3 + 3$

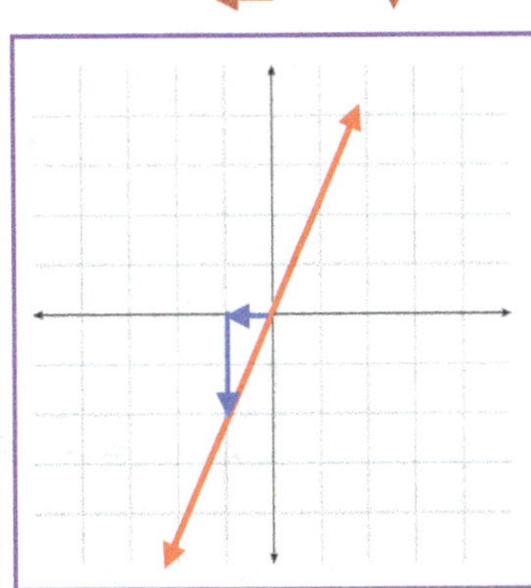
$g(x) = 2(x + 1) - 2$

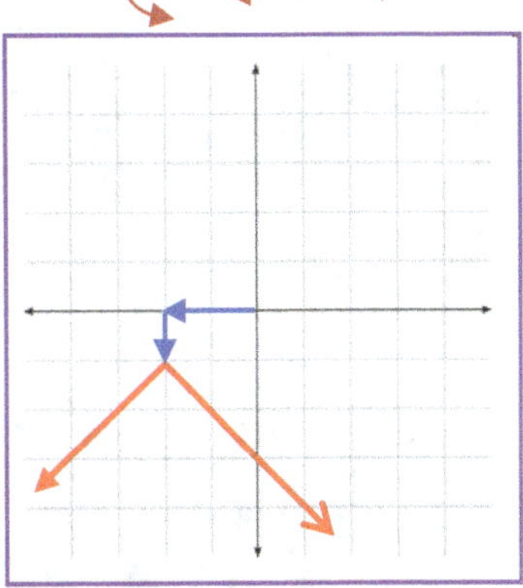
$h(x) = -|x + 2| - 1$

Example 2 Sketch the graphs $f(x) = \sqrt{x + 2} + 1$, $g(x) = -3x - 2$, $h(x) = (x + 1)^2 + 2$

SOLUTION ▶

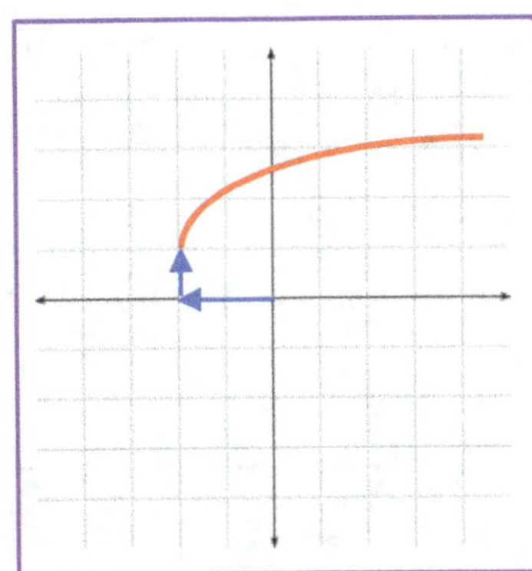
$f(x) = \sqrt{x + 2} + 1$

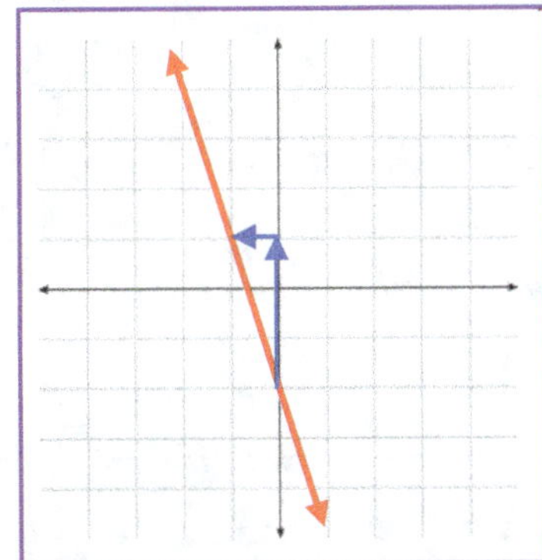
$g(x) = -3x - 2$

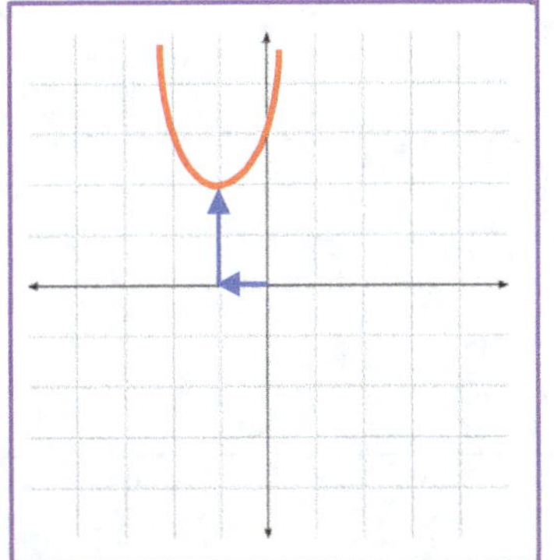
$h(x) = (x + 1)^2 + 2$

Sketch the graph of each **function.**

1. $y = -x - 2$

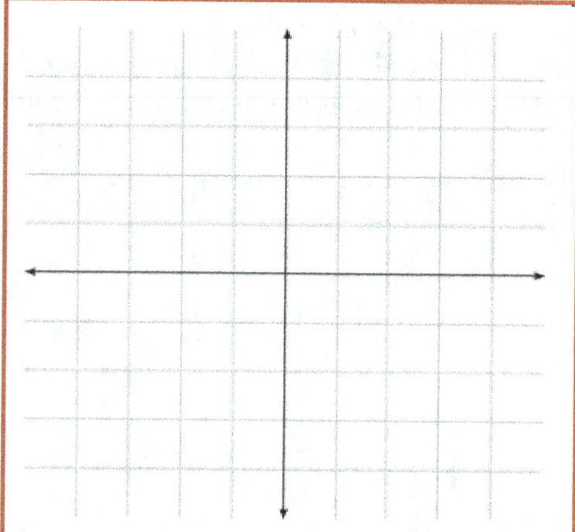

2. $y = -|x + 2| + 1$

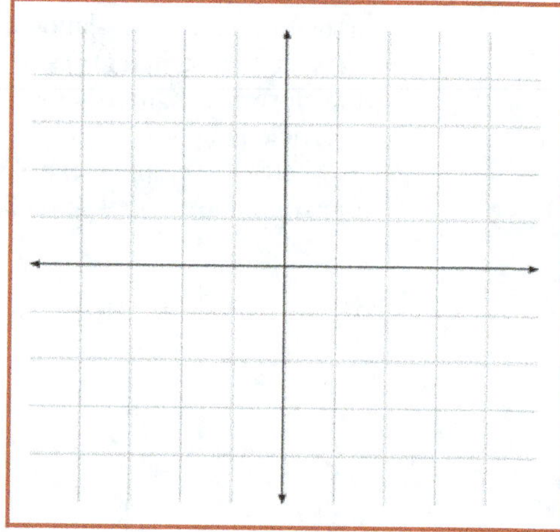

3. $y = (x - 2)^2 + 2$

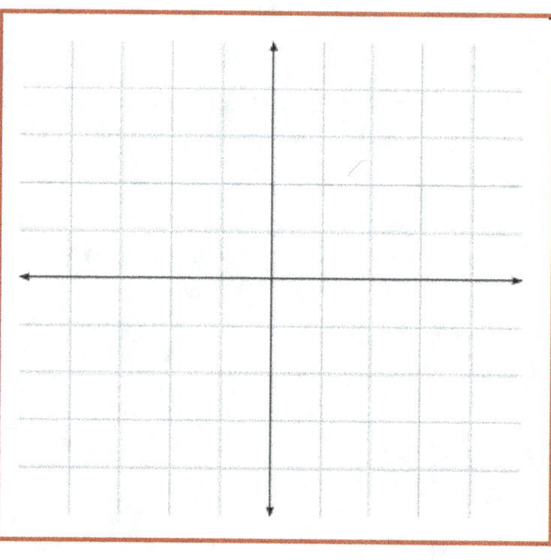

4. $y = -(1 - x)^3$

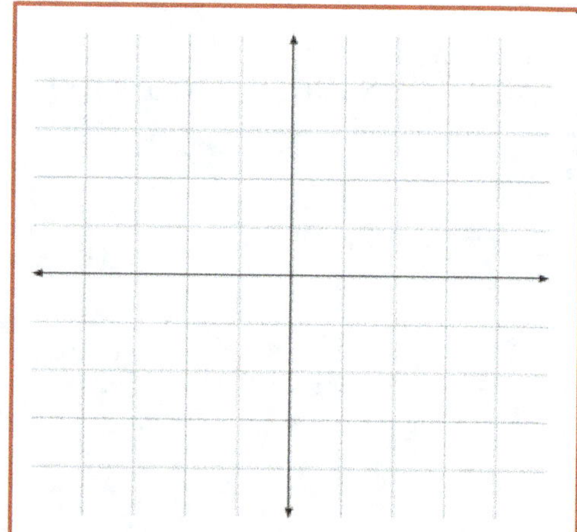

7.3. Polynomial and Fractional of Functions

- POLYNOMIAL FUNCTIONS
- FRACTIONAL FUNCTIONS

■ POLYNOMIAL FUNCTIONS

We will use the following definition of a **polynomial function**.

Definition ▶ A **polynomial function with the power n** is a function

$$y = P_n(x) = a_n x^n + a_{n-1} x^{n-1} + ... + a_1 x + a_0$$

where $a_n \neq 0$, and $a_n, a_{n-1}, ..., a_1, a_0 \in R, n \in N.$ [*]

- If **n** is an even number (**n = 2k**) the function is called the polynomial function with even power

$$y = P_{2k}(x) = a_{2k} x^{2k} + a_{2k-1} x^{2k-1} + ... + a_1 x + a_0$$

where $a_{2k} \neq 0$, and $a_{2k}, a_{2k-1}, ..., a_1, a_0 \in R, \; n \in N.$

- If **n** is an odd number (**n = 2k+1**) the function is called the polynomial function with odd power

$$y = P_{2k+1}(x) = a_{2k+1} x^{2k+1} + a_{2k} x^{2k} + ... + a_1 x + a_0$$

where $a_{2k+1} \neq 0$, and $a_{2k+1}, a_{2k}, ..., a_1, a_0 \in R, n \in N.$

[*] **R** is the set of the real numbers and **N** is the set of the natural numbers.

Examples The basic algebraic function $y = x$ as well as any linear function $y = mx + b$ are polynomial functions with odd power **1**.
The basic algebraic function $y = x^2$ as well as any quadratic function $y = ax^2 + bx + c$ are polynomial functions with even power **2**.
The functions $y = x^3$, $y = -2x^3$, $y = x^3 - 2$, $y = x^3 - x$, and $y = x^3 + 3x^2 + 2x$ are the polynomial functions with odd power **3**.
The functions $y = x^4$, $y = -x^4$, $y = -x^4 + 2$, $y = x^4 - x^2$, and $y = x^4 - 3x^3$ are the polynomial functions with even power **4**.
The functions $y = x^5$, $y = -x^5$, $y = x^5 - 2$, $y = x^5 - 4x^3$, and $y = x^5 + x^4$ are the polynomial functions with odd power **5**.

END BEHAVIORS OF EVEN AND ODD POWER FUNCTIONS

The end behaviors of **even** and **odd** power polynomial functions at $x \to \pm\infty$ are shown below.

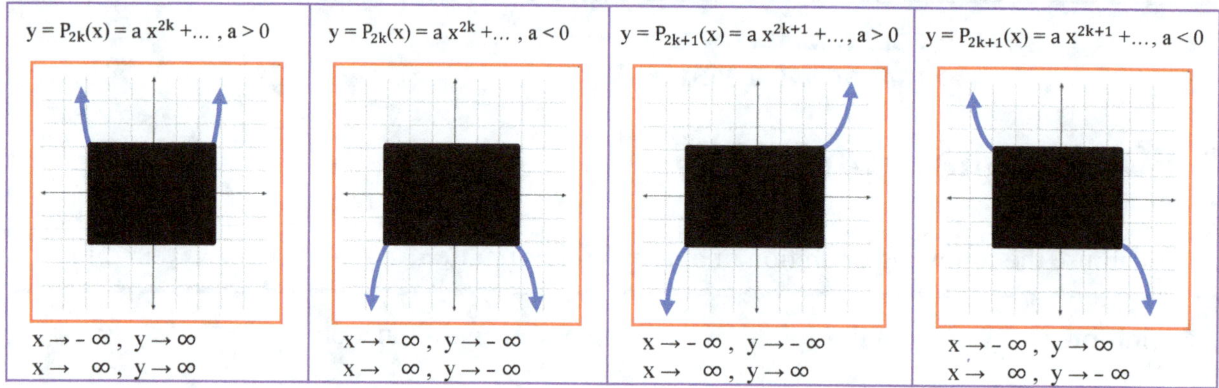

There are different curves with different numbers of x- and y-intercepts inside the black boxes depended on the natures of other non-leading terms of the polynomial functions.

The end behaviors and curves inside the black boxes of several basic functions are shown below.

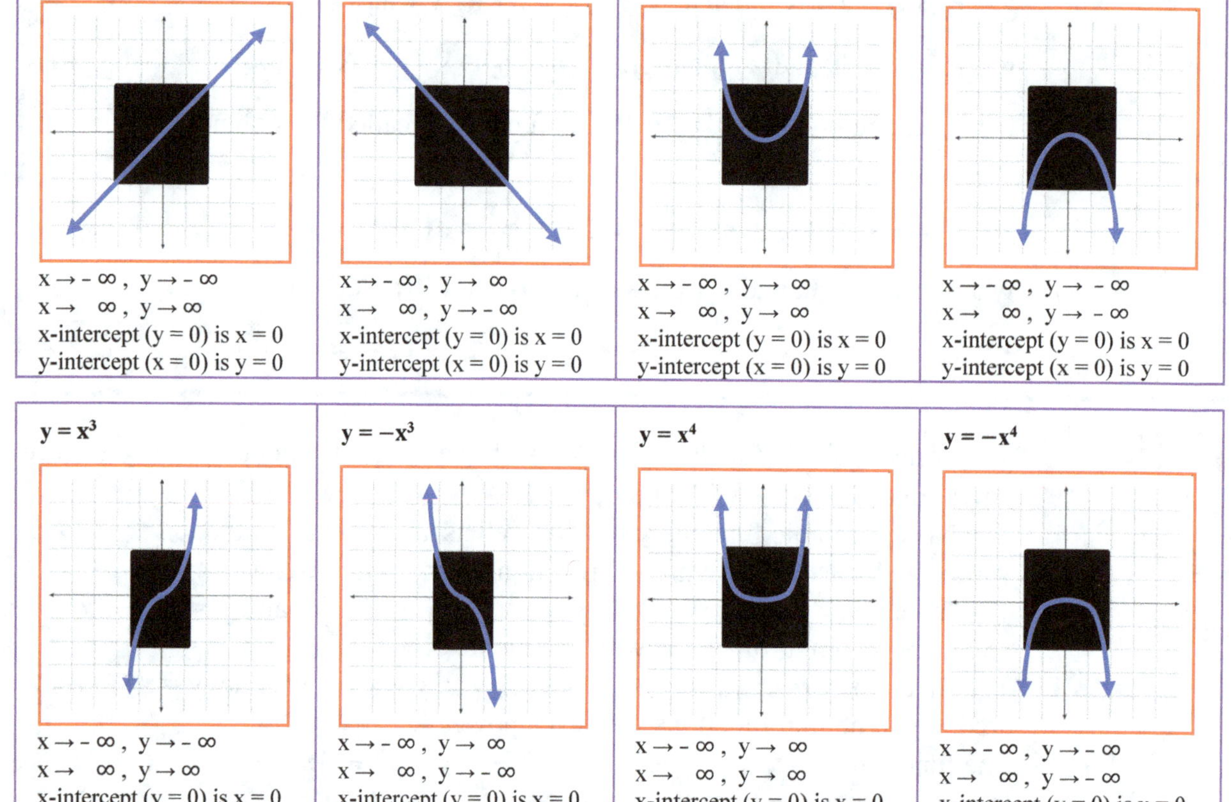

SKETCHING GRAPHS OF POLYNOMIAL FUNCTIONS

The recipe to sketch a graph of any polynomial function is:
1. Find x- and y-intercepts of the given polynomial function.
2. Use end behaviors of the given function.
3. Connect x- and y-intercepts and use corresponding end behaviors of the given polynomial function to sketch the graph of that function.

Examples Sketch the graphs of the functions $y = x^3 - x$, $y = x^4 - x^2$, and $y = x^5 - 5x^3 + 4x$.

SOLUTION ▶ To sketch the graph $y = x^3 - x$ we use:

1. **x-intercepts** (is same as **y = 0**): **x = 0, 1, –1**
   $y = 0 \rightarrow x^3 - x = 0 \rightarrow x(x^2 - 1) = 0 \rightarrow x(x - 1)(x + 1) = 0 \rightarrow x = 0, 1, -1$
2. **y-intercepts** (is same as **x = 0**): **y = 0** $(x = 0 \rightarrow y = 0)$
3. **The end behaviors** of the polynomial function with odd power and positive leading coefficient (power of function is **3**, leading coefficient is **1**).

To sketch the graph $y = x^4 - x^2$ we use:

1. **x-intercepts:** **x = 0, 0, 1, –1** (0 with multiplicity 2)
   $y = 0 \rightarrow x^4 - x^2 = 0 \rightarrow x^2(x^2 - 1) = 0 \rightarrow x^2(x - 1)(x + 1) = 0 \rightarrow x = 0, 1, -1$
2. **y-intercepts:** **y = 0** $(x = 0 \rightarrow y = 0)$
3. **The end behaviors** of polynomial function with even power and positive leading coefficient (power of function is **4**, leading coefficient is **1**).

To sketch the graph $y = x^5 - 5x^3 + 4x$ we use:

1. **x-intercepts:** **x = 0, 1, –1, 2, –2** $(y = 0 \rightarrow x^5 - 5x^3 + 4x \rightarrow$
   $x(x^4 - 5x^2 + 4) = 0 \rightarrow x(x^2 - 4)(x^2 - 1) = 0 \rightarrow x = 0, 1, -1, 2, -2)$
2. **y-intercepts:** **y = 0** $(x = 0 \rightarrow y = 0)$
3. **The end behaviors** of polynomial function with even power and positive leading coefficient (power of function is **4**, leading coefficient is **1**).

$y = x^3 - x$	$y = x^4 - x^2$	$y = x^5 - 5x^3 + 4x$
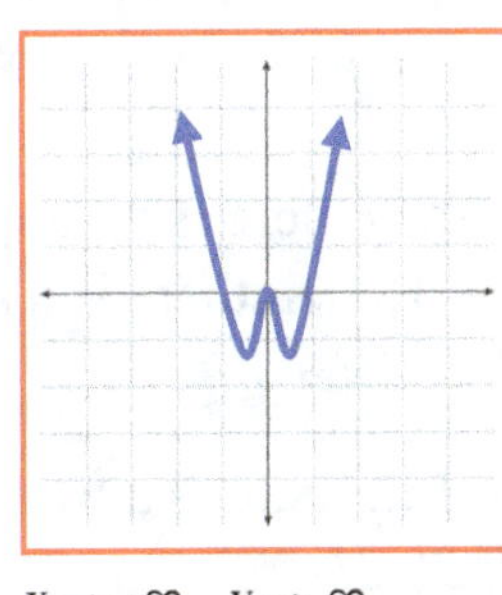	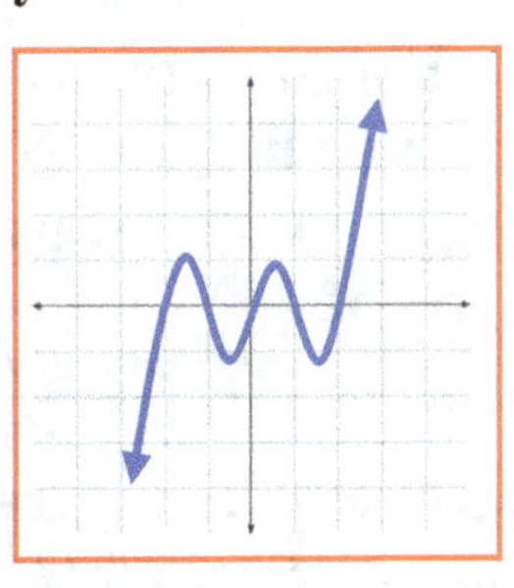	
$x \rightarrow -\infty, \ y \rightarrow -\infty$	$x \rightarrow -\infty, \ y \rightarrow \infty$	$x \rightarrow -\infty, \ y \rightarrow -\infty$
$x \rightarrow \infty, \ y \rightarrow \infty$	$x \rightarrow \infty, \ y \rightarrow \infty$	$x \rightarrow \infty, \ y \rightarrow \infty$
x-intercept (y = 0) is x = 0, 1, -1	x-intercept (y = 0) is x = 0, 0, 1, -1	x-intercept (y = 0) is x = 0, 1, -1, 2, -2
y-intercept (x = 0) is y = 0	y-intercept (x = 0) is y = 0	y-intercept (x = 0) is y = 0

■ FRACTIONAL FUNCTIONS

We will use the following definition of a single-variable **fractional (or rational) function**.

> **Definition** ▶ A **single-variable fractional function** is a function
>
> $$y = \frac{P(x)}{Q(x)}$$
>
> where $Q(x) \neq 0$, $P(x)$ and $Q(x)$ are any polynomial expressions.

Examples The basic algebraic function $y = \dfrac{1}{x}$ is a fractional function.

The functions $y = \dfrac{1}{x+2}$; $y = \dfrac{x-1}{x+1}$; $y = \dfrac{x-1}{x-4}$; $y = \dfrac{x-1}{x-4}$; $y = \dfrac{9x+1}{-3x+6}$; $y = \dfrac{x^2-2x}{x^2-1}$;

$y = \dfrac{x^2-1}{x^2-4}$; $y = \dfrac{x^2-4}{x^2-1}$; $y = \dfrac{2x^2-4x+5}{x^2-4x+4}$ also are fractional/rational functions.

THE RECIPE TO SKETCH THE GRAPHS OF FRACTIONAL FUNCTIONS

The recipe to sketch a graph of any fractional function is:
1. **Find vertical asymptote (s):** solve equation $Q(x) = 0$.
2. **Find horizontal or slant asymptote:** compare the powers of leading terms $P(x)$ and $Q(x)$.
3. **Find x- and y-intercepts of the given polynomial function:** put $y = 0$ and $x = 0$.
4. **Analyze end behaviors of the given function:** analyze behaviors of the function on left and right sides of vertical asymptote(s) and end behaviors if $x \to -\infty$ and if $x \to \infty$.
5. **Connect x- and y-intercepts and use corresponding end behaviors** of the given function and sketch the graph of that function.

To find the vertical asymptote(s) we have to solve the equation $\mathbf{Q(x) = 0}$ (denominator is zero). Vertical line(s) corresponds to solution(s) of equation $\mathbf{Q(x) = 0}$ is (are) vertical asymptote (s) for the given function.

To find the Horizontal or slant asymptote(s) we have to compare the powers of the leading terms of the expressions $\mathbf{P(x)}$ and $\mathbf{Q(x)}.$
- If **degree of $P(x)$ > degree of $Q(x)$,** then there is no any horizontal asymptote.
- If **degree of $P(x)$ = degree of $Q(x)$ + 1,** then there is a **slant asymptote line** with the equation $y = \dfrac{coeficient\ of\ leading\ term\ of\ P(x)}{coeficient\ of\ leading\ term\ of\ Q(x)}\,\mathbf{x.}$
- If **degree of $P(x)$ = degree of $Q(x)$,** then there is a **horizontal asymptote line** with the equation $y = \dfrac{coeficient\ of\ leading\ term\ of\ P(x)}{coeficient\ of\ leading\ term\ of\ Q(x)}.$
- If **degree of $P(x)$ < degree of $Q(x)$,** then there is a **horizontal asymptote line $y = 0$.**

Below we will study the sketching of the graphs of the several fractional functions according to this recipe.

Example 1 Sketch the graph of the function $y = \dfrac{1}{x}$

SOLUTION▶

1. The vertical asymptote is a line **x = 0**.
2. The horizontal asymptote is a line **y = 0**, because the degree of the numerator is **0** which is less than the degree of the denominator, that is **1**.
3. There is no **x**-intercept(s) because numerator is **1** $(\neq 0)$, so **y** can't be zero. There is no y-intercepts because x can't be zero (denominator of fraction can't be 0).
4. End behaviors will be:

$$x \to -\infty, \; y \to \frac{1}{-\infty} \to 0^-$$

$$x \to \;\;\infty, \; y \to \frac{1}{\infty} \to 0^+$$

$$x \to 0^-, \quad y \to \frac{1}{0^-} \to -\infty$$

$$x \to 0^+, \quad y \to \frac{1}{0^+} \to \infty$$

5. The graph of the function $y = \dfrac{1}{x}$ is shown below

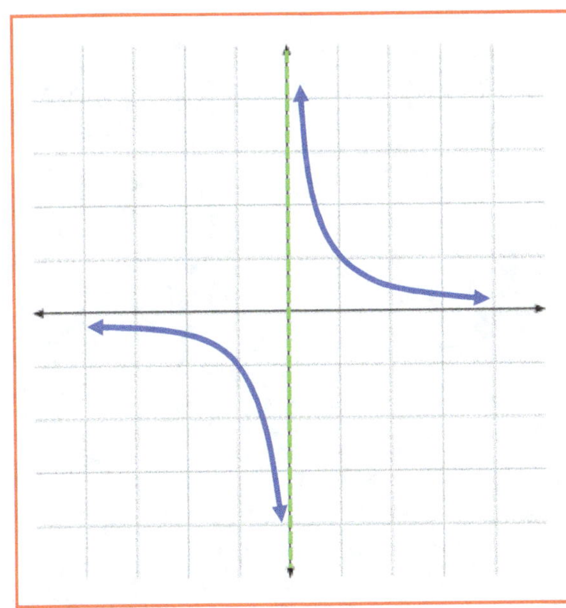

  Sketch the graph of the function $y = \dfrac{4x + 8}{-2x + 2}$

SOLUTION ▶

1. The vertical asymptote is a line $x = 1$ ($-2x + 2 = 0 \rightarrow x = 1$)
2. The horizontal asymptote is a line $y = -2$ (the degree of the numerator is **1**, the degree of the denominator is **1**, so horizontal asymptote is $y = \dfrac{4}{-2} \rightarrow y = -2$)
3. The x-intercept is $x = -2$ ($y = 0 \rightarrow 4x + 8 = 0 \rightarrow 4x = -8 \rightarrow x = -2$)

    The y-intercept is $y = 4$ ($x = 0 \rightarrow y = \dfrac{8}{2} \rightarrow y = 4$)
4. End behaviors will be:

    $$x \rightarrow -\infty, y \rightarrow \frac{4(-\infty)+8}{-2(-\infty)+2} \rightarrow -2^{+}$$

    $$x \rightarrow \infty, y \rightarrow \frac{4(\infty)+8}{-2(\infty)+2} \rightarrow -2^{-}$$

    $$x \rightarrow 1^{-}, \quad y = \frac{4x+8}{-2x+2} \rightarrow \infty$$

    $$x \rightarrow 1^{+}, \quad y = \frac{4x+8}{-2x+2} \rightarrow -\infty$$

5. The graph of the function $y = \dfrac{4x + 8}{-2x + 2}$ is shown below

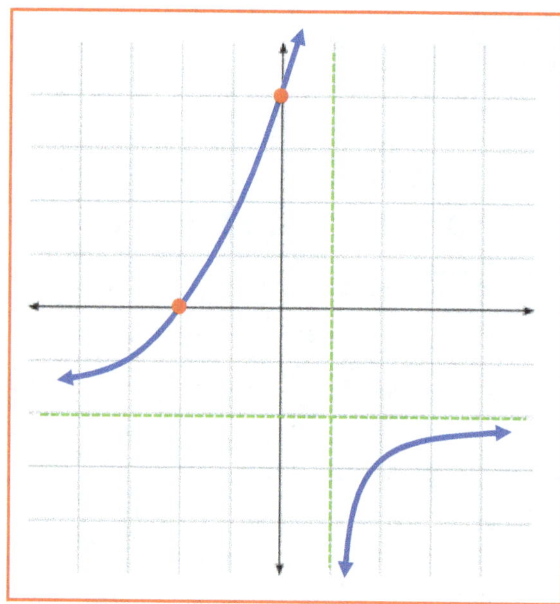

Example 3 Sketch the graph of the function $y = \dfrac{x-2}{x-1}$

SOLUTION ▶

1. The vertical asymptote is a line $x = 1$ $(x - 1 = 0 \rightarrow x = 1)$

2. The horizontal asymptote is a line $y = 1$ (the degree of the numerator is **1**, the degree of the denominator is **1**, so horizontal asymptote is $y = \dfrac{1}{1} \rightarrow y = 1$)

3. The x-intercept is $x = 2$ $(y = 0 \rightarrow x - 2 = 0 \rightarrow x = 2)$

   The y-intercept is $y = 2$ $(x = 0 \rightarrow y = \dfrac{-2}{-1} \rightarrow y = 2)$

4. End behaviors will be:

   $$x \rightarrow -\infty, \ y \rightarrow \dfrac{-\infty - 2}{-\infty - 1} \rightarrow 1^{+}$$

   $$x \rightarrow \ \infty, \ y \rightarrow y \rightarrow \dfrac{\infty - 2}{\infty - 1} \rightarrow 1^{-}$$

   $$x \rightarrow 1^{-}, \quad y \rightarrow \dfrac{x-2}{x-1} \rightarrow \infty$$

   $$x \rightarrow 1^{+}, \quad y \rightarrow \dfrac{x-2}{x-1} \rightarrow -\infty$$

5. The graph of the function $y = \dfrac{x-2}{x-1}$ shown below

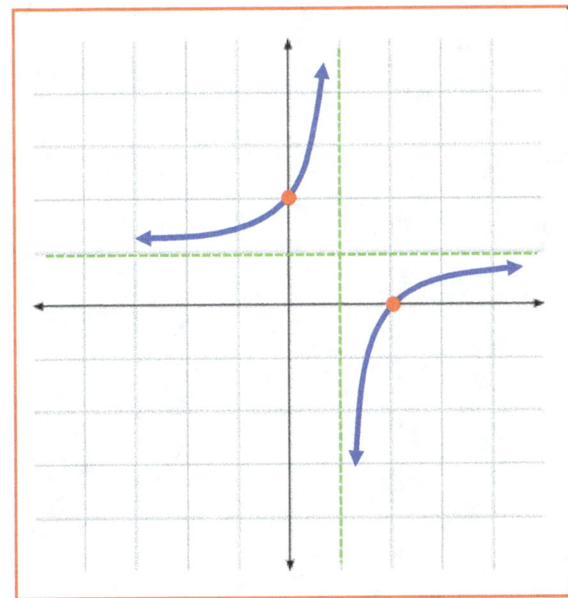

<table><tr><td>**Example 4**</td><td>Sketch the graph of the function $y = \dfrac{x^2 - 1}{x^2 - 4}$</td></tr></table>

1. The vertical asymptotes are lines $x = -2$ and $x = 2$ ($x^2 - 4 = 0 \rightarrow x = -2, 2$)

2. The horizontal asymptote is a line $y = 1$ (the degree of the numerator is **2**, the degree of the denominator is **2**, so horizontal asymptote is $y = \dfrac{1}{1} \rightarrow y = 1$)

3. The x-intercepts are $x = -1$ and $x = 1$ ($y = 0 \rightarrow x^2 - 1 = 0 \rightarrow x = -1, 1$)

   The y-intercept is $y = \dfrac{1}{4}$ ($x = 0 \rightarrow y = \dfrac{-1}{-4} \rightarrow y = \dfrac{1}{4}$)

4. End behaviors will be:

$$x \rightarrow -\infty, \quad y = \frac{x^2 - 1}{x^2 - 4} \rightarrow \frac{\infty - 1}{\infty - 4} \rightarrow 1^+$$

$$x \rightarrow \infty, \quad y = \frac{x^2 - 1}{x^2 - 4} \rightarrow \frac{\infty - 1}{\infty - 4} \rightarrow 1^+$$

$$x \rightarrow 2^-, \quad y = \frac{(x-1)(x+1)}{(x-2)(x+2)} = \frac{(+)(+)}{(-)(+)} \rightarrow y \rightarrow -\infty$$

$$x \rightarrow 2^+, \quad y = \frac{(x-1)(x+1)}{(x-2)(x+2)} = \frac{(+)(+)}{(+)(+)} \rightarrow y \rightarrow \infty$$

$$x \rightarrow -2^-, \quad y = \frac{(x-1)(x+1)}{(x-2)(x+2)} = \frac{(-)(-)}{(-)(-)} \rightarrow y \rightarrow \infty$$

$$x \rightarrow -2^+, \quad y = \frac{(x-1)(x+1)}{(x-2)(x+2)} = \frac{(-)(-)}{(-)(+)} \rightarrow y \rightarrow -\infty$$

5. The graph of the function $y = \dfrac{x-2}{x-1}$ is shown below

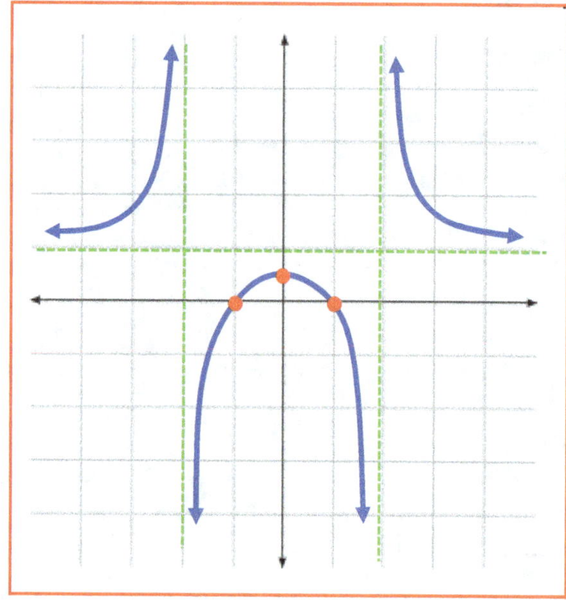

Example 5   Sketch the graph of the function $y = \dfrac{x^2 - 2x - 3}{x - 1}$

SOLUTION▶

1. The vertical asymptote is a line $x = 1$ ($x - 1 = 0 \rightarrow x = 1$)
2. The slant asymptote is a line $y = x$ (the degree of the numerator is **2**, the degree of the denominator is **1**, so slant asymptote is $y = x$)
3. The x-intercepts are $x = -1$ and $x = 3$ ($y = 0 \rightarrow x^2 - 2x - 3 = 0 \rightarrow (x - 3)(x + 1) = 0$)

   The y-intercept is $y = 3$ ($x = 0 \;\rightarrow\; y = \dfrac{-3}{-1} \;\rightarrow y = 3$)
4. End behaviors will be:

$$x \rightarrow -\infty, \quad y = \frac{x^2 - 2x - 3}{x - 1} \;\rightarrow\; x^+$$

$$x \rightarrow \;\;\infty, \quad y = \frac{x^2 - 2x - 3}{x - 1} \;\rightarrow\; x^-$$

$$x \rightarrow 1^-, \quad y = \frac{x^2 - 2x - 3}{x - 1} = \frac{(x - 3)(x + 1)}{x - 1} = \frac{(-)(+)}{(-)} \;\rightarrow y \rightarrow \infty$$

$$x \rightarrow 1^+, \quad y = \frac{x^2 - 2x - 3}{x - 1} = \frac{(x - 3)(x + 1)}{x - 1} = \frac{(-)(+)}{(+)} \;\rightarrow y \rightarrow -\infty$$

5. The graph of the function $y = \dfrac{x^2 - 2x - 3}{x - 1}$ is shown below

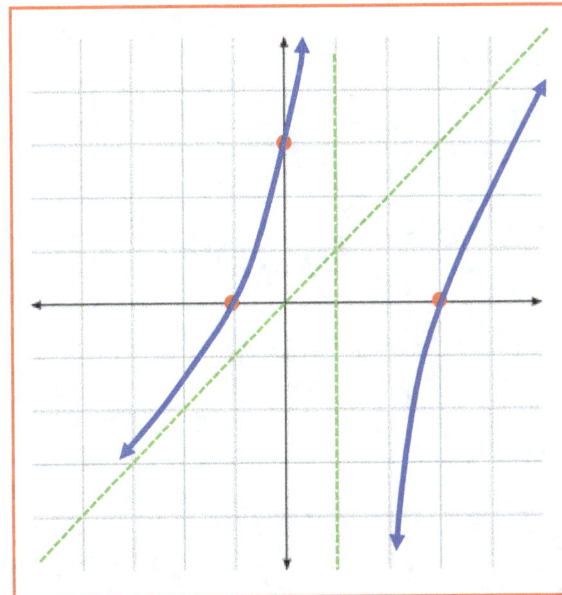

Sketch the graphs of the following functions.

1. $y = x^3 - 2x^2 - 3x$

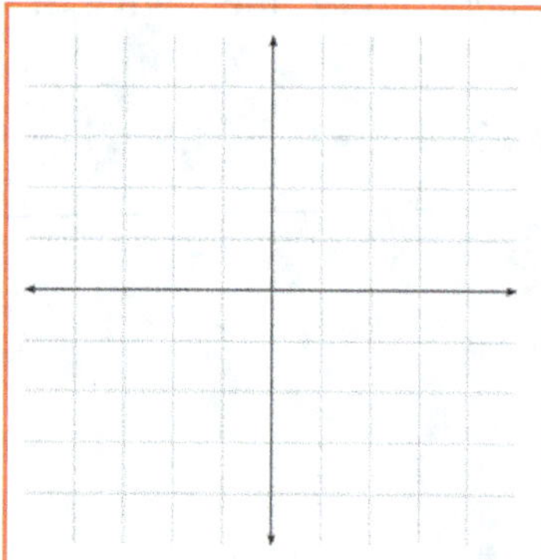

2. $y = \dfrac{1}{x-1}$

3. $y = x^4 - 2x^3 + x^2$

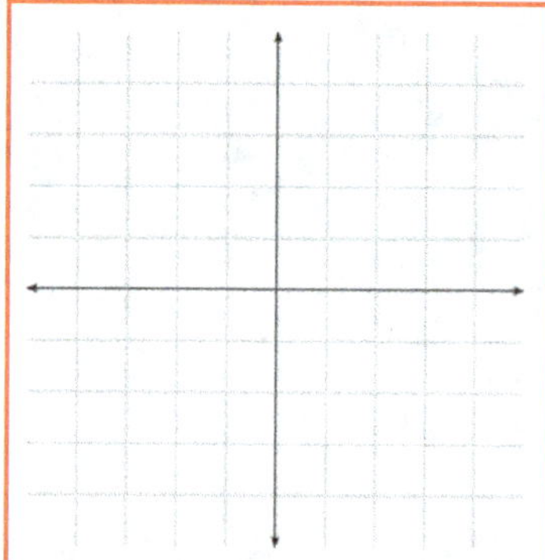

4. $y = \dfrac{x+2}{x-2}$

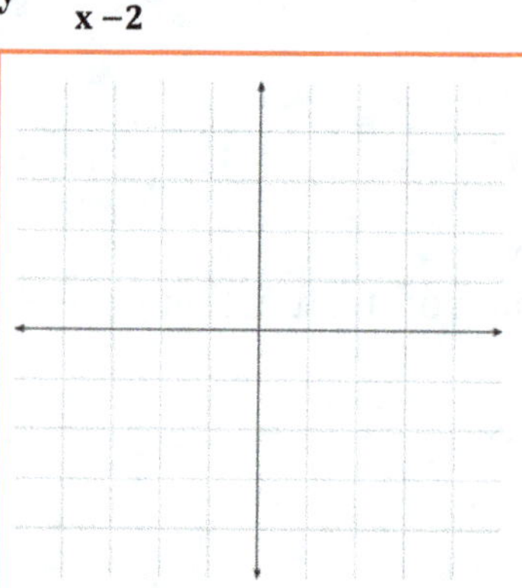

5. $y = x(x+1)(x-1)(x+2)(x-2)$

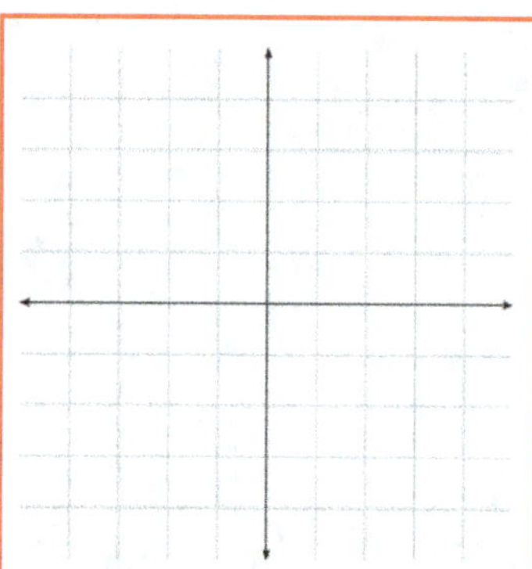

6. $y = \dfrac{x^2-4}{x^2-1}$

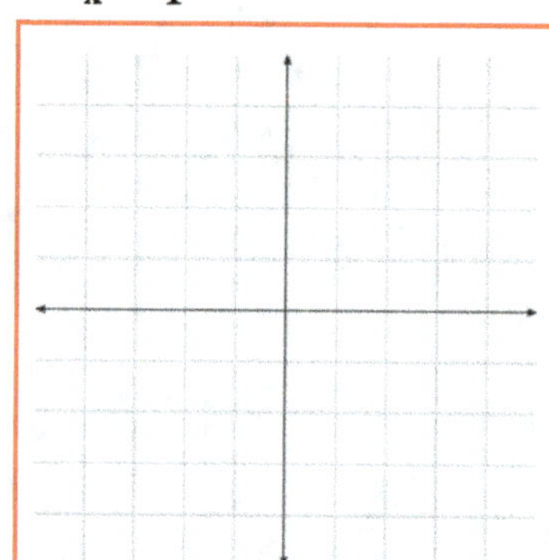

1. Decide whether the given relation is a function.
   If the given relation is a function, give the domain and the range.

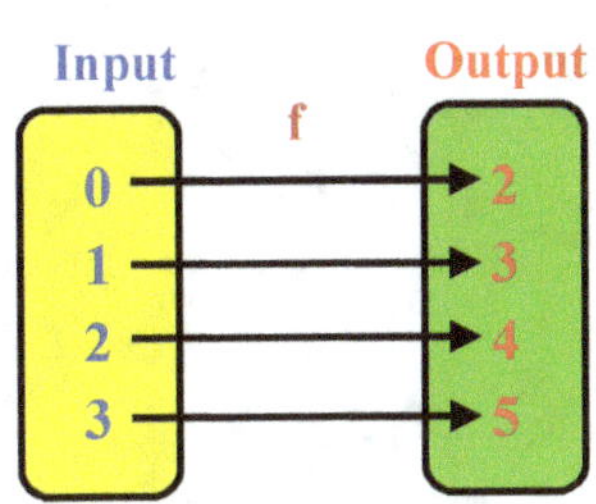
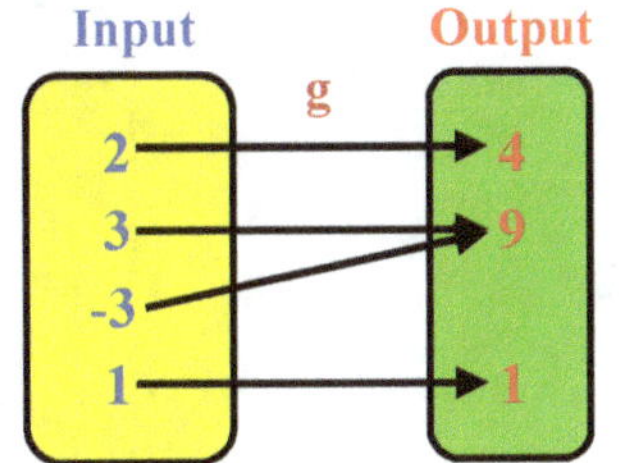
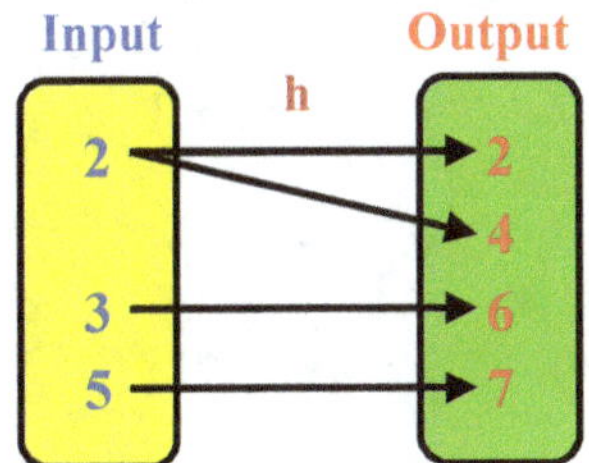

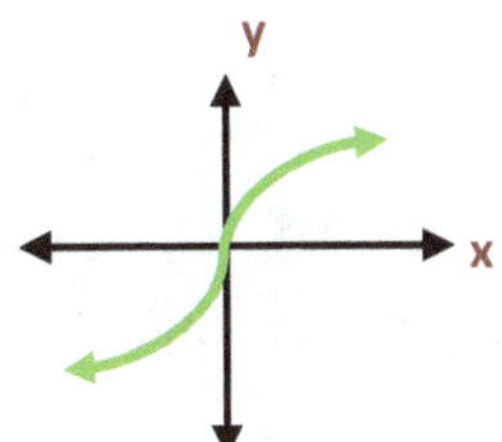
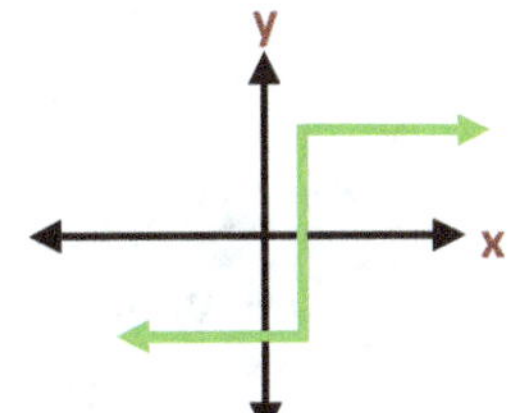
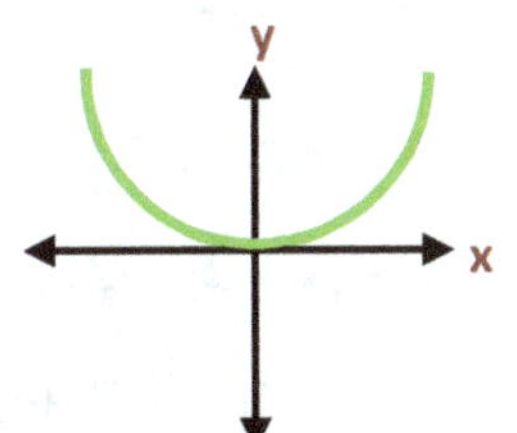

2. Find the domain of the functions

   a) $f(x) = \sqrt{x-1}/(x-4)$
   b) $g(x) = 1/x\sqrt{x-2}$
   c) $h(x) = x^5 + x^3 - 2$
   d) $k(x) = 1/\sqrt{x+1} + 1/(x-3)$

3. Find the domain and the range of the functions

   a) $f(x) = (x-1)^3 - 1$
   b) $g(x) = 1/[(x-1)\sqrt{x+2}\,]$
   c) $h(x) = x^6 - 1$
   d) $k(x) = 1/\sqrt{1-x} + 1/(x+2)$

4. Classify functions as even, odd, or neither

   a) $f(x) = 2|x|$
   b) $g(x) = x^3 - x$
   c) $h(x) = x^2 - 3x$
   d) $k(x) = x^4 - x^6$

5. Let $f(x) = x + 2$, $g(x) = 3x - 3$.
   Find and graph the functions

   a) $f^{-1}(x)$
   b) $g^{-1}(x)$

6. Let $f(x) = x$, $g(x) = x + 1$.
   Find the following

   a) $f(x) + g(x) + 2f(x)g(x) + \dfrac{f(x)}{g(x)+1}$
   b) $f(x) \circ f(x) + f(x) \circ g(x)$
   c) $g(x) \circ f(x) - g(x) \circ g(x)$
   d) $f(x) \circ f^{-1}(x) + f^{-1}(x) \circ f(x)$
   e) $f^{-1}(x) \circ f^{-1}(x)$
   f) $f^{-1}(x) \circ g(x)$
   g) $f^{-1}(x) \circ g^{-1}(x)$
   h) $g(x) \circ f^{-1}(x)$
   i) $g(x) \circ g^{-1}(x) - g^{-1}(x) \circ g(x)$
   j) $g^{-1}(x) \circ g^{-1}(x)$
   k) $g^{-1}(x) \circ f^{-1}(x)$

Sketch the graph of each function using transformation.

7. $y = |-x| + 1$

8. $y = -|x| - 1$

9. $y = |x - 2|$

10. $y = -|x + 2|$

11. $y = |x - 2| + 1$

12. $y = |x + 2| - 3$

13. $y = -2|x - 1| + 1$

14. $y = -x^2 - 1$

15. $y = -x^2 + 2$

16. $y = x^2 + 1$

17. $y = -2x^2 - 1$

18. $y = -2(x - 1)^2$

19. $y = (x - 2)^2 - 3$

20. $y = (x + 3)^2 + 1$

21. $y = -(x - 1)^2$

22. $y = -(x - 3)^2 + 1$

23. $y = (x - 2)^3 + 3$

24. $y = \sqrt{x + 1} - 1$

25. $y = -2\sqrt{x - 2} + 1$

26. Write the basic function corresponded to the graph bellow.

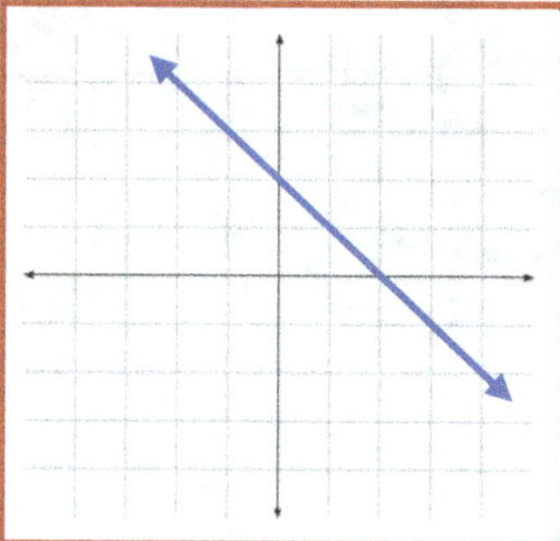

27. Write the basic function corresponded to the graph bellow.

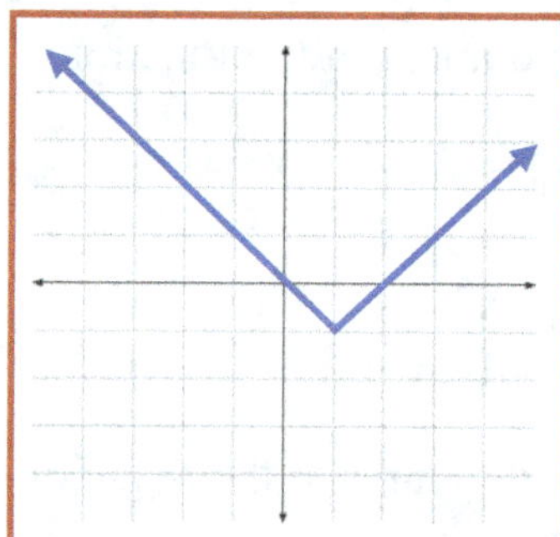

28. Write the basic function corresponded to the graph bellow.

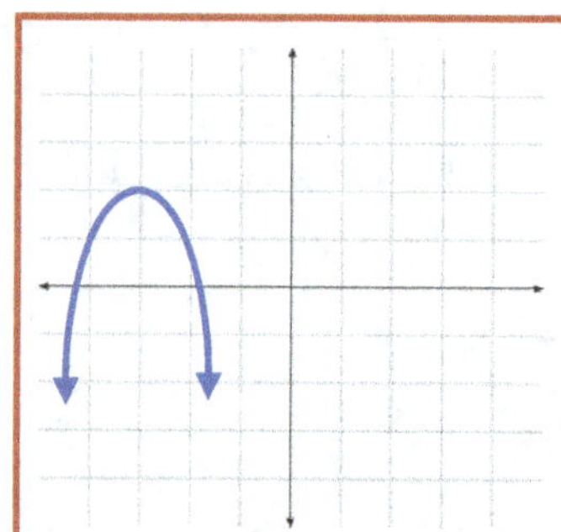

Sketch the graph of each function.

29. $y = x^3 - 3x^2$

30. $y = x^3 - 4x$

31. $y = -2x^3 - 2x$

32. $y = -x^3 - 3x^2 + x + 3$

33. $y = x(x + 1)(x + 3)$

34. $y = x(x - 1)(x + 2)$

35. $y = -x(x - 2)(x + 4)$

36. $y = x^2(x - 1)$

37. $y = x^4 - 4x^2$

38. $y = x^4 - 5x^2 + 4$

39. $y = x^4 - 10x^2 + 9$

40. $y = x^4 + 3x^3 - x^2 - 3x$

41. $y = x(x-1)(x + 2)(x + 3)$

42. $y = x(x+1)(x - 3)(x - 4)$

43. $y = x^2(x - 3)(x + 2)$

44. $y = x^5 + 2x^4 - x^3 - 2x^2$

45. $y = x(x+1)(x +2)(x - 1)(x - 2)$

46. $y = -x^2(x + 1)^2(x - 2)$

47. $y = \dfrac{1}{x + 2}$

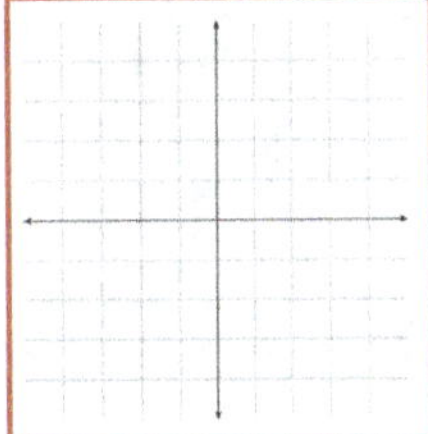

48. $y = \dfrac{x - 3}{x + 1}$

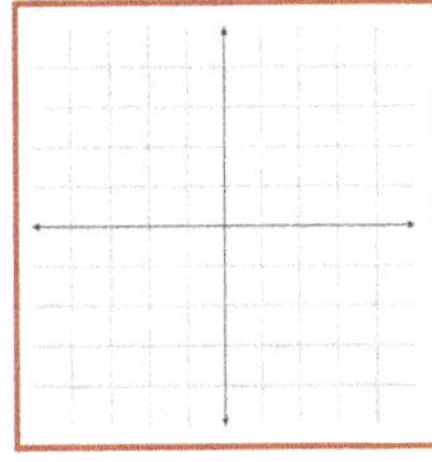

49. $y = \dfrac{x^2 - 1}{x^2 - 4}$

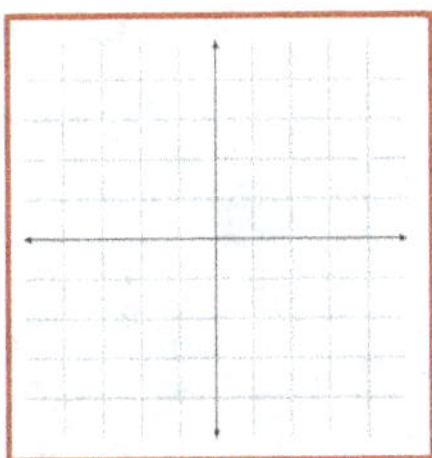

50. $y = \dfrac{x^2 - 1}{x + 2}$

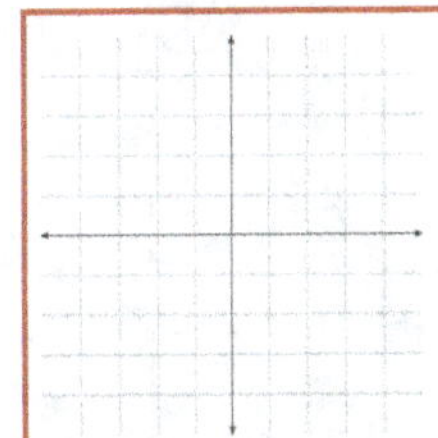

EXTRA PRACTICE
Functions and Graphs

Hayk Yegoryan & Rubik Yegoryan

INTERMEDIATE ALGEBRA

Sketch the graphs of the following functions.

51. $y = x^3 - 2x^2$

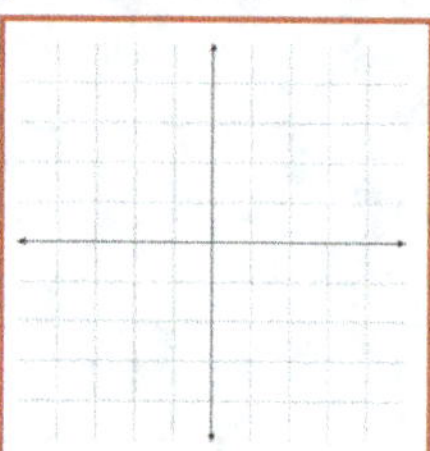

52. $y = -x^3 - 4x$

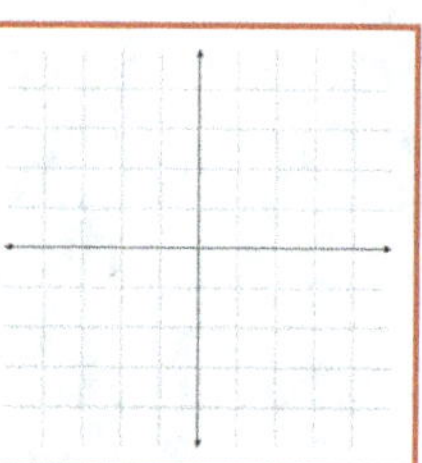

53. $y = -x^4 + 9x^2$

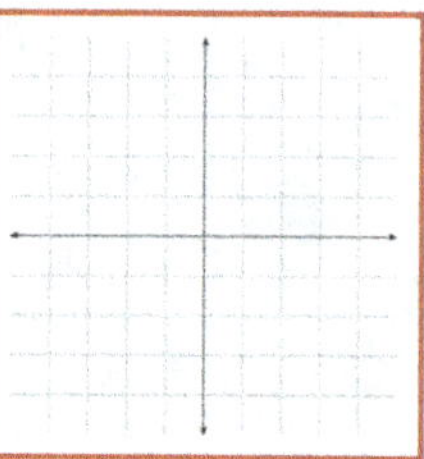

54. $y = x^2(x-1)(x+1)(x+2)$

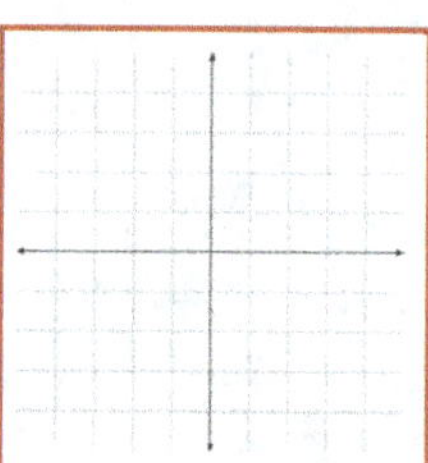

55. $y = \dfrac{1}{x-3}$

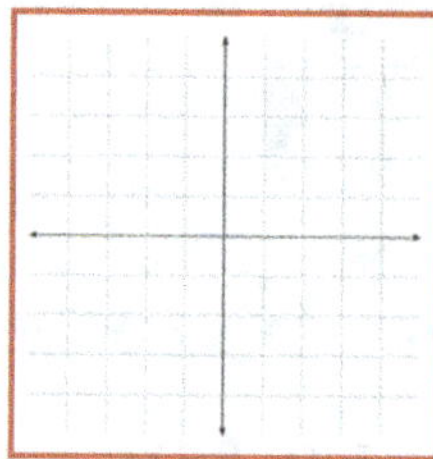

56. $y = \dfrac{x-2}{x-4}$

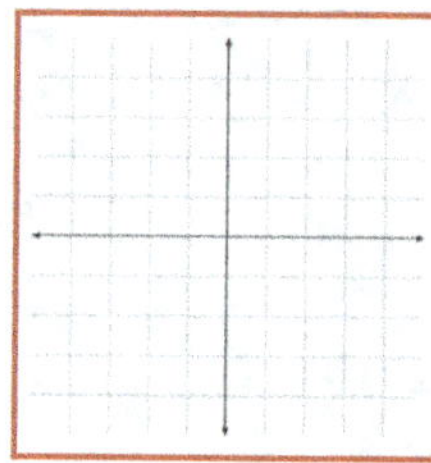

57. $y = \dfrac{x^2-9}{x^2-1}$

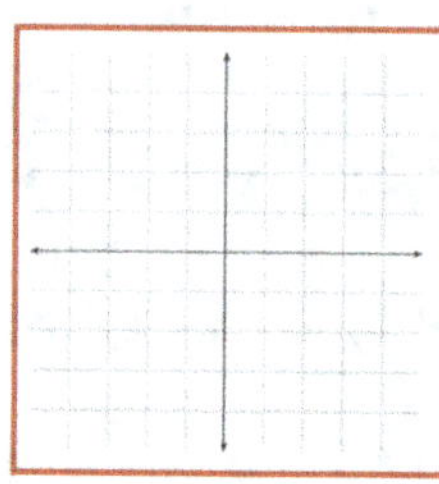

58. $y = \dfrac{x^2-4}{x+2}$

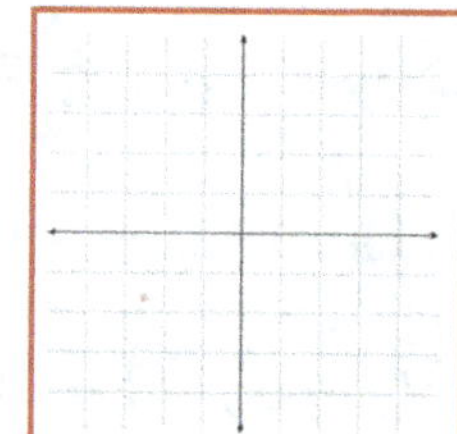

1. Let **f(x) = x + 1** and **g(x) = x − 2**. Find the following.

    a) $f^{-1}(x)$
    b) $g^{-1}(x)$
    c) $f^{-1}(x) \circ g^{-1}(x)$
    d) $g^{-1}(x) \circ f^{-1}(x)$
    e) $g(x) \circ f(x)$
    f) $f(x) \circ g(x)$
    g) $g^{-1}(x) \circ g(x)$
    h) $f^{-1}(x) \circ f^{-1}(x)$

2. Sketch the graph of functions using transformations.

    a) **f(x) = −|x + 3| − 1**
                                     b) **g(x) = (x + 2)² + 1**

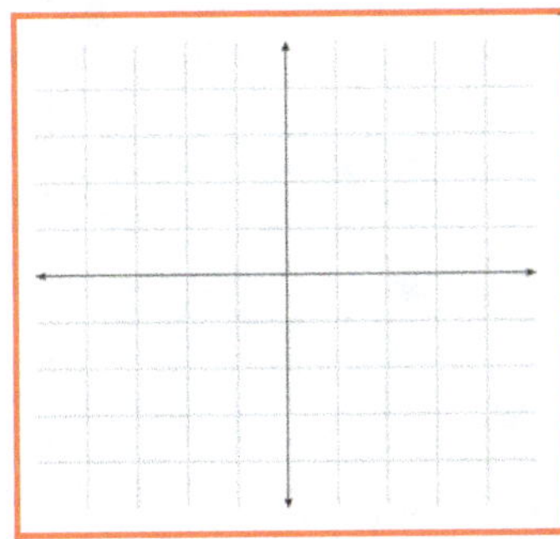
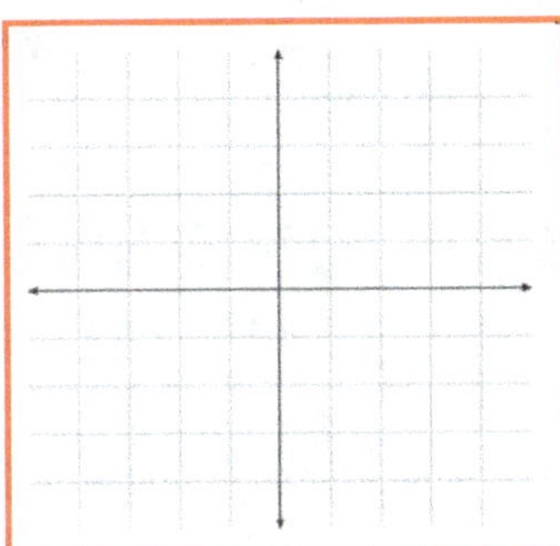

3. Sketch the graph of polynomial and fractional functions.

    a) **f(x) = x(x − 1)²(x + 1)(x + 2)²**
                                 b) **g(x) = $\dfrac{x-2}{x+1}$**

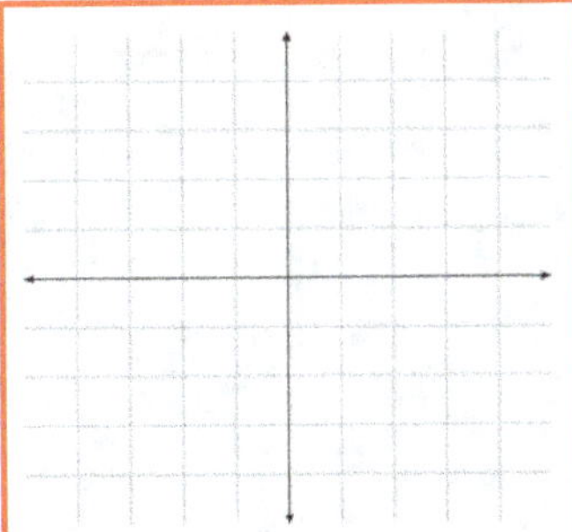
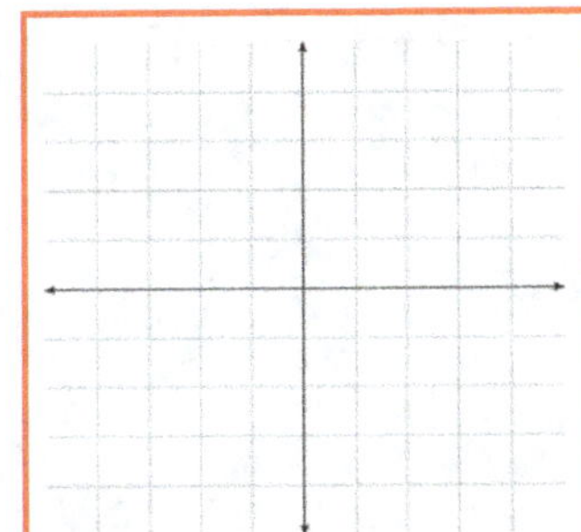

1. Let **f(x) = x + 3** and **g(x) = 2x − 1**. Find the following.

    i) $f^{-1}(x)$
    j) $g^{-1}(x)$
    k) $f^{-1}(x) \circ g^{-1}(x)$
    l) $g^{-1}(x) \circ f^{-1}(x)$
    m) $g(x) \circ f(x)$
    n) $f(x) \circ g(x)$
    o) $g^{-1}(x) \circ g(x)$
    p) $f^{-1}(x) \circ f^{-1}(x)$

2. Sketch the graph of functions using transformations.

    b) **f(x) = −|x −3| − 3**
    b) **g(x) = − (x − 2)²**

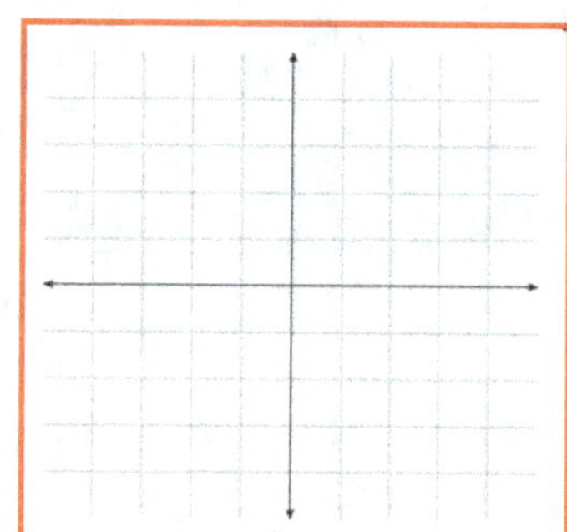
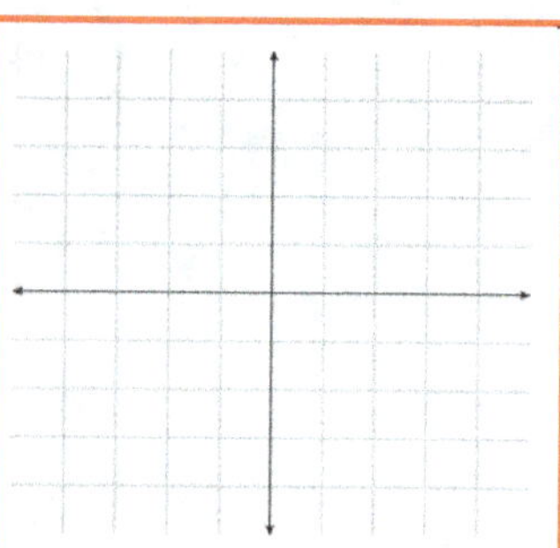

3. Sketch the graph of polynomial and fractional functions.

    b) **f(x) = x²(x−1)(x + 1)²(x + 2)³**
    b) $g(x) = \dfrac{x - 3}{x - 1}$

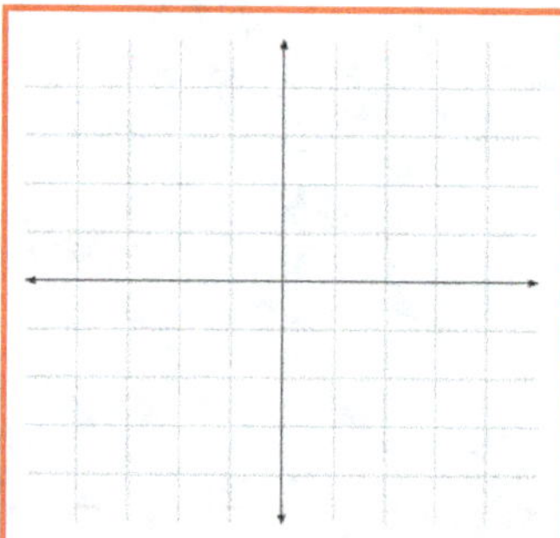
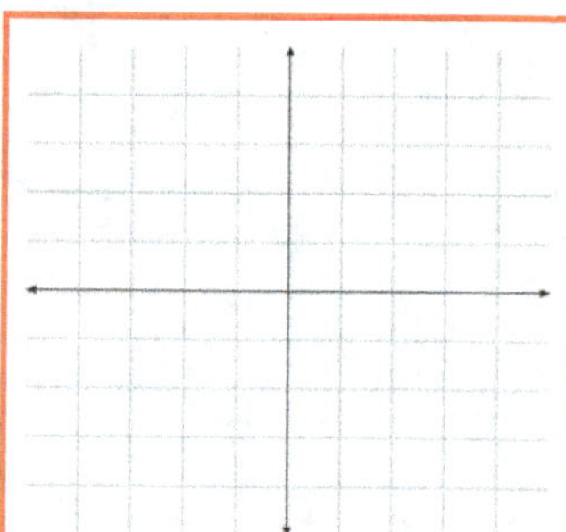

1. Decide whether the given relation is a function.

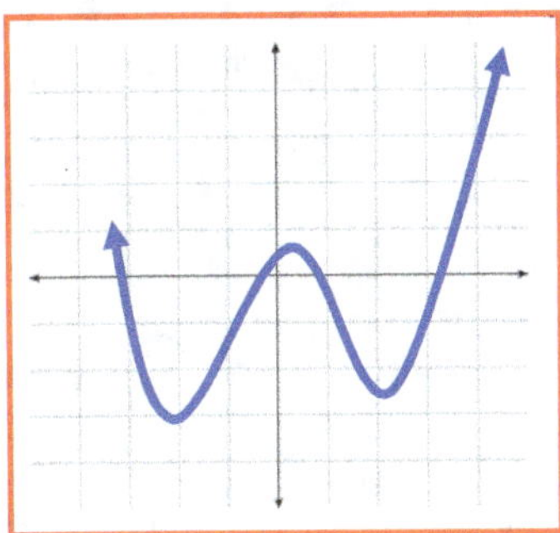

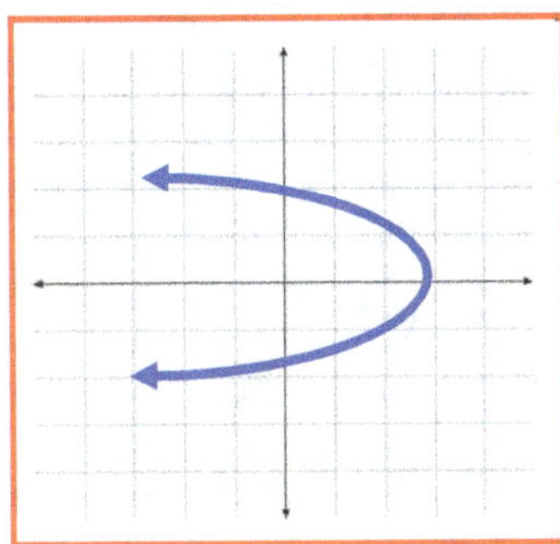

2. Find the domain and range of each function.

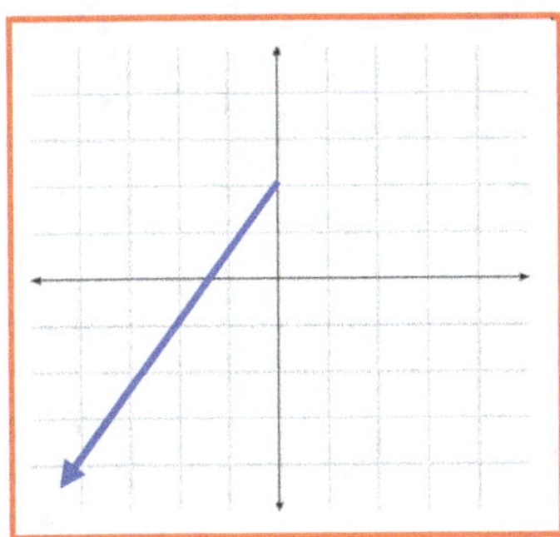

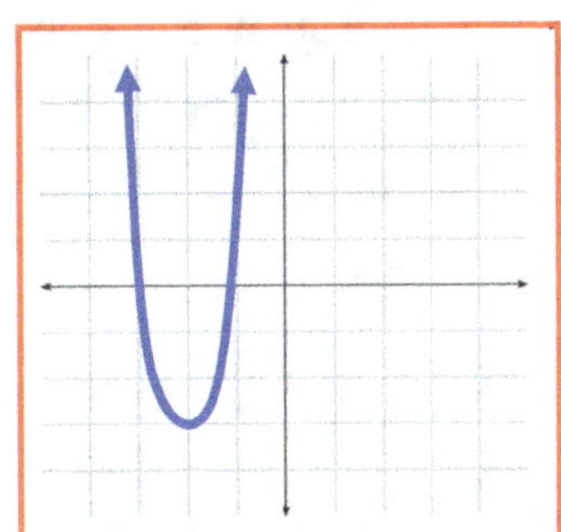

3. Classify functions as even, odd, or neither.

a) $y = -x^3$ b) $y = -3x^2$ c) $y = x^4$ d) $y = 5x^5 + 1$

4. Let $f(x) = x - 1$, $g(x) = \dfrac{x+1}{3}$. Find and graph the functions $f^{-1}(x)$ and $g^{-1}(x)$.

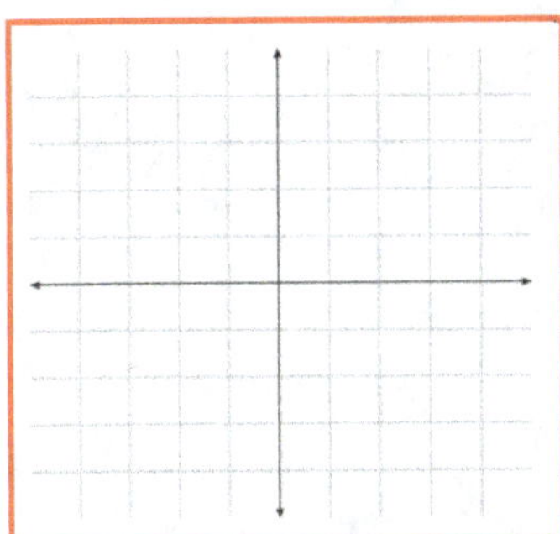

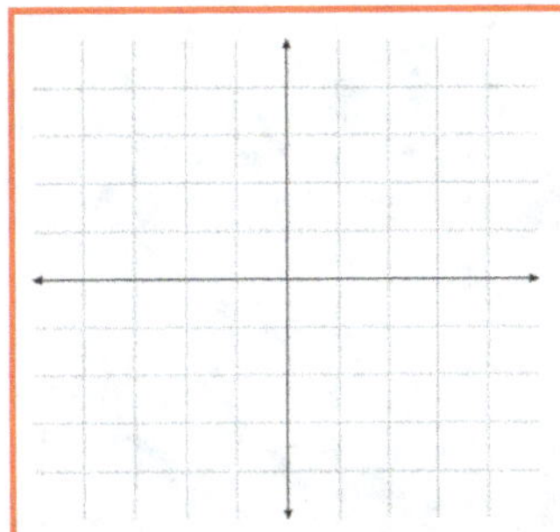

5. Let $f(x) = x$ and $g(x) = -x + 2$. Find the following.

    q) $f^{-1}(x)$

    r) $g^{-1}(x)$

    s) $f^{-1}(x) \circ g^{-1}(x)$

    t) $g^{-1}(x) \circ f^{-1}(x)$

    u) $g(x) \circ f(x)$

    v) $f(x) \circ g(x)$

    w) $g^{-1}(x) \circ g(x)$

    x) $f^{-1}(x) \circ f^{-1}(x)$

6. Sketch the graph of functions using transformations.

    c) $f(x) = -2|x + 2| - 3$ b) $g(x) = -2(x - 1)^2 - 1$

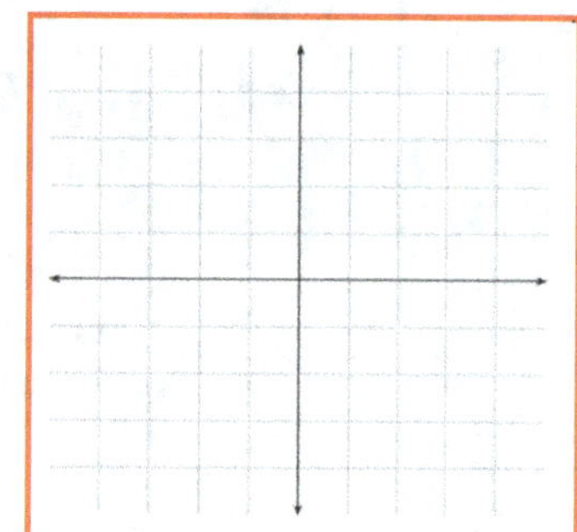 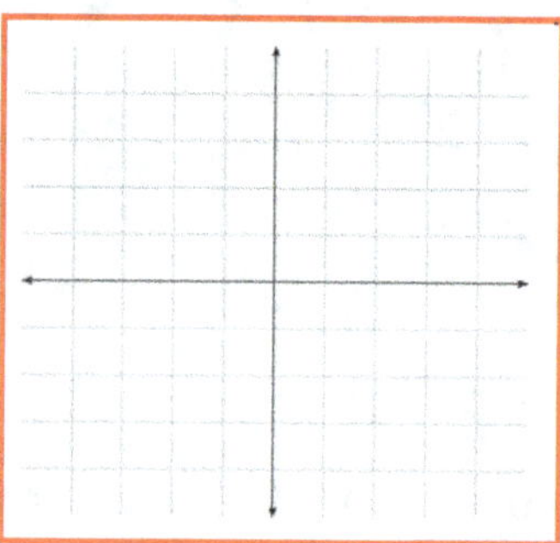

7. Sketch the graph of polynomial and fractional functions.

    c) $f(x) = x(x-1)(x+1)(x+2)$ b) $g(x) = \dfrac{x - 4}{x + 2}$

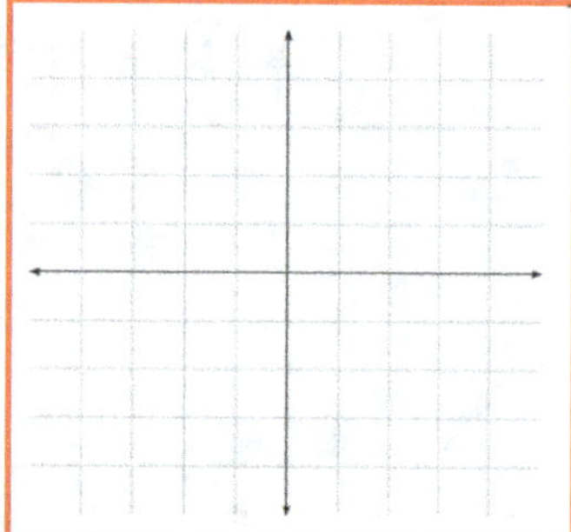 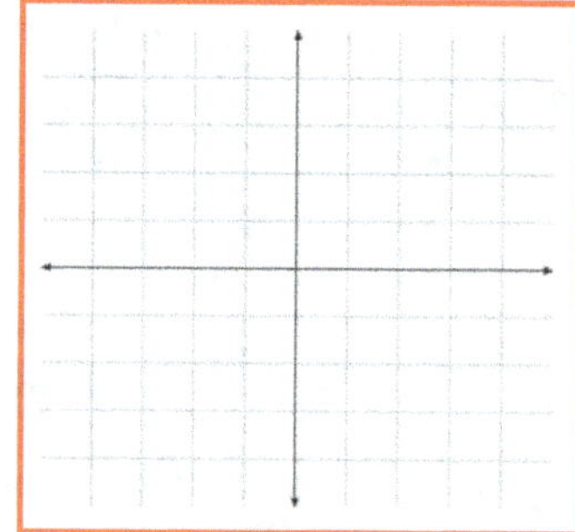

Extra Part 5 Bank of Tests

$x(x-2)=0$	$x=0$ or $x=2$			$\{0,2\}$
$(x-4)(x+3)>0$	$x<-3$ or $x>4$		$x\in(-\infty,-3)\cup(4,\infty)$	$\{x/x<-3,\ x>4\}$

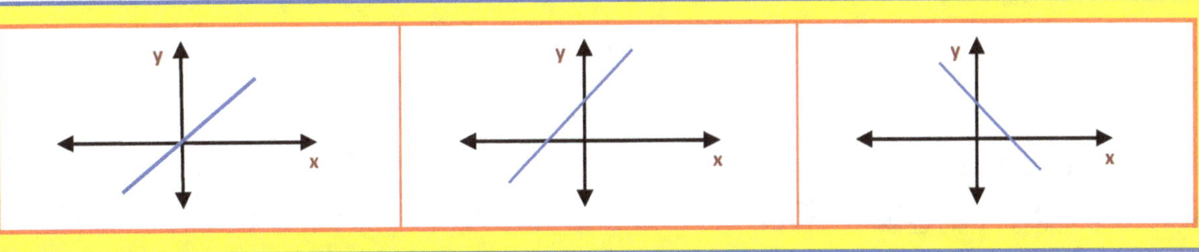

$$\det A = \begin{vmatrix} a_1 & b_1 & c_1 \\ a_2 & b_2 & c_2 \\ a_3 & b_3 & c_3 \end{vmatrix} = (a_1b_2c_3 + b_1c_2a_3 + c_1a_2b_3) - (a_3b_2c_1 + b_3c_2a_1 + c_3a_2b_1)$$

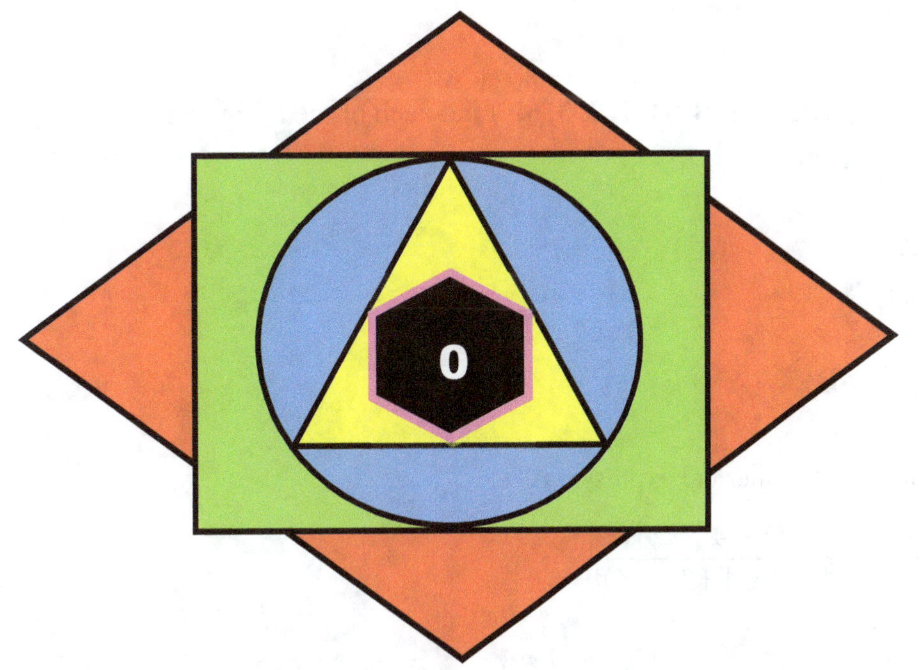

1. In the set $\{-6, -4.\overline{3}, -\frac{1}{3}, 0, \sqrt{5}, 3, \sqrt{16}, 8, 12, 15.5\}$ list all of the:

- Natural Numbers
- Whole Numbers
- Integers
- Even Numbers
- Odd numbers
- Prime Numbers
- Rational Numbers
- Irrational Numbers
- Real Numbers

2. Evaluate the numerical expression

$$-1 + (-2) \cdot (-3) \cdot (-5) \div (-10)$$

3. Let $A = \{1, 2, 4, 6\}$, $B = \{-1, 2, 3, 6\}$. Find each of the following.

a) $A \cup B$

b) $A \cap B$

c) $A \cap A$

d) $(A \cap B) \cup (A \cup B)$

4. Evaluate the numerical expression.

$$\frac{|3 - 6| + |-(3 - 4) - 4 \div 2|}{|1 - 4| + |-8 \div 4 + (5 - 4)|}$$

5. Use PEMDAS + L→R to evaluate the expression.

$$(-5 + 4) \cdot \{[-(-3)^2 + (-3)] + (2 - 4)\} - 9 \div (-3) + 1$$

6. Find the result of operations.

$$\frac{x^5 y^2}{x^6} \div \frac{x^{-2} y^{-3}}{x^{-6} y^{-4}} \cdot (x^{-2} y^3)$$

7. Write the expression in scientific notation form.

$$5(4 \cdot 10^{10})(7 \cdot 10^{-7})$$

8. Classify the polynomial by number of terms, by degree, and as binomial or trinomial.

$$x^5 + 3x^7 - 5x^8$$

9. Combine like terms in the expression and write the answer in standard form.

$$(2x^4 + x - 2x^2 + x^3) - (2x^3 + 2x^4 - 2x^2 + x)$$

10. Multiply the following polynomials. Write the answer in simplest form.

$$(x + 1) \cdot (x^2 - x + 1)$$

11. Divide a polynomial by using **long division** and **synthetic division**.

$$\frac{y^2 + 2y - 3}{y + 2}$$

12. Factor

$$x^5 - x^2$$

13. Factor

$$x^5 - x^4 - 20x^3$$

14. Factor

$$2x^2 - 3x - 5$$

15. Simplify

$$\sqrt{24} - \sqrt{18} + \sqrt{2}$$

16. Simplify

$$\left(\frac{4x^4}{y}\right)^{1/2} \cdot \frac{\sqrt[4]{y^2}}{2x^2} + \frac{y^3\sqrt{y^2}+\sqrt[3]{x}}{x^{1/3}+y^{5/2}}$$

17. Rationalize the denominator

$$\frac{\sqrt{7}}{\sqrt{7}+1}$$

1. In the set $\{-8, -4.1, -1.25, -\frac{1}{4}, 0, \sqrt{3}, 3, \pi, 5\frac{2}{3}, 8, 10.\overline{3}, 14.5\}$ list all of the:

   - Natural Numbers

   - Whole Numbers

   - Integers

   - Even Numbers

   - Odd numbers

   - Prime Numbers

   - Rational Numbers

   - Irrational Numbers

   - Real Numbers

2. Evaluate the numerical expression

   $$-5 + 4 \cdot (-2) \cdot (-1)$$

3. Let $A = \{2, 3, 5, 6\}$, $B = \{-1, 2, 3, 5, 7\}$. Find each of the following.

   a) $A \cup B$
   b) $A \cap B$
   c) $A \cap A$
   d) $(B \cup B) \cap A$
   e) $(A \cap B) \cup A$
   f) $\{x/\ x \in A,\ x \notin B\}$
   g) $\{x/\ x \in B,\ x \geq 3\}$

4. Evaluate the numerical expression.

   $$\frac{|4 - 5| + |-(3 - 9) - 9 \div 3|}{|1 - 2| + |-4 \div 2 + (5 - 6)|}$$

5. Evaluate the variable expression for $x = 2$, $y = 4$.

   $$\frac{x + 3y}{3x + 1} + 3$$

6. Use PEMDAS + L➔R to evaluate the expression.

$$(-5) \cdot \{[-(-1)^2 + 4] + 1 \} - 12 \div (-3) + 6$$

7. Find the result of operations.

$$\frac{x^2 y^3}{x^7} \div \frac{x^{-9} y^{-2}}{x^{-3} y^{-5}} \cdot (x y^5)$$

8. Write expression in scientific notation form.

$$8(4 \cdot 10^{10})(5 \cdot 10^{-9})$$

9. Classify the polynomial by number of terms, by degree, and as binomial or trinomial.

$$x^3 + 3x^9 - 5x^5$$

10. Evaluate the polynomial for **x = 2, y = 2, z = −2.**

$$2x^3 + xy^3 z^2 + 2z + y - 3x$$

11. Combine like terms in the expression and write the answer in standard form.

$$(x^4 + x - 3x^2 + x^3) - (2x^3 + x^4 - 2x^2 + x)$$

12. Multiply the following polynomials. Write the answer in simplest form.

$$(x + 2) \cdot (x^2 - 2x + 4)$$

13. Divide a polynomial by using **long division** and **synthetic division**.

$$\frac{y^2 - 4y + 4}{y - 2}$$

14. Simplify.

$$\frac{x^2 - 4y^2}{x + 2y} - x + 2y$$

15. Factor

$$x^6 - x^2$$

16. Factor

$$x^3 - x^2 - 12x$$

17. Factor

$$2x^2 - 5x - 7$$

18. Simplify

$$\sqrt{27} - \sqrt{48} + 4\sqrt{3}$$

19. Simplify

$$\left(\frac{x^3}{y^2}\right)^{1/3} \cdot \frac{\sqrt[3]{y^2}}{x^2}$$

20. Simplify

$$\frac{x\sqrt[3]{x^2} + \sqrt[3]{y}}{y^{1/3} + x^{5/2}}$$

21. Rationalize the denominator

$$\frac{\sqrt{3}}{\sqrt{3} + 3}$$

22. Rationalize the numerator

$$\frac{\sqrt{5} - \sqrt{3}}{2}$$

1. Solve

   a) $\quad 4(x-2)+5-3x=-4$

   b) $\quad \dfrac{4x}{3}-1=\dfrac{5x}{6}$

   c) $\quad \dfrac{x}{2x-1}=\dfrac{1}{3}$

2. Solve

   a) $\quad 4x-1>-(x-3)+1$

   b) $\quad 1\le 2+x\le 7$

   c) $\quad 2x<3x-1\le 11$

   d) $\quad 2x-1\le 5 \quad \text{or} \quad -x+2<-4$

   e) $\quad -x+1>0 \ \text{ and } \ 2+x<6+2x$

3. Solve

   a) $\quad |2-x|+4=8$

   b) $\quad |x+2|+5=1$

   c) $\quad |x+1|=|2x-3|$

4. Solve

   a) $\quad |1+x|\ge 3$

   b) $\quad |2x-3|\le 3$

   c) $\quad |x+2|>2 \ \text{ and } \ |x|\le 5$

   d) $\quad |x-1|\le 4 \ \text{or} \ |x+1|>6$

5. Solve

   a) $x^2 + 4x = 8x$

   b) $2x^2 + 7x - 9 = 0$

   c) $x^2 + 3x - 10 = 0$

   d) $25x^2 - 9 = 0$

6. Solve

   a) $x^2 - 64 > 0$

   b) $9x^2 + 4x - 5 \geq 0$

   c) $x^2 + 2x - 15 < 0$

   d) $3x^2 - 27 \leq 0$

7. Solve

   a) $x^4 + 5x^3 = 0$

   b) $6x^7 + 3x^6 - 9x^5 = 0$

   c) $x^5 + x^4 - 6x^3 = 0$

   d) $x^4 - 1 = 0$

8. Solve

   a) $x^4 - 16 > 0$

   b) $4x^5 + x^3 - 3x^2 \geq 0$

   c) $x^4 + 2x^3 - 8x^2 < 0$

   d) $2x^4 - 8x^2 \leq 0$

9. Solve

   a) $\dfrac{x+1}{x+3} = \dfrac{1}{2}$

   b) $\dfrac{x+1}{x+3} < 0$

   c) $\dfrac{1}{x+3} = 1 + \dfrac{1}{x-3}$

   d) $\dfrac{1}{x-1} \geq \dfrac{1}{2x+1}$

10. Solve

   a) $\sqrt{4 - 3x} = x$

   b) $\sqrt{x-3} = \sqrt{x+2} - 1$

   c) $\sqrt{4x-3} < x$

   d) $\sqrt{4x-3} \leq \sqrt{x} + \sqrt{3x-3}$

11. Graph these lines using **the point's method.**

a) $x + 3y = 9$

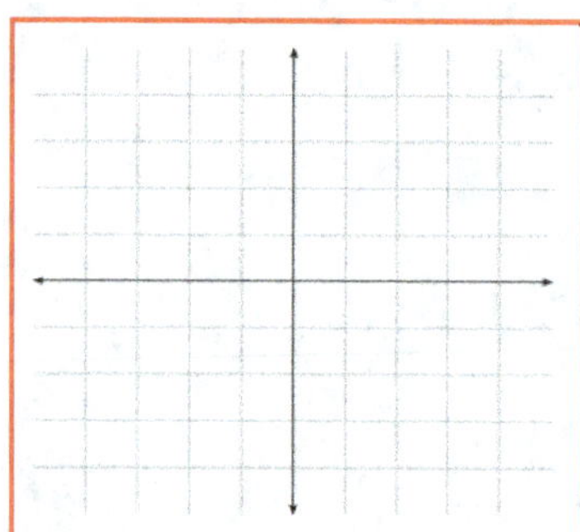

b) $y = -2 + 4x$

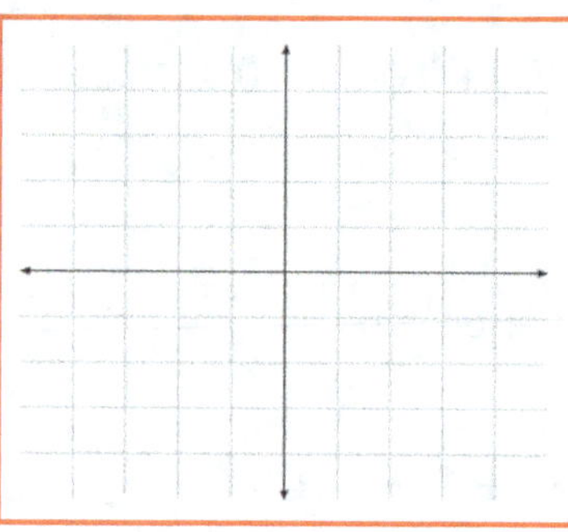

12. Graph these lines using **x- and y-intercepts method.**

a) $4x + 6y = 12$

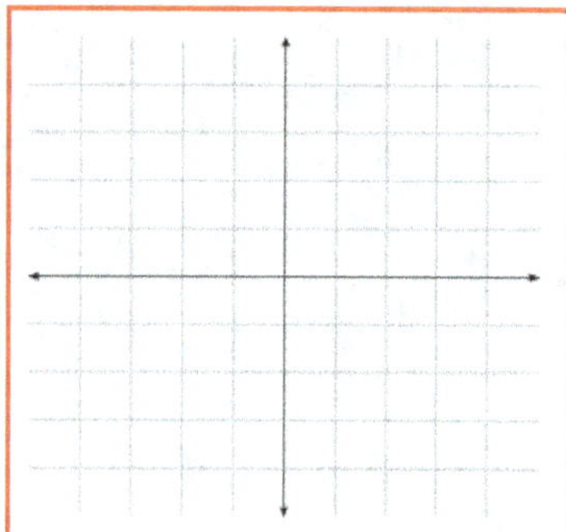

b) $y = 4x - 2$

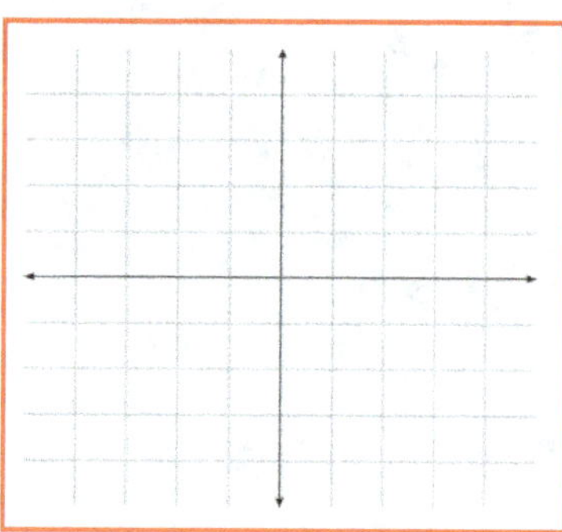

13. Graph these lines using **the slope – y– intercept method**.

a) $x + y = 4$

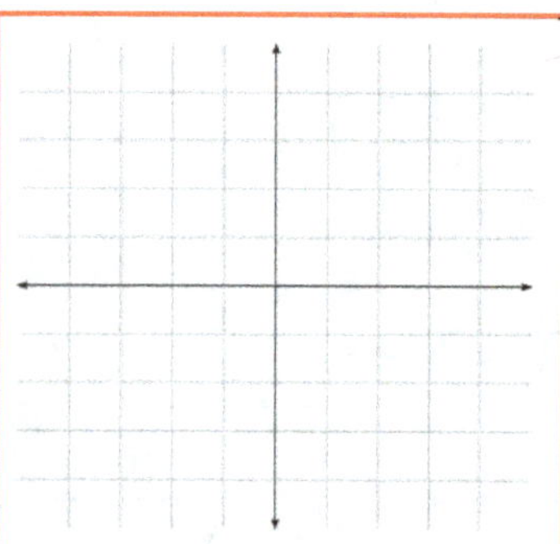

b) $y = \dfrac{1}{3}x - 2$

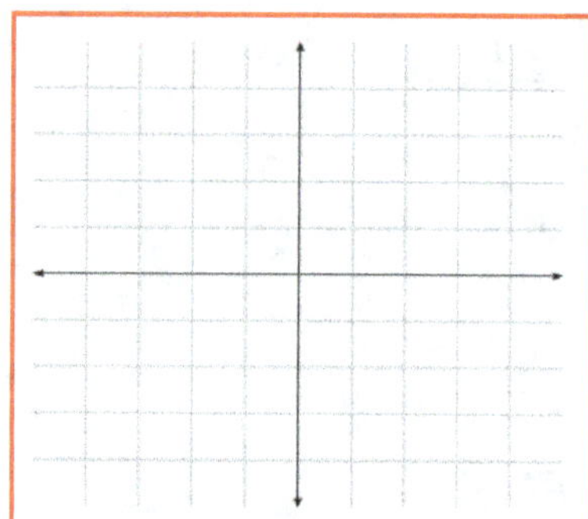

14. Graph the solution sets of the following inequalities.

a) $x + 3y > 6$

b) $2x + 2y \leq 4$

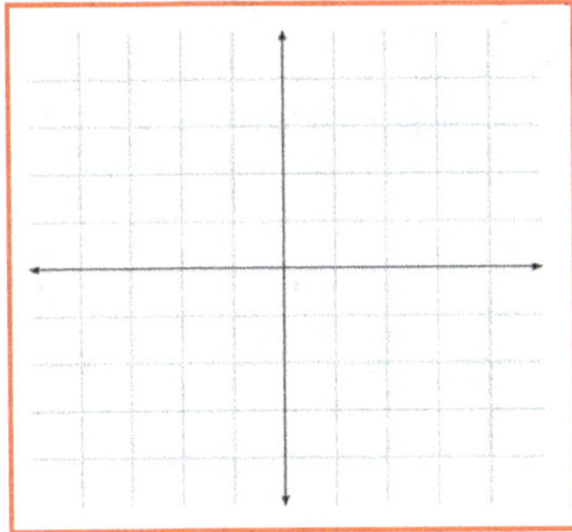

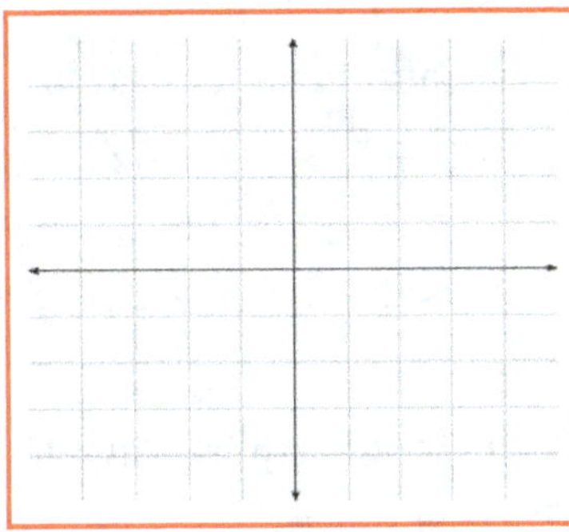

15. Graph the solution sets of the following inequalities.

a) $y < -3x + 4$

b) $y \geq -2x$

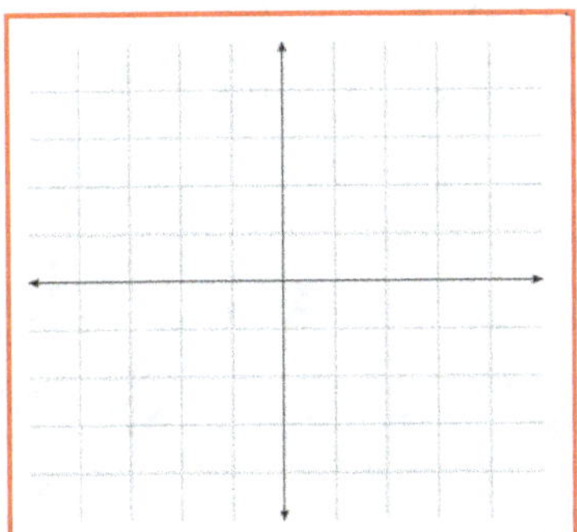

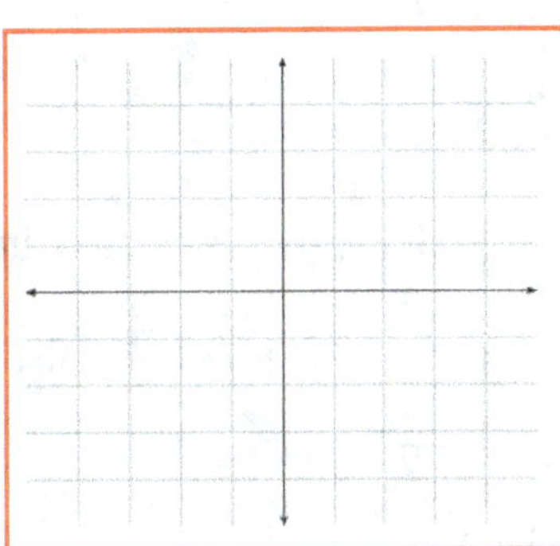

16. Graph the solution sets of the following inequalities.

a) $y \geq -3$

b) $x < 3$

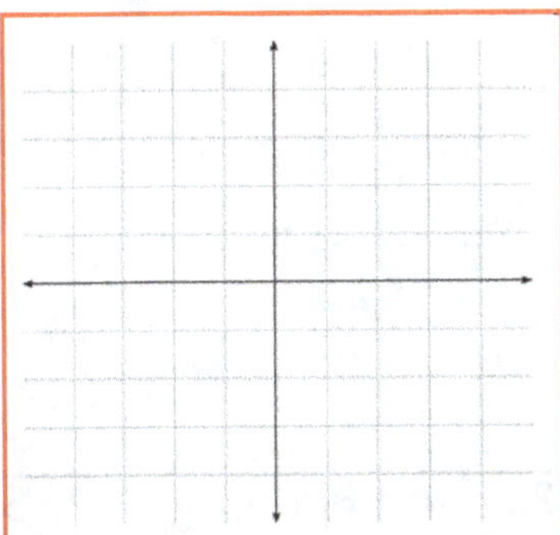

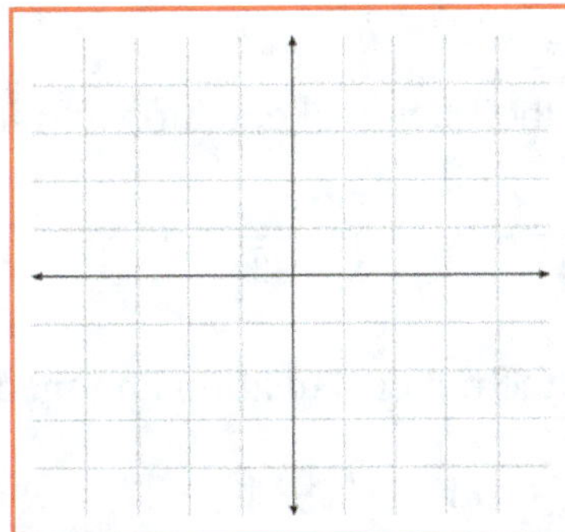

<table><tr><td>

PART 2   **TEST 2B**
Equations, Inequalities, and Graphs

</td><td>

Hayk Yegoryan & Rubik Yegoryan

INTERMEDIATE ALGEBRA

</td></tr></table>

1. **Solve each linear equation and inequality**

   a) $2(x - 1) + 5x = -4 + 6x$

   b) $\dfrac{x}{2} - 1 = \dfrac{x}{4} + \dfrac{1}{6}$

   c) $2x - 1 \le 2 + x$ or $-2x + 8 < -2$

   d) $-3x + 1 < 4$ and $1 + 4x < 3(1 + x)$

2. **Solve each absolute value equation and inequality**

   a) $2|1 - x| + 4 = 8$

   b) $|x - 2| + 2 = 6$

   c) $|1 - x| + 8 \ge 4$

   d) $|3x - 1| + 3 < 5$

3. **Solve each quadratic equation and inequality**

   a) $x^2 + 3x = 0$

   b) $x^2 + 3x - 10 = 0$

   c) $x^2 - 64 > 0$

   d) $8x^2 + 3x - 5 \ge 0$

4. **Simplify each expression and write all the answers in a + bi form**

   a) $(1 + \sqrt{-16}\,)(1 + \sqrt{-4}\,) + i^5$

   b) $\dfrac{1 + \sqrt{-2}}{1 + \sqrt{-1}}$

5. **Solve each polynomial equation and inequality**

   a) $x^4 - 64x^2 = 0$

   c) $x^4 + x^3 - 12x^2 = 0$

   b) $4x^7 + 2x^6 - 6x^5 \le 0$

   d) $2x^5 - 8x^3 < 0$

6. **Solve the fractional equation and inequality**

   a) $\dfrac{1}{x + \sqrt{3}} + \dfrac{1}{x - \sqrt{3}} = 1$

   b) $\dfrac{1}{x} - \dfrac{1}{x + 2} < 1$

7. **Solve the radical equation and inequality**

   a) $\sqrt{x - 1} = \sqrt{x + 1} - 2$

   b) $\sqrt{2x - 4} \le \sqrt{x + 1} - \sqrt{x - 5}$

8. Graph these lines using **the point's method.**

    a) $x + 3y = 6$ b) $y = 1 - 2x$

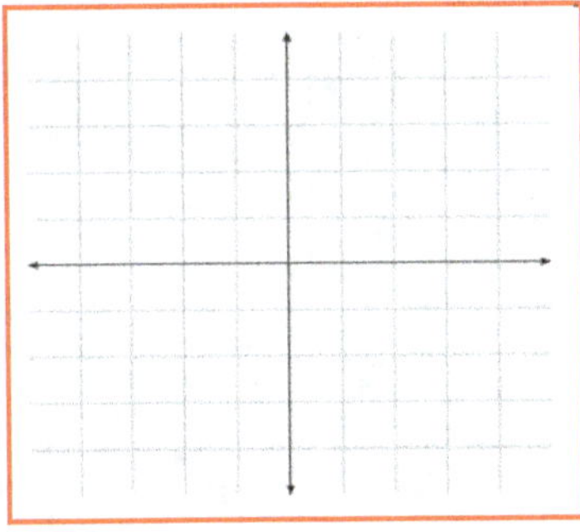
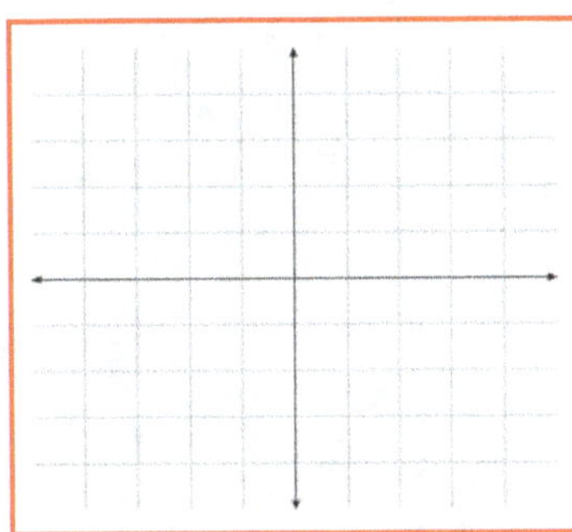

9. Graph these lines using **x- and y-intercept's method.**

    a) $4x + 3y = 12$ b) $y = 3x - 3$

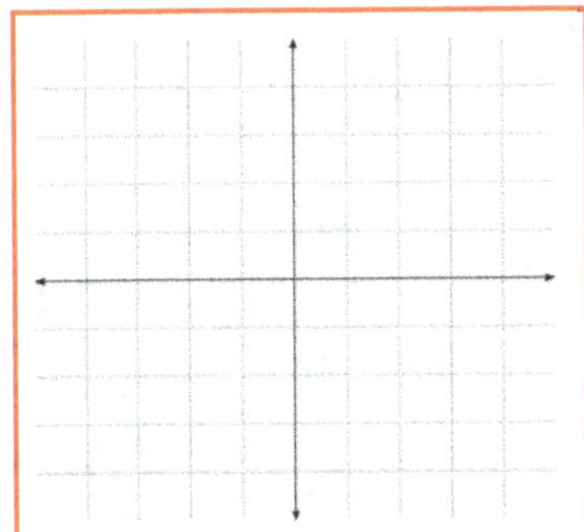
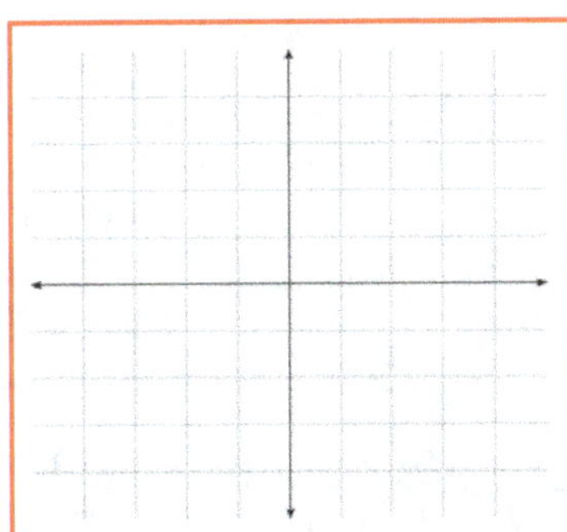

10. Graph these lines using **the slope – y– intercept's method.**

    a) $x + y = 3$ b) $y = \dfrac{1}{2}x - 2$

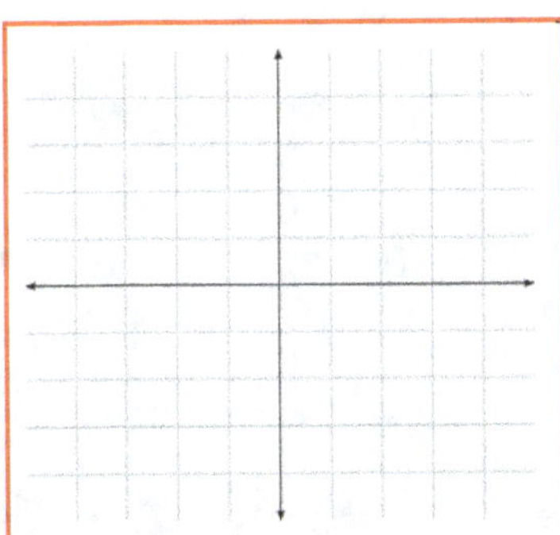
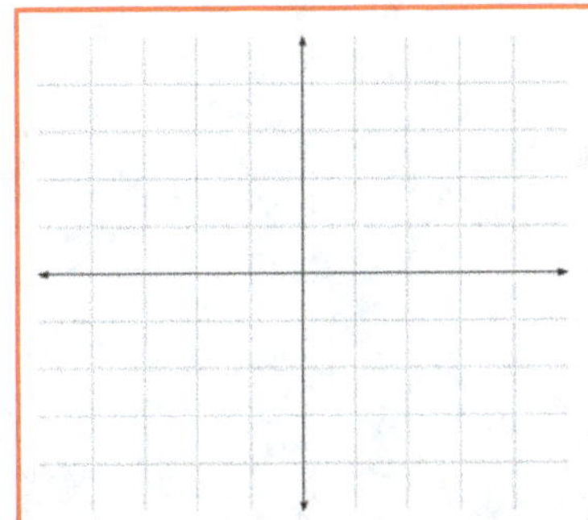

11. Graph the solution sets of following inequalities.

a) $x + 3y > 6$

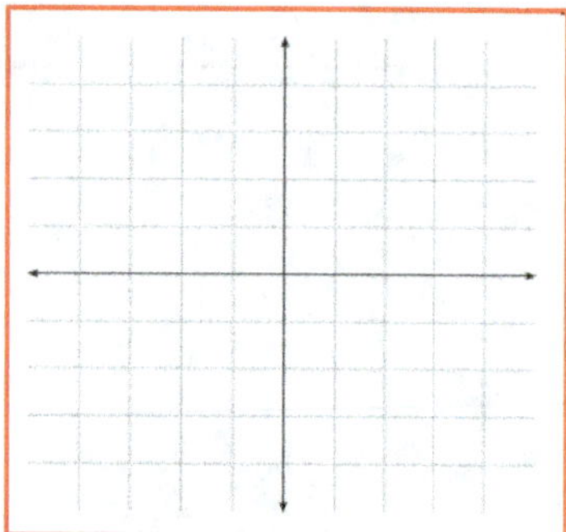

b) $2x + y \leq 1$

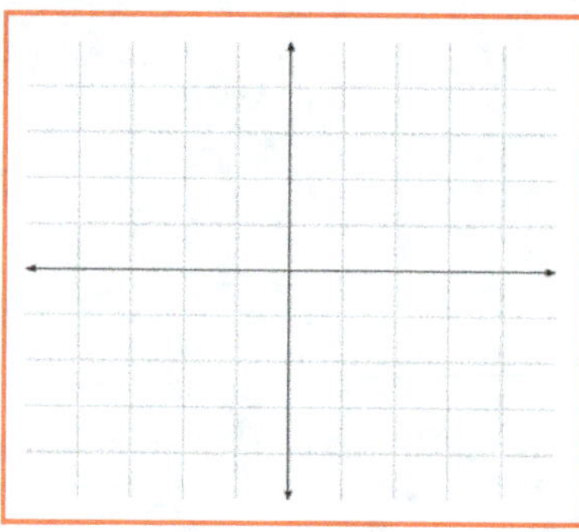

12. Graph the solution sets of following inequalities.

a) $y < -x + 4$

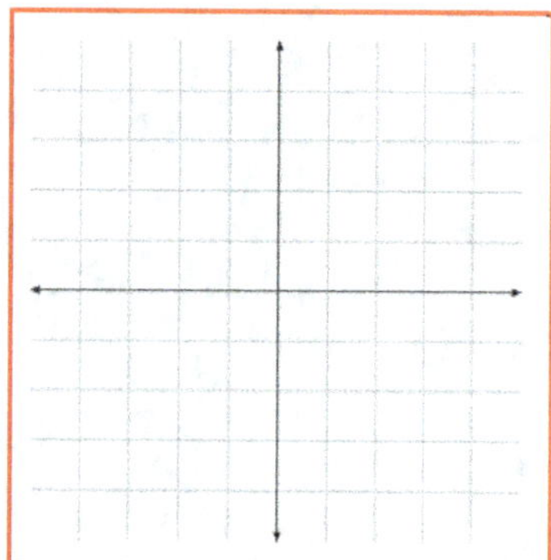

b) $y \geq 3x$

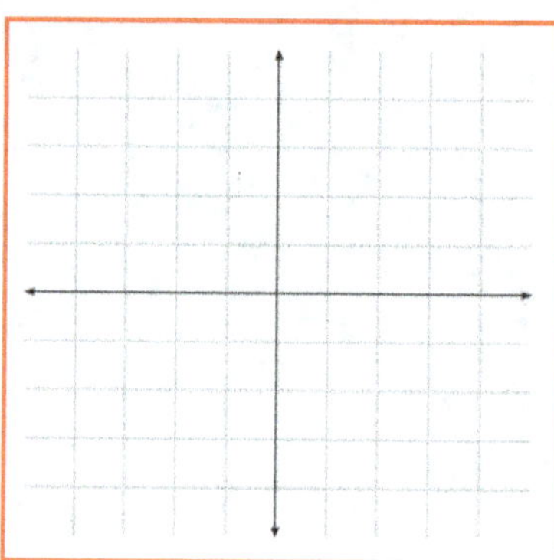

13. Graph the solution sets of following inequalities.

a) $y \geq 3$

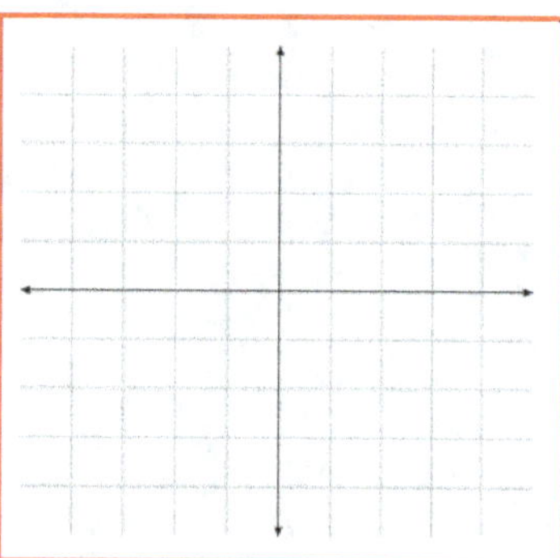

b) $x < -2$

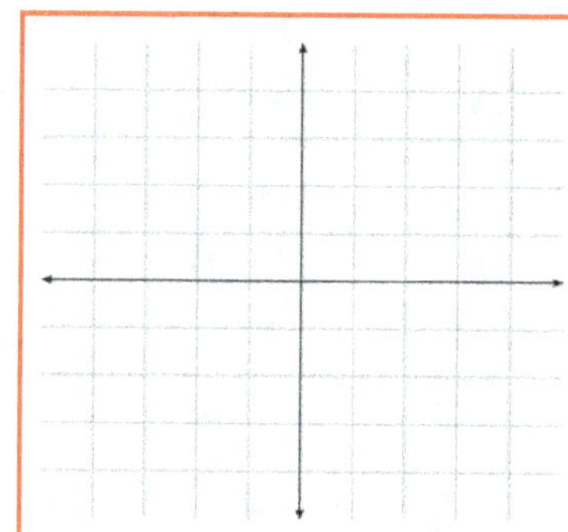

1. Solve the system by substitution method.

$$\begin{cases} x + y = 3 \\ 3x + 2y = 8 \end{cases}$$

2. Solve the system by graph method.

$$\begin{cases} 2x - y = -1 \\ 3x + y = 6 \end{cases}$$

3. Solve the system by elimination method.

$$\begin{cases} x - y + z = 1 \\ x + y + 2z = 4 \\ 2x + y - 2z = 1 \end{cases}$$

4. Solve each system by corresponding Cramer's Rule.

$$\begin{cases} x + y = 3 \\ x + 2y = 5 \end{cases} \qquad \begin{cases} x - y + z = 0 \\ x + y + z = 2 \\ 2x + 3y - z = 5 \end{cases}$$

5. Sketch the graph of the set of solutions for each system of inequalities.

a) $\begin{cases} x - y > -2 \\ y \geq -x + 2 \end{cases}$
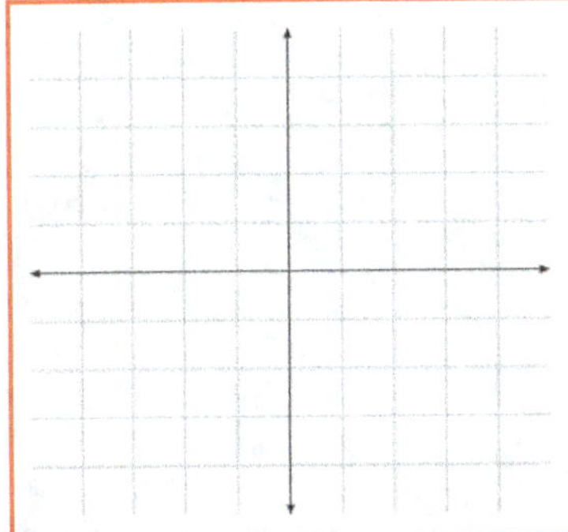

b) $\begin{cases} -x + 2y > -4 \\ y \leq -3x \end{cases}$
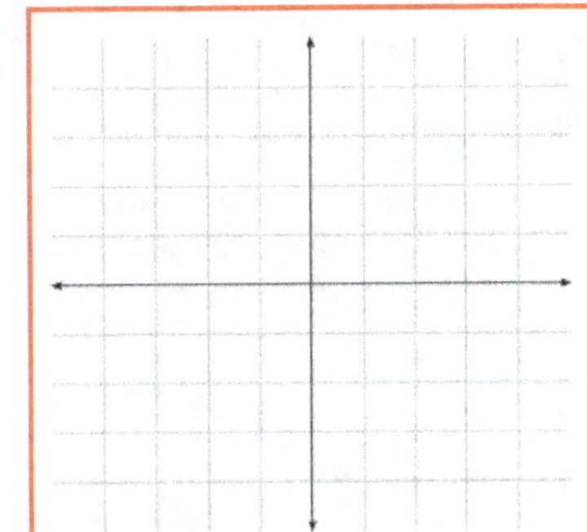

1. Solve the system by substitution method.

$$\begin{cases} x + y = 2 \\ 2x + 5y = 7 \end{cases}$$

2. Solve the system by graph method.

$$\begin{cases} 2x - y = -3 \\ x + y = 6 \end{cases}$$

3. Solve the system by elimination method.

$$\begin{cases} x - y + z = 2 \\ x + y + 2z = 3 \\ 2x + 3y - z = 1 \end{cases}$$

4. Solve each system by corresponding Cramer's Rule.

$$\begin{cases} x + y = 4 \\ x + 2y = 7 \end{cases} \qquad \begin{cases} x - 2y + z = 0 \\ x + 2y + z = 4 \\ 2x + y - z = 2 \end{cases}$$

5. Sketch the graph of the set of solutions for each system of inequalities.

a) $\begin{cases} 2x - y > -3 \\ y \geq -x + 2 \end{cases}$

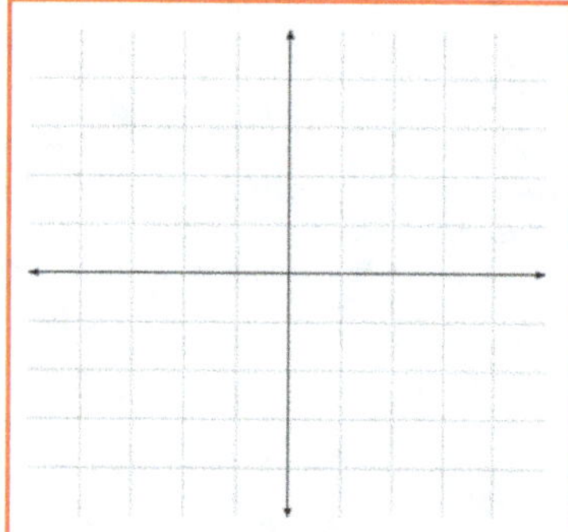

b) $\begin{cases} x - y > -1 \\ y \leq -x + 1 \end{cases}$

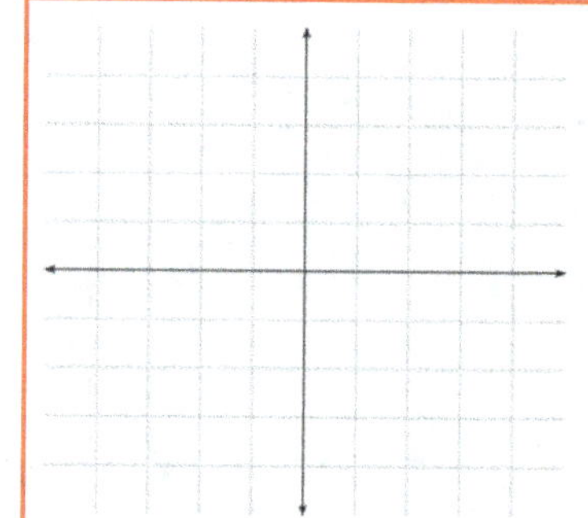

1. Decide whether the given relation is a function.

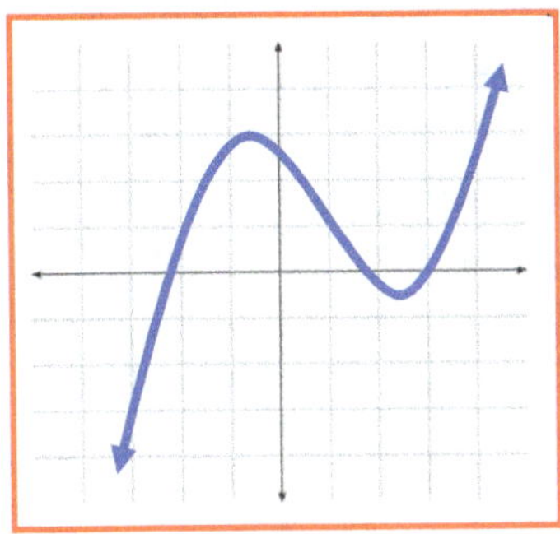   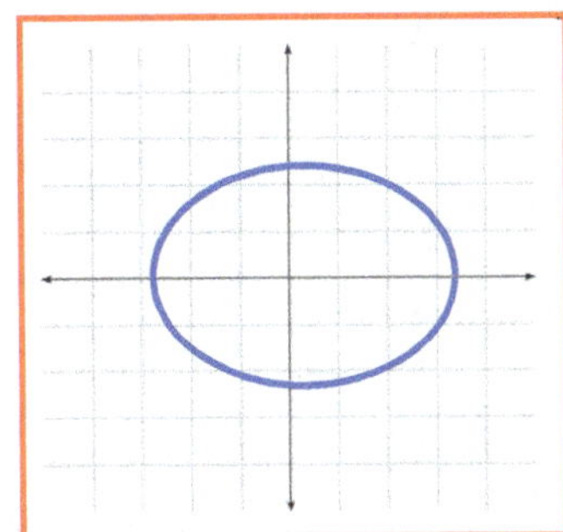

2. Find the domain and range of each function.

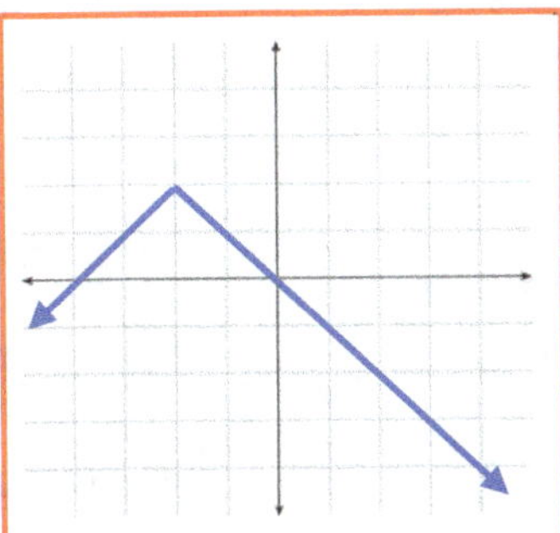   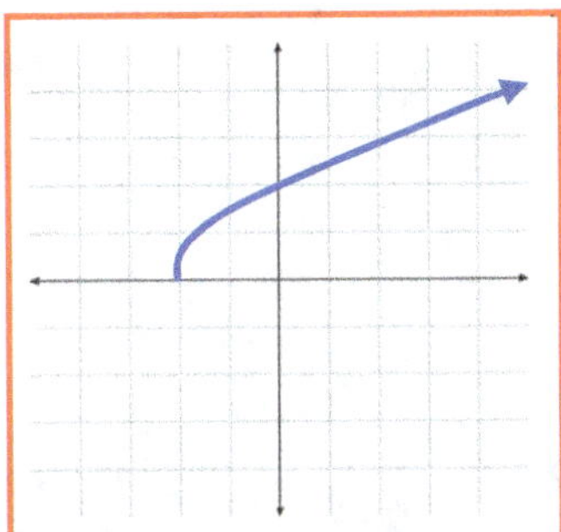

3. Classify functions as even, odd, or neither.

    a) $y = x + x^5$ b) $y = -2x^4$ c) $y = x^2 + x^3$ d) $y = 3|x| + x^2$

4. Let $f(x) = 3x - 2$, $g(x) = \frac{1}{2}x + 1$. Find and graph the functions $f^{-1}(x)$ and $g^{-1}(x)$.

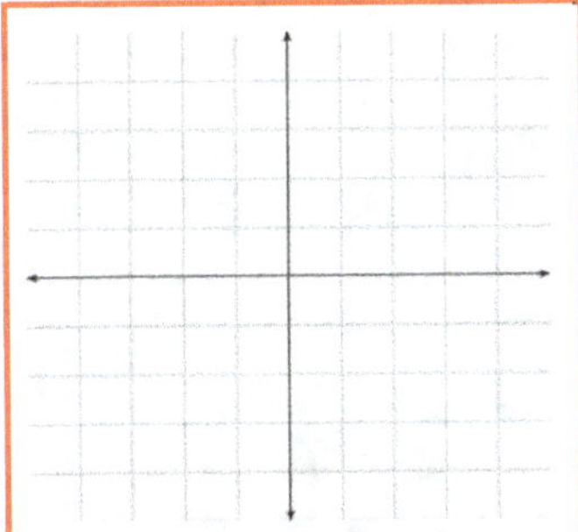   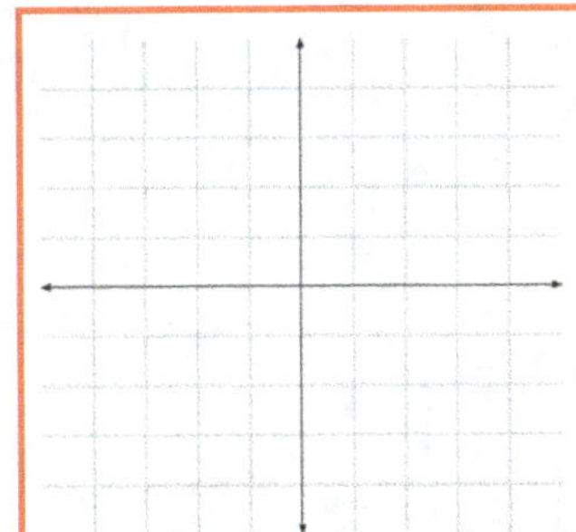

5. Let $f(x) = 2x$, $g(x) = -x + 1$. Find the following.

    a) $f^{-1}(x)$

    b) $g^{-1}(x)$

    c) $f^{-1}(x) \circ g^{-1}(x)$

    d) $g^{-1}(x) \circ f^{-1}(x)$

    e) $g(x) \circ g^{-1}(x)$

    f) $f(x) \circ f^{-1}(x)$

    g) $g^{-1}(x) \circ g(x)$

    h) $f^{-1}(x) \circ f^{-1}(x)$

6. Sketch the graph of functions using transformations.

    a) $f(x) = 2|x + 1| - 3$ b) $g(x) = -(x - 1)^2 + 3$

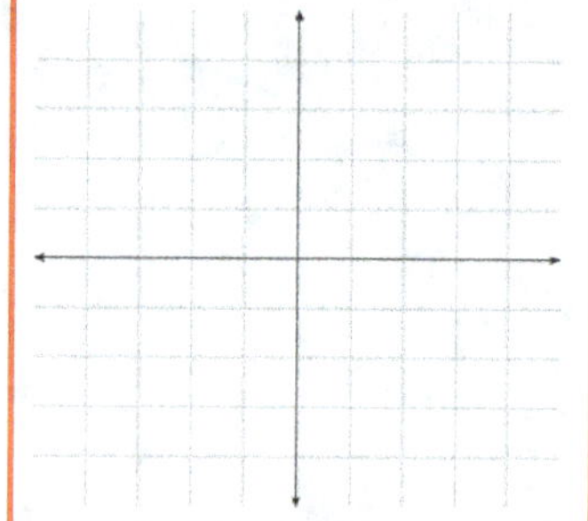 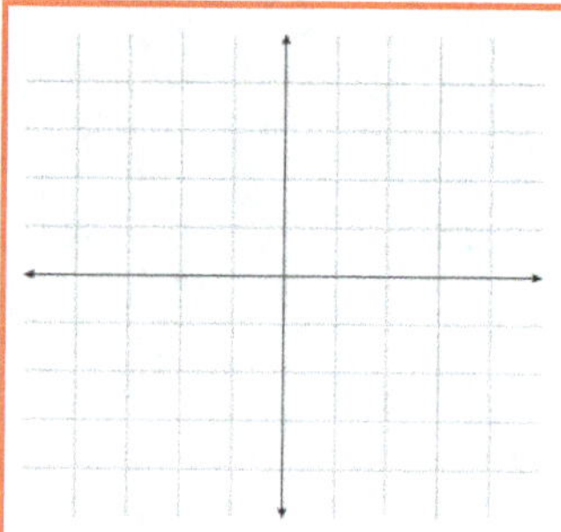

7. Sketch the graph of polynomial and fractional functions.

    a) $f(x) = x - x^3$ b) $g(x) = \dfrac{x + 2}{x - 2}$

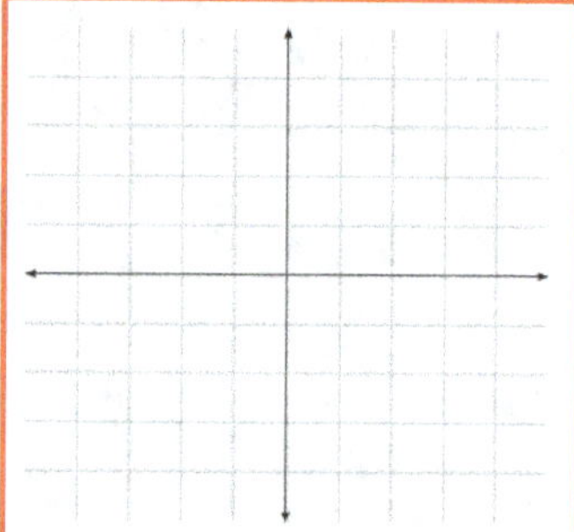 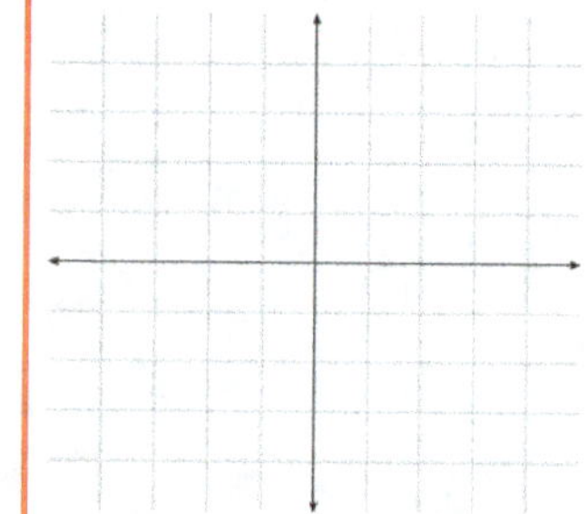

1. Decide whether the given relation is a function.

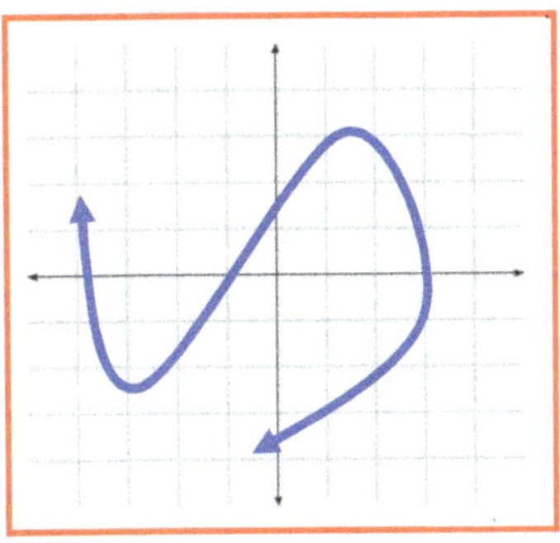

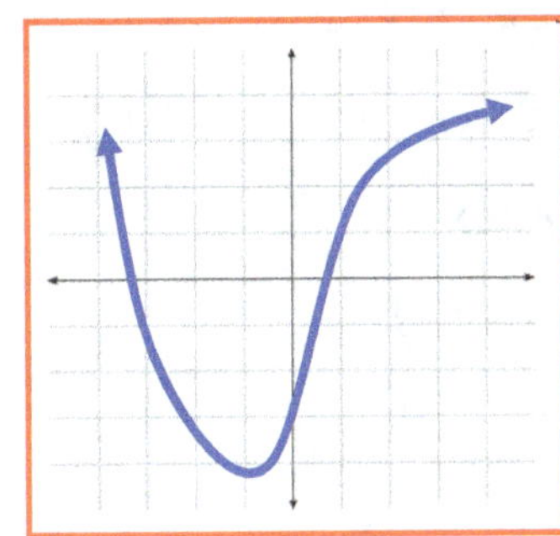

2. Find the domain and range of each function.

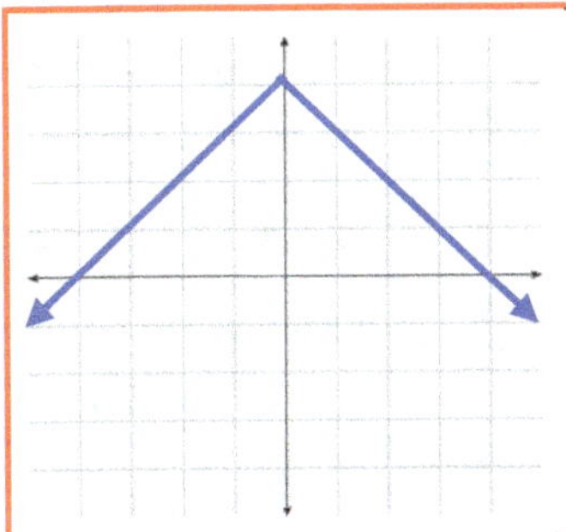

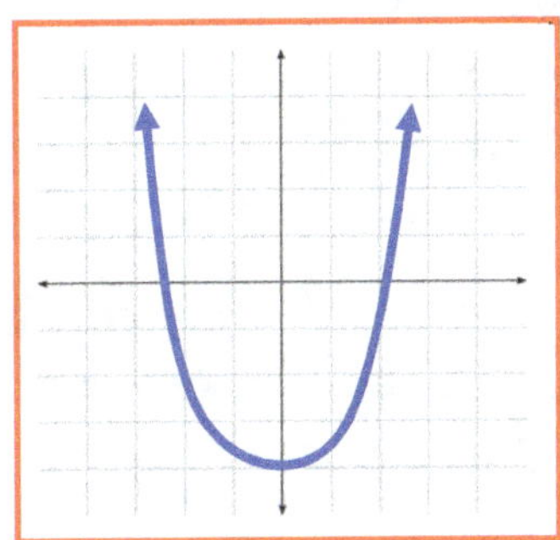

3. Classify functions as even, odd, or neither.

a) $y = 2x - x^3$ b) $y = -3x^2 + |x|$ c) $y = x^3 + 1$ d) $y = 3x^2 + 1$

4. Let $f(x) = x - 2$, $g(x) = \frac{1}{3}x$. Find and graph the functions $f^{-1}(x)$ and $g^{-1}(x)$.

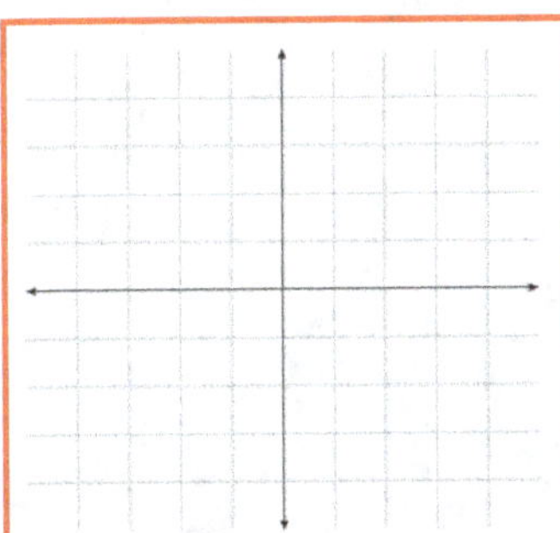

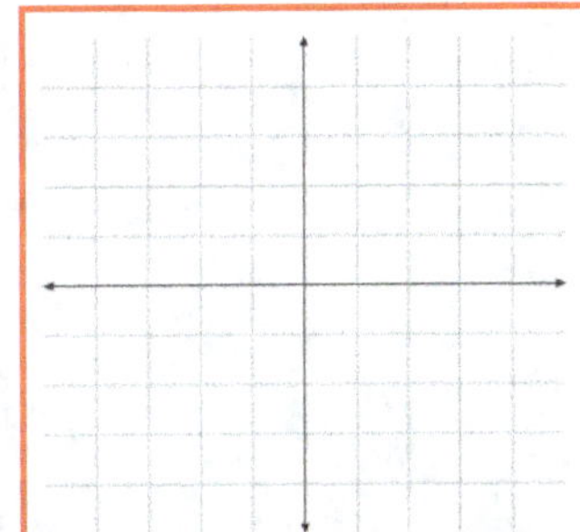

5. Let $f(x) = x + 1$, $g(x) = -2x$. Find the following.

   a) $f^{-1}(x)$

   b) $g^{-1}(x)$

   c) $f^{-1}(x) \circ g^{-1}(x)$

   d) $g^{-1}(x) \circ f^{-1}(x)$

   e) $g(x) \circ f(x)$

   f) $f(x) \circ g(x)$

   g) $g^{-1}(x) \circ g(x)$

   h) $f^{-1}(x) \circ f^{-1}(x)$

6. Sketch the graph of functions using transformations.

   a) $f(x) = -\,|x - 2| + 3$ b) $g(x) = (x + 1)^2 - 2$

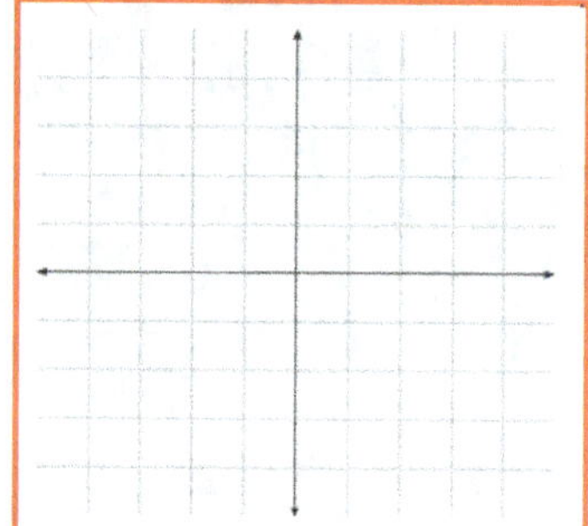

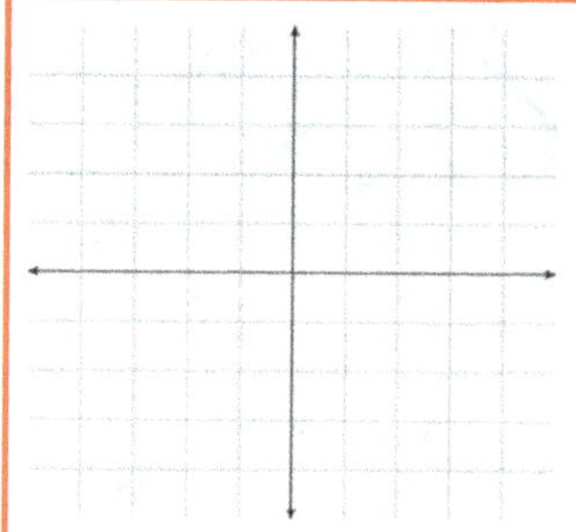

7. Sketch the graph of polynomial and fractional functions.

   a) $f(x) = -x(x-1)(x+2)$ b) $g(x) = \dfrac{x + 3}{x - 2}$

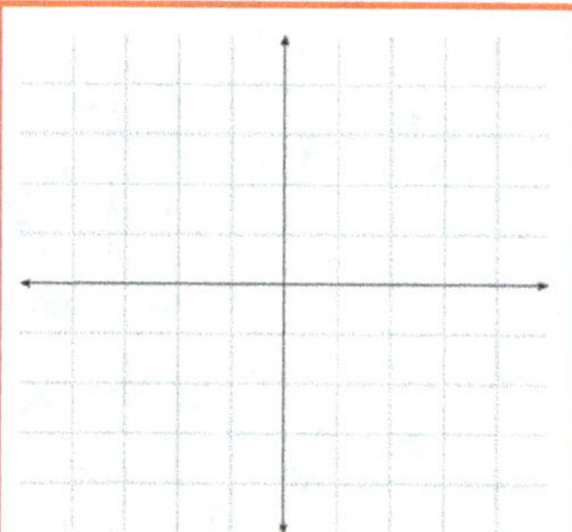

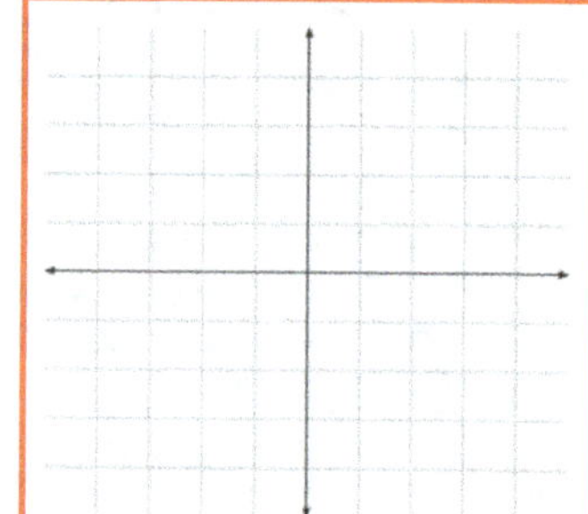

FINAL TEST 1

Hayk Yegoryan & Rubik Yegoryan

INTERMEDIATE ALGEBRA

1. Evaluate the numerical expression

$$1 + (-5) \cdot (-2) \cdot (-4) \div (-20)$$

2. Let A = {1, 2, 3, 5}, B = {−1, 2, 3, 5, 6}. Find each of the following.

e) $A \cup B$

f) $A \cap B$

g) $A \cap A$

h) $(A \cap B) \cup (A \cup B)$

3. Evaluate the numerical expression.

$$\frac{|1 - 4| + |-(2 - 5) - 2 \div 2|}{|1 - 2| - 8 \div 4 + |1 - 7|}$$

4. Use PEMDAS + L→R to evaluate the expression.

$$(-2 + 4) \cdot \{[-(-1)^2 + (-2)] + (2^3 - 3) + 6 \div (-2)\} + 1$$

5. Simplify.

$$\frac{x^3 y^7}{x^2} \div \frac{x^{-4} y^{-6}}{x^{-5} y^{-2}} \cdot (x^{-5} y^{-3})$$

6. Write expression in scientific notation form.

$$5(8 \cdot 10^{12})(5 \cdot 10^{-6})$$

7. Combine like terms in each expression and write the answer in standard form.

$$(5x^4 + 2x - 2x^3 + x^5) - (-2x^3 + 5x^4 - 2x^2 + 2x)$$

8. Multiply the following polynomials. Write the answer in simplest form.

$$(x - 2) \cdot (x^2 + 2x + 4)$$

9. Divide a polynomial by using **long division** and **synthetic division**.

$$\frac{y^2 + 3y - 4}{y + 4}$$

10. Factor

$$x^6 - x^2$$

11. Factor

$$x^6 - x^5 - 12x^4$$

12. Factor

$$3x^2 + 2x - 5$$

13. Solve

$$2x^2 + 5x - 7 = 0$$

14. Simplify

$$2\sqrt{18} - \sqrt{50} + \sqrt{2}$$

15. Simplify

$$\left(\frac{16x^8}{y^2}\right)^{1/4} \cdot \frac{\sqrt[4]{y^2}}{2x^2} + \frac{y^2\sqrt[6]{y^3}+\sqrt[3]{x}}{x^{1/3}+y^{5/2}}$$

16. Rationalize the denominator

$$\frac{\sqrt{2}+1}{\sqrt{2}-1}$$

17. Solve

a) $-(x-2)+x-3=-(1-x)$

b) $\dfrac{x-1}{2}=\dfrac{x}{3}$

18. Solve

a) $3x - 2 > -(1 - x) + 1$

b) $4 \leq 3x + 1 \leq 7$

c) $x - 1 \leq -4 \quad \text{or} \quad -x + 1 < -6$

d) $-x + 1 > 0 \quad \text{and} \quad 5 + x < 1 + 2x$

19. Solve

a) $|2 - x| + 3 = 4$

b) $|x - 3| + 1 = 1$

c) $|x - 2| + 1 \leq 4$

d) $|3x - 1| + 5 > 3$

20. Graph these lines using **the points method**

a) $x + y = 4$

b) $y = -4x + 2$

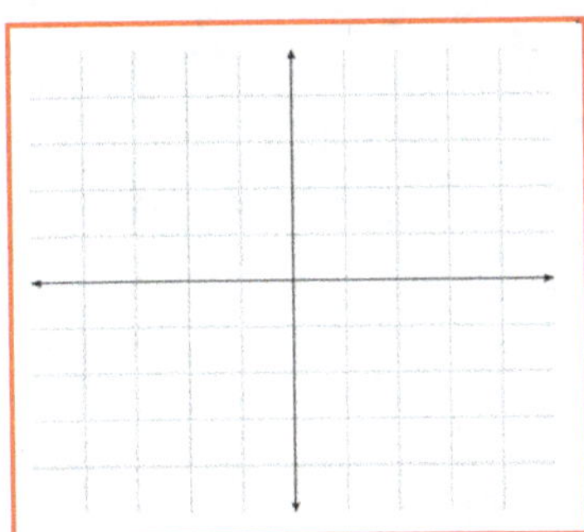

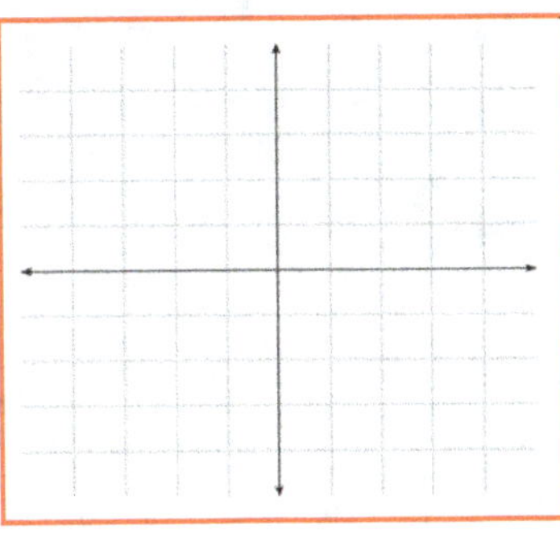

21. Graph these lines using **x- and y-intercepts method**

a) $2x + 3y = 6$

b) $y = 2x - 2$

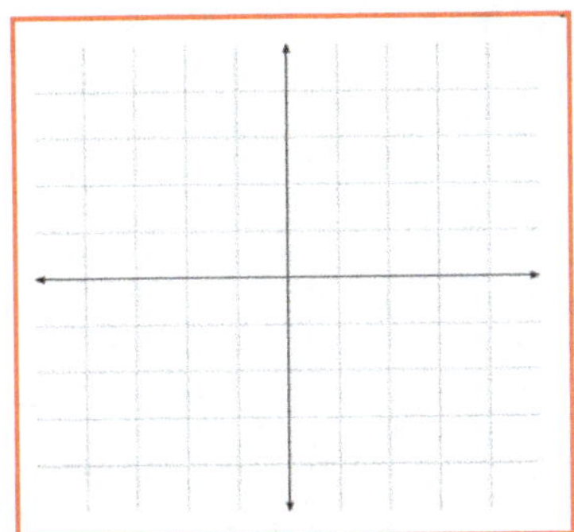

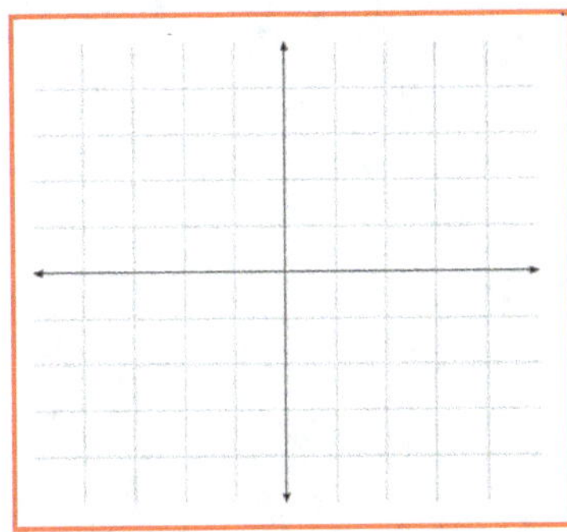

22. Graph these lines using **the slope – y– intercept method**

a) $x + 3y = 3$

b) $y = \dfrac{2}{5}x - 3$

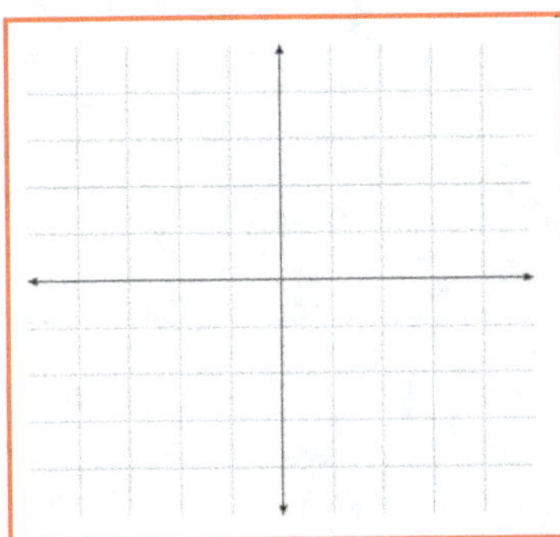

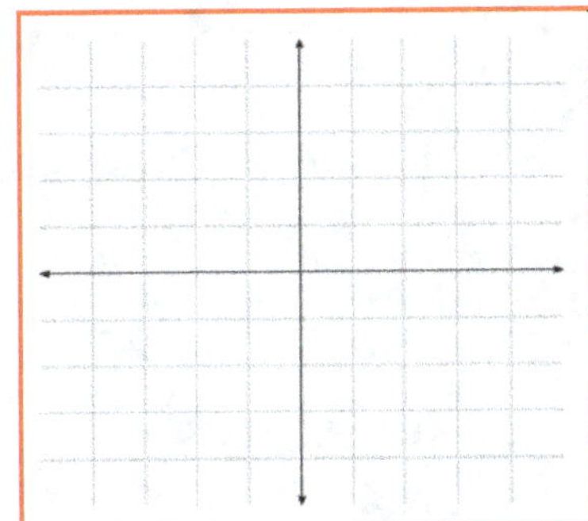

23. Graph the solution sets of the following inequalities

a) $x + 3y > 3$

b) $4x + 2y \leq 8$

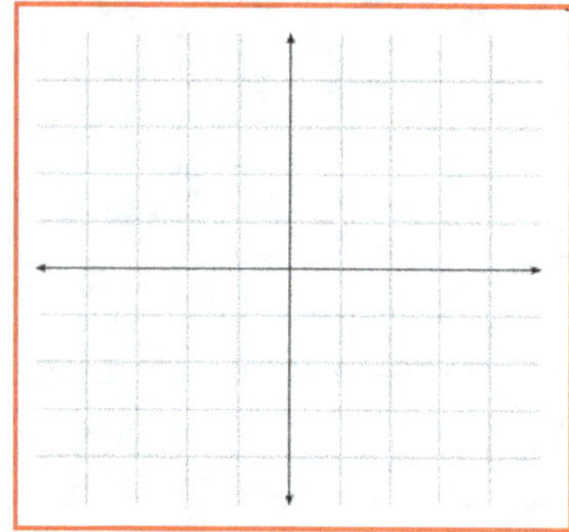

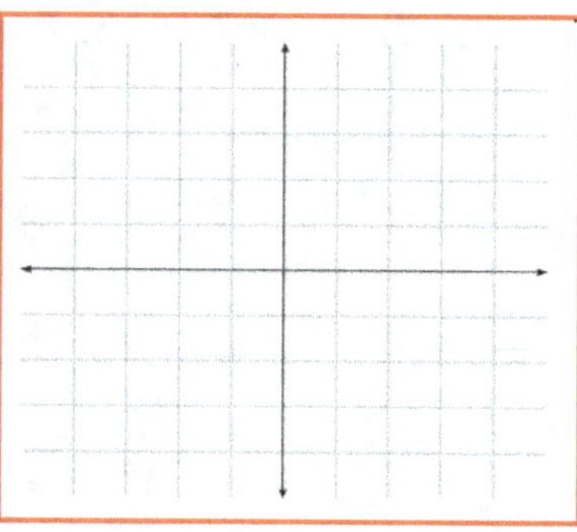

24. Graph the solution sets of the following inequalities

a) $y > -x$

b) $y \geq -2x + 4$

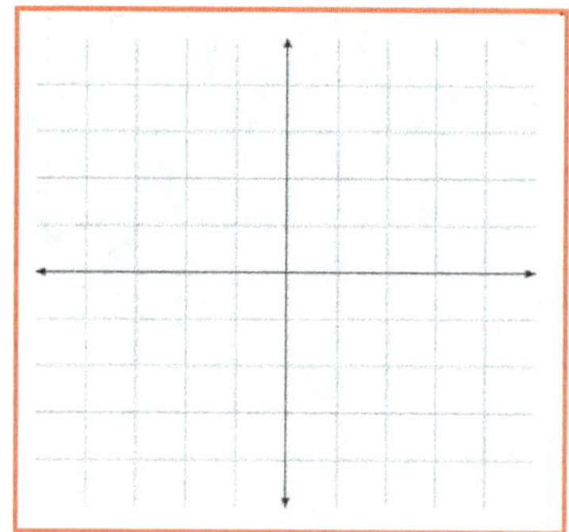

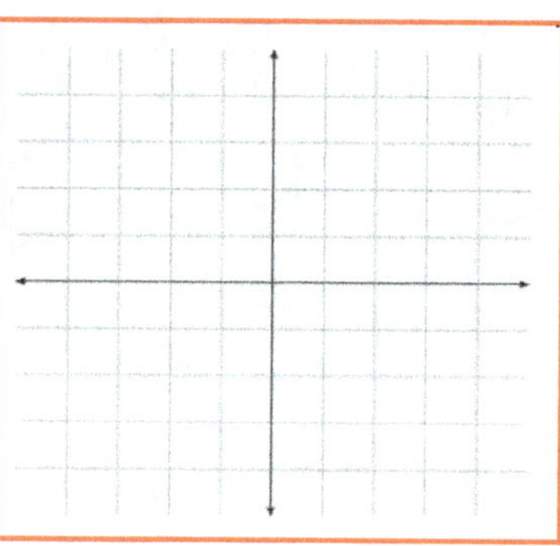

25. Graph the solution sets of the following inequalities

a) $y \geq 0$

b) $x < 0$

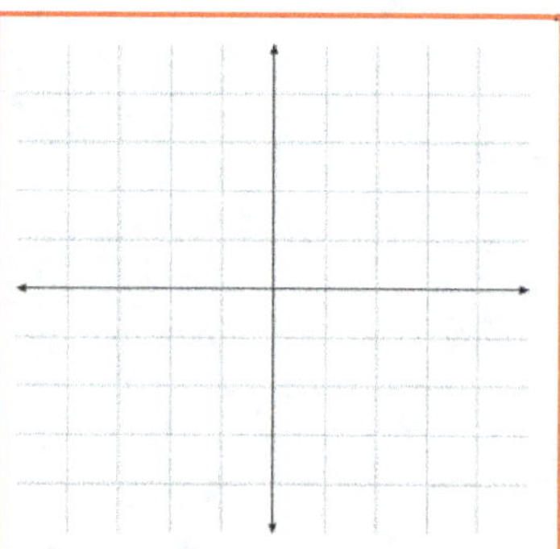

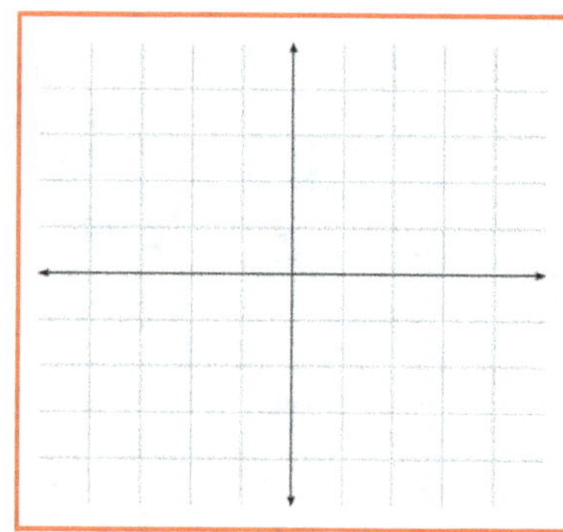

26. Solve each system

a) $\begin{cases} x + y = 4 \\ 2x + y = 6 \end{cases}$

b) $\begin{cases} x - y + z = 2 \\ x + y + 2z = 3 \\ 2x + y - z = 1 \end{cases}$

27. Solve each system by corresponding Cramer's Rule.

a) $\begin{cases} x + y = 5 \\ 2x + y = 7 \end{cases}$

b) $\begin{cases} x - y + z = 2 \\ x + y + z = 6 \\ x + 2y - z = 4 \end{cases}$

28. Sketch the graph of the set of solutions for each system of inequalities.

a) $\begin{cases} x - 3y > -3 \\ y \geq -x + 2 \end{cases}$

b) $\begin{cases} -x + y > -2 \\ y \leq x + 2 \end{cases}$

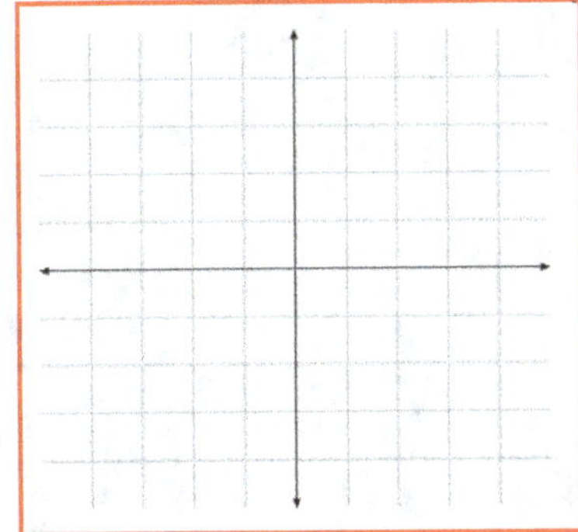

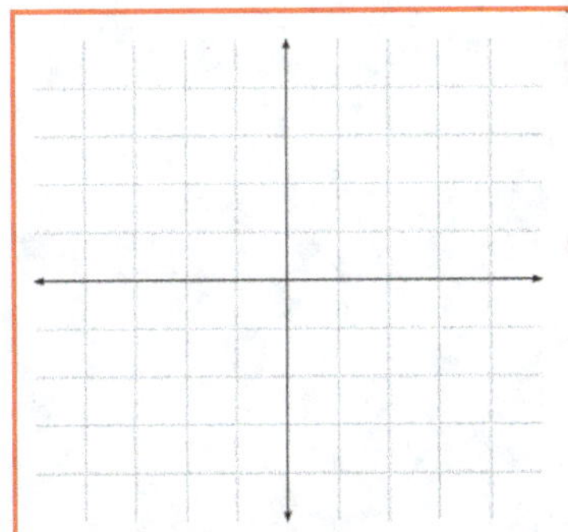

29. Let $f(x) = x$ and $g(x) = -x + 2$. Find the following.

   a) $f^{-1}(x) \circ g^{-1}(x)$
   b) $g^{-1}(x) \circ f^{-1}(x)$
   c) $g(x) \circ f(x)$
   d) $f(x) \circ g(x)$
   e) $g^{-1}(x) \circ g(x)$
   f) $f^{-1}(x) \circ f^{-1}(x)$

30. Sketch the graph of functions using transformations.

   a) $f(x) = -2|x + 2| - 3$

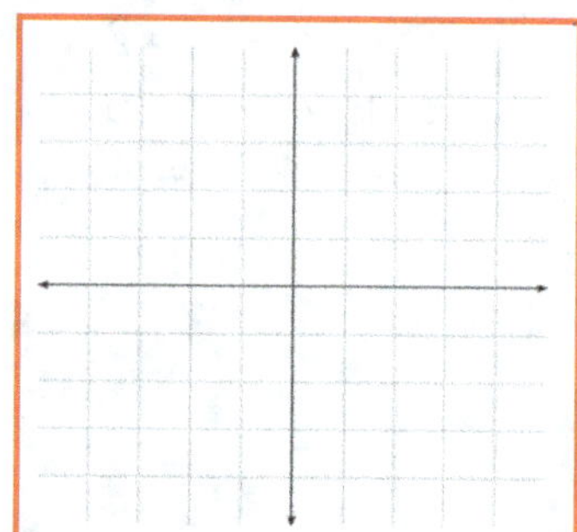

   b) $g(x) = -2(x - 1)^2 - 1$

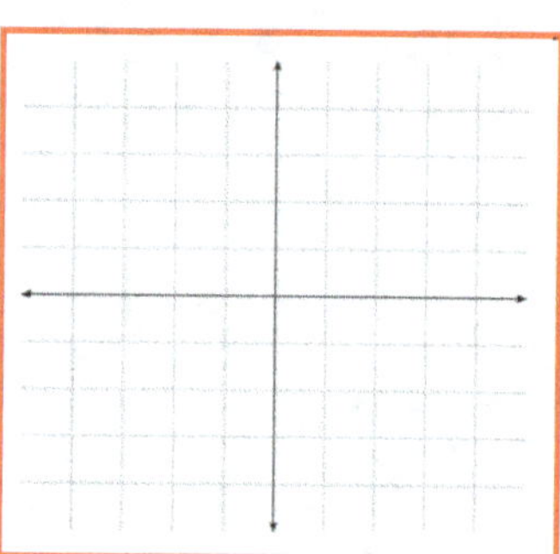

31. Sketch the graph of polynomial and fractional functions.

   a) $f(x) = x(x-1)(x + 1)(x + 2)$

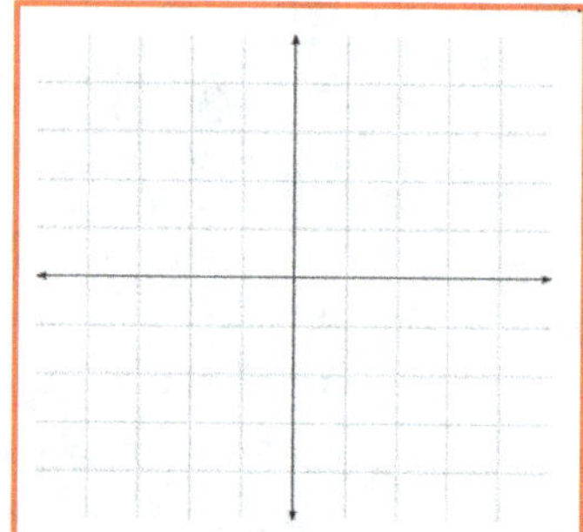

   b) $g(x) = \dfrac{x - 4}{x + 2}$

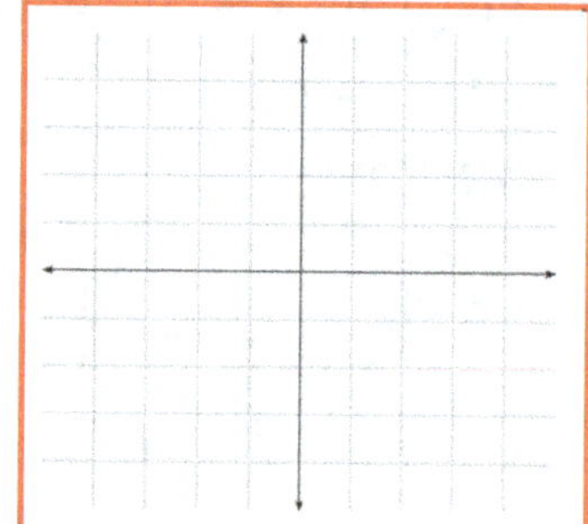

1. Evaluate the numerical expression

$$1 + (-5 + 1) \div (-2) \cdot (-4) - 8$$

2. Let $A = \{0, 1, 2, 3, 4, 5\}$, $B = \{-2, 0, 2, 4, 6\}$. Find each of the following.

a) $A \cup B$

b) $A \cap B$

c) $A \cap A$

d) $(A \cup B) \cup (A \cap B)$

3. Evaluate the numerical expression.

$$\frac{|8 - 5| + |-(1 - 4) - 3 \div 3|}{|4 - 3| - 6 \div 3 + |1 + 5|}$$

4. Use PEMDAS + L→R to evaluate the expression.

$$(-3) \cdot \{[-(-2)^3 + (-3)] + (2^2 - 3) + 8 \div (-2)\} + 6$$

5. Simplify.

$$\frac{x^4 y^2}{x^3 y} \cdot \frac{x^2 y^5}{x^2} \div \frac{x^{-2} y^{-3}}{x^{-4} y^{-3}} \cdot (x^{-3} y^{-2})$$

6. Write the expression in scientific notation form.

$$5(8.4 \cdot 10^7)(4 \cdot 10^{-9})$$

7. Combine like terms in each expression and write the answer in standard form.

$$(3x^5 + x - 4x^2 + x^4) - (-2x^5 + x^4 - 4x^2 + x)$$

8. Multiply the following polynomials. Write the answer in simplest form.

$$(x + 3) \cdot (x^2 + 3x + 9)$$

9. Divide a polynomial by using **long division** and **synthetic division**.

$$\frac{x^2 + x - 12}{x - 3}$$

10. Factor

$$x^5 - x^2$$

11. Factor

$$x^3 - 2x^2 - 8x$$

12. Factor

$$4x^2 + 3x - 7$$

13. Solve

$$x^2 + 5x - 6 = 0$$

14. Simplify

$$\sqrt{25} - \sqrt{18} + 3\sqrt{2}$$

15. Simplify

$$\left(\frac{27x^6}{y}\right)^{1/3} \cdot \frac{\sqrt[6]{y^2}}{x^2} + \frac{y\sqrt[4]{y^2} + \sqrt[4]{x}}{x^{1/4} + y^{3/2}}$$

16. Rationalize the denominator

$$\frac{\sqrt{7} + 2}{\sqrt{7} - 1}$$

17. Solve

a) $\quad 4(x - 2) + x - 3 = -4(1 - x)$

b) $\quad \dfrac{x - 4}{2} + 1 = \dfrac{x}{6}$

18. Solve

a) $3x - 1 > -(1 - 2x) + 1$

b) $1 \leq 1 + 2x \leq 5$

c) $2x - 1 \leq 5$ or $-x + 2 < -4$

d) $-x + 1 > 0$ and $2 + x < 6 + 2x$

19. Solve

a) $|1 - 2x| + 5 = 6$

b) $|x - 3| + 4 = 1$

c) $|x - 1| + 2 \leq 3$

d) $|x - 1| + 2 > 5$

20. Graph these lines using **the points method**

    a) $x + 2y = 6$

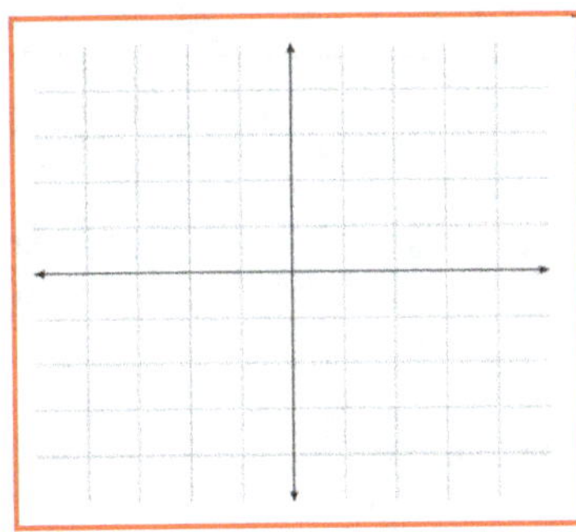

    b) $y = -2x + 3$

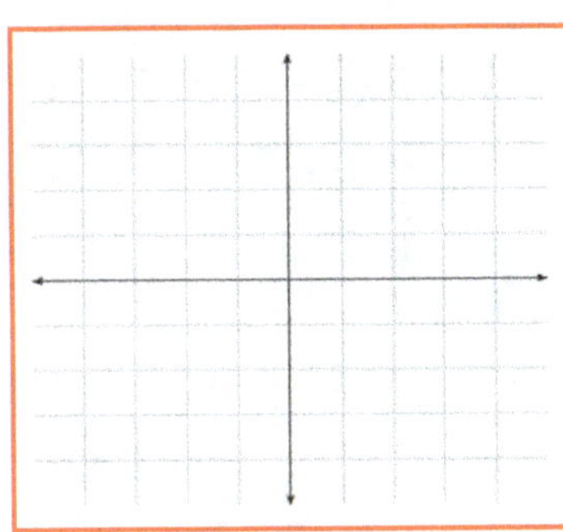

21. Graph these lines using **x- and y-intercepts method**

    a) $2x + 3y = 6$

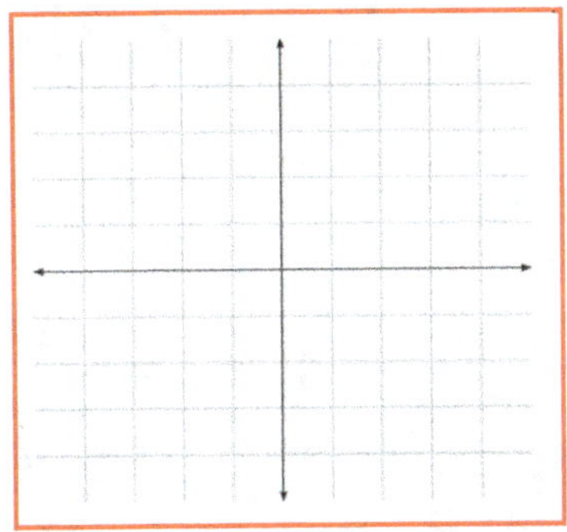

    b) $y = 2x - 2$

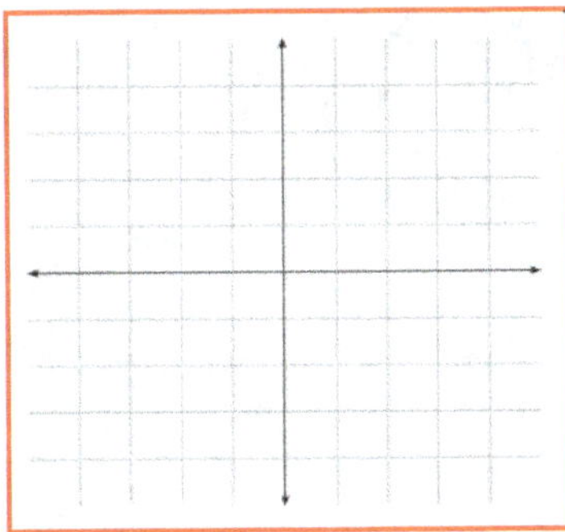

22. Graph these lines using **the slope – y– intercept method**

    a) $2x + y = 3$

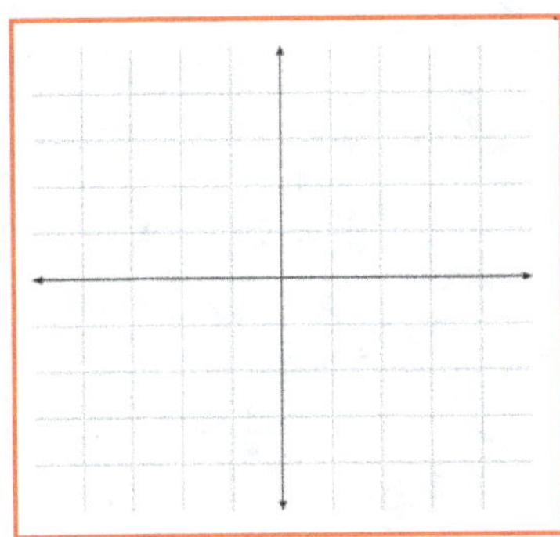

    b) $y = \dfrac{2}{3}x - 4$

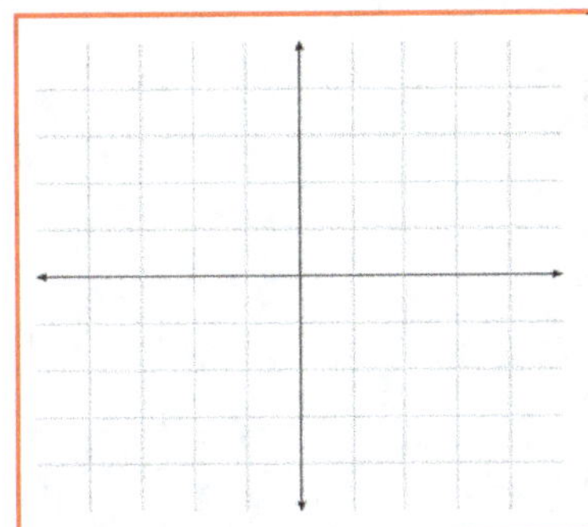

23. Graph the solution sets of the following inequalities

   a) $x + y > 3$

   b) $3x + 2y \le 6$

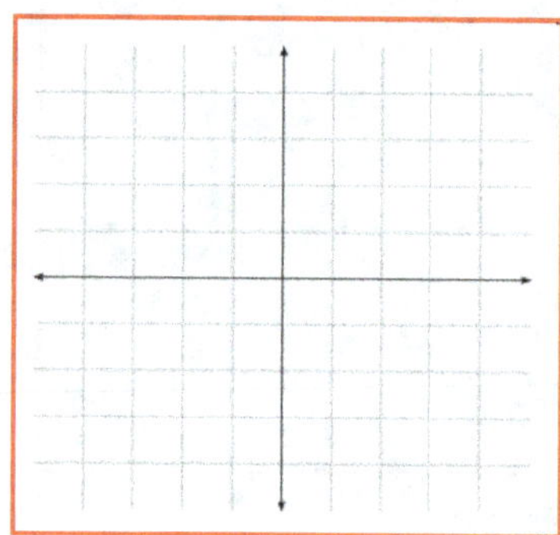 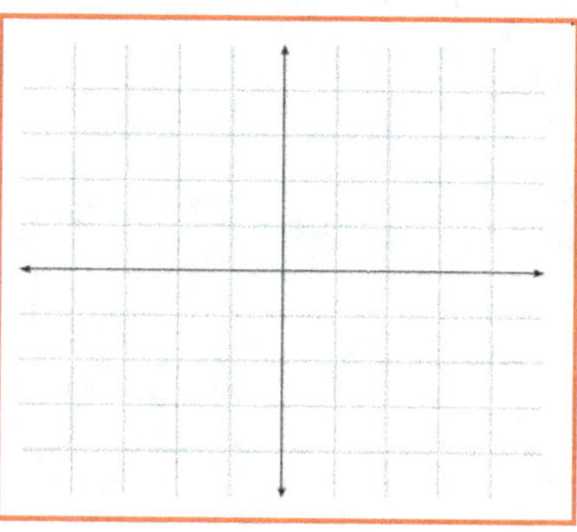

24. Graph the solution sets of the following inequalities

   a) $y < -x + 3$

   b) $y \ge -x$

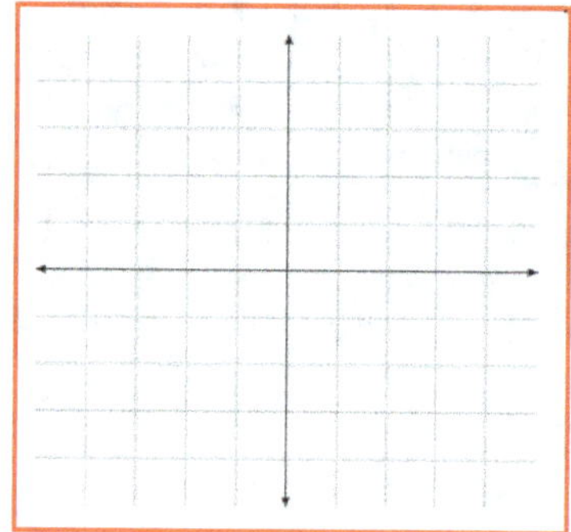 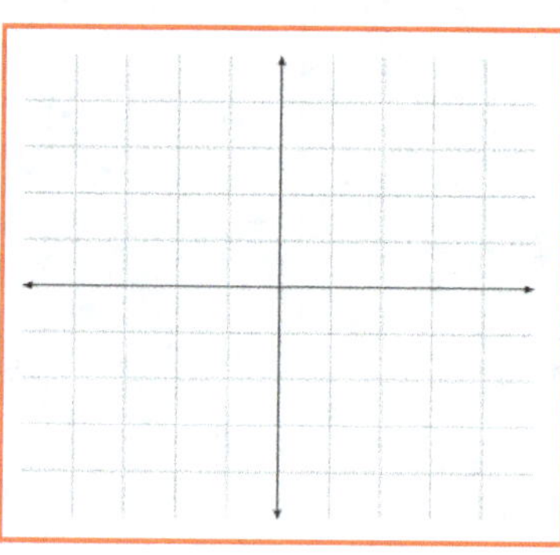

25. Graph the solution sets of the following inequalities

   a) $y \ge -1$

   b) $x < -2$

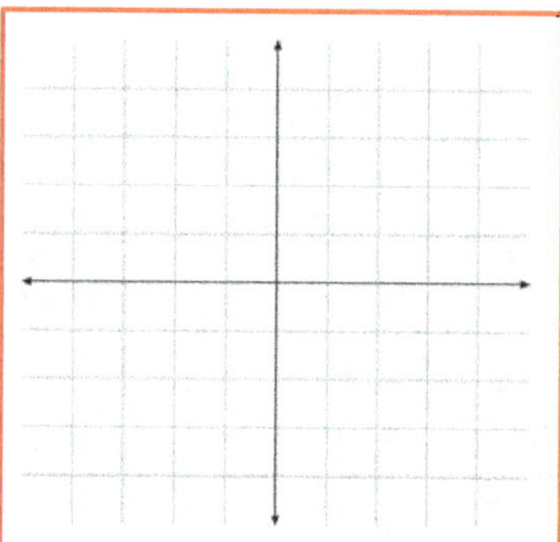 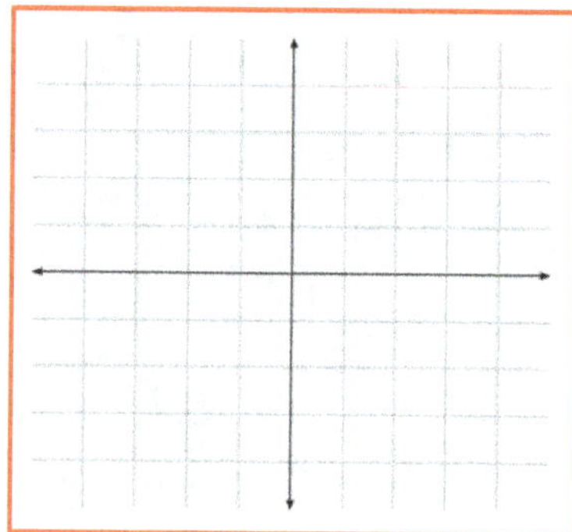

26. Solve each system

a) $\begin{cases} x + y = 2 \\ 2x + 4y = 6 \end{cases}$

b) $\begin{cases} x - y + z = 2 \\ x + y + 2z = 6 \\ x + y - z = 0 \end{cases}$

27. Solve each system by corresponding Cramer's Rule.

a) $\begin{cases} x + y = 3 \\ 2x + 3y = 8 \end{cases}$

b) $\begin{cases} x - y + z = 1 \\ x + y + 2z = 4 \\ x + 2y - 3z = 0 \end{cases}$

28. Sketch the graph of the set of solutions for each system of inequalities.

a) $\begin{cases} 2x - y > -3 \\ y \geq -2x + 1 \end{cases}$

b) $\begin{cases} -x + y > -4 \\ y \leq -32x \end{cases}$

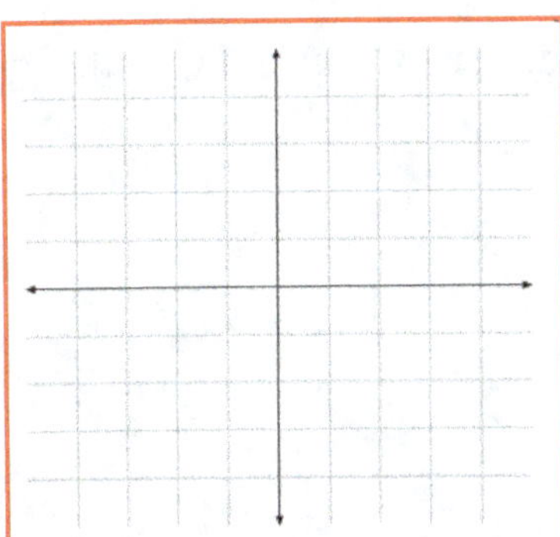

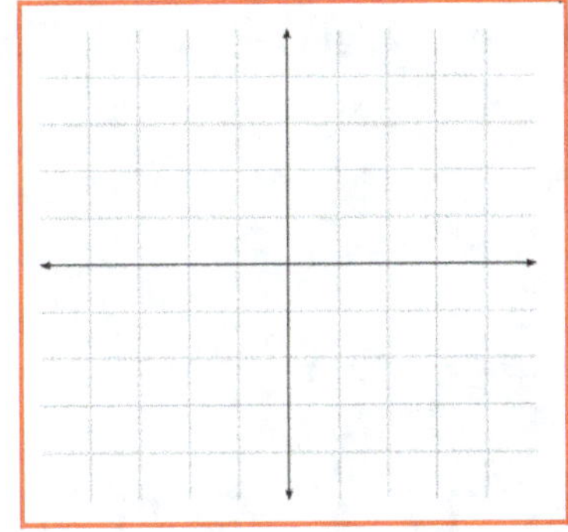

29. Decide whether the given relation is a function.

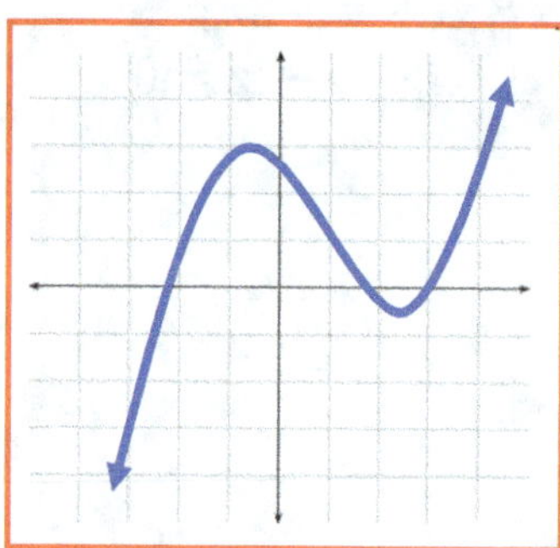 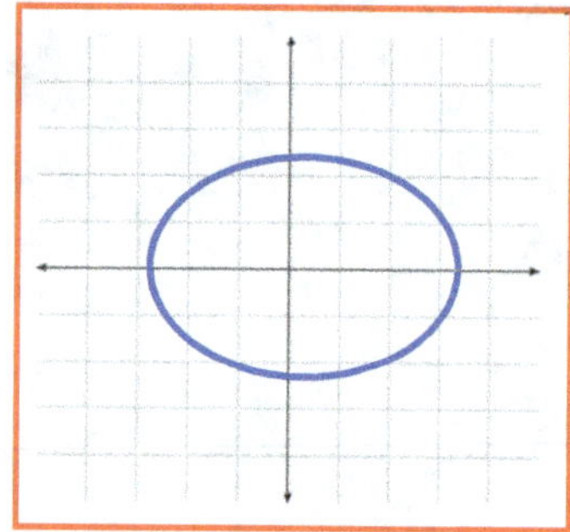

30. Find the domain and range of each function.

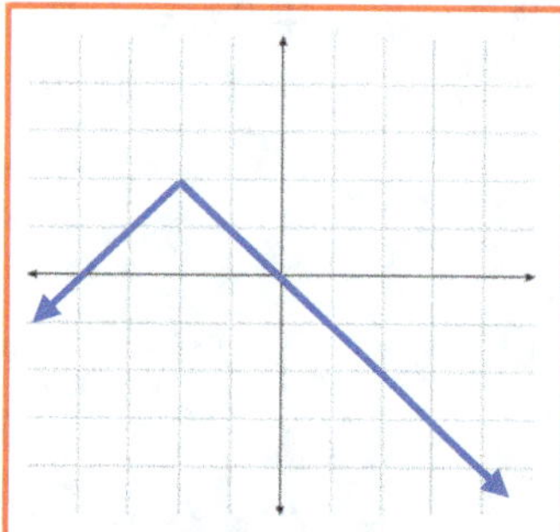 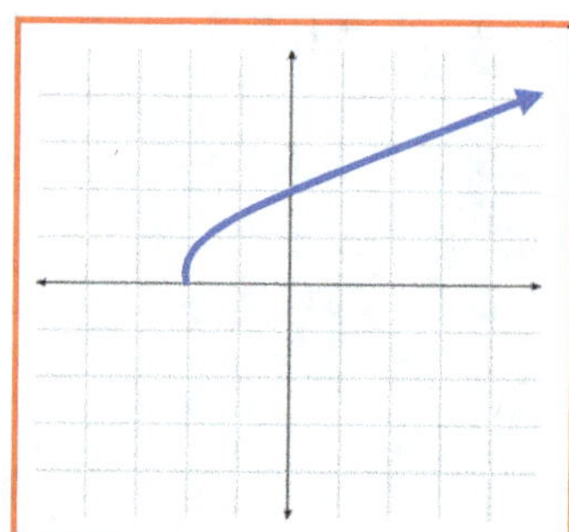

31. Classify functions as even, odd, or neither.

e) $y = x + x^5$ f) $y = -2x^4$ g) $y = x^2 + x^3$ h) $y = 3|x| + x^2$

32. Let $f(x) = 3x - 2$, $g(x) = \frac{1}{2}x + 1$. Find and graph the functions $f^{-1}(x)$ and $g^{-1}(x)$.

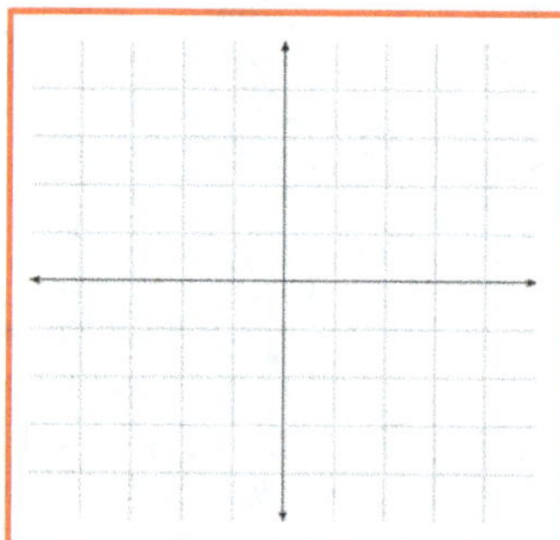 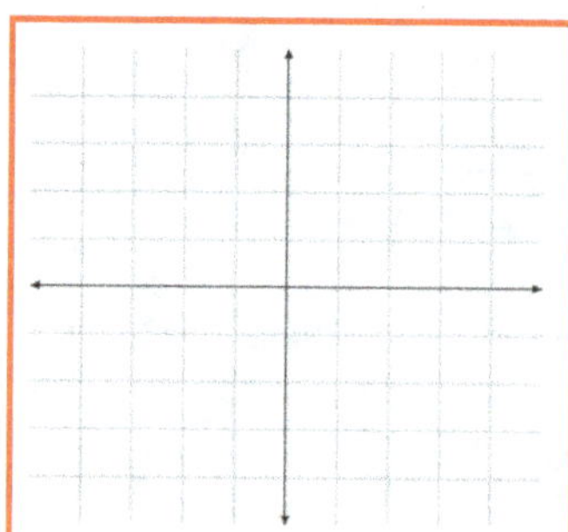

33. Let $f(x) = 2x$, $g(x) = -x + 1$. Find the following.

    i) $f^{-1}(x)$

    j) $g^{-1}(x)$

    k) $f^{-1}(x) \circ g^{-1}(x)$

    l) $g^{-1}(x) \circ f^{-1}(x)$

    m) $g(x) \circ g^{-1}(x)$

    n) $f(x) \circ f^{-1}(x)$

    o) $g^{-1}(x) \circ g(x)$

    p) $f^{-1}(x) \circ f^{-1}(x)$

34. Sketch the graph of functions using transformations.

    b) $f(x) = 2|x + 1| - 3$ b) $g(x) = -(x - 1)^2 + 3$

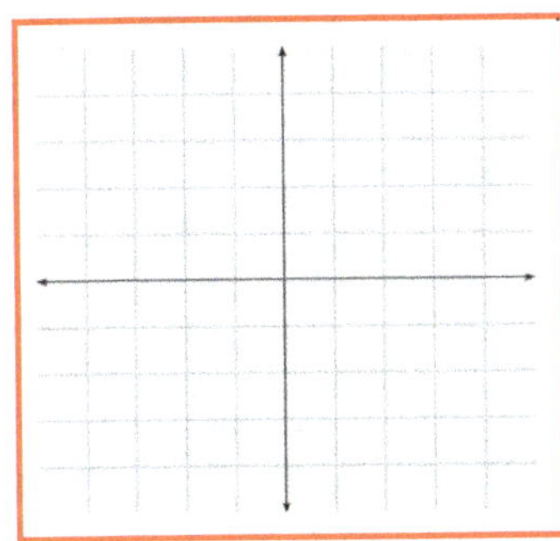 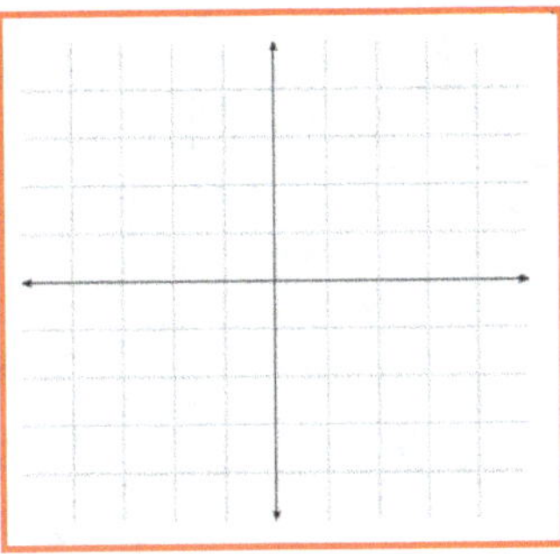

35. Sketch the graph of polynomial and fractional functions.

    b) $f(x) = x - x^3$ b) $g(x) = \dfrac{x + 2}{x - 2}$

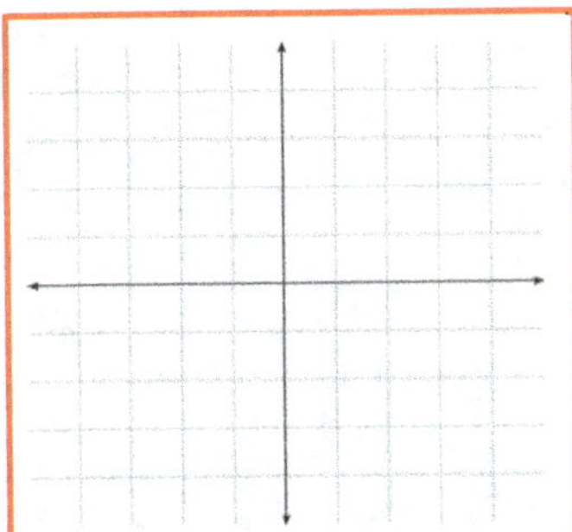 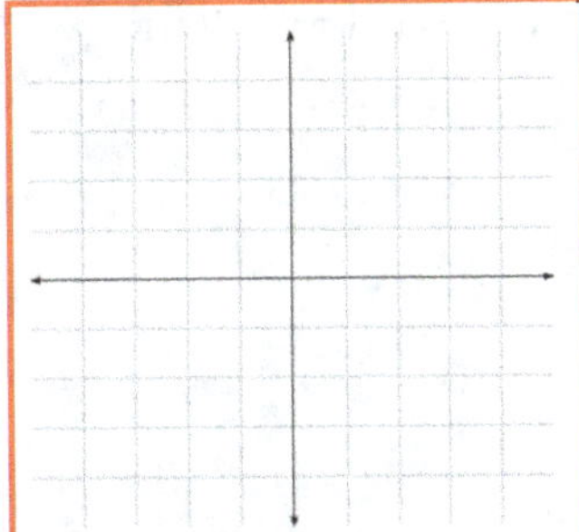

<table>
<tr><td>PART
1-4</td><td>FINAL TEST 3</td><td>Hayk Yegoryan & Rubik Yegoryan
INTERMEDIATE ALGEBRA</td></tr>
</table>

PROBLEM	ANSWER
1. Simplify $2(3-5)^2 - (2-4)^2 - 4$	1. ________
2. Simplify $\dfrac{2\left|1-5\right| - 2\left|2-3\right|}{-2\left|-3+2\right| - 4\left|6-5\right|}$	2. ________
3. Simplify $3(2x - 3y + 5z) - 12\left(-\dfrac{3}{4}y + \dfrac{1}{2}x + z\right)$	3. ________

4. Solve the equation

$$\frac{1}{2}x - 1 = \frac{1}{3}x$$

4. _____________

---

5. Solve the equation

$$4(2x - 1) + 6\left(\frac{1}{3}x - 2\right) = 8(x - 2) + 2$$

5. _____________

---

6. Use the symbols > or < for ☐ and compare expressions

$$9 - 2^3(5 + 1) \div 4 \quad \boxed{} \quad -28 \div 4 - (-8 \div 2) + 1$$

6. _____________

7. **Solve the inequality and graph**

$$4x - 1 \leq 2x + 7$$

7. __________

8. **Solve the inequality and graph**

$$-6(x - 1) + 8 \leq -4x + 16$$

8. __________

9. **Solve the inequality and graph**

$$3 \leq 2x + 6 \leq 12$$

9. __________

10. Solve and graph

$x + 3 < -8$ or $x - 2 \geq 9$

11. Solve and graph

$2x \leq 8$ and $3x + 5 > -1$

12. Graph these lines

a) $-6x + 3y = 12$

b). $y = \frac{1}{2}x - 1$

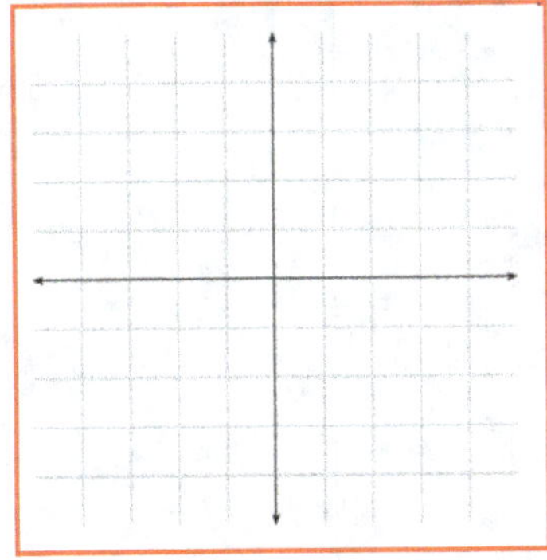

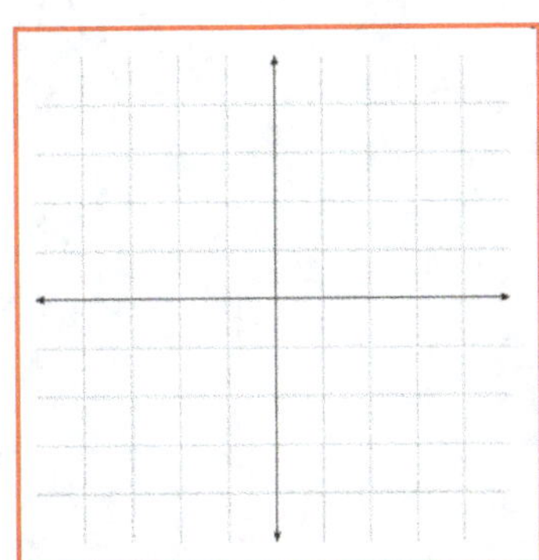

13. Are lines $3x+2y = 1$ and $y = \frac{2}{3}x - 1$ *parallel, perpendicular, or neither?*

    A.   parallel
    B.   perpendicular
    C.   neither

14. Graph these lines

a) $y = -3$ b) $x = 4$

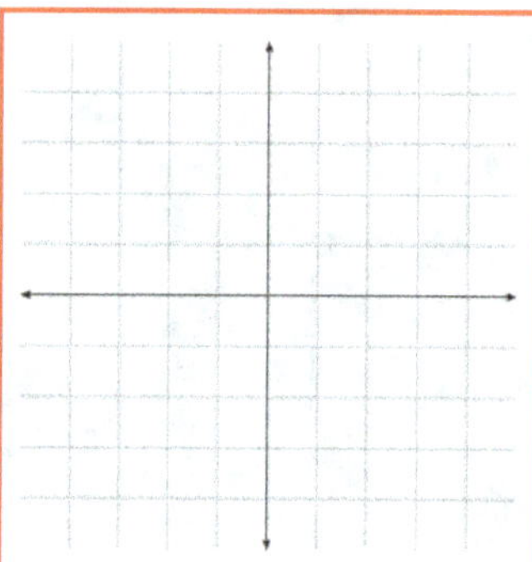
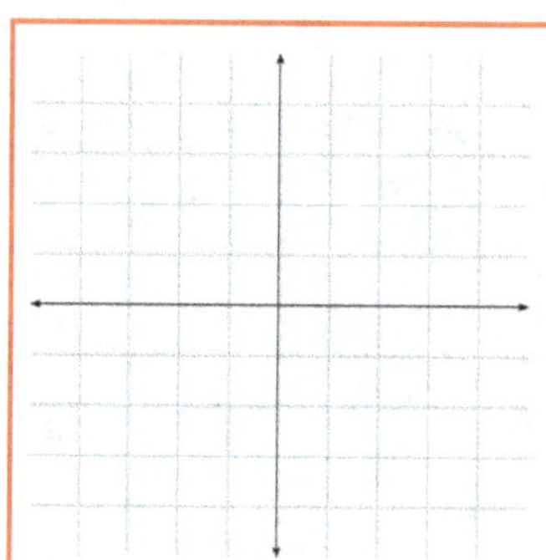

15. Write the equation of the line which passes through the given points

a) (-2,0) & (0,3)   b) (-1,0) & (0,-3)

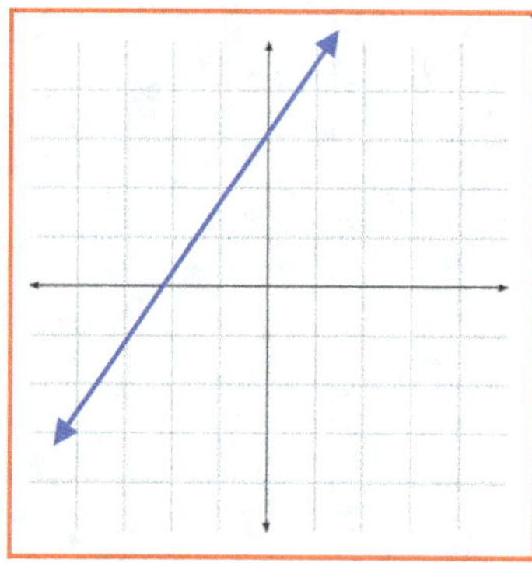
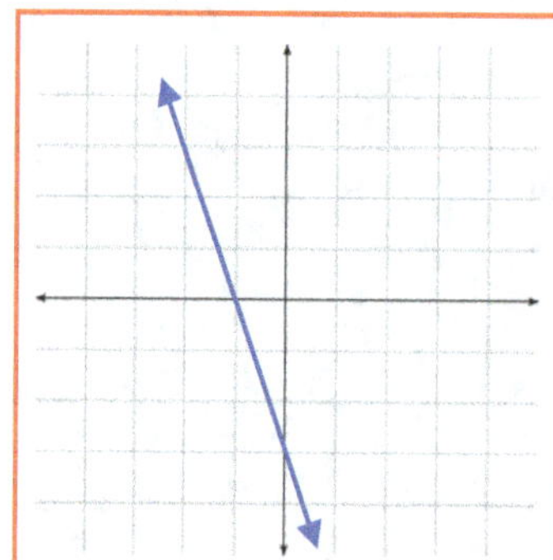

16. Graph the following inequalities

a) $x + 3y > -3$

b) $y \leq -\dfrac{1}{3}x + 2$

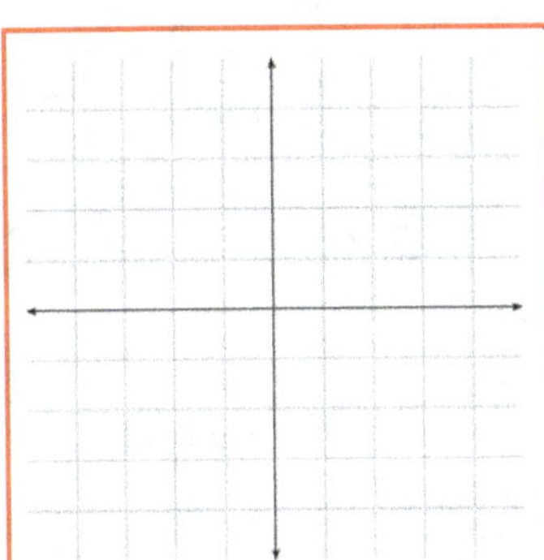 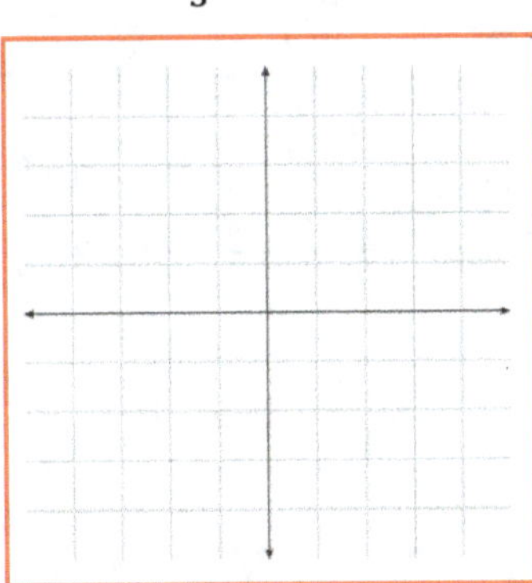

16. __________

17. Graph the following inequalities

a) $2x + y > -2$

b) $y > \dfrac{2}{4}x - 3$

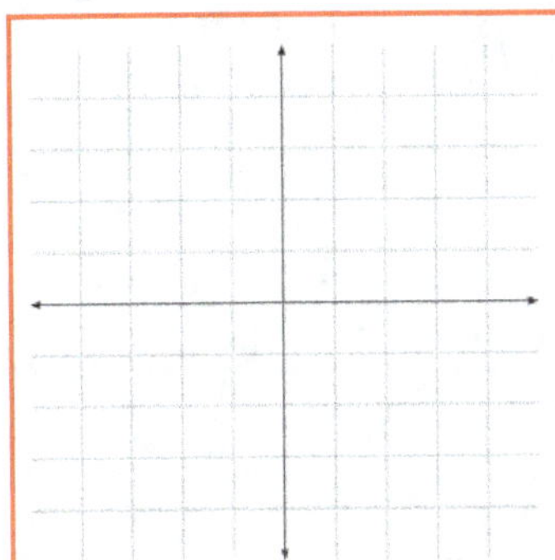 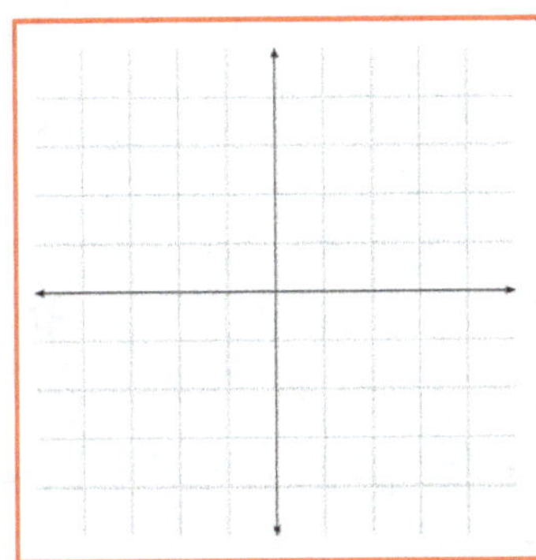

17. __________

18. Graph the following inequalities

a) $x > -3$

b) $y \leq +2$

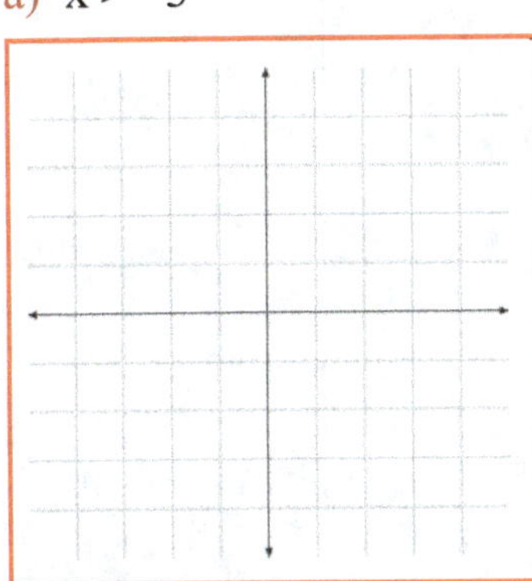 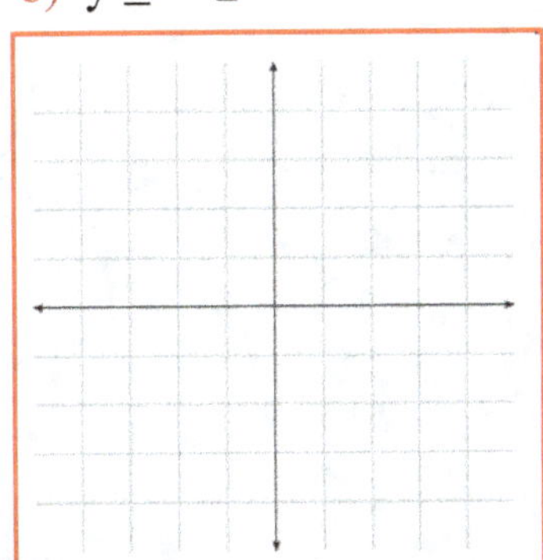

18. __________

19. Solve the system by graphing

$$\begin{cases} 2x + y = 2 \\ 4x + 2y = 8 \end{cases}$$

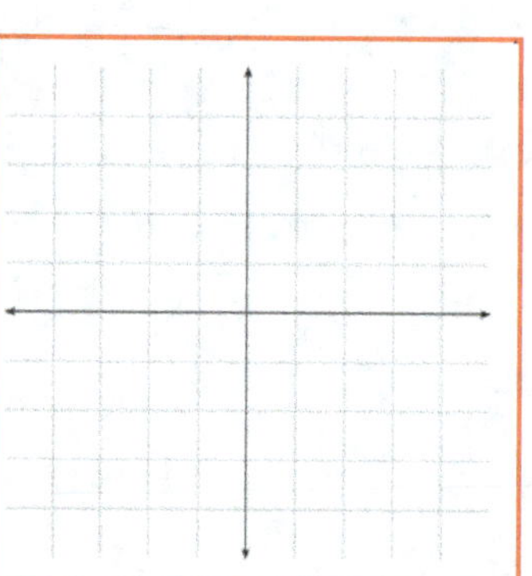

19. ____________

20. Solve the system by elimination method

$$\begin{cases} 2x + y = 5 \\ x - y = 1 \end{cases}$$

20. ____________

21. Solve the system by substitution method

$$\begin{cases} x - 2y = 2 \\ 3x - 5y = 7 \end{cases}$$

21. ____________

FINAL TEST 4

Hayk Yegoryan & Rubik Yegoryan

INTERMEDIATE ALGEBRA

PROBLEM	ANSWER
1. Simplify $$\dfrac{x^2 - 5x + 6}{x^2 - 4}$$	————
2. Simplify $$\dfrac{x^2 - 1}{x^2 + 2x + 1} \div \dfrac{x - 1}{x + 1}$$	————
3. Simplify $$\dfrac{1 - \dfrac{2}{x}}{x - \dfrac{4}{x}}$$	————

4. **Solve**

$$\frac{1-x}{x} + \frac{2}{x} = 2$$

---

5. **Solve**

$$\frac{2x}{x+1} = \frac{2x-1}{x+3}$$

---

6. **Rationalize the denominator**

$$\frac{\sqrt{5}-2}{\sqrt{5}+2}$$

7. Simplify

a) $\sqrt{18x^5y^2}$

b) $\dfrac{\sqrt[3]{x^6 y^9}}{x\sqrt{y^4}}$

____

8. Solve

a) $\sqrt{2x + 3} = x$

b) $\sqrt{x + 3} = x + 1$

____

9. Solve

a) $2x^2 + 4x = 0$

b) $4x^2 - 9 = 0$

____

10. Solve

a) $x^2 + 6x - 16 = 0$

b) $5x^2 + 3x - 2 = 0$

___

11. Simplify

a) i^{15}

b) $\sqrt{-4} + i^9 + \dfrac{2}{i}$

___

12. Sketch the graphs

a) $y = \sqrt{x - 3} - 1$

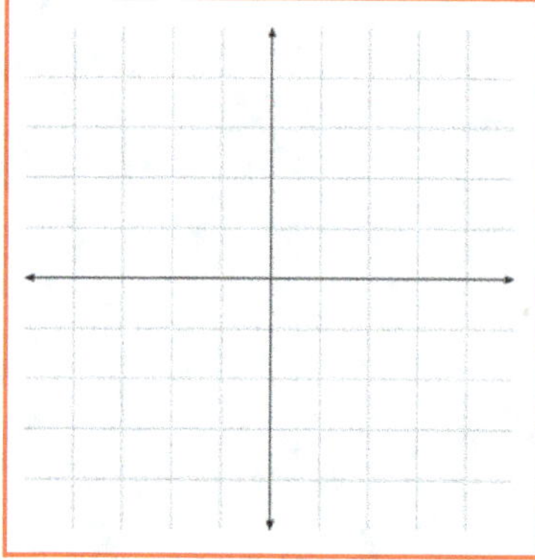

b) $y = -\sqrt{x + 1} + 2$

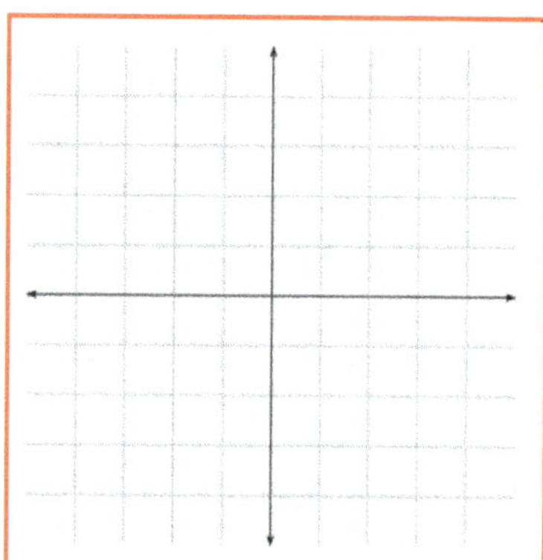

13. Sketch the graphs

a) $y = (x - 3)^2 + 2$

b) $y = x^2 - 2x - 1$

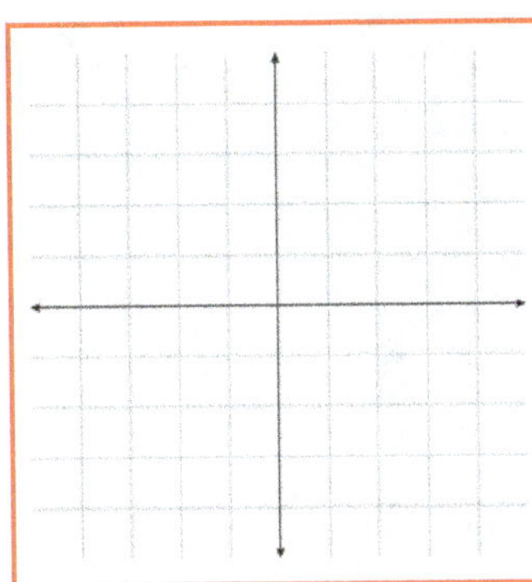

14. Graph these lines using the points method

a) $x + y = 3$

b) $y - 2x = 1$

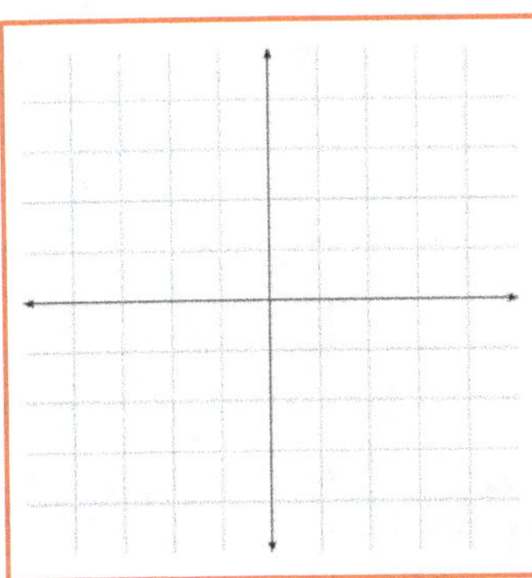

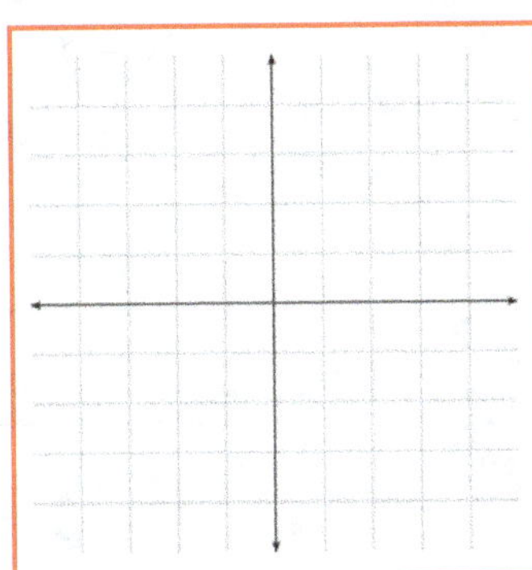

15. Graph these lines using the x– and y– intercepts method

a) $2x + 3y = 6$

b) $y = 2x - 4$

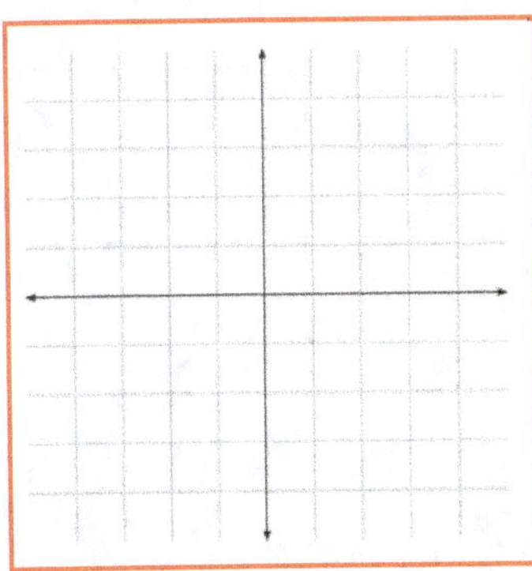

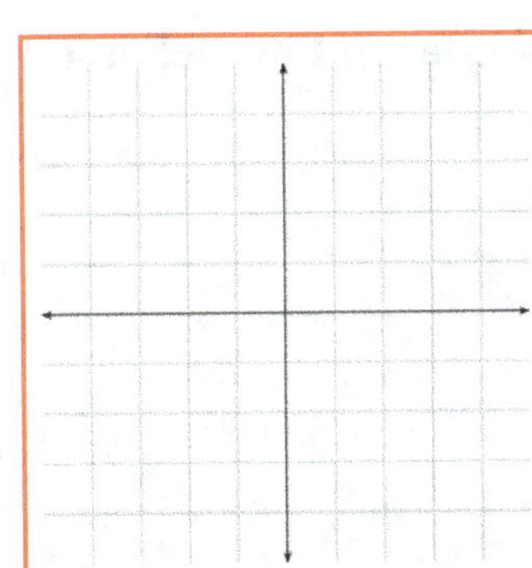

16. **Graph these lines using the slope – y–intercept method**

  a) $3x + 2y = 4$ b) $y = (2/3)\,x + 1$

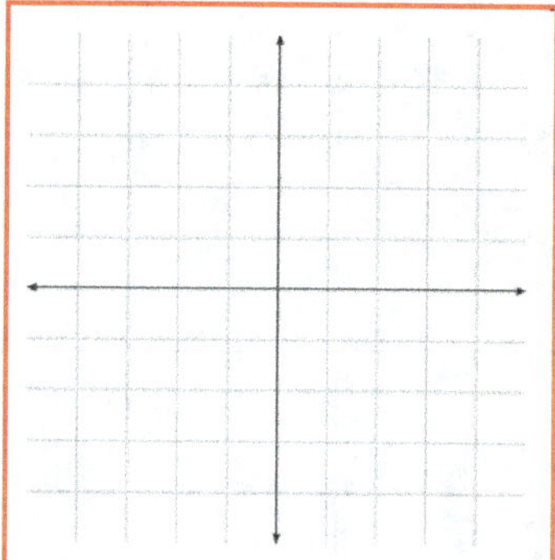 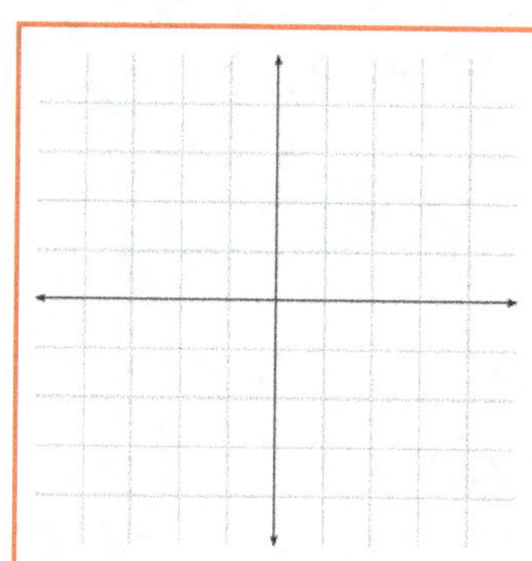

17. **Graph the lines**

  a) $y = -3$ b) $x = 2$

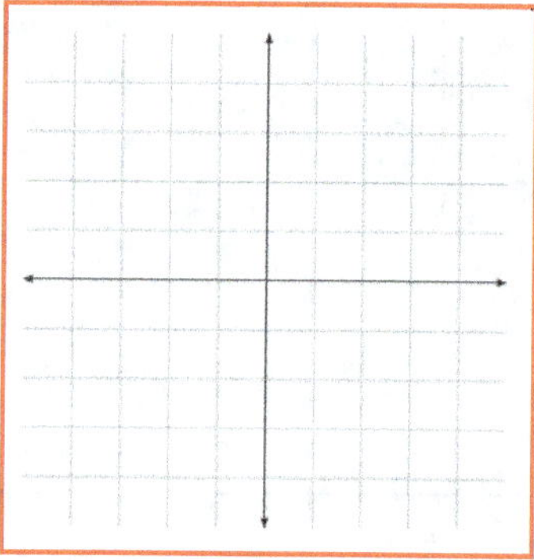 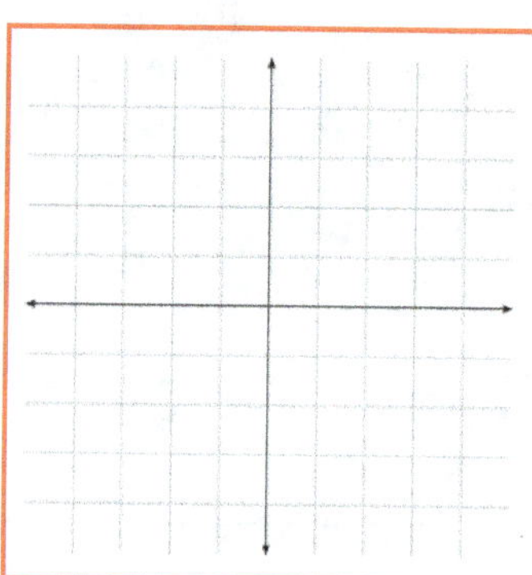

18. **Find the slope and y–intercept of each line**

  a) $5x + 4y = 12$ b) $y = -3 + 2x$

19. Write the equation of the line with the given slope and y-intercept

$m = 2, \ b = -3$

____

20. Write the equation of the line with the given point and slope

$(-2, 0), \ m = -3$

____

21. Write the equation of the line which passes through the points
(0, 3) **and** (1, 5)

____

22. Are the graphs of $3x + 2y = 1$ and $y = \dfrac{2}{3}x - 1$ *parallel,*
perpendicular, or neither?

A. parallel
B. perpendicular
C. neither

____

23. **Graph the following inequalities**

a) $x + 3y > -3$

b) $y \leq -\dfrac{1}{3}x + 2$

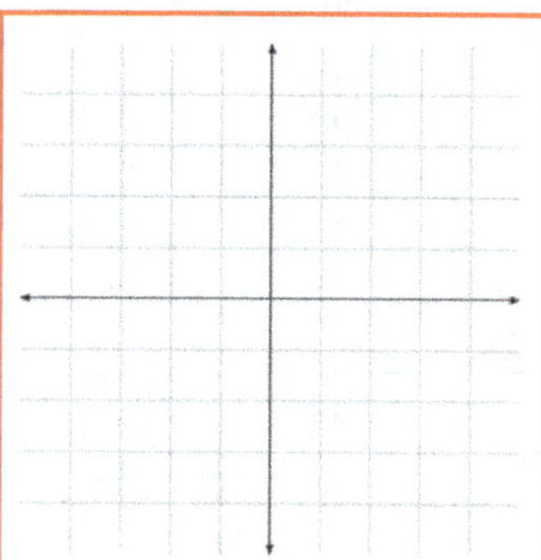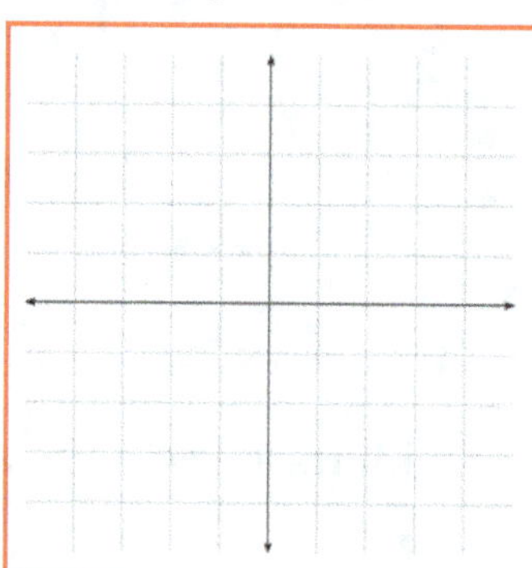

24. **Solve the system by graphing**

$$\begin{cases} 2x + y = 2 \\ 4x + 2y = 8 \end{cases}$$

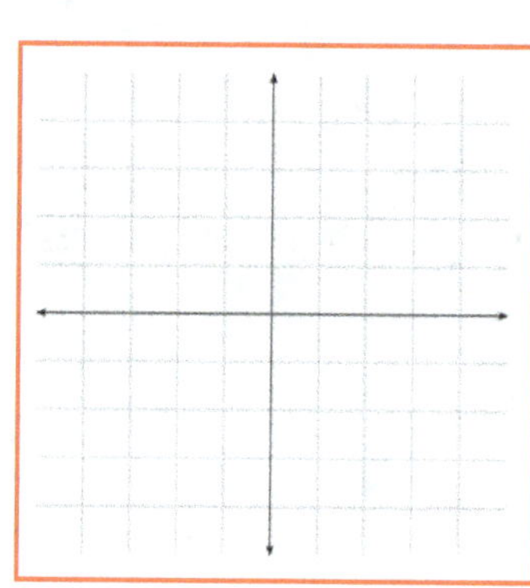

____

25. **Solve system by elimination method**

$$\begin{cases} 2x + y = 5 \\ x - y = 1 \end{cases}$$

____

26. **Solve by substitution method**

$$\begin{cases} x - y = 7 \\ 4x - 5y = 25 \end{cases}$$

____

27. **Solve by Cramer's rule**

$$\begin{cases} 3x - y = 5 \\ 2x + 4y = 8 \end{cases}$$

____

$$-x + 5 > 0 \text{ and } 2 + x < 7 + 2x$$

$$-4 + \left|\,\left|\,|x-2| + 3\,\right| - 5\,\right| + 5 = 12$$

$$\det A = \begin{vmatrix} a_1 & b_1 & c_1 \\ a_2 & b_2 & c_2 \\ a_3 & b_3 & c_3 \end{vmatrix} = \begin{vmatrix} a_1 & b_1 & c_1 \\ a_2 & b_2 & c_2 \\ a_3 & b_3 & c_3 \end{vmatrix} \begin{matrix} a_1 & b_1 \\ a_2 & b_2 \\ a_3 & b_3 \end{matrix} = (a_1b_2c_3 + b_1c_2a_3 + c_1a_2b_3) - (a_3b_2c_1 + b_3c_2a_1 + c_3a_2b_1)$$

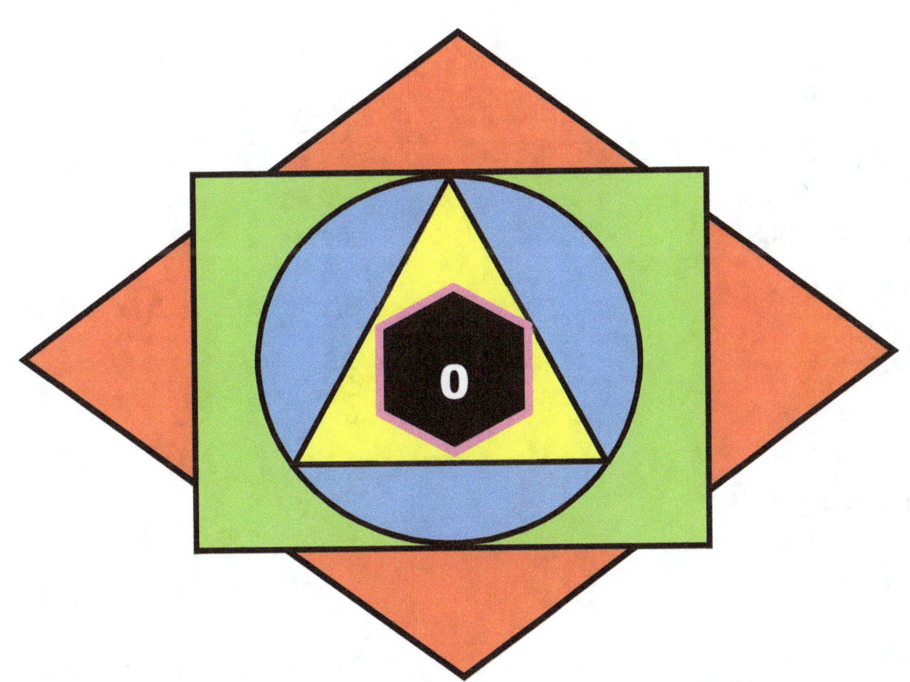

BANK OF EXTRA EXERCISES

Hayk Yegoryan & Rubik Yegoryan

INTERMEDIATE ALGEBRA

Solve the linear equations and inequalities.

1. $x + 2 = 3$

2. $4 + x = 7$

3. $x + \dfrac{1}{4} = \dfrac{1}{2}$

4. $\dfrac{1}{3} + x = \dfrac{1}{6}$

5. $x + 0.5 = -3$

6. $4.5 = 1 + 2.5$

7. $x - 4 = 7$

8. $3 - x = 2$

9. $x - \dfrac{1}{4} = \dfrac{1}{6}$

10. $\dfrac{1}{2} - x = \dfrac{1}{8}$

11. $x - 0.7 = 10.8$

12. $4.5 = 1 - 2.5$

13. $2x = 6$

14. $\dfrac{1}{2}x = \dfrac{1}{8}$

15. $0.5x = 4.5$

16. $\dfrac{3}{2}x = 6$

17. $0.5 + 2x = 4.5$

18. $3x + 5 = 20$

19. $4 - 3x = \dfrac{1}{6}$

20. $\dfrac{1}{3}x - \dfrac{1}{2} = -\dfrac{1}{6}$

21. $-3(5 - y) = 3y - 12$

22. $0.2(x - 5) + 2.5 = -2x + 0.5(x + 1)$

23. $\dfrac{x + 1}{2} = 2x + 5$

24. $\dfrac{1}{3}(1 - x) + \dfrac{1}{6} = \dfrac{1}{4} + \dfrac{1}{2}(x + 2)$

25. $\dfrac{2x - 2}{3} = \dfrac{3x + 4}{4}$

26. $x + 3 > 3$

27. $2 + x < 7$

28. $x + \dfrac{1}{2} \leq \dfrac{1}{6}$

29. $\dfrac{1}{3} \leq \dfrac{1}{6} + x$

30. $5(x + 3) + 4 \geq 2(2x - 1) + 15$

31. $\dfrac{1}{6} < 2x - 3 \leq \dfrac{3}{2}$

32. $x < 3x + 4 \leq 4$

33. $2x \leq 2 + 4x \leq 5x$

34. $x - 2 \leq 3 \quad$ or $\quad -x + 1 < 3$

35. $3x - 1 \leq 3 \quad$ or $\quad x + 1 > 4$

36. $x - 1 \leq 3x \quad \cup \quad 2x - 8 < -4$

37. $-x + 2 > 2 \quad$ and $\quad 5 + 4x > 5 + 3x$

Solve the linear equations and inequalities.

1. $2x + 1 = 5$

2. $4x - 1 = 7$

3. $\dfrac{1}{4}x - \dfrac{1}{2} = -\dfrac{1}{8}$

4. $-2(6 + y) = -2y - 12$

5. $0.2(x - 5) = -1 + 0.3(x + 1)$

6. $\dfrac{5x + 1}{2} = x + 2$

7. $\dfrac{2}{3}(1 - x) + \dfrac{1}{2} = \dfrac{1}{4} + \dfrac{1}{3}x$

8. $\dfrac{x + 1}{3} = \dfrac{x + 4}{4}$

9. $x < 2x - 1 \le 11$

10. $2 \le 2 + x \le 9$

11. $2x - 1 \le 3 \ $ or $\ -x + 2 < -4$

12. $4x - 1 \le 3x \ \cup \ -2x - 8 < -2$

13. $-x + 2 > 0 \ $ and $\ 2 + x < 6 + 3x$

14. $x - 2 \le 3x \ \cap \ 2x - 8 < -2$

15. $\begin{cases} x + 2 > -8 \\ 4x + 2 \le 3(x + 1) \\ -x - 1 < -5 \end{cases}$

16. $\begin{cases} x - 1 > 1 \\ 3x - 2 \le 2(x + 1) \\ x - 1 < 6 \end{cases}$

17. Find x,y, m∡CED, m∡DEB, m∡BEA, m∡AEC

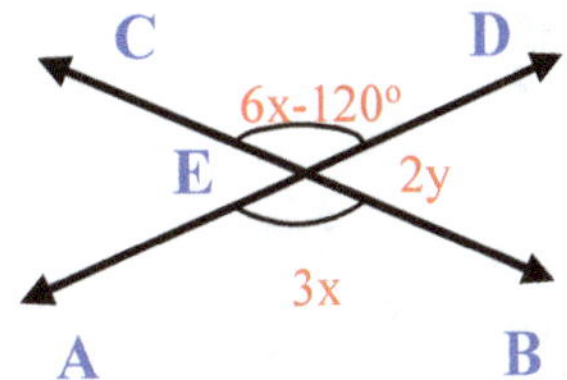

18. Find x, m∡1, m∡2, and m∡3.

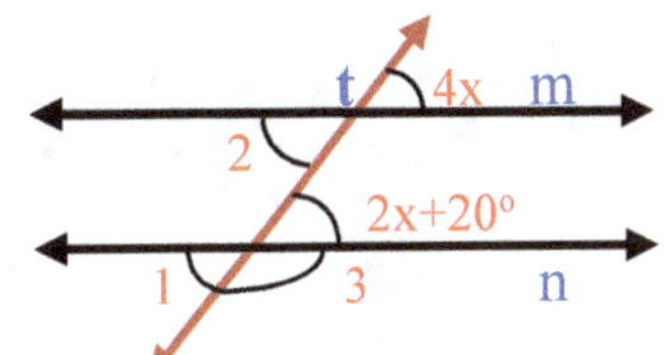

19. Find x and m∡AOB.

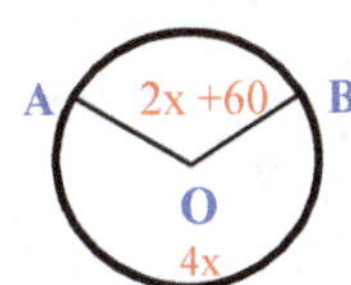

20. Find x, m∡A, m∡B, m∡C, m∡D, m∡E, and m∡F

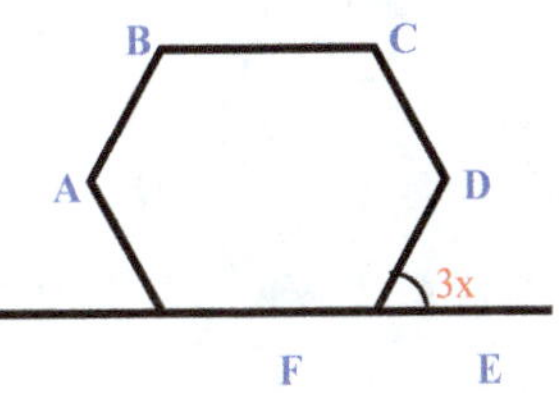

21. Find the range of possible values of x.

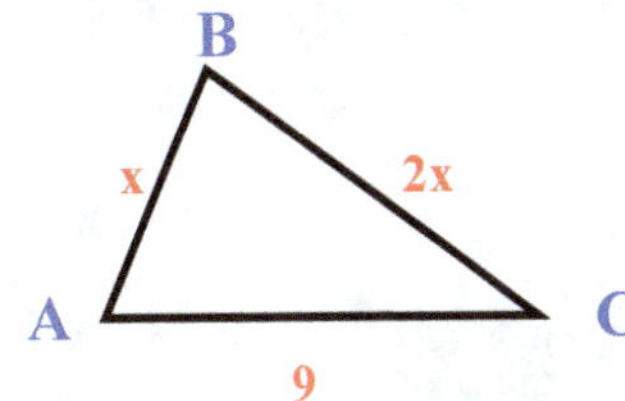

Solve the absolute value equations and inequalities.

1. $|4 - 2x| = 4$

2. $\left|x - \dfrac{3}{2}\right| = -2$

3. $\dfrac{1}{3}|x - 2| = 2$

4. $5|6x - 3(x - 4)| - 1 = 3$

5. $|1 - 2x| + 5 = 2$

6. $|4 - 2x| + 3 = 3$

7. $|||x + 1| + 1| + 1| = 7$

8. $|||x - 1| - 3| - 7| = 1$

9. $|||x - 2| + 3| - 5| = 11$

10. $||x - 3| - 2| = 4$

11. $|x| = x$

12. $|x| = -3x$

13. $|-x + 2| = 2x$

14. $|4 - x| = -3x$

15. $2|x - 1| = 2x - 2$

16. $|3 + x| = |9 - x|$

17. $|x + 3| = |x - 6|$

18. $|1 - 2x| = |x - 2| - 2$

19. $|x| < 5$

20. $|x| \geq -4$

21. $|x + 3| > 2$

22. $|x - 1| < 1$

23. $|4 - x| \geq 2$

24. $-1 \leq |1 + x|$

25. $|x| + 0.5 > 0.3$

26. $-0.2 > |n| - 2$

27. $-|x - 1| \geq 0$

28. $|x + 3| > 5$

29. $|3 - x| > -1$

30. $|x + 3| \leq 6 + 2x$

31. $2|x + 1| \geq 4x$

32. $|x - 1| \leq 3 \text{ or } -|x + 2| < -4$

33. $1 \leq |1 - x| \leq 9$

34. $x < |x - 2| \leq 11$

35. $x - 1| \leq 5 \text{ or } -|x - 3| < -2$

36. $|x + 1| > 4 \text{ and } |x| \leq 6$

37. $|x - 4| \geq 0 \cap x < |1 + 2x|$

Factor the polynomials.

1. $x^4 - 4$
2. $x^4 - 49x^2$
3. $x^5 - 4x^3$
4. $x^4 - 12x^2 - 64$
5. $x^4 + 4x^2 + 4$
6. $x^4 - 2x^2 + 1$
7. $x^4 - 3x^2 - 4$
8. $x^3 - 27$
9. $x^3 + 125$
10. $x^5 - 243$
11. $2x^6 - 32x^2$
12. $x^8 - 16x^4$
13. $x^5 - 9x^3$
14. $x^7 + 5x^5$
15. $x^4 - 9x^2 + 2x^2 - 18$
16. $x^3 - x^2 + 4x - 4$
17. $x^3 + 6x^2 + 5x$
18. $x^4 + 3x^2 + 2$
19. $x^5 + x^4 - 81x^3 - 81x^2$
20. $3x^3 + 2x^2 - 5x$
21. $5x^4 - 4x^3 - x^2$

22. $x^4 - 25$
23. $3x^4 - 27x^2$
24. $x^6 - 25x^4$
25. $x^4 - 2x^2 - 4$
26. $x^4 + 4x^2 + 4$
27. $x^4 - 2x^2 + 1$
28. $x^4 - 5x^2 - 36$
29. $x^5 - 32$
30. $x^6 + 8$
31. $x^5 - 1$
32. $5x^6 - 45x^4$
33. $x^8 + 5x^4$
34. $x^5 - 4x^3$
35. $x^9 + x^5$
36. $x^4 - 4x^2 + 3x^2 - 12$
37. $x^3 + x^2 - x - 1$
38. $x^3 + 5x^2 + 4x$
39. $x^4 + 7x^2 + 6$
40. $x^5 + 3x^4 - 16x^3 - 48x^2$
41. $7x^3 + 3x^2 - 4x$
42. $2x^4 - 4x^3 - 6x^2$

<table>
<tr><td>

6

</td><td>

BANK OF EXTRA EXERCISES

</td><td>

Hayk Yegoryan & Rubik Yegoryan

INTERMEDIATE ALGEBRA

</td></tr>
</table>

Factor and solve.

1. $12x - 4 = 0$	2. $8 - 6x = 0$
3. $x^2 + 6x = 0$	4. $3x^2 - 6x = 0$
5. $x^2 = -10x$	6. $4x^2 = 8x$
7. $x^2 + 9x + 18 = 0$	8. $x^2 + 6x - 16 = 0$
9. $x^2 + x - 12 = 0$	10. $x^2 + x - 42 = 0$
11. $x^2 + 5x = 6$	12. $x^2 + 2x - 3 = 0$
13. $x^2 - 6 = -x$	14. $4x + 8 = 4x^2$
15. $x^2 + 8x + 7 = 0$	16. $x^2 + 9x + 20 = 0$
17. $x^2 - 49 = 0$	18. $x^2 + 8x = 0$
19. $2x^2 + 5x + 3 = 0$	20. $2x^2 + x - 3 = 0$
21. $3x^2 - x - 4 = 0$	22. $4x^2 - 2x - 6 = 0$
23. $7x^2 + 9x + 2 = 0$	24. $6x^2 + 8x + 2 = 0$
25. $x^2 - 1 = 0$	26. $x^2 - 4 = 0$
27. $x^2 - 49 = 0$	28. $x^3 - 9x = 0$
29. $x^4 - 81 = 0$	30. $2x^6 - 32x^2 = 0$
31. $81x^4 - 16 = 0$	32. $27x^4 - 48 = 0$
33. $x^2 - 7x - 18 = 0$	34. $3x^2 + 8x - 11 = 0$
35. $-x^2 + 5x - 14 = 0$	36. $-4x^2 + 25 = 0$
37. $x^2 - 3 = 0$	38. $3x^4 - 2x^2 = 0$
39. $x^2 - 7x - 18 = 0$	40. $3x^2 + 8x - 11 = 0$
41. $81x^4 - a^4 = 0$ (solve for x)	42. $27x^4 - 48b^4 = 0$ (solve for x)

Write the equation of each line in **slope-y-intercept** form and then in **standard** form.

1. $m = 2, b = 3$
2. $m = 4, b = -1$
3. $m = -1, b = 0$
4. $m = -1/5, b = 2$
5. $m = 2/3, b = -2$
6. $m = 0, b = 3$
7. $m = 2, (1, 2)$
8. $m = 3, (4, 0)$
9. $m = 1/3, (0, 5)$
10. $m = -2, (0, 3)$
11. $m = -1/2, (0, 0)$
12. $m = -2/3, (3, 1)$
13. $m = 2, (5, 3)$
14. $m = 1/3, (6, 1)$
15. $(2, 1)\ (4, 5)$
16. $(2, 6)\ (3, 4)$
17. $(-1, 1)\ (2, 7)$
18. $(-1, 2)\ (-1, 7)$
19. $(-2, 4)\ (-3, 4)$
20. $(-3, 1)\ (-1, 3)$
21. $(-3, -1)\ (-4, -3)$
22. $(-2, 4)\ (-5, 3)$
23. $(-3, -1)\ (-4, -3)$
25. $(-2, 4)\ (-5, 3)$

24.

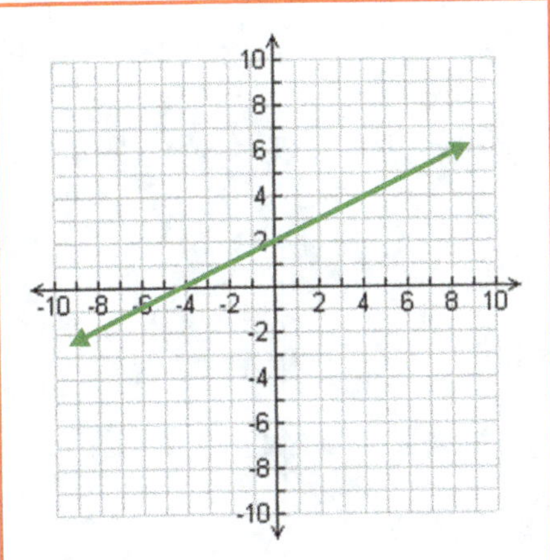

26.

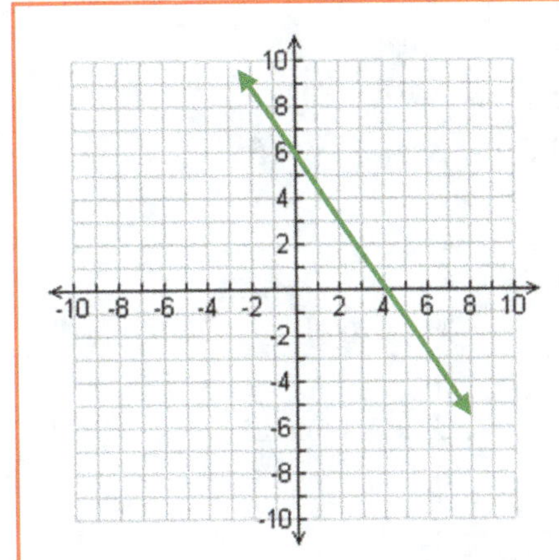

Sketch the graphs of the following functions.

1. $y = -2x - 3$ & $y = \dfrac{1}{3}x + 2$

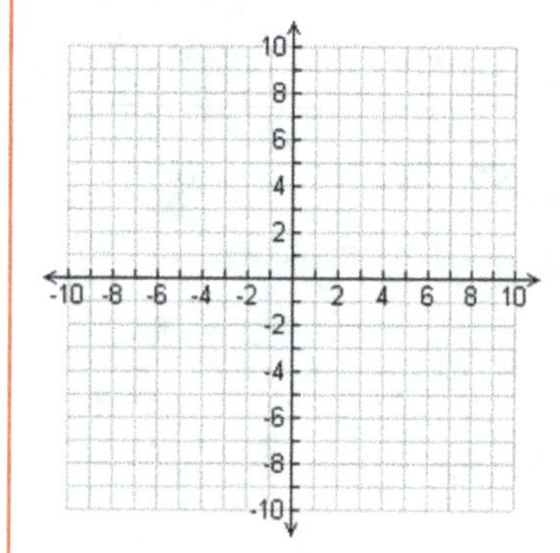

4. $y = x - 2$ & $y = \dfrac{2}{3}x - 3$

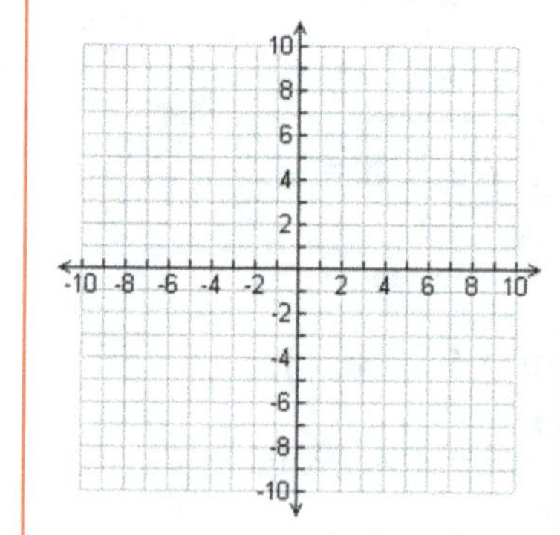

2. $y = x - 2$ & $y = -\dfrac{3}{4}x + 3$

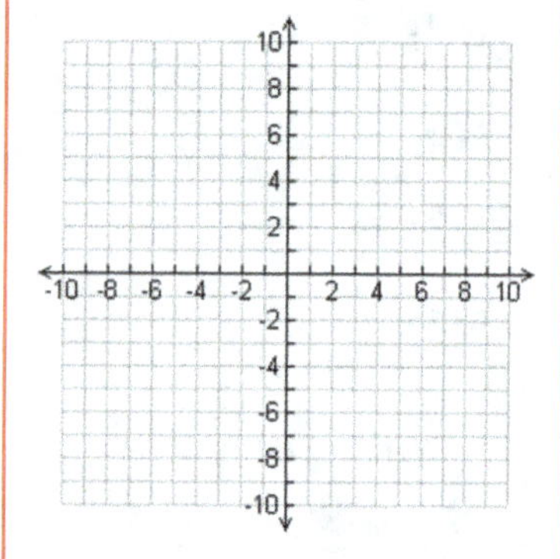

5. $y = -2x + 3$ & $y = \dfrac{1}{2}x + 6$

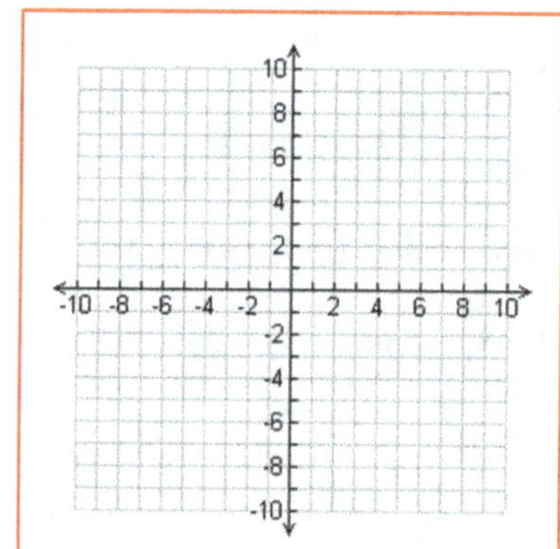

3. $y = -4x$ & $y = 5x$

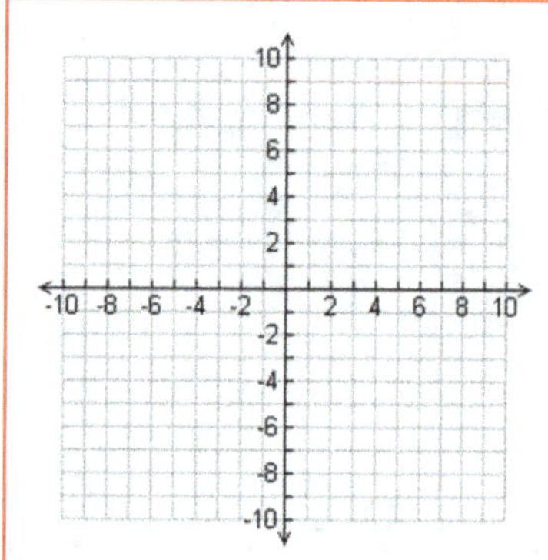

6. $y = x$ & $y = x + 4$

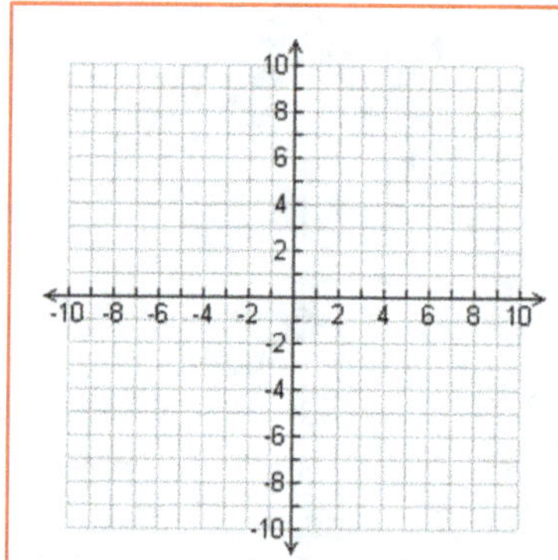

Sketch the graphs of the following functions.

1. $y = x - 3$ & $y = -\dfrac{3}{5}x + 4$

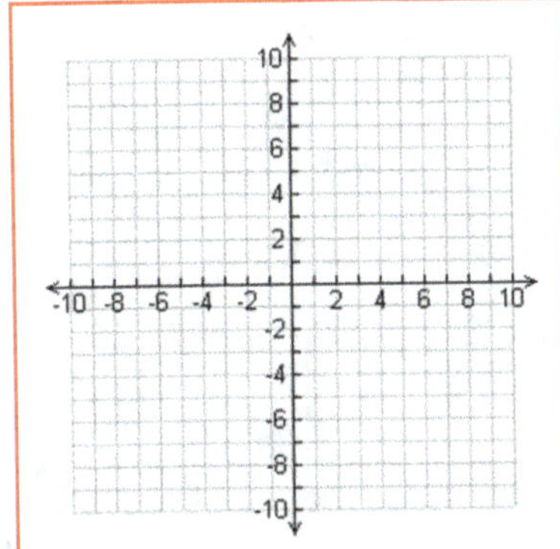

4. $y = -5x + 3$ & $y = \dfrac{5}{6}x - 2$

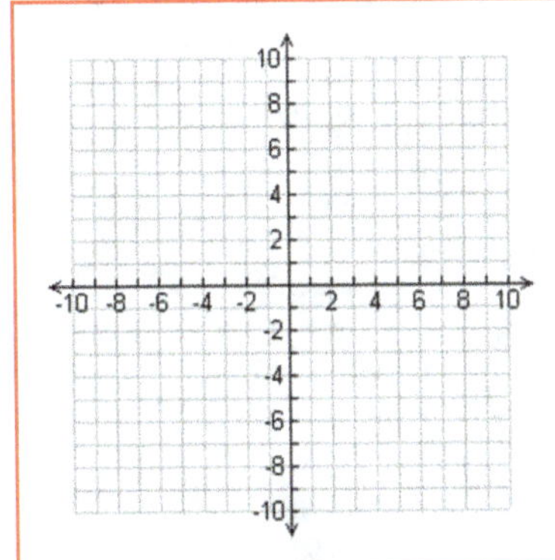

2. $y = 4x - 4$ & $y = -\dfrac{3}{7}x + 5$

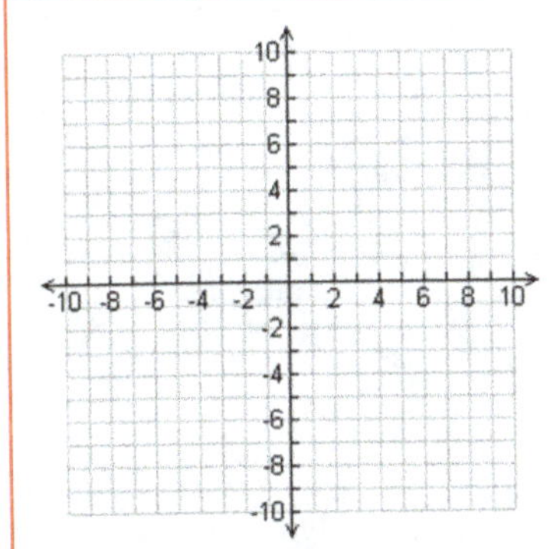

5. $y = -3x + 4$ & $y = \dfrac{3}{2}x + 1$

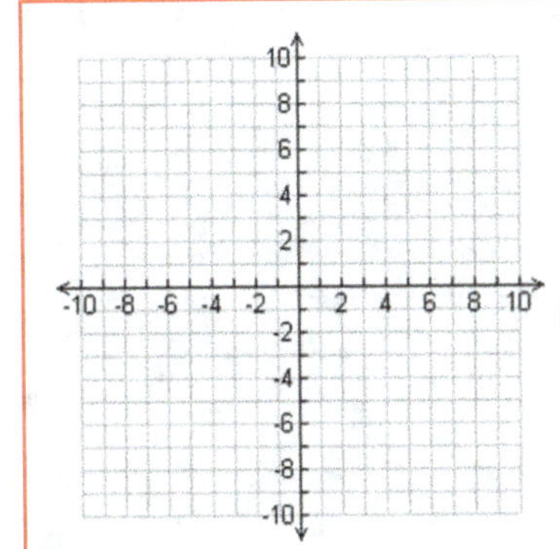

3. $y = -x$ & $y = \dfrac{5}{3}x$

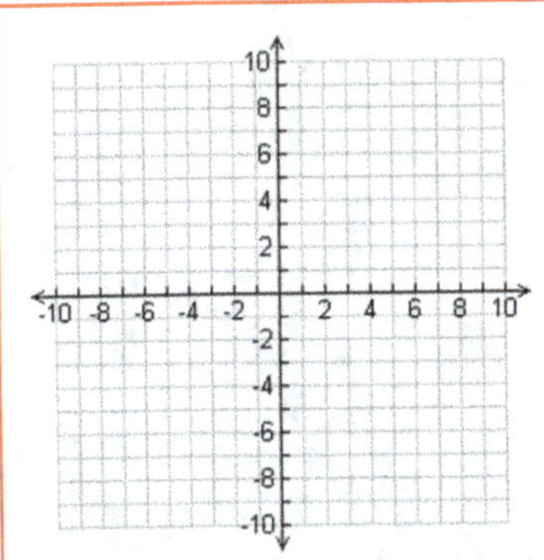

6. $y = -2x$ & $y = -\dfrac{3}{?}x$

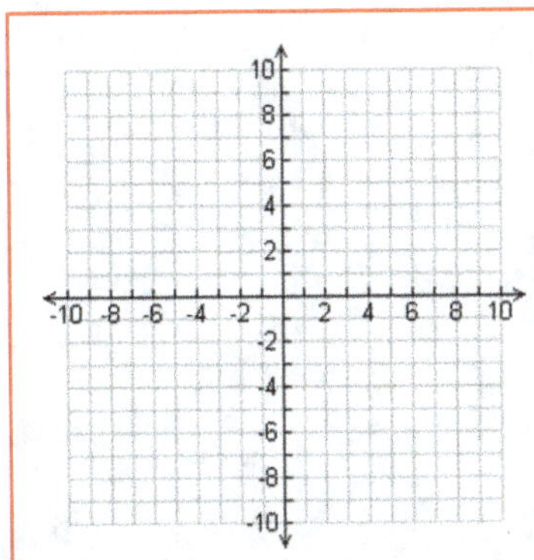

Sketch the graphs of the following functions.

1. $x + y = -4$ & $-x + 2y = 2$

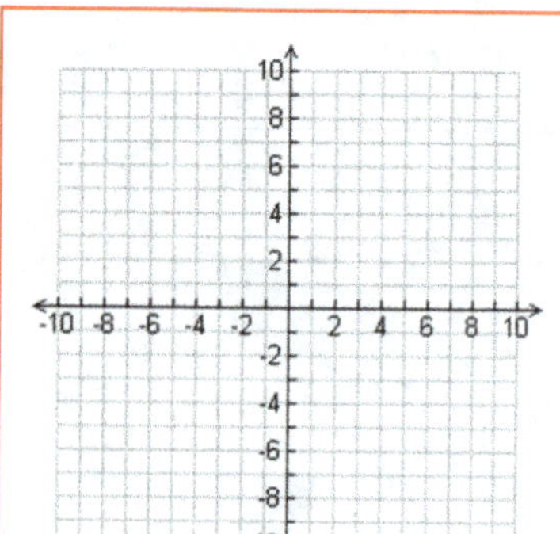

4. $y = x - 2$ & $y = -3$

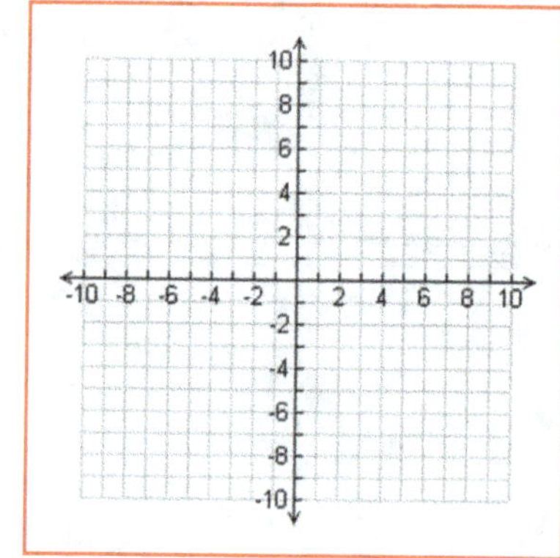

2. $x = 2$ & $2y + 3x = 4$

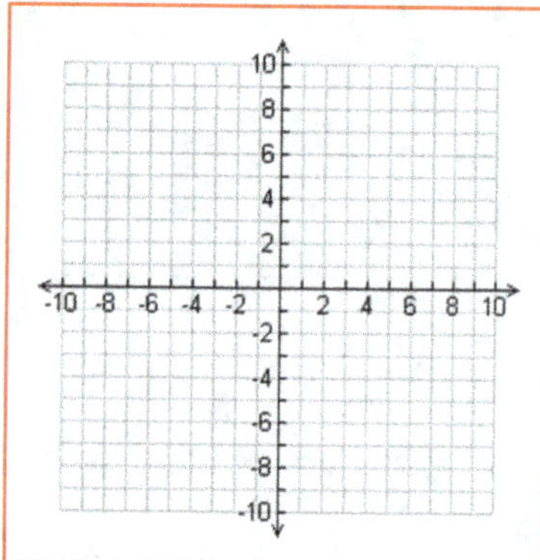

5. $y - 2x = 3$ & $x + 2y = 2$

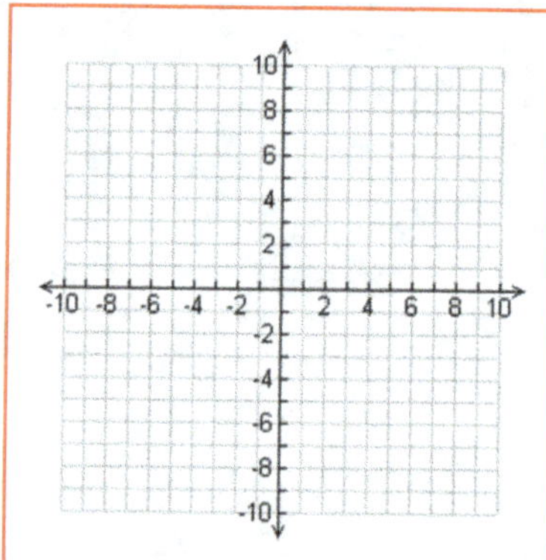

3. $y - 2x = 0$ & $4x - 2y = 6$

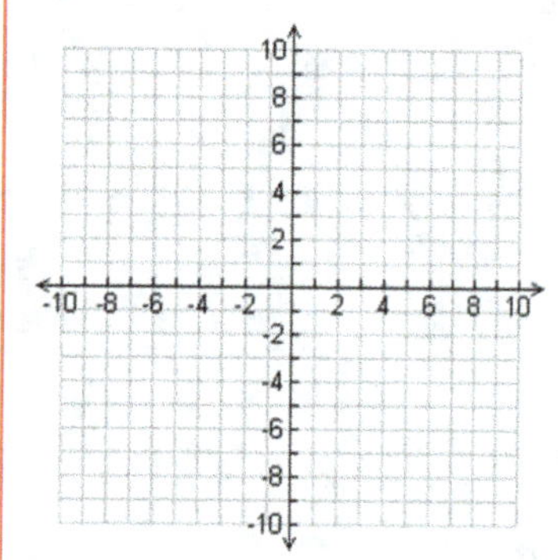

6. $y = x$ & $5x + 3y = 9$

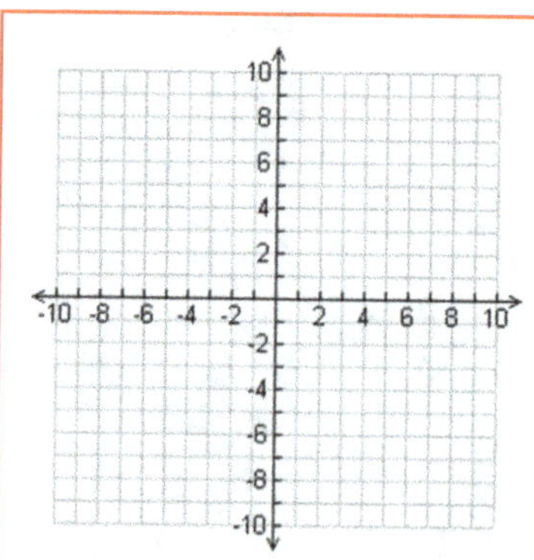

Sketch the graphs of the following functions.

1. $2x + 3y = 6$ & $4y - 3x = 12$

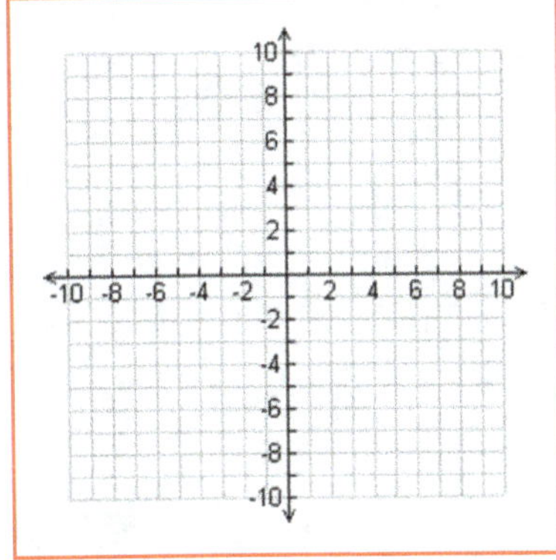

2. $x + 4y = 8$ & $5y + x = 5$

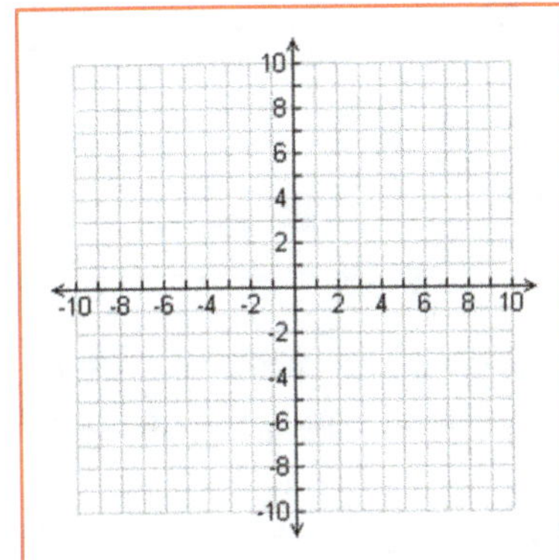

3. $3y - x = 0$ & $3x - 2y = 12$

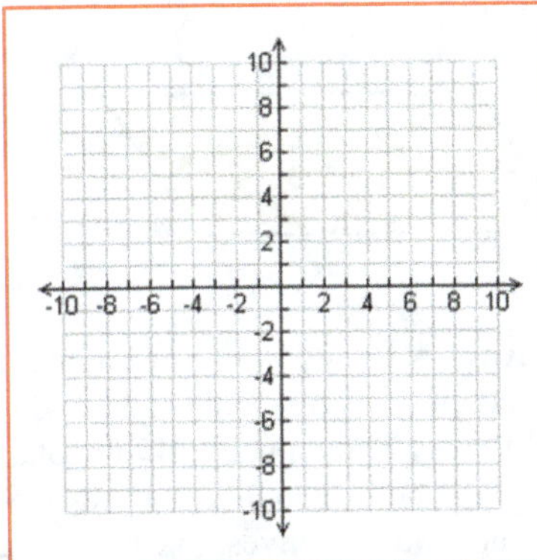

4. $y + 3 = 2x$ & $y - 3x = 3$

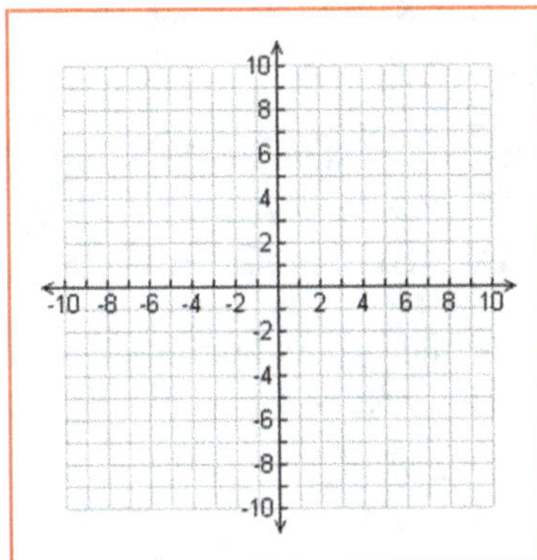

5. $y - 2x = 4$ & $4x + 2y = 8$

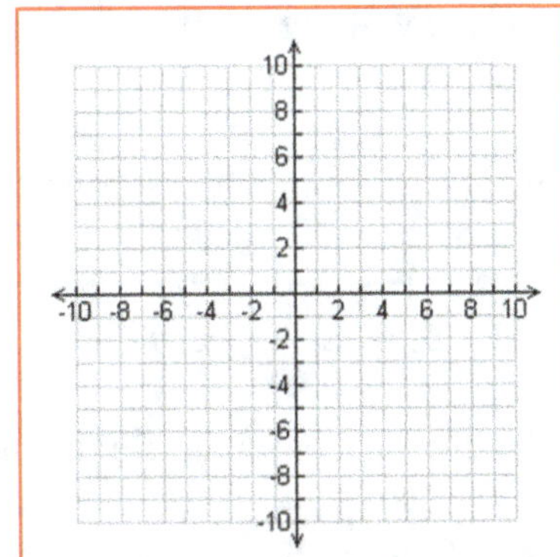

6. $y + x = 0$ & $2x + 5y = 10$

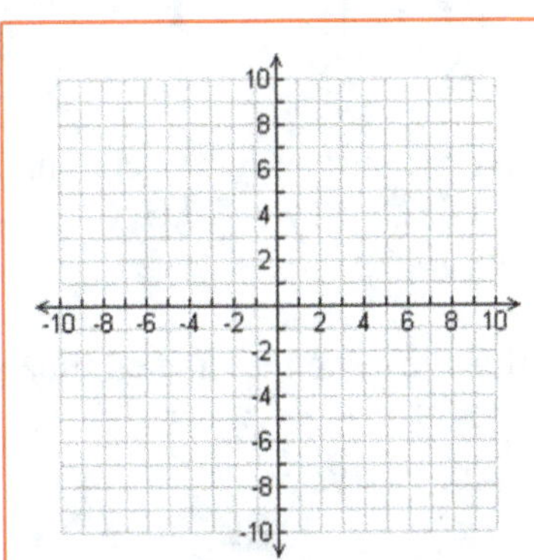

1. Solve the system by the substitution method

$$\begin{cases} y = x + 1 \\ 3x + y = 5 \end{cases}$$

2. Solve the system by the substitution method.

$$\begin{cases} x - 3y = 2 \\ 4x - 8y = 12 \end{cases}$$

3. Solve the system by the substitution method.

$$\begin{cases} 4y - 8x = 8 \\ x = \dfrac{y}{2} - 1 \end{cases}$$

4. Solve the system by the substitution method.

$$\begin{cases} 3x - \quad z = 3 \\ x + 2y + 3z = 9 \\ 2x - y + z = 8 \end{cases}$$

5. Solve the system by the substitution method.

$$\begin{cases} x + y \quad = 3 \\ 2x + y + z = 7 \\ x - 3y + z = -2 \end{cases}$$

6. Solve the system by the elimination method.

$$\begin{cases} y - x = 2 \\ 2x + y = 5 \end{cases}$$

7. Solve the system by the elimination method.

$$\begin{cases} 2x - 3y = -1 \\ 5x - 2y = 3 \end{cases}$$

8. Solve the system by the elimination method.

$$\begin{cases} 12x - 8y = 16 \\ 3x - 2y = 4 \end{cases}$$

9. Solve the system by the elimination method.

$$\begin{cases} x + y + z = -2 \\ 2x + 3y - z = 1 \\ x - y + 2z = -7 \end{cases}$$

10. Solve the system by the graph method.

$$\begin{cases} y = 3x + 1 \\ y = -2x + 6 \end{cases}$$

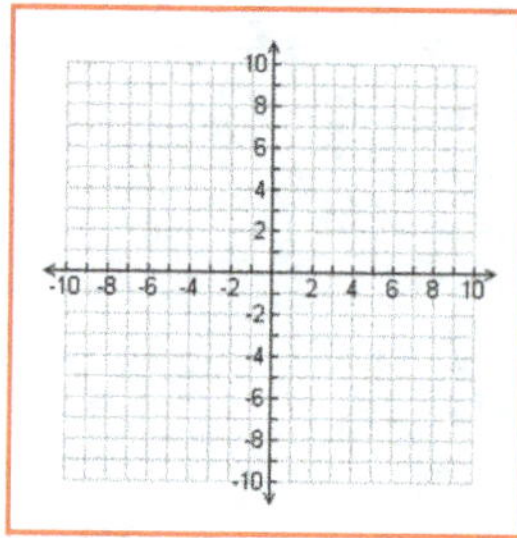

11. Write the corresponding system.

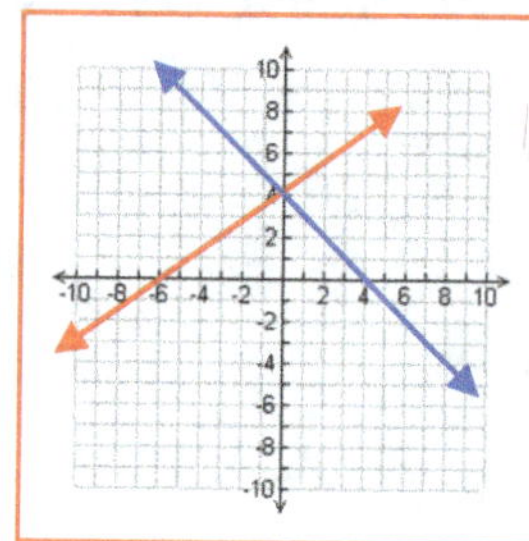

12. Solve the system by any method.

$$\begin{cases} \dfrac{x}{3} + \dfrac{y}{6} = \dfrac{1}{2} \\ \dfrac{x}{6} + \dfrac{2y}{8} = \dfrac{5}{12} \end{cases}$$

13. Find x and y.

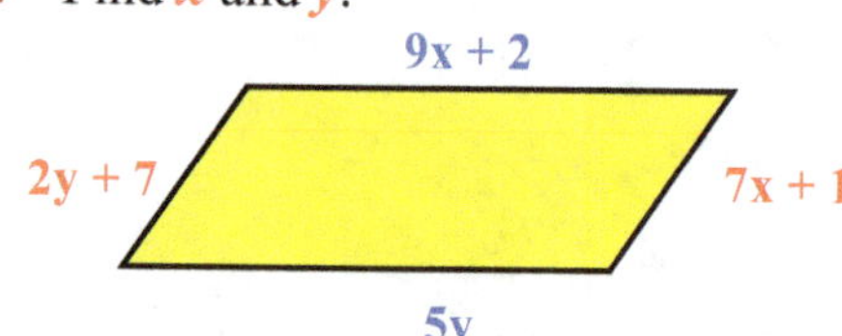

14. Bob invests $ 10,000 in two accounts. The first account earns 8% and the second account earns 7% annually. Bob's total earned from both accounts in a year is $770. How much did Bob invest in each account?

Solve each of the following systems using **Cramer's rule**

1. $\begin{cases} x - y = 0 \\ x + y = 6 \end{cases}$

2. $\begin{cases} x + y = 2 \\ 2x - y = 1 \end{cases}$

3. $\begin{cases} 2x - y = 5 \\ x + 3y = 6 \end{cases}$

4. $\begin{cases} 3x - 2y = 1 \\ 2x + 5y = 7 \end{cases}$

5. $\begin{cases} 4x - 2y = 6 \\ -2x - y = -1 \end{cases}$

6. $\begin{cases} 2y = 2x - 4 \\ x - y = 2 \end{cases}$

7. $\begin{cases} 3x - 4y = 1 \\ -6x + 8y = -2 \end{cases}$

8. $\begin{cases} 2x = y - 3 \\ y - 3x = 2 \end{cases}$

9. $\begin{cases} 0.3x - 0.2y = 0.6 \\ 0.2x + 0.7y = 0.4 \end{cases}$

10. $\begin{cases} \frac{1}{2}x = \frac{1}{3}y \\ \frac{1}{4}x + \frac{1}{2}y = 4 \end{cases}$

11. $\begin{cases} \frac{1}{5}x - \frac{1}{4}y = -\frac{1}{20} \\ \frac{1}{10}x - \frac{1}{8}y = -\frac{1}{30} \end{cases}$

12. $\begin{cases} \frac{1}{5}x - \frac{1}{3}y = \frac{1}{15} \\ \frac{1}{3}x + \frac{1}{4}y = \frac{11}{12} \end{cases}$

13. $\begin{cases} x + y = 1 \\ y - z = 1 \\ x + z = 0 \end{cases}$

14. $\begin{cases} x + y = 1 \\ y + z = 1 \\ x + z = 2 \end{cases}$

15. $\begin{cases} x + y = 1 \\ y + z = 2 \\ x + z = 1 \end{cases}$

16. $\begin{cases} x + y - z = 1 \\ 3x + y - 2z = 2 \\ x - 2y + 2z = 1 \end{cases}$

17. $\begin{cases} x + y - 2z = 0 \\ x + y + z = 3 \\ 3x - 3y + 2z = 2 \end{cases}$

18. $\begin{cases} x + y + z = 3 \\ x - y + z = 1 \\ 5x + 5y + 5z = 15 \end{cases}$

19. $\begin{cases} x + y + z = 3 \\ -x - y = z - 3 \\ x - y = 0 \end{cases}$

20. $\begin{cases} \frac{1}{2}x + \frac{1}{3}y + \frac{1}{2}z = \frac{5}{6} \\ \frac{1}{3}x + \frac{2}{3}y - \frac{1}{3}z = \frac{1}{3} \\ \frac{1}{4}x + \frac{1}{2}y - \frac{1}{4}z = \frac{1}{4} \end{cases}$

21. $\begin{cases} \frac{1}{6}x + \frac{1}{4}y - \frac{1}{8}z = 1 \\ \frac{1}{3}x - \frac{1}{2}y - \frac{2}{3}z = 3 \\ \frac{1}{3}x + \frac{1}{2}y - \frac{1}{4}z = 2 \end{cases}$

Solve each of the following systems by any method.

1. $\begin{cases} x + y = 2 \\ 4x - y = 3 \end{cases}$

2. $\begin{cases} x - y = 0 \\ x + y = 4 \end{cases}$

3. $\begin{cases} x - 2y = 0 \\ x + 3y = 5 \end{cases}$

4. $\begin{cases} 4x - y = 6 \\ 3x + y = 8 \end{cases}$

5. $\begin{cases} 5x - 2y = 3 \\ -2x + y = -1 \end{cases}$

6. $\begin{cases} -6x - 2y = 0 \\ 9x + 3y = 2 \end{cases}$

7. $\begin{cases} 3x - 4y = 1 \\ -6x + 8y = -2 \end{cases}$

8. $\begin{cases} x - 2y = 1 \\ -2x + 4y = 3 \end{cases}$

9. $\begin{cases} 3x - 2y = 13 \\ x + 0.5y = 2 \end{cases}$

10. $\begin{cases} \frac{1}{3}x - \frac{1}{4}y = \frac{1}{12} \\ \frac{1}{4}x + \frac{1}{2}y = \frac{3}{4} \end{cases}$

11. $\begin{cases} \frac{1}{5}x - \frac{1}{4}y = -\frac{1}{20} \\ \frac{1}{10}x - \frac{1}{8}y = \frac{1}{30} \end{cases}$

12. $\begin{cases} \frac{1}{3}x - \frac{1}{2}y = \frac{1}{6} \\ \frac{1}{2}x + \frac{1}{4}y = \frac{5}{4} \end{cases}$

13. $\begin{cases} x - 2y = 0 \\ y + 2z = 4 \\ x + 3z = 5 \end{cases}$

14. $\begin{cases} x - y = 0 \\ y + z = 2 \\ x + z = 2 \end{cases}$

15. $\begin{cases} x - 2y = z - 1 \\ 2y + 2z = x + 1 \\ 3x + 4y - 5z = 7 \end{cases}$

16. $\begin{cases} x + y - z = 1 \\ 3x + y - 2z = 2 \\ x - 2y + 2z = 1 \end{cases}$

17. $\begin{cases} x - y - 2z = 2 \\ x + y + z = 4 \\ 3x - 3y - 6z = 0 \end{cases}$

18. $\begin{cases} x + y + z = 3 \\ x + 3y - 2z = 17 \\ 2x + 3y + 7z = 0 \end{cases}$

19. $\begin{cases} 2x + 2y + 4z = 8 \\ -x - 2z = y - 4 \\ x - y = 0 \end{cases}$

20. $\begin{cases} \frac{1}{2}x + \frac{1}{4}y + \frac{1}{8}z = \frac{7}{8} \\ \frac{1}{3}x + \frac{2}{3}y - \frac{1}{3}z = \frac{2}{3} \\ \frac{1}{5}x + \frac{1}{3}y - \frac{1}{5}z = \frac{1}{3} \end{cases}$

21. $\begin{cases} x + 2y + z - w = 5 \\ 2x + y + z + w = 3 \\ 3x - y - z - w = 0 \\ x - y + 3z + w = 2 \end{cases}$

<table>
<tr><td>

6 **BANK OF EXTRA EXERCISES**

</td><td>

Hayk Yegoryan & Rubik Yegoryan

INTERMEDIATE ALGEBRA

</td></tr>
</table>

Solve each of the following systems by any method.

1. $\begin{cases} x - y = 1 \\ x + y = 1 \end{cases}$

2. $\begin{cases} 2x + y = 3 \\ 3x - y = 2 \end{cases}$

3. $\begin{cases} x - 2y = 4 \\ x + 2y = 0 \end{cases}$

4. $\begin{cases} 3x - y = 8 \\ x + 2y = 5 \end{cases}$

5. $\begin{cases} 3x - 2y = 3 \\ -2x + 3y = -2 \end{cases}$

6. $\begin{cases} 3x - y = 4 \\ 6x - 2y = 2 \end{cases}$

7. $\begin{cases} 2x - y = 3 \\ -x + y = -2 \end{cases}$

8. $\begin{cases} 2x - 4y = 4 \\ -2x + 4y = -4 \end{cases}$

9. $\begin{cases} 0.6x - 0.2y = 0.4 \\ 0.1x + 0.6y = 0.7 \end{cases}$

10. $\begin{cases} \frac{1}{3}x - \frac{1}{2}y = \frac{1}{2} \\ \frac{1}{2}x + \frac{1}{3}y = \frac{11}{6} \end{cases}$

11. $\begin{cases} \frac{1}{2}x - \frac{1}{4}y = -\frac{1}{2} \\ \frac{1}{3}x - \frac{1}{5}y = -\frac{1}{3} \end{cases}$

12. $\begin{cases} \frac{1}{5}x - \frac{1}{2}y = -1 \\ \frac{1}{3}x - \frac{1}{4}y = -\frac{1}{2} \end{cases}$

13. $\begin{cases} 3x - 2y = 5 \\ 4x - 3y = 6 \end{cases}$

14. $\begin{cases} x + y = 2 \\ y + z = 2 \\ x + z = 2 \end{cases}$

15. $\begin{cases} x - 2y = 1 \\ y + 2z = 5 \\ 2x - z = 4 \end{cases}$

16. $\begin{cases} x + y - z = 1 \\ 3x + y - 2z = 2 \\ x - 2y + 2z = 1 \end{cases}$

17. $\begin{cases} x - y - 2z = 2 \\ x + y + z = 4 \\ 3x - 3y - 6z = 0 \end{cases}$

18. $\begin{cases} x + 2y + 3z = 6 \\ x - y + z = 1 \\ 2x + y + 2z = 5 \end{cases}$

19. $\begin{cases} x + y + z = 3 \\ -x - y = z - 3 \\ x - y = 0 \end{cases}$

20. $\begin{cases} \frac{1}{3}x + \frac{1}{2}y + \frac{1}{3}z = \frac{5}{2} \\ \frac{1}{3}x + \frac{2}{3}y - \frac{1}{3}z = \frac{2}{3} \\ \frac{1}{2}x - \frac{1}{2}y + \frac{1}{3}z = 2 \end{cases}$

21. $\begin{cases} \frac{1}{2}x - \frac{1}{2}y - \frac{1}{3}z = -\frac{1}{3} \\ \frac{1}{3}x - \frac{1}{3}y - \frac{2}{3}z = -\frac{2}{3} \\ \frac{1}{4}x - \frac{3}{4}y - \frac{1}{2}z = -\frac{3}{2} \end{cases}$

BANK OF EXTRA EXERCISES

Hayk Yegoryan & Rubik Yegoryan

INTERMEDIATE ALGEBRA

Solve each of the following systems by any method.

1. $\begin{cases} x - y = 0 \\ x + y = 2 \end{cases}$

2. $\begin{cases} 2x + y = 2 \\ 3x - y = 3 \end{cases}$

3. $\begin{cases} 2x - y = 0 \\ x + 2y = 5 \end{cases}$

4. $\begin{cases} 4x - y = 4 \\ 2x + y = 2 \end{cases}$

5. $\begin{cases} 5x - 2y = 3 \\ -2x + y = -1 \end{cases}$

6. $\begin{cases} -6x - 2y = 0 \\ 9x + 3y = 2 \end{cases}$

7. $\begin{cases} 3x - 4y = 1 \\ -6x + 8y = -2 \end{cases}$

8. $\begin{cases} x - 2y = 1 \\ -2x + 4y = 3 \end{cases}$

9. $\begin{cases} 0.3x - 0.2y = 0.1 \\ 0.2x + 0.7y = 0.9 \end{cases}$

10. $\begin{cases} \frac{1}{2}x - \frac{1}{3}y = -\frac{1}{6} \\ \frac{1}{4}x + \frac{1}{2}y = \frac{5}{4} \end{cases}$

11. $\begin{cases} \frac{1}{5}x - \frac{1}{4}y = -\frac{1}{20} \\ \frac{1}{10}x - \frac{1}{8}y = \frac{1}{30} \end{cases}$

12. $\begin{cases} \frac{1}{6}x + \frac{1}{2}y = \frac{2}{3} \\ \frac{1}{3}x - \frac{1}{4}y = \frac{1}{12} \end{cases}$

13. $\begin{cases} 2x - 3y = 3 \\ 2x + 3y = 9 \end{cases}$

14. $\begin{cases} x + y = 1 \\ y + z = 1 \\ x + z = 2 \end{cases}$

15. $\begin{cases} x - 2y = 1 \\ 4y + 2z = 10 \\ x - z = 1 \end{cases}$

16. $\begin{cases} x + y - z = 1 \\ 3x + y - 2z = 2 \\ x - 2y + 2z = 1 \end{cases}$

17. $\begin{cases} x - y - 2z = 2 \\ x + y + z = 4 \\ 3x - 3y - 6z = 0 \end{cases}$

18. $\begin{cases} x + y + z = 3 \\ x - y + z = 1 \\ 5x + 5y + 5z = 15 \end{cases}$

19. $\begin{cases} x + y + z = 3 \\ -x - y = z - 3 \\ x - y = 0 \end{cases}$

20. $\begin{cases} \frac{1}{2}x + \frac{1}{2}y + \frac{1}{2}z = \frac{3}{2} \\ \frac{1}{3}x + \frac{2}{3}y - \frac{1}{3}z = \frac{2}{3} \\ \frac{1}{4}x + \frac{1}{2}y - \frac{1}{4}z = \frac{1}{2} \end{cases}$

21. $\begin{cases} \frac{1}{3}x - \frac{1}{4}y - \frac{1}{3}z = 1 \\ \frac{2}{3}x - \frac{1}{2}y - \frac{2}{3}z = 2 \\ \frac{1}{2}x + \frac{3}{4}y - \frac{1}{2}z = -3 \end{cases}$

Extra Part 7 Bank of Nonstandard Problems

2/5 - □/5 = 1/5	1/□ - 1/5 = 2/15	1/□ - 2/7 = □/14
□/5 - 3/4 = 1/20	□/6 - 3/□ = 7/30	□/7 - 1/□ = 5/14
2/3 + □/□ = 11/12	1/3 - □/□ = 1/21	1/3 - 1/□ = 1/□

$$\square\square \frac{\square\square}{\square\square} - \square\square \frac{\square}{\square} = 47$$

$$\square = \square - \square = \square + \square \quad \square - \square$$

$$\square\square \cdot \square + \square = 11 \qquad \square\square \div \square - \square = 98 \qquad \square\square \cdot \square - \square = 1$$

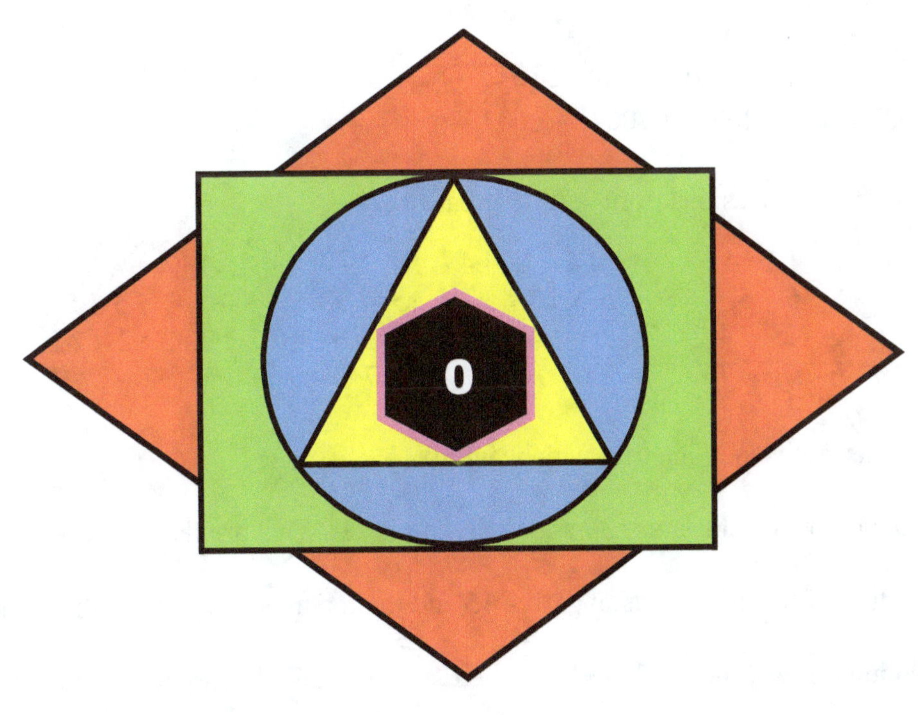

Hayk Yegoryan & Rubik Yegoryan

ELEMENTARY ALGEBRA

1. Use addition, subtraction, multiplication, and division operations between five 2s to get the numbers 0, 1, 2, 3, 4, 5, 6, 7, 8, 9, or 10. For example, $2 + 2 + 2 - 2 \div 2 = 5$.

2. Use addition, subtraction, multiplication, and division operations between six 9s to get 100.

Find the sum of the first **100**, **102**, and **150** terms of each sequence.

3. 1, 2, 3, 1, 2, 3, 1, 2, 3,...
4. 1, 2, 3, 4, 1, 2, 3, 4, 1, ...
5. 1, 2, 3, 4 -5, 1, 2, 3, 4, -5,...
6. –1, 0, 1, 2, –1, 0, 1, 2,...
7. –2, –1, 0, 1, 2, –2, –1, 0, 1, 2, ...

Find the value of each expression.

8. $(-3)^4 - (-3)^3 + (-3)^2 - (-3)$
9. $(-2)^5 - (-2)^4 + (-2)^3 + (-2)^2 + (-2)^1 + (-2)^0$
10. $(-1)^{2006} + (-1)^{2007}$
11. $(-5)^3 + (-5)^{-3}$
12. $(-10)^2 - (-10)^2 + (-10)^3 - (-10)^3$
13. $\{[(1957 \cdot 1957) + 1957] \div 1957\} - 1957$
14. $\{[(2007 \cdot 2007) - 2007] \div 2007\} - 2007$

List the values of expressions from least to greatest.

15. $4 + 4 \div 4 + 4$, $\quad 4 \cdot 4 \div 4 + 4$, $\quad 4 \div 4 \cdot 4 \div 4$, $\quad 4 \cdot 4 - 4 \cdot 4$, $\quad 4 \div 4 \cdot 4$, $\quad 4 \div 4 + 4 \cdot 4$, $4 \cdot 4 - 4 \div 4$
16. $5 \cdot 5 \cdot 5 \div 5$, $\quad 5 \cdot 5 \cdot 5 + 5$, $\quad 5 \cdot 5 \div 5 + 5$, $\quad 5 \div 5 + 5 \div 5$, $\quad 5 \div 5 + 5 - 5$, $\quad 5 \cdot 5 \cdot 5 \cdot 5$, $\quad 5 \div 5 + 5 \cdot 5$
17. $2 + 2 + 2 + 2$, $\quad 2 - 2 - 2 - 2$, $\quad 2 + 2 - 2 - 2$, $\quad 2 \cdot 2 + 2 \cdot 2$, $\quad 2 \cdot 2 \cdot 2 \cdot 2$, $\quad 2 \div 2 - 2 \cdot 2$, $\quad 2 \cdot 2 \div 2 \div 2$
18. $7 - 7 \cdot 7 \cdot 7$, $\quad 7 + 7 \cdot 7 \cdot 7$, $\quad 7 \cdot 7 + 7 \cdot 7$, $\quad 7 - 7 \cdot 7 + 7$, $\quad 7 \div 7 - 7 \cdot 7$, $\quad 7 \cdot 7 \div 7 \cdot 7$, $\quad 7 - 7 \div 7 \cdot 7$

19. Find **3** consecutive odd integers which sum to **15**.

20. The product of two integers is **30** and their sum is **11.** What are those integers?

21. The product of two integers is **negative 15** and their sum is **2.** What are those integers?

22. The product of two integers is **132** and their sum is **–23.** What are those integers?

23. The product of two positive integers is **75** and their difference is **10.** What are those integers?

24. What is the sum of the prime factors of **48**?

25. What are two factors of **24,** whose product is **36**?

26. What is the sum of all the factors of **12**?

27. Find exponent $9 \cdot 3^{\square} = 243$.

28. What is the **3rd** term of the sequence if the **10th** term is **60** and the difference between any two consecutive terms is **5?**

29. Let the sum of all even natural numbers less than **1,000** be **E**. What is the sum of all odd natural numbers less than **1,000?**

30. Let the sum of all odd natural numbers less than **10,000** be **B**. What is the sum of all even natural numbers less than **10,000?**

31. How many numbers less than **101** are divisible by **11?**

32. Find the number the double of which equals to **2** more than the number.

33. Find the largest **6**-digit number that is divisible by **117.**

34. Find the largest **5**-digit number that is divisible by **136.**

35. Find the smallest **6**-digit number that is divisible by **121.**

36. Find the smallest **5**–digit number that is divisible by **87.**

37. Find all natural numbers that, if multiplied by **27,** yield a **3**-digit number, and if multiplied by **30,** yield a **4**-digit number.

38. Find the product of all integers from **–7** to **2,** inclusive

39. What month will be the **485th** month from now, if now it is July?

40. What day will be the **501st** day from now, if today is a Sunday?

41. What day and what time will it be **129** hours from now, if now is **8p.m.** on a **Monday?**

42. If **A** is the set of integers between **–20** and **20,** inclusive, and **B** is the set of squares of the integers of the set **A,** how many elements of **B** are also elements of the set **A?**

43. What is the product of **12** and **18** rounded to the nearest ten?

44. If **A** is the set of integers between **−30** and **30,** inclusive, and **B** is the set of cubes of the integers of the set A, how many elements of **B** are also elements of **A?**

45. Diana was **25** years old in **2000,** when her son, Bob, was **1** year old. In what year was Diana exactly **4** times as old as Bob?

46. Divide the **16**[th] term of the first sequence by the **16**[th] term of the second:
     2, 4, 8, 16, 32, 64,…
     2, 4, 6, 8, 10, 12,…

47. If **3** times a number is **48,** then what is that number divided by **4?**

48. Find the common multiples of **5** and **7** that are less than **196.**

49. Find the difference between the sum of integers from **51** to **100,** inclusive, and the sum of integers from **1** to **50,** inclusive.

50. Find a positive integer the square of which is twice that integer.

51. Prove that $10^4 + 10^3$ is divisible by **11.**

52. Prove that $5^{2006} + 5^{2005}$ is divisible by **6.**

53. Prove that $13^{14} - 13^{13}$ is divisible by **4.**

54. Prove that $16^5 - 4^9$ is divisible by **3.**

55. Prove that $7^9 - 7^8 + 7^7$ is divisible by **43.**

56. Prove that $3^{45} + 3^{44} + 3^{43} + 3^{42}$ is divisible by **120.**

57. Prove that $16^4 + 4^6 + 2^8$ is divisible by **7.**

58. Prove that $27^4 - 9^5 + 3^9$ is divisible by **5**

59. The sum of five consecutive even integers is **740.** What is the largest of these integers?

60. Find the natural number that is the sum of all the natural numbers preceding it.
    Find the natural number that is twice less than the sum of all its preceding natural numbers.

61. If $O - \Delta = 10$ and $O + \Delta = 2.$ What are the values of O and Δ ?

62. Find the natural number that is 5 times less than the sum of all the numbers preceding it.

63. Prove that any 3-digit number that has the same digit (e.g. 111) throughout is divisible by 37.

64. Find a 2-digit number that is 4 times greater than the sum of its digits and twice greater than their product

65. Find a 2-digit number that is 4 times greater than the sum of its digits and 3 times greater than their product.

66. Prove that any two consecutive odd numbers greater than 1 are relatively prime numbers.

67. Prove that the product of any five consecutive natural numbers is divisible by 5.

68. Prove that the product of any three consecutive natural numbers is divisible by 6.

69. Prove that the sum of any four consecutive odd numbers is divisible by 8.

70. The sum of two consecutive odd numbers equals 24,568. Find those numbers.

71. The sum of two numbers equals 145. Find those numbers.

72. The sum of the squares of the digits of a 2-digit even number is 25. Find the 2-digit number.

73. The sum of the squares of the digits of a 2-digit odd number is 25. Find the 2-digit number.

74. Find three consecutive even integers so that the sum of the squares of the first two equals to the square of the third.

75. Put the signs + and − between numbers and find the missing number, so that all three equations are *true.* The signs in all the equations are the same at the respective places.

8 ☐ 15 ☐ 1 ☐ 7 = 15

4 ☐ 24 ☐ 2 ☐ 9 = 17

3 ☐ 17 ☐ 5 ☐ 6 = ☐

76. Find the sum of the first **17** numbers of the sequence.

1, 3, 3, 3, 5, 5, 5, 5, 5,..

77. Find the next **4** numbers that follow the pattern.

1, 3, 7, 13, ☐ , ☐ , ☐ , ☐ ,...

78. Write the following numbers in the pattern up to the **10th**, inclusive.

1, 2, 5, 10, 25, 50, ☐ , ☐ , ☐ , ☐ ,…

79. Write the missing number on the vertex of the third triangle.

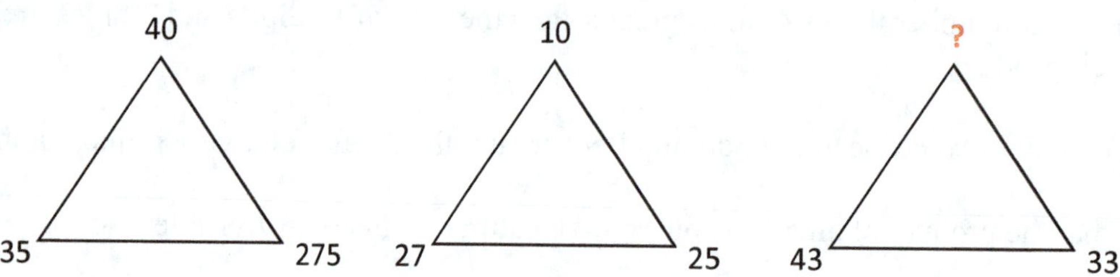

80. Write the missing number on the vertex of the third triangle.

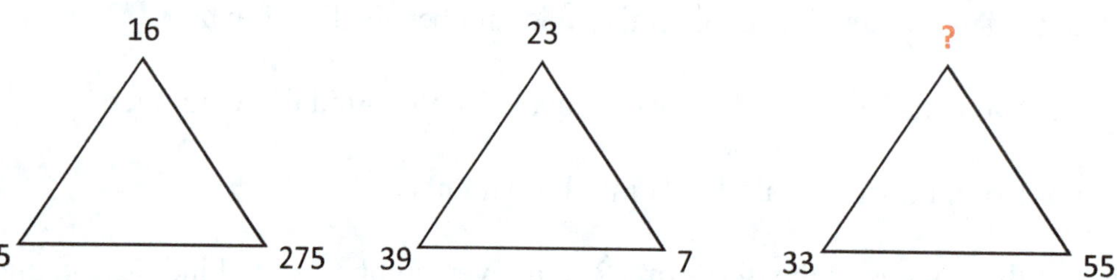

81. Write the missing number on the vertex of the third triangle.

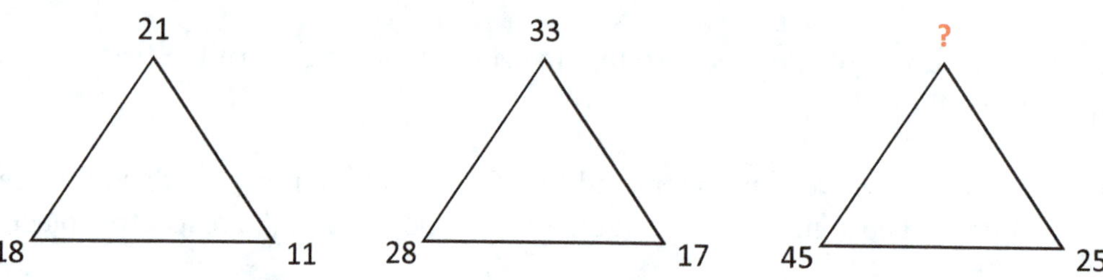

82. Write the missing number in the parentheses.

17   (25)   42
21   ( )   58

83. Write the missing number in the parentheses.

12   (51)   39
35   ( )   47

84. Write the missing number in the parentheses.

13   (11)   143
15   ( )   135

85. Find the next **2** numbers of the pattern.

1, 5, 13, 25, 41, ☐ , ☐ , …

86. Find the next **3** numbers of the pattern.

4, 9, 15, 22, 30, ☐ , ☐ , ☐ , …

87. Find the next number of the sequence.

145, 876, 514, 587, 651, ☐ , …

88. Find the pattern and replace the question mark. Move along a spiral from the bottom-left corner of the square to the question mark.

26	23	30	27
19	36	33	24
22	29	?	31
15	32	25	28

89. Replace question marks with numbers that satisfy the pattern inside each square.

2	10	5
8	?	9
4	28	7

6	9	3
25	?	18
8	13	5

5	5	25
7	?	49
8	12	96

9	25	34
7	?	18
12	11	16

90. Replace the question marks to numbers that satisfy the pattern.

56	7	45	2
17	9	20	?
21	13	?	14
18	29	12	24

91. Change the question marks to numbers.

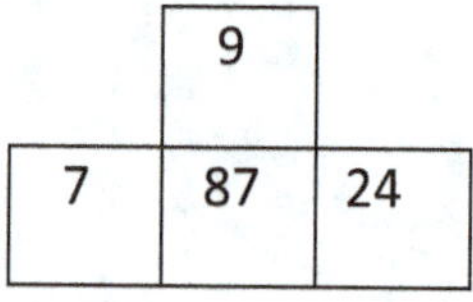

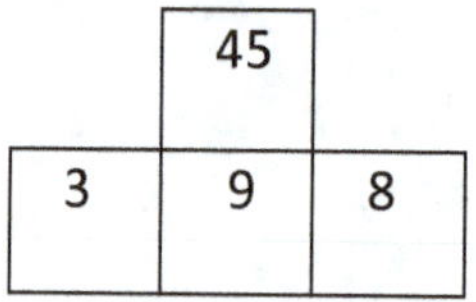

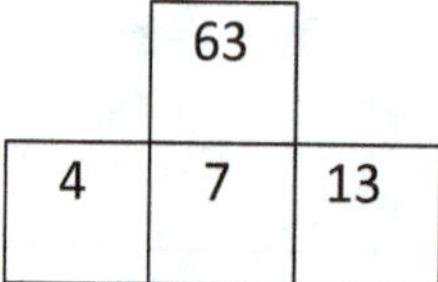

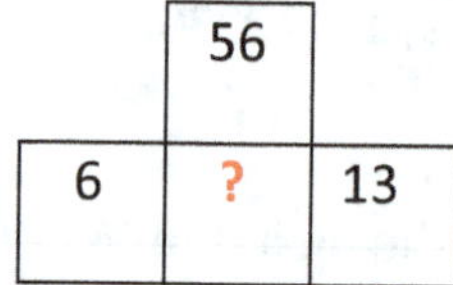

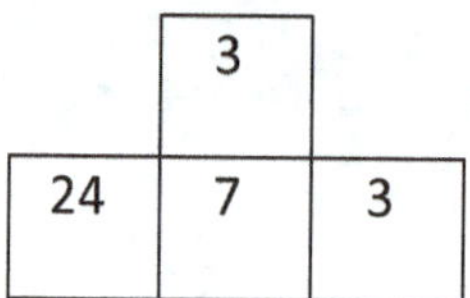

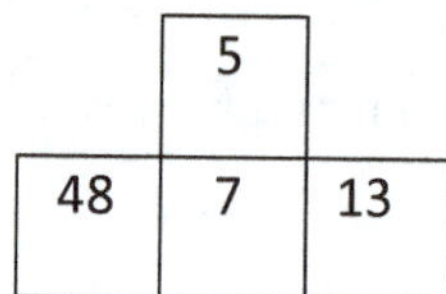

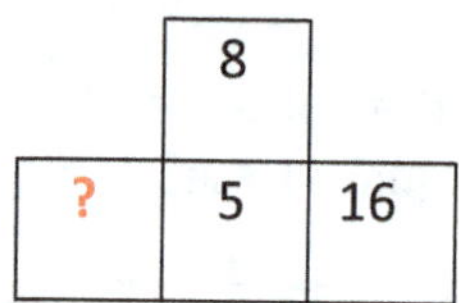

92. Write the signs + or − between the numbers in the first equation and copy them to their respective places in the second equation. Then, find the missing number in the second equation so that both equations end up being true.

1 ☐ **5** ☐ **9** ☐ **13 = 2**
2 ☐ ☐ ☐ **8** ☐ **11 = 5**

93. Write the missing positive integers.

☐☐ · ☐ + ☐ = 11   ☐☐ ÷ ☐ − ☐ = 98   ☐☐ · ☐ − ☐ = 1

94. Write the missing positive integers.

☐ 7 + ☐ = ☐☐ 1   ☐☐ + ☐☐ = ☐ 97   ☐☐ ÷ 9 = ☐ 0

95. Put the integers **0, 1, 2, 3, 4, 5, 6, 7, 8, 9** (use each only once) in the following expressions, so as to get two true statements.

☐☐ · ☐ = ☐☐
☐☐ · ☐ = ☐☐

96. Write the missing positive integers.

□ **8** + □ = □ □ **5**   □ **2·** □ = **8** + □   □ □ **6** − □ = □ **8**

97. Put the integers **1, 2, 3, 4, 5, 6, 7, 8, 9** (use each only once) in the correct spaces to get a true statement.

□ + □ = □ + □ = □ + □ = □ + □ = **2·** □

98. Put the integers **1, 2, 3, 4, 5, 6, 7, 8, 9** (use each only once) in the correct spaces to get a true statement.

□ + □ = □ - □ = □ **·** □ = □ □ ÷ □

99. Put the integers **0, 1, 2, 3, 4, 5, 6, 7, 8, 9** (use each only once) in the correct spaces so as to get a true statement.

□ ÷ □ + □ = □

x □

□ □ = □ + □ − □

100. Replace the spaces with the same natural number to get a true statement

□ = **2 · {[(** □ **+1) − 2] ÷ 3 + 1}**

101. Replace the spaces with the same natural number to get a true statement

2 · □ = **3 · {[(** □ **+ 2) − 1] ÷ 4 + 1}**

102. Put the signs +, −, x, and ÷ between the numbers, so that the following equations are true. They don't have to be the same from one equation to another.

2 □ **3** □ **4** □ **5** □ **6 = 1**
7 □ **8** □ **9** □ **10** □ **11 = 100**
12 □ **13** □ **14** □ **15** □ **16 = − 70**

103. Write the missing positive integer.

$$[(\boxed{} + 2) \div 6 \cdot 3 + 1] \div 4 \cdot 5 = 5$$

104. Put the integers **1, 2, 3, 4, 5, 6, 7, 8, 9** (use each only once) in the correct squares to get a true statement.

$$\boxed{} = \boxed{} + \boxed{} = \boxed{} + \boxed{}$$

with the top branch $= \boxed{} - \boxed{}$ and the bottom branch $= \boxed{} - \boxed{}$

105. Replace the question mark with a number to follow the pattern.

8		5
	42	
6		9

7		3
	13	
5		4

12		10
	46	
11		13

14		10
	?	
17		15

106. Replace the question mark with a number to follow the pattern.

3		2
	10	
4		1

7		6
	26	
8		5

11		10
	?	
12		9

107. Put the integers **2, 4** and **6** in the small squares so as to get a *magic square*.

2		
	4	
		6

108. Replace the question mark with a number to follow the pattern.

3		2
	10	
4		1

7		6
	26	
8		5

11		10
	?	
12		9

109. Replace the question mark with a number to follow the pattern.

63		5
	9	
45		7

56		7
	7	
49		8

72		8
	?	
64		9

110. Replace the question mark with a number to follow the pattern.

	5	
9	62	3
	7	

	6	
8	44	4
	2	

	10	
1	?	3
	12	

111. How can you take exactly 7 gallons of water from a well having only 3 and 8 gallon canisters?

112. How can you take exactly 10 gallons of water from a well having only 9 and 11 gallons canisters?

113. How can you take exactly 2 gallons of water from a well with only 7 and 11 gallons canisters?

114. How can you take exactly 4 gallons of water from a well having only 3 and 5 gallon canisters?

115. How can you pour in 2 liters of wine into one 5 liter jug and one 7 liter jug, having a barrel with 12 liters of wine in it?

116. Find the **missing digits** in each question below.

$$2/5 - \Box/5 = 1/5 \qquad 1/\Box - 1/5 = 2/15 \qquad 1/\Box - 2/7 = \Box/14$$
$$\Box/5 - 3/4 = 1/20 \qquad \Box/6 - 3/\Box = 7/30 \qquad \Box/7 - 1/\Box = 5/14$$
$$2/3 + \Box/\Box = 11/12 \qquad 1/3 - \Box/\Box = 1/21 \qquad 1/3 - 1/\Box = 1/\Box$$

117. Find the **missing digits** in each question below

```
    5 □ 4 . □ 9 0          □ 3 . 3 □ 2            9 □ , 5 2 3
  -  □ 5 □ . 2 3 □       -  7 2 . □ 5 □         -    7 , 1 □ □
  ─────────────────       ───────────────       ─────────────
    1 9 2 . 9 5 5          1 □ . 9 2 6            □ 6 , □ 8 4
```

118. Use **Gaussian Method** to find the result.

$$1 + 2 + 3 + 4 + \ldots + 20 = \Box$$
$$1 + 2 + 3 + 4 + \ldots + 1000 = \Box$$
$$-1 - 2 - 3 - 4 - \ldots - 51 = \Box$$
$$-1 - 2 - 3 - 4 - \ldots - 1003 = \Box$$

119. Put the digits **1, 2, 3, 4, 5, 6, 7, 8, 9** (use each only once) in the boxes to get a true statement.

$$\Box - \Box$$
$$\| \qquad \|$$
$$\Box = \Box - \Box = \Box + \Box\Box - \Box$$

120. Put the digits **0, 1, 2, 3, 4, 5, 6, 7, 8, 9** (use each only once) in the boxes to get a true statement.

$$\Box\Box\frac{\Box\Box}{\Box\Box} - \Box\Box\frac{\Box}{\Box} = 47$$

121. Put the digits **0, 1, 2, 3, 4, 5, 6, 7, 8, 9** (use each only once) in the boxes to get a true equation
½ - ½ = 0. Find out all possibilities.

$$\frac{\Box\Box}{\Box\Box} - \frac{\Box\Box\Box}{\Box\Box\Box} = 0$$
$$½ \quad - \quad ½ \quad = 0$$

122. **R**eplace the question marks with a number to follow the pattern.

60	20	40		80	?	40		100	?	?

123. Solve the equation.

$$-\left(-\left(-\left(-\left(-(x)\right)\right)\right)\right) = x$$

124. Solve the equation.

$$\left(-\left(-\left(-\left(-(x^3)\right)\right)\right)\right) = -x^2$$

125. Solve the equation.

$$\frac{1}{x} + 1 = \frac{1}{2x}$$

126. Solve the equation for x in terms of other variables.

$$\frac{1}{x} + 2 = \frac{3z}{xy}$$

127. Find the largest integer that makes the inequality true.

$$4(x - 2) + 1 < 17$$

128. Find the smallest integer that makes the inequality true.

$$3(x - 1) - 5 > 16$$

129. Find the smallest and largest integers that make the double inequality true.

$$3 < \frac{x + 6}{7} \le 8$$

130. Tell whether the statement is *sometimes* true, *always* true, or *never* true, if **x** and **y** are real numbers and $0 < x < y$.

a) $x + 2y < 2x + y$ b) $y^2 < x$ c) $x^2 y < xy^2$ d) $\dfrac{x}{y} < \dfrac{y}{x}$ e) $xy > x + y$

131. What is the relationship between **p** and **q**, if double inequality $p < x < q$ has no solution.

132. Solve the equations.

a) $|-x| = -|x|$ b) $|-x| + 3 = |x| + 3$ c) $|-x + 1| = |x + 1|$ d) $||-x| - 2| = |-x| - 2$

e) $|3 - x| = \dfrac{x}{2}$ f) $\left|\dfrac{x}{3} - \dfrac{1}{3}\right| = \dfrac{1}{2}$ g) $|-x| + |x| = 2x$ h) $|2 + x| = |x| + 2$

133. Solve the inequalities.

a) $|x| \le -x$ b) $|-x| > -x$ c) $|-x + 2| > x - 2$ d) $||-x| - 2| < |-x| - 2$

134. Write the equation of the line with the following x-intercepts and slopes:

a) x-intercept = 2 and m = $-\dfrac{3}{2}$ b) x-intercept = -3 and m = $\dfrac{4}{3}$ c) x-intercept = 0 and m = 2

135. Write the equation in point-slope, in slope-y-intercept, and in standard of the line that passes through the given points.

a) (a, b), (–a, 3b)   b) (2a, –b), (a, a –b)   c) (a, b + 2a), (3a, b)

136. Determine whether the lines are perpendicular.

a) $y = 10x$; $y = -0.1x + 2$ b) $y = -0.25x + 1$; $y = 4x + 3$ c) $y = -0.5 + x$; $y = 2x$

137. Find the coordinates of vertex D of the rectangle ABCD if coordinates of vertices A, B, and C are given: A (1,3), B (4,1), C(0, – 5) .

138. Use the relation $x^2 + y^2 = 25$ and find all possible values of y if x = 4.

139. Measures of three sides of a right triangle are consecutive numbers. Find that numbers.

140. The square ABCD is inscribed in the circle and the perimeter of ABCD 16. Find the perimeter/circumference of circle.

NOTES